FRAMINGHAM STATE COLLEGE
3 3014 00123 1143

D1002269

CHEMISTRY OF HETEROCYCLIC COMPOUNDS IN FLAVOURS AND AROMAS

CHEMISTRY OF HETEROCYCLIC COMPOUNDS IN FLAVOURS AND AROMAS

Editor:
G. VERNIN
Maitre de Recherche
Organic Chemistry
National Centre of Scientific Research
Marseilles, France

ELLIS HORWOOD LIMITED
Publishers · Chichester

Halsted Press: a division of
JOHN WILEY & SONS
New York · Brisbane · Chichester · Toronto

First published in 1982 by

ELLIS HORWOOD LIMITED
Market Cross House, Cooper Street, Chichester, West Sussex, PO19 1EB, England

The publisher's colophon is reproduced from James Gillison's drawing of the ancient Market Cross, Chichester.

Distributors:

Australia, New Zealand, South-east Asia:
Jacaranda-Wiley Ltd., Jacaranda Press,
JOHN WILEY & SONS INC.,
G.P.O. Box 859, Brisbane, Queensland 40001, Australia

Canada:
JOHN WILEY & SONS CANADA LIMITED
22 Worcester Road, Rexdale, Ontario, Canada.

Europe, Africa:
JOHN WILEY & SONS LIMITED
Baffins Lane, Chichester, West Sussex, England.

North and South America and the rest of the world:
Halsted Press: a division of
JOHN WILEY & SONS
605 Third Avenue, New York, N.Y. 10016, U.S.A.

British Library Cataloguing in Publication Data
Chemistry of heterocyclic compounds in flavours and aromas. –
(Ellis Horwood series in chemical science)
1. Heterocyclic compounds
2. Aroma and Flavouring compounds
Analysis Synthesis Precursors
Legislation, Foods
3. Maillard reaction
Computor application
I. Vernin, G.
597'.595045'9 QD400.3

Library of Congress Card No. 82-3034 AACR2

ISBN 0-85312-263-6 (Ellis Horwood Ltd., Publishers)
ISBN 0-470-27336-4 (Halsted Press)

Typeset in Press Roman by Ellis Horwood Ltd.
Printed in Great Britain by Unwin Brothers Ltd., of Woking

Table of Contents

Preface 13
J. METZGER, University of Aix-Marseilles, France

Introduction 15
G. VERNIN, University of Aix-Marseilles, France

Chapter I **Heterocyclic Aroma Compounds Precursors**
J. GARNERO, Robertet Company, Grasse, France

I–1 Introduction 17
I–2 Meat and Related Products: Beef, Pork, Lamb and Chicken . . . 18
2.1 Primary Precursors and their Hydrolysis Products 18
2.2 Intermediate Precursors in Meat Flavours 20
I–3 Nut Products 23
3.1 Peanuts 23
3.2 Hazelnuts 25
I–4 Non-alcoholic Beverages: Tea, Cocoa, Coffee 25
4.1 Tea Aroma Precursors 26
4.2 Cocoa Flavour Precursors 32
4.2.1 Precursors involved during the Fermentation Step 32
4.2.2 Precursors involved during Roasting 34
4.3 Coffee Flavour Precursors 34
I–5 Fruits and Vegetables 40
5.1 Tomato 40
5.2 Asparagus 43
5.3 Potato 46
I–6 Bread and Wheat Flour 48
6.1 Bread Processing 48
6.2 Precursors 50

I–7 Milk and Dairy Products 52
7.1 Composition 52
7.2 Precursors 54
I–8 Alcoholic Beverages: Wine 59
I–9 References 62

Chapter II Heterocyclic Aroma Compounds in Foods: Occurrence and Organoleptic Properties

Gaston VERNIN and Genevieve VERNIN, University of Aix-Marseilles, France

II–1 Introduction 72
II–2 Five-membered Heterocyclic Systems Containing One, Two, or more Heteroatoms 74
2.1 Furans and Reduced Systems 74
2.1.1 Structure and Occurrence 74
2.1.2 Sensory Properties and Patents 83
2.2 Furanones 85
2.2.1 3-(2*H*)-Furanones 85
2.2.2 4,5-Dihydro-2 (3*H*)-furanones (or γ-lactones) and 2-(5*H*)-furanones 87
2.3 Thiophenes and Thiophenones 92
2.4 Pyrroles and Reduced Systems 97
2.5 Dioxolanes, Dithiolanes, Trithiolanes, Imidazoles, and Pyrazoles 101
2.6 Oxazoles, Oxazolines, and Oxazolidines 103
2.7 Thiazoles and Reduced Systems 106
2.7.1 Structure and Occurrence 106
2.7.2 Sensory Properties and Patents 110
II–3 Six or more Membered Heterocyclic Systems containing one, two, or more Heteroatoms 112
3.1 Pyrans, Thiapyrans, and Derivatives 112
3.2 Pyridines and Derivatives 116
3.3 1,4-Dioxanes, Oxathianes, Dithianes, Dithiins, Trithianes, Oxadithianes, and Dioxathianes 120
3.4 Pyrazines 120
3.5 Miscellaneous: Cyclic Polysulphides and Macrocyclic Lactones 131
II–4 Fused Ring Heterocyclic Compounds 132
4.1 Fused Heterocyclic Derivatives from Five-membered Heterocyclic Compounds 132
4.2 Fused Heterocyclic Derivatives from Six-membered Heterocyclic Compounds 136

4.3 Polyfused Heterocyclic Compounds 140
II–5 References 140

Chapter III Mechanisms of Formation of Heterocyclic compounds in Maillard and Pyrolysis Reactions

Gaston VERNIN, University of Aix-Marseilles, France
Cyril PAŘKÁNYI, The University of Texas at El Paso, U.S.A.

III–1 Introduction. 151
III–2 Immediate Precursors of Heterocyclic Compounds in Foods: The various steps of the Maillard Reactions 152
2.1 *N*-Glycosylamines 152
2.2 Amadori and Heyns Intermediates 153
2.3 Rearrangement of Amadori and Heyns Intermediates: Reductones and Dehydroreductones 154
2.4 Retro-aldolization of Rearranged Amadori and Heyns Intermediates 154
2.5 Aldehydes 154
2.5.1 Strecker Oxidative Degradation of Amino Acids 156
2.5.2 Hydrolysis of *D*-Glucose in the Presence of Amino Acids . 158
2.5.3 Oxidative Degradation of Fatty Acids. 158
2.5.4 Aldol Condensation. 159
2.5.5 Miscellaneous. 159
2.6 Hydrogen Sulphide and Alkyl Sulphides 160
III–3 Occurrence and Formation of Heterocyclic Compounds in Model Systems 161
3.1 Furans and Furanones 161
3.2 Thiophenes 171
3.3 Pyrroles. 175
3.4 Oxazoles and Oxazolines. 181
3.5 Imidazoles 182
3.6 Thiazoles 184
3.7 Pyrans and Pyrones 186
3.8 Pyridines 189
3.9 Pyrazines 192
3.10 Cyclic Sulphides and Polysulphides, and Related Substances . 198
III–4 Conclusions 200
III–5 References 201

Chapter IV General Synthetic Methods for Heterocyclic Compounds used for flavourings

Gaston VERNIN, University of Aix Marseilles, France
A. K. EL-SHAFFEI, Assiut University, Sohag, Egypt

IV–I Introduction. 208
IV–2 Five-membered Rings. 208

2.1 Five-membered Ring with one Heteroatom: Furans, Thiophenes, Pyrroles, and Related Reduced Systems 208
2.1.1 Formation of C–X Bonds 209
2.1.1.1 From 1,4-Diketones 209
2.1.1.2 From 1,4-Disubstituted Butanes or 2-Butenes 210
2.1.1.3 From Pyrolysis of Alkanes 211
2.1.2 Formation of the C_3–C_4 Bond 212
2.1.2.1 From β-Dicarbonyl Compounds and α-Aminoketones 212
2.1.2.2 From α-Halogenoketones and β-Ketoesters 212
2.1.2.3 From α-Mercaptoketones 213
2.1.3 Formation of Other Bonds 213
2.1.4 Formation from Other Heterocyclic Compounds 213
2.1.5 Variously Substituted Derivatives from Parent Compounds 214
2.2 Five-membered Rings with two Heteroatoms in the 1,3 positions: Oxazoles, Thiazoles, and Imidazoles 215
2.2.1 From α-Halogenoketones 216
2.2.2 From (α-Acylamino Acid Esters) and (α-acylaminoketones) 216
2.2.3 From α-Aminoketones or α-Dicarbonyl Compounds 217
2.2.4 From Acyl Derivatives of α-Hydroxy- and α-Mercaptoketones 217
2.2.5 From α-Thiocyanoketones 218
2.2.6 From β-Hydroxy-, β-Amino-, and β-Mercapto Acylamines . 218
2.2.7 From 1,2-difunctionnal ethanes 219
2.2.8 Miscellaneous 219
2.3 Five-membered Rings with two or more Heteroatoms in the 1,2-, and 1,2,4-positions, respectively 219
2.3.1 Isoxazoles and Pyrazoles 220
2.3.2 Isothiazoles 221
2.3.2.1 From 1,2-Dithiolanes 221
2.3.2.2 From β-Thioxo Imines 222
2.3.2.3 From α-(*O*-Tosyloximino) Nitriles 222
2.3.2.4 From Isoxazoles 222
2.3.2.5 From β-Crotonic Esters and Thiophosgene 222
2.3.3 1,2-Dithiolanes 223
2.3.4 1,2,4-Trithiolanes 223
IV–3 Six-membered Rings 223
3.1 Six-membered Rings with one Heteroatom: Pyrans, Thiopyrans, Pyridines and Related Reduced Systems 223
3.1.1 From 1,5-Dicarbonyl Compounds and Derivatives 224
3.1.2 From β-Ketoesters, Aldehydes and Ammonia 228
3.1.3 From 1,3,5-Tricarbonyl Compounds 228
3.1.4 Pyridines from Other Sources 229

3.2 Six-membered Rings with two Heteroatoms in the 1,4-positions. 230
3.2.1 Pyrazines. 230
3.2.1.1 By Direct Ring Closure 231
3.2.1.2 From Cyclized Compounds 232
3.2.2 Dioxanes, Dithianes, Piperazines and 1,4-Oxathianes 234
3.3 Six-membered Rings with two Heteroatoms in the 1,2-, and 1,3- positions 235
3.3.1 1,2-Oxazines, 1,2-Pyridazines and 1,2-Dithiins 235
3.3.2 Pyrimidines 236
3.3.3 1,3-Dioxanes, 1,3-Thioxanes, 1,3-Dithianes, Reduced Pyrimidines, 1,3-Oxazines and 1,3-Thiazines 237
3.4 Six-membered Rings with more than two Heteroatoms. 237
IV–4 Fused Ring Systems 237
4.1 Benzofurans, Benzothiophenes, and Indoles 237
4.1.1 General Methods. 237
4.1.2 The *Fischer* Indole Synthesis. 238
4.1.3 The *Reissert* Indole Synthesis 238
4.1.4 The *Madelung* Indole synthesis 238
4.1.5 The *Bischler* Indole Synthesis 239
4.1.6 Benzothiophenes Synthesis 239
4.1.7 Acyl Derivatives 239
4.2 Benzoxazoles, Benzothiazoles, Benzimidazoles, Indazoles and Benzisothiazoles 240
4.3 Quinolines and Isoquinolines 241
4.4 Benzopyrans and Derivatives 243
4.5 Quinoxalines and Cyclopentapyrazines 244
4.6 Miscellaneous 245
IV–5 References 246

Chapter V Computer Application of Non-Interactive Program of Simulation of Organic Synthesis in Maillard's Reaction: A Proposition for New Heterocyclic Compounds for Flavours

R. BARONE and M. CHANON, University of Aix-Marseilles, France

V–1 Introduction. 249
V–2 Principle of the Method 250
2.1 Representation of Molecules 250
2.2 Representation of Reactions 250
2.3 How does the Program Work? 251
V–3 Results 253
3.1 From 4-Hydroxy-3 (2*H*)-Furanones, Hydroxyacetone, Glucose, Maltol and Isomaltol 253
3.2 From Furfural-Ammonia-Hydrogen Sulphide Model System . 258

V–4 Conclusion 260
V–5 References 261

Chapter VI Recent Techniques in the Analysis of Heterocyclic Aroma Compounds in Foods
L. PEYRON, Lautier-Aromatiques Company, Grasse, France
VI–1 Introduction 262
VI–2 Isolation of Heterocyclic Compounds from Complex Media . . 264
2.1 General 264
2.2 Isolation and Purification 267
2.2.1 Physical Methods 267
2.2.2 Chemical Methods 271
2.3 Examples of Isolation of Heterocyclic Compounds from Complex Flavouring Materials 272
2.3.1 Oxygen-Containing Heterocyclic Compounds 272
2.3.2 Sulphur-Containing Heterocyclic Compounds 273
2.3.3 Nitrogen-Containing Heterocyclic Compounds 273
VI–3 Identification, Determination, and Characterization 276
3.1 General 276
3.2 Identification and Determination 277
3.2.1 Identification 277
3.2.1.1 Organoleptic Methods 278
3.2.1.2 Physical Methods 281
3.2.1.3 Chemical and Biochemical Methods 296
3.2.1.4 Combination of Physical and Chemical Methods: Derivitization Methods 296
3.2.2 Determination 297
3.3 Characterization 297
3.3.1 Physical and Chemical Analytical Methods used to establish a Likely Structure 298
3.3.2 Confirmation by Total Synthesis 298
3.3.3 Examples 298
VI–4 Conclusions 299
VI–5 References 300

Chapter VII Mass Spectrometry of Heterocyclic Compounds used for flavouring
Gaston VERNIN and **Michel PETITJEAN**, University of Aix-Marseilles, France
VII–1 Introduction 305
VII–2 Furans, Thiophenes and Pyrroles 306

VII–3 Oxazoles, Thiazoles, Imidazoles, 1,3-Dioxolanes and Isothiazoles . . . 313
VII–4 Lactones, Thiolactones and Lactams . . . 319
VII–5 Pyrans, Thiapyrans, Pyridines and Pyrazines . . . 321
VII–6 Polysulphur-containing Ring Systems . . . 330
VII–7 *Bis*-Heteroaryl Compounds and Fused Heterocyclic Systems . 332
VII–8 Bank Mass Spectra Data: An indispensable tool for a fast identification of heterocyclic aroma compounds . . . 338
VII–9 References . . . 340

Chapter VIII The Legislation of Flavours
Gaston VERNIN, University of Aix-Marseilles, France
VIII–1 Introduction . . . 343
VIII–2 Toxicity of Heterocyclic Compounds . . . 344
VIII–3 References . . . 362

Index . . . 363

In memory of L. C. Maillard, a French chemist who lived at the beginning of the twentieth century and to whom we are indebted for the discovery of the role of sugars and amino acids in food flavours

Preface

Given their major importance in food flavours, heterocyclic compounds justified a systematic approach from the double point of view of analysis and synthesis.

Dr Gaston Vernin is, to date, one of the most competent experts of this complex field for writing a comprehensive review on such an interesting topic. He is at the same time a distinguished synthetic heterocyclic chemist and a most advanced researcher in the field of analytical chemistry. The seven other contributors to the eight chapters of the work are also distinguished researchers from universities and industry.

The major role of heterocyclic compounds in food flavours has been only recently recognized, owing to the recent progress in the techniques of analysis and identification of molecules (GC, GC/MS, HPLC, High resolution NMR). It seems therefore worth while reviewing our present knowledge in this most topical field exemplified by the prominent place occupied by nature-identical compounds in the flavour industry.

Gone are the times when the consumer could get his food products directly from the farm, since now most of them are processed on a large scale, and people's taste has become more uniform through the use of synthetic flavours.

Dr Vernin's work is very exhaustive, taking into account all the various aspects of the origin, the occurrence, the methods of separation, the methods of identification, and finally the methods of synthesis of heterocyclic components of aroma compounds.

Among the original aspects of the work, I have especially appreciated the use of computer-assisted synthesis of heterocyclic compounds in its simplified form by R. Barone and M. Chanon. Indeed this method allows the description of all the heterocyclic molecules likely to result from non-enzymatic reaction of a given sugar with a given amino acid. This original approach to Maillard reaction provides evidence for the extreme complexity of the mixtures responsible for the flavour of our foods.

I noted also the very useful data on mass fragmentation patterns of heterocyclic molecules reported in the last chapter, which may bring most valuable assistance to their identification.

This book will undoubtedly be appreciated by a large population of chemists both from academic and industrial laboratories. In the various chapters they will find all the necessary information for understanding the complex reactions of food chemistry used to identify the heterocyclic components responsible for flavour, and to develop new synthetic pathways for the preparation of the most significant. For these reasons I am quite confident of the success of Dr Vernin's most illuminating book.

JACQUES V. METZGER

Marseilles, France
March 1981

Introduction

Within the last few years, our knowledge concerning the composition of flavour compounds in foods has considerably improved, and this is due to the increasingly refined analytical methods. Among the volatile components of flavours, up to a thousand heterocyclic compounds have been identified so far in various processed foods, including: meats and fats, fish, vegetables, fruits, cereals, vegetable oils, nut and oilseed products, milk and dairy products, alcoholic and nonalcoholic beverages, tobacco, and essential oils to mention only a few.

All the categories of oxygen-, sulphur- and/or nitrogen-containing five- or six-membered rings or fused rings are present in trace amounts in flavours. They are mainly formed in non-enzymatic browning reactions known as Maillard reactions. Owing to their extraordinarily high odour strength, they are now highly appreciated by the food industry as ideal flavouring ingredients. Processed foods to which they can be added include baked goods, meat sauces and soups, condiments, pickles, snacks, butter, margarine, gelatins and puddings, ice creams, icings, candy, chewing gum, nonalcoholic beverages, tobacco, perfumes etc.

A combination of our interest in the exceptional olfactory properties of these newly discovered flavouring substances as well as our experience in heterocyclic chemistry have prompted us to write this book.

The first chapter is devoted to the precursors of heterocyclic compounds in food flavours. The role of carbohydrates, amino acids and peptides, vitamins, lipids and other specific components of each class of foods are discussed in detail. It is clear from the results of thermal and/or oxidative degradation studies of these food components that the smaller size molecules which are formed can act as the immediate precursors of heterocyclic compounds. The role of enzymes and that of ribonucleotides in the formation of food flavours is also emphasized.

In the next chapter we have attempted to develop the occurrence and the olfactory properties of heterocyclic compounds resulting mainly from technological processes. All the known substances have been classified according to their ring size, and in each category according to the number of O-, S- and/or N heteroatoms. Their olfactory properties are also reported.

In Chapter III, heterocyclic compounds identified in Maillard reactions and related model systems are classified as described above. The mechanisms of their formation from precursors are also discussed.

Chapter IV entitled 'General Synthetic Methods', is devoted to the main pathways for the preparation of each class of heterocyclic aroma compounds. Owing to the wide extent of this topic, our task has not been easy.

Then a short but important chapter deals with the computer-assisted synthesis applied to Maillard reactions. This original approach can lead not only to the heterocyclic compounds previously described, but also to new flavouring substances which will probably be identified in processed foods in the near future.

The next chapter describes the methods for isolation, separation, and identification of heterocyclic compounds in flavours. Among them, the essential role of chromatographic and sensory techniques is emphasized. We call special attention to mass spectrometry in combination with gas chromatography. This sensitive analytical tool is certainly the best one. For this reason, mass fragmentation patterns in heterocyclic chemistry are treated in chapter VII.

Finally, heterocyclic substances authorized as flavourings in the USA and/or in the European Community are given in chapter VIII as well as the toxicity of some heterocyclic compounds.

Each topic is so extensive that only highlights have been covered.

Increasing interest in the flavour of foodstuffs is evident from the continuously growing number of papers, reviews, books, patents, and symposia.

We hope that this book will contribute to a better understanding of the importance of heterocyclic compounds in food flavours. We are convinced that the identification of many more of them will be forthcoming. Last, but not least, if this book can in any way stimulate research in the above discussed areas, its main purpose will have been served.

ACKNOWLEDGEMENTS

Firstly, I wish to express my thanks to Professor J. Metzger for kindly agreeing to write the Preface of this book and for his valuable encouragement.

I wish to thank also all the contributors.

I am particularly indebted to C. Párkányi, Professor of Organic Chemistry at the University of Texas at El Paso, who corrected some chapters and translated Chapter VI.

I must also thank Mr B. Dorison, who translated Chapter I.

Mr J. Hariel, Manager of the ORIL Co Society, has given me the benefit of his views and criticism of the legislation of food flavours.

My thanks are also due to Mrs J. de Caseneuve who typed some parts of the manuscript.

I should like to thank my wife, and I want to express my deep gratitude for her valuable assistance in the artwork and in all stages of production of the book.

Finally, thanks are due to the University of Aix-Marseilles III, for financial support and to my publisher Ellis Horwood for his encouragement and assistance.

G. VERNIN March 1981

CHAPTER I

Heterocyclic Aroma Compounds Precursors

J. GARNERO, Robertet Company, Grasse, France

I–INTRODUCTION

During the last twenty years, our knowledge of the chemical composition of food aromas has made considerable progress mainly because of the GC/SM coupling. Among a total of about 10 000 substances occuring in aromas, heterocyclic compounds occupy a prominent position, as a result of their quite exceptional olfactory properties. Indeed, some of these substances exhibit such a low olfactory perception threshold that it is possible to detect the equivalent of one milligram in five hundred tons of water (e.g. 2-isobutyl-3-methoxypyrazine).

The origin of these compounds, whether they are aliphatic or heterocyclic, can be considered in two ways:

a) The enzymatic and microbiological processes to which the fermentation processes can be linked, since they consist of a series of enzymatic reactions. The main point to be observed in the course of these processes is the formation of carbon chain aliphatic compounds, substituted or not, for functional groups containing O, S, N heteroatoms, terpenes, but few heterocyclic compounds.

b) The non-enzymatic processes resulting from thermal treatments such as cooking, roasting, and especially coffee roasting and so on. These reactions which give rise to the formation of heterocyclic compounds are known as non-enzymatic browning reactions or *Maillard* reactions.

The first of these two pathways plays a significant role in the formation of certain aromas such as those from alcoholic beverages (wines, brandies, beer), dairy products (cheese, yoghurts. . .), non-alcoholic beverages (tea) and various species of *Allium* and *Brassicas*. The processes involved are complex and specific to each particular foodstuff.

The second pathway plays an important role in the formation of aromas from meat, coffee, and nuts. For bread and cocoa, both pathways come into play in order to produce the final aroma. Whatever the foodstuff under consideration, the role of precursors is essential to the formation of aroma. But the latter will vary according to the techniques used to make the foodstuff fit for consumption.

In the case of raw food, the initial precursors are then only to be found in the processed foodstuffs, the aroma resulting from biotransformations and complex chemical reactions involving these initial precursors. The partial or total destruction of the latter leads to reactive intermediates directly responsible for the formation of aromatic substances.

Two kinds of precursors are to be considered according as to whether they already exist in the fresh food or are formed in the course of biochemical and/or chemical processes.

I–2 MEAT AND RELATED PRODUCTS : BEEF, PORK, LAMB, AND CHICKEN

The meats we usually consume are composed of a tissue with a high glycogen content, the main component of muscles, and they are also rich in adipose tissue consisting of triglycerides. The presence of glycoproteins is also worth noting (as they give rise to peptides, amino acids, and sugars), fibrous proteins (actin, myosin), and phospholipids to a smaller amount.

Among various *post mortem* phenomena described by several authors [104], the conversion of nucleotides into sugar-phosphate and into sugars (ribose) can be mentioned. Fatty acids result from phospholipids and triglycerides hydrolysis. As for proteins, they are only slightly hydrolysed [65].

2.1 Primary precursors and their hydrolysis products

In the course of the various processes of conservation or cooking of meats, a certain number of reactions lead to peptides, amino acids, and sugars. Some authors have studied the main water-soluble precursors contributing to the overall flavour of cooked meat including beef, pork, lamb, and chicken [42,111, 220] (*see Table I–1*).

Table I–1 – Water-soluble meat flavour precursors [111]

Glycopeptides	Peptides
Nucleic acids	α-Amino acids
Free nucleotides	Sugar phosphate
Peptide-bound nucleotides	Amino-sugars
Nucleosides	(Sugar-amine)
Nucleotide sugar	Free sugars
Nucleotide sugar-amine	Amines
Nucleotide acetylsugar-amine	Organic acids

During heating in aqueous medium, these precursors generate peptides, amino acids, and sugars whose composition is reported in *Table I–2* [110].

Table I–2 – The main α-amino acids, peptides and sugars arising from the primary degradation of initial precursors [110]

α-Amino acids		
Taurine	Methionine sulphoxide	Tyrosine
α-Alanine	Asparagine	Phenylalanine
Cysteine	Aspartic acid	Hydroxylysine
β-Alanine	Threonine	Lysine
Glutamic acid	Serine	Arginine
Tryptophan	Proline	Valine
Methionine	Glycine	
Peptides	**Sugars**	
Anserine	Ribose	
Carnosine	Glucose	
Glutathione	Fructose	
	Glucosamine	

The ratios of the different free amino compounds in red meats vary according to the species as shown in *Table I–3* [110].

Table I–3 – Ratios of free amino compounds in red meats [110]

	Percent (%)		
Amino compounds	Beef	Lamb	Pork
Neutral amino acids	14.4	18.1	9.4
Hydroxy amino acids	5.5	6.2	3.4
Acidic amino acids	3.4	5.9	2.9
Sulphur-containing amino acids	9.6	27.0	13.9
Basic amino acids	9.8	20.2	7.1
Aromatic amino acids	2.0	2.9	1.0
Dipeptides (anserine + carnosine)	55.8	19.7	61.1

The two dipeptides : anserine and carnosine are by far the most abundant free amino components in red meats, followed by sulphur-containing amino acids (cysteine and cystine), neutral α-amino acids and other α-amino acids. The analysis of dialysable and non-dialysable fractions of aqueous beef extracts shows that the main α-amino acids are in decreasing order of importance: glutamic and aspartic acids followed by lysine, leucine, alanine, and valine, respectively [139,140]. Nucleotides present in red meats (*Table I–4*) are enzymatically decomposed to ribose-5′-phosphate [198] which on heating gives a meat flavour.

Table I–4 – Nucleotide content in red meat [110] a)

Meat	CMP	UMP	IMP	GMP	AMP
Beef	12.0	13.0	150.0	8.0	17.0
	1.0	1.6	107.0	2.1	6.6
Pork	1.9	1.6	123.0	2.5	7.6
Mutton	1.9	0.6	83.5	5.1	6.8

a) in mg/100 g; CMP: cystosine-5′-monophosphate; UMP: Uridine-5′-monophosphate; IMP: Inosine-5′-monophosphate; GMP: Guanidine-5′-monophosphate; AMP: Adenosine-5′-monophosphate.

Lipids in meat can be classified as intermuscular (mainly in connective tissues) and as intramuscular or muscle tissue lipids. They exist in close association with proteins and contain a large proportion of the total phospholipids. These latter compounds are highly susceptible to hydrolysis and yield linoleic and arachidonic acids. It must be pointed out that some free amino acids, peptides, sugar-amine, sugar-phosphate and organic acids exist in bovine meat [82].

2.2 Intermediate precursors in meat flavours

α-Amino acids and sugars – These two most important chemical categories participate in the various steps of the *Maillard* reaction. This very important flavour-producing chemical reaction involves oxidative deamination of an amino acid molecule with the formation of an aldehyde with one carbon less than the original acid. This has been the subject of several reviews [49,74,111,212] and will be detailed in *Chapter III*. Under roasting conditions (220° to 270°C) amino acids undergo a thermal degradation [105] leading to the formation of very complex mixtures (*Table I–5*). Similarly, sugars give furans on heating to 250°C. In the case of glucose, more than a hundred compounds have been identified. Among them were furan itself and its 2-acyl-, α-dicarbonyl and 2-alkyl derivatives (*Chapter II*).

Sulphur-containing amino acids – During preservation or simmering of chicken and beef meats, cysteine, cystine, and methionine produce ammonia and sulphur-containing small molecules such as hydrogen sulphide, methyl mercaptan, and methional. These simple compounds along with acetaldehyde and cysteamine (β-mercaptoethylamine) appear to be important precursors of many volatile aroma compounds.

Thermal degradation of thiamin – This reaction yields furans and sulphur-containing molecules such as hydrogen sulphide, thiophenes and thiazoles with meat-like flavours [41].

Table I–5 – Thermal decomposition products of some α-amino acids [105]

Valine	Leucine	Isoleucine	Alanine	β-Alanine
NH_3	NH_3	NH_3	NH_3	NH_3
CO_2	CO_2	CO_2	CO_2	CO_2
CO	CO	CO	CO	CO
Propane	Isobutane	Butane	Ethane	Ethane
Propylene	Isobutylene	Butylene	Ethylene	Propylene
Isobutylene	Isopentane	Isopentane	Propylene	Acetonitrile
Acetone	3-Methyl-1-isobutene	2-Methyl-1-butene	2-Butene	Acetone
Isobutanal	Acetone	2-Butanone	Acetaldehyde	Pyridine
Isobutylamine	Isobutanal	2-Methylbutanal	Ethylamine	β-Picoline
N-Isobutylidene isobutylamine	Isopentanal	2-Methylbutylamine	*N*-Ethylidene-ethyl amine	2,4-Lutidine
	Isobutylamine	2-*N*-Methylbutylamine		3,5-Lutidine
Diisobutylamine	*N*-Isobutylidene isoamylamine	*Bis*-(2-methylbutyl amine)	Pyridine	2,3,5-Trimethyl pyridine
	N-Isoamylidene diisoamylamine		*N*-Ethylpropionamide	

Ribose-5′-phosphate – As mentioned earlier, ribonucleotides decompose enzymatically to give ribose-5′-phosphate. This latter compound on heating is converted in beef broth into 4-hydroxy-5-methyl-2,3-dihydrofuran-3-one [198,199]. The reaction of this furanone with hydrogen sulphide gives 3-mercaptofurans and 3-mercaptothiophenes in various degrees of saturation (*Chapter III*).

Unsaturated fatty acids – Fatty acids oxidation leads to the formation of saturated, mono- and polyunsaturated aldehydes. They play an important role in cooked chicken flavour. As aldehydes arising from *Strecker* degradation of α-amino acids, they can react with either hydrogen sulphide and/or ammonia to give important flavour components. For example, methylthioethanethiol is a key compound of roast beef aroma [177,178].

Carbonyl- and nitrogen-containing compounds – The thermal degradation of sugars (hexoses, pentoses) and the *Strecker* degradation of amino acids are the main sources of carbonyl compounds in meats. Nitrogen-containing products arise from amino sugars, α-amino acids, and amines. It is from these various immediate precursors and hydrogen sulphide that heterocyclic compounds and other aroma compounds are formed. The main groups of precursors and types of reaction responsible for meat flavour formation can be summarized according to *Scheme 1a* [56,207,212].

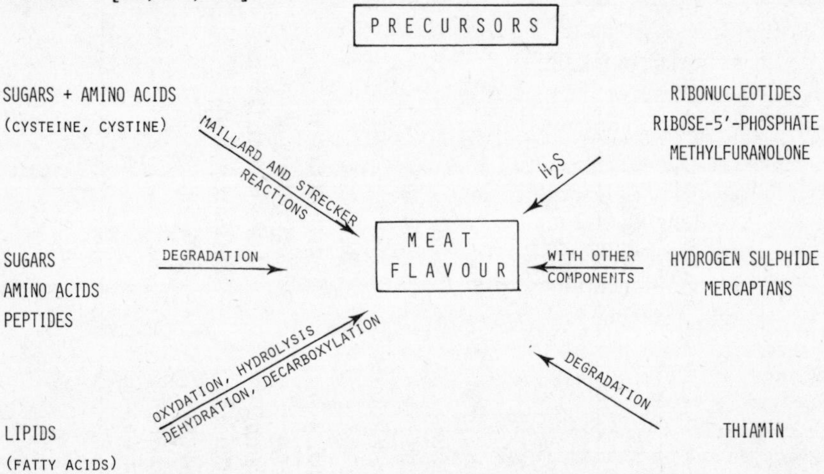

PRECURSORS

SUGARS + AMINO ACIDS
(CYSTEINE, CYSTINE)
MAILLARD AND STRECKER REACTIONS
RIBONUCLEOTIDES
RIBOSE-5′-PHOSPHATE
METHYLFURANOLONE
H_2S
SUGARS
AMINO ACIDS
PEPTIDES
DEGRADATION
MEAT FLAVOUR
WITH OTHER COMPONENTS
HYDROGEN SULPHIDE
MERCAPTANS
OXYDATION, HYDROLYSIS
DEHYDRATION, DECARBOXYLATION
DEGRADATION
LIPIDS
(FATTY ACIDS)
THIAMIN

Scheme 1a – Flavour precursors and types of reactions leading to meat flavour [56,207,212]

Flavour-enhancing products – The overall flavour sensation of meat may be divided into sensation due to 'taste' and 'aroma'. All *D*-amino acids have a sweet taste, while free *L*-glutamic acid and some of its dipeptide derivatives have a well-known brothy taste. The presence of these amino acids as well as that of ribonucleotides (IMP, GMP), sugars, and salts, have flavour-enhancing effects [101,191,228,229]. Solms [187] points out the flavour sensation as shown in *Scheme 1b*.

FLAVOUR SENSATION	Low boiling compounds and high boiling compounds with odour effects	ODOUR
	Potentiators and synergists	
	Non-volatile compounds with taste and tactile effects	TASTE

Scheme 1b – Flavour sensation [187]

I–3 NUT PRODUCTS

3.1 Peanuts

The peanut (*Arachis hypogaea L.*) is a leguminous New World plant that has been adapted to many other parts of the world, particularly Africa. They are very sensitive to cool weather and will not survive any frost. Although some peanuts are consumed as food with little processing (only cooking, roasting, and grinding to peanut butter) and some are used directly for animal feeding, most of the peanut crop is processed into oil and meal for animal feeds.

The volatile aroma of roasted peanut which numbers over 230 components is particuarly rich in (i) aliphatic and aromatic compounds and (ii) heterocyclic compounds (furans, pyrazines, thiophenes, and pyrroles). Carbonyl compounds consist of alkanals among which hexanal (in raw peanut) and 2-methylpropanal and butanal (mainly in roasted peanuts), 2-alkanones, 2-alkenals and 2,4-alkadienals [12,13]. Pyrazines play an important part in the aroma quality. Koehler [99] found 6 mg/kg of 2-methylpyrazine and 11 mg/kg of 2,5-dimethylpyrazine in peanut aroma [86,87,185].

All the above-mentioned compounds arise from proteins and amino acids as well as from reducing sugars or their mutual interactions. Peanut seed composition is indicated in *Table I–6*. Its main components are proteins and amino acids, lipids and fatty acids, carbohydrates and vitamins.

Table I–6 – Peanut content [138]

	g/100 g	Vitamins	
Water	4.5		
Sugars	3.1	Nicotinic acid	16 ng/100 g
Starch	5.5	Vitamin E	8.1 ng/100 g
Total nitrogen	4.5	(γ-Tocopherol)	
Proteins	24.3	Free folic acid	28 μg/100 g
Fatty oils	49.0	Total folic acid	110 μg/100 g
Carbohydrates	8.6	Pantothenic acid	2.7 mg/100 g

Proteins and α-amino acids. According to François [52] the average protein content in peanut cattle-cakes reaches 28%. These proteins are located in the seed cotyledons in globular protein form. Apparently flavour originates from rather specific types of micromolecules rather than from general macromolecular cellular components such as the large globular protein and the starches. Essential α-amino acids identified from peanuts are reported in *Table I–7*.

Table I–7 – Essential α-amino acids content in peanuts [35]

	g/100 g of proteins		g/100 g of proteins
Tryptophan	1.1	Cysteine	1.6
Lysine	3.5	Glycine	5.4
Threonine	2.9	Alanine	2.9
Leucine	6.2	Serine	6.6
Isoleucine	4.2	Proline	5.1
Valine	5.0	Arginine	10.6
Phenylalanine	5.0	Histidine	2.1
Methionine	1.0	Aspartic acid	14.1
Tyrosine	3.0	Glutamic acid	20.0

Young [231b] studied the amino acid composition of three commercial varieties of American peanuts. Glutamic acid (6.40%), arginine (3.48%) and aspartic acid are also the three most abundant amino acids. Free arginine can be used as a criterion to assess the seed maturity and the quality of the peanut aroma.

Mason *et al*. [118] have studied amino acid and peptide concentrations in peanuts before and after roasting. The biogenesis of peptides generating amino acids during roasting, as well as phenylalanine, coincides with the seed-ripening process and is necessary to favourable aroma development.

Sugars. The main sugars to be found in peanut seed are sucrose, fructose, glucose, inositol, and an unknown carbohydrate [128]. Newell *et al*. [128] have studied sugar and amino acid alterations in various samples before and after roasting. They postulate the existence of a formation mechanism for roasted peanut involving these precursors.

Lipids and fatty acids. The oil is one of the main sources of carbonyl compounds in peanut seed. The composition of the fatty acids of this oil is well known. It is a mixture of triglycerides, with palmitic, stearic, oleic, and linoleic acids. The effects of genotypes and habitat on the composition in fatty acids, on the oil quantity and that of proteins, have already been studied [26,75]. The carbonyl compounds occurring in peanut have been thoroughly studied. They are important seed aroma constituents before and after roasting. About 90 aldehydes and

ketones have been identified in the volatile fraction of roasted peanut [13,86, 136,219]. Some are obtained from an enzymatic pathway during seed-ripening. Hexanal appears to be a characteristic aroma component. According to Brown *et al.* [12] a whole range of aldehydes such as hexanal, octanal, nonanal, and 2-nonen-al seems to be responsible for the green or beany peanut aroma. The storage or drying conditions of the seed have been studied too [137,231].
Pyrazines. Over forty pyrazines have been identified in roasted peanut aroma (*see Chapter III*). They arise from amino acids and sugars in *Maillard* and *Strecker* reactions [128].

Aspartic acid, asparagine, glutamic acid, and phenylalanine contribute highly to the nutty note of this aroma. Some authors [231a] laid stress on the importance of diketoses and the influence of roasting on pyrazine formation.

3.2 Hazelnuts

Hazelnut (*Corylus avellana*) as other nuts (peanut, almond, pecan, walnut etc.) is a seed which is consumed before or after roasting. The main components of the seed [35,138] are water, lipids with polyunsaturated fatty acids, proteins, and carbohydrates (*Table I–8*). The roasted hazelnut aroma has been studied by several workers [98,112,183]. Among nearly two hundred components identified were pyrazines (≙40), furans (≙20), pyrroles (≙12), thiazoles, pyridines, thiophenes, and carbonyl compounds. Both the composition and the origin of these aroma compounds are very similar to those reported for roasted peanuts. Storage, drying, and roasting conditions also influence the production and the quality of the hazelnut flavour.

Table I–8 – Hazelnut content [35,138]

	g/100 g	g/100 g
Water	41.4	6
Sugars	4.7	–
Starch	2.1	–
Total nitrogen	1.4	–
Proteins	7.6	12.7
Fat	36.0	60.9
Carbohydrates	6.8	18.0
Fibre	–	3.5
Polyunsaturated fatty acids	–	23.0

I–4 NON-ALCOHOLIC BEVERAGES : TEA, COFFEE, COCOA

Tea, coffee, and cocoa are similar in character both in their composition and in the various steps leading to the production of the corresponding beverages. They contain alkaloids of the xanthine group. Caffeine is the main component of coffee and tea, whereas theobromine is present in relatively large amounts in cocoa bean.

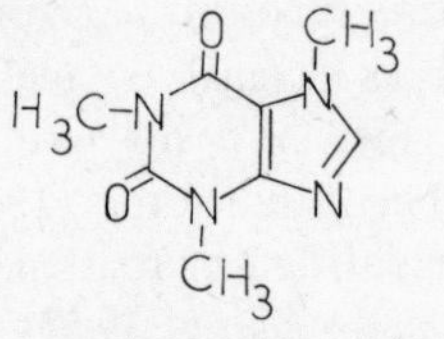

Caffeine

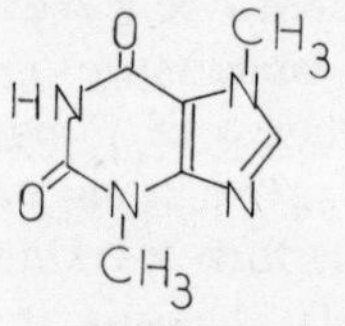

Theobromine

The technology involved in stimulant beverages processing has been summarized by Reymond [153]. The first step involves fermentation and enzymatic degradations giving rise to flavour precursors. This step is followed by firing (tea) or drying (cocoa, coffee) in driers. In the case of coffee and cocoa, another step of roasting ($T < 150°$ for cocoa and $> 180°$ for coffee) is carried out during which primary and intermediate precursors undergo important thermal degradations. These *Maillard* reactions lead to the formation of a great number of volatile substances responsible for the coffee aroma (≙ 650 components).

Several reviews showed the olfactive importance of these compounds in tea [216a,227], coffee [217] and cocoa [208,216b].

4.1 Tea aroma precursors

The black tea aroma is a typical example of the important changes induced by fermentation. Three groups of substances are concerned as flavour precursors. These important contributors are: polyphenols (flavanols), carotenoids, unsaturated fatty acids (e.g. mainly linoleic acid). Other usual components of living matter which contribute to thermal degradation, may also act as precursors. In *Table I–9* and *I–10*, the total analysis of Indian tea leaves [138] and the composition of tea leaves are reported, respectively [46].

Table I–9 – Indian tea leaves content [138]

	g/100 g
Water	9.3
Sugars	3.0
Total nitrogen	4.08[a)]
Proteins	19.6
Fatty oil	2.0
Carbohydrates	3.0

a) From which 0.95 g of nitrogen contained in alkaloids.

Table I–10 – Constituents from tea leaves [46]

Phenols
Flavanols
Flavonol glycosides and others
Leucoanthocyanins
Acids and depsides (theogallins)
Non-phenolic materials
Caffeine, theobromine, theophylline
Amino acids (theanine)[a)] and proteins
Carbohydrates, sugars and polysaccharides
Organic acids
Lipids
Chlorophylls and carotenoid
Volatile constituents
Vegetable fibres
Ash
Enzymes
Polyphenol oxydase
Pectin methylesterase
Phosphatases

a) For formula see Chapter III, Fig. 2, No. **116**.

Polyphenols. They play an important role in black tea preparation. It is a well-known fact that from oldest Chinese antiquity, green tea has been highly appreciated. This is owing to the disactivation during processing of a dimerase enzyme, the polyphenoloxidase which acts on flavanols such as epigallocatechin and epicatechin [167,168,221]. This enzyme occurs in the formation of the characteristic flavour of the infusion prepared from black tea. In tea leaves freshly harvested the following flavanols: (-)-epicatechin **1a**, (-)-epicatechin-3-gallate **1b**, (-)-epigallocatechin **1c**, (-)-epigallocatechin-3-gallate **1d**, (+)-catechin **2a**, and (+)-gallocatechin **2b** were present (*Fig. 1*). The total amount of these flavanols will vary from about 15 to 25% (dry weight basis). In the course of the fermentation step, the polyphenol oxidase comes into contact with flavanols.

An oxidation of the flavanols **1** and **2** and gallic acid **3** by coupled oxidation leads to the formation of the *bis*-flavanols A, B and C (**5a, 5b, 5c**), theaflavin **6a**, theaflavin gallates A and B (**6b, 6c**), theaflavin digallate **6d**, epitheaflavic acid **7a**, 3′-galloyl epitheaflavic acid **7b**, and thearubigins **8**. Many chemical changes take place during the tea manufacturing process, especially during tea fermentation and the subsequent firing (drying) step that are essential to the formation of the aroma characteristic of tea. Millin *et al.* [120] have established that polymerization products of flavanols play an outstanding role in the mouth feel of black tea infusions. They isolated four fractions by means of gel permeation chromatography and by dialysis. An intermediate molecular weight fraction including

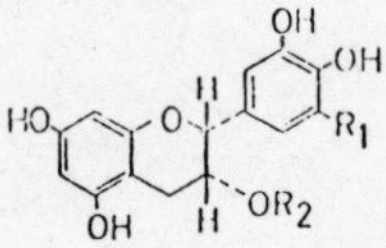

1; a) $R^1 = R^2 = H$
b) $R^1 = H$, $R^2 = \underline{4}$
c) $R^1 = OH$, $R^2 = H$
d) $R^1 = OH$, $R^2 = H$

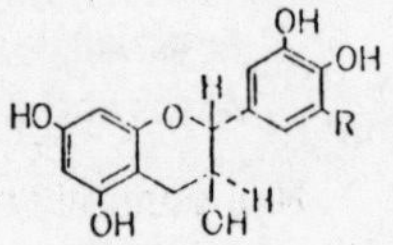

2; a) R = H
b) R = OH

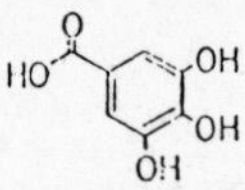

3

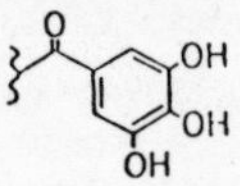

4
(Galloyl group)

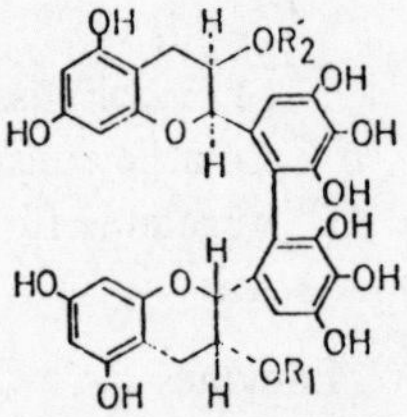

5; a) $R^1 = R^2 = \underline{4}$
b) $R^1 = H$, $R^2 = \underline{4}$
c) $R^1 = R^2 = H$

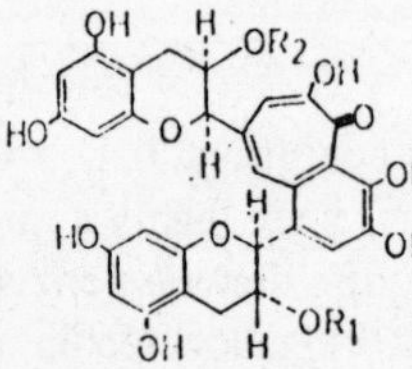

6; a) $R^1 = R^2 = H$
b) $R^1 = H$, $R^2 = \underline{4}$
c) $R^1 = \underline{4}$, $R^2 = H$
d) $R^1 = R^2 = \underline{4}$

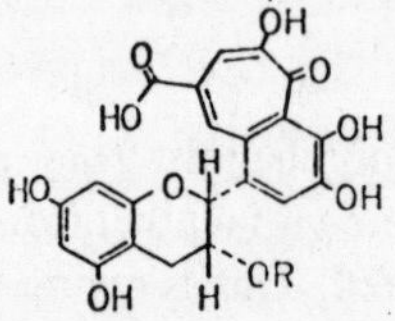

7; a) R = H
b) R = $\underline{4}$

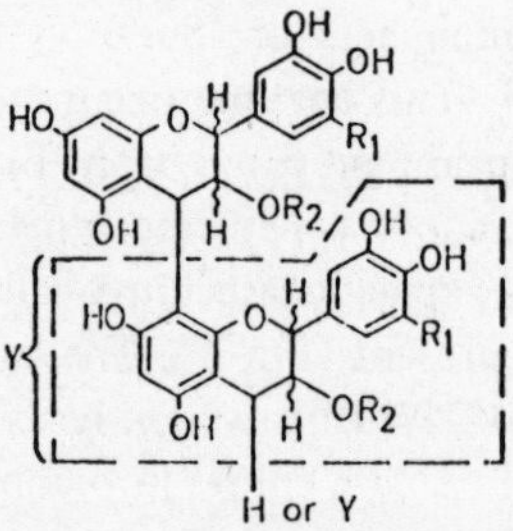

H or Y

8; a) $R^1 = R^2 = H$
b) $R^1 = H$, $R^2 = \underline{4}$
c) $R^1 = OH$, $R^2 = H$
d) $R^1 = OH$, $R^2 = \underline{4}$

Fig. 1 – Polyphenols in tea [168].

theaflavins, leucoanthocyanins, and theogallin is described as having brisk, astringent, and strong taste. A high molecular weight fraction mainly consisting of polysaccharides and phenolic substances possesses softness and flatness taste.

Carotenoids. Suggested by several workers, the carotene oxidation constitutes one of the important factors contributing to tea flavour [125]. Carotenoids present in fresh tea leaves have been separated and identified by Sanderson *et al.* [166]. They are: theoxanthin, violaxanthin, lutein, and β-carotene. Amounts of carotenoids decrease during the fermentation step. The model system studied by these authors [166] shows the formation of β-ionone derivatives such as dihydroactinidiolide, 2,2,6-trimethylcyclohexanone, 2,2,6-trimethyl-6-hydroxycyclohexanone, and 5,6-epoxyionone. Some of these substances exhibit very low thresholds of perception.

The analysis of the black tea aroma reveals the presence of a great number of substances arising from carotene oxidation. These latter compounds and those resulting from transformation of these primary oxidation products are reported in *Fig. 2*. Among them theaspirone **16** and dihydroactinidiolide **15** play a preponderant role in black tea aroma [8,150].

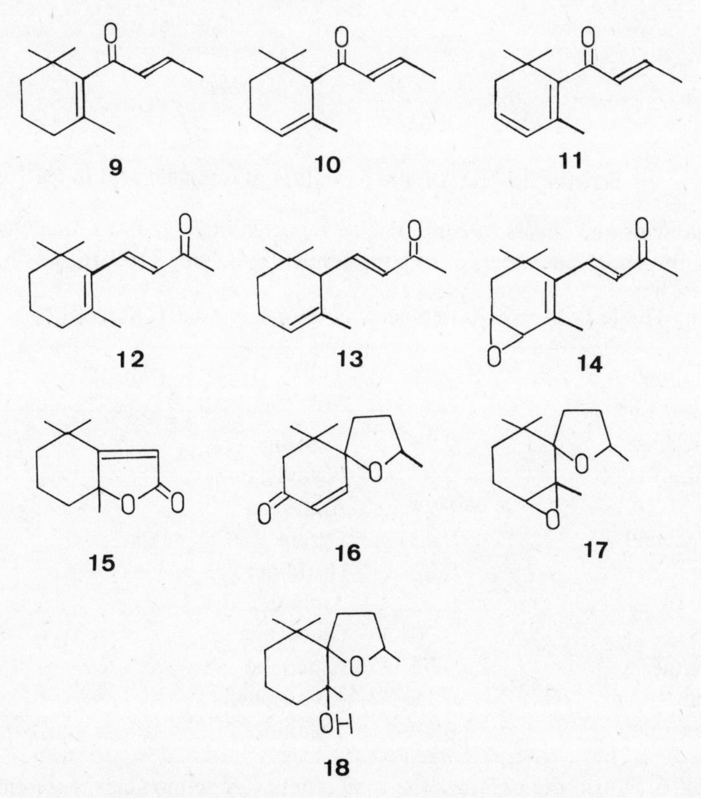

Fig. 2 – Volatiles resulting from carotene oxidation in black tea.

The model used by Sanderson & Graham [167] gives also an explanation for the presence of monoterpenoids such as linalool, *cis* and *trans* linalool oxides in tea aroma.

Lipids. The carbonyl compound formation under the action of lipoxygenase upon C–18 unsaturated fatty acids of lipids contained in leaves and fruits of various plants has been the subject of important work [28,181,182]. Hatanaka *et al*. [69] using tea leaf chloroplasts showed that ^{14}C–labelled linolenic acid is converted into 12-oxo(*E*)-dodecen-10-oic acid. They proposed the following scheme to explain the formation of (*E*)-2-hexenal found in tea aroma [59]. It is described as having a coarse vegetable-like aroma.

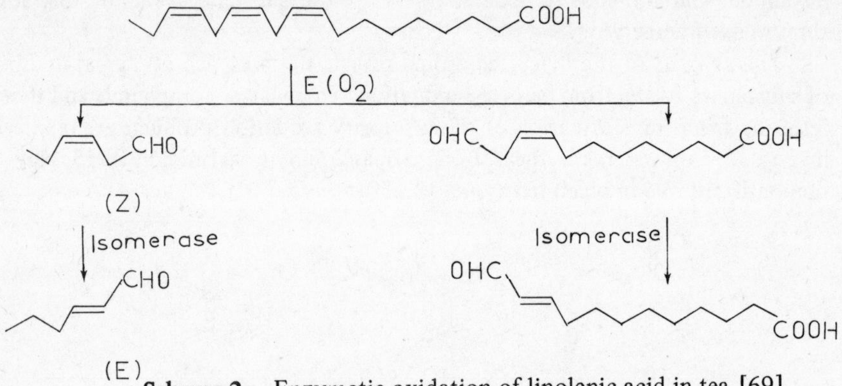

COOH

$E(O_2)$

CHO

OHC COOH

(Z)

Isomerase

Isomerase

CHO

OHC

COOH

(E)

Scheme 2 – Enzymatic oxidation of linolenic acid in tea [69]

α-Amino acids and sugars. α-Amino acids reported in *Table I–11* are considered as being terpenoid precursors during the fermentation step [167,168].

Table I–11 – α-Amino acids in the tea extract [167,168] a)

Amino acids	(%)	Amino acids (continued)	(%)
Aspartic acid	0.39	Tyrosine	0.15
Threonine	0.07	Phenylalanine	0.16
Serine	0.24	Ammonia	0.13
Glutamic acid	0.42	Lysine	0.04
Glycine	0.02	Histidine	0.002
Alanine	0.12	Arginine	0.03
Valine	0.20	Glutamine	0.19
Methionine	0.02	Aspargine	0.24
Isoleucine	0.18	Tryptophan	0.09
Leucine	0.19	Theanine	3.4

a) Total 6.29%; in the beverage the total amount of amino acids represents 2.1% (theanine 1.1%)

Wickremasinghe [222] proposed the following scheme (*see Scheme 3*) for the biogenesis of terpenoids, carotenoids, and steroids from leucine in tea leaf chloroplasts. The validity of this scheme is based on the fact that leucine concentrations in black tea are very weak as compared to the quantity that occurs in tea leaves extract.

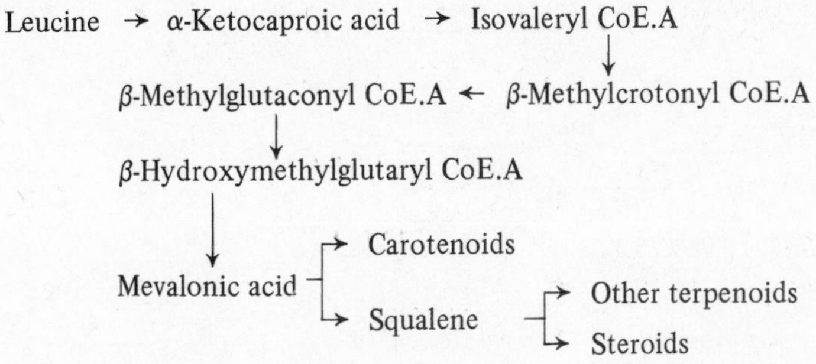

Leucine → α-Ketocaproic acid → Isovaleryl CoE.A
↓
β-Methylglutaconyl CoE.A ← β-Methylcrotonyl CoE.A
↓
β-Hydroxymethylglutaryl CoE.A
↓
Mevalonic acid → Carotenoids
Mevalonic acid → Squalene → Other terpenoids
Squalene → Steroids

Scheme 3 – Terpenoids, carotenoids, and steroids biogenesis from leucine [222] (CoE.A = coenzyme A)

Also, the production of α-ketocaproic acid is particularly significant during the flavour development step. This is due to the joint action of an enzyme: the acetoglutarate leucine, a transaminase of coenzyme A and manganese [4]. According to *Strecker*, the changes occurring during the fermentation process may account for the formation of aldehydes by amino acid degradation [25, 162]. During the next firing step, sugar and amino acid interactions are responsible for the formation of the following heterocyclic compounds: furans, pyrroles, pyridines, pyrazines, oxazoles, thiazoles, and quinolines [216a]. In the solid extract of tea, 6.52% of total sugars are present including fructose (2.0%), glucose (1.9%), *meso*-inositol (0.5%), sucrose (1.6%), maltose (0.1%), and raffinose (0.3%) [168].

Hydrogen sulphide and dimethyl sulphide. These two important reactive compounds which have been found to be present in tea aroma, contribute to the formation of sulphur-containing heterocyclic compounds (thiophenes, thiazoles, dithiolanes etc.). Their formation may be explained by the presence of methylmethionine sulphonium salt in the extract.

$$(CH_3)_2\overset{\oplus}{S}CH_2CH_2CH(COOH)NH_2 \rightarrow (CH_3)_2S + HO(CH_2)_3CH(COOH)NH_2$$

(Homoserine)

Methanethiol, methional, and thiophene have also been found in the black tea aroma.

Alkaloids. Caffeine, which accounts for 7% of the solid extract, is not practically affected during the preparation of the black tea. On the other hand, it has a significant effect on the astringency taste of the infusion. The galloyl groups on the black tea polyphenols are the specific sites involved in the complexation with caffeine [168].

The various steps of the formation of the black tea aroma from the above precursors can be summarized as shown in *Scheme 4* [167].

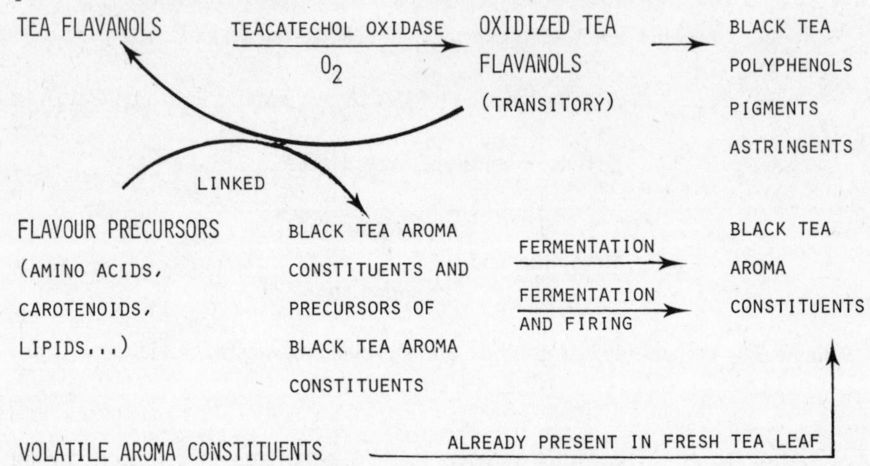

Scheme 4 – Proposed scheme for the formation of black tea aroma during fermentation and firing [167]

4.2 Cocoa flavour precursors

As with tea. cocoa flavour precursors interfere during the various manufacturing stages. The latter range from the cocoa bean to the edible food, either in cocoa powder form or in chocolate form. The two main steps are the fermentation which produces important changes in the structure and the composition of cotyledons, followed by roasting at a temperature not exceeding 150°C for 40 minutes. The precursors participating in the fermentation step are: enzymes, polyphenols and aromatic acids, lipids, and amino acids. The precursors involved in the second stage are sugars, proteins, peptides, amino acids, and theobromine.

4.2.1 Precursors involved during the fermentation step. They have been reviewed by several authors [121,152,157b].

Enzymes. In the course of the fermentation step, biodegradations take place which lead to the formation of (a) α-amino acids and peptides from proteins, (b) reducing sugars through polysaccharide and anthocyanin hydrolysis, and (c) polyphenol oxidation products. The enzymes involved during these biotransformations are: polyphenoloxidase, catalase, peroxidase, decarboxylase and coenzyme A [144]. On the other hand, the role played by bacteria has been studied by Ostovar & Keeney [131] who have identified ten different families of microorganisms in Trinidad cocoa beans.

Significant differences were clearly shown in cocoa bean composition before and after fermentation [156b]. These differences concern flavonoids, sugars, and amino acids. The latter are to be found in a larger amount after the fermentation step.

Polyphenols and phenol acids. Flavonols in cocoa from Ghana and Trinidad were examined by Rohan *et al.* [156b] and Griffith [61]. The main components belonging to this class are leucoanthocyanin (flavylogenes 1, 2, 3, 4), flavonols (quercetin and quercitrin) and three phenol acids: coumaric, caffeic, and chlorogenic acids. A comprehensive study of aromatic acids present in fermented cocoa was published by Quesnel [145]. He showed that protocachetic, phenylacetic, *o*-hydroxyphenylacetic, phloretic acids and esculin arise in the course of the fermentation process.

Sugars. The following sugars have been identified in cocoa beans: *D*-fructose, *D*-glucose, sucrose, raffinose, stachyose, *D*-galactose, melibiose, mannitriose, and mesoinositol [21]. The limit values of total and reducing sugars fluctuate between 0.34 and 1.38% for the former and 0.39 and 3.48 for the latter [157a]. During the fermentation step, sucrose is almost completely removed, thus increasing the amounts of reducing sugars (fructose and glucose) [157a]. According to the varieties under examination, an increase in reducing sugars from 2- to 16-fold can be observed [148].

Peptides and amino acids. They are formed in cocoa beans during fermentation in the course of proteolytic reactions [121,156a]. The amino acid composition undergoes important quantitative modifications as shown in *Table I–12* [156a].

Table I–12 – α-Amino acid content in the dialysis fraction of cocoa beans before and after fermentation [156a]

α-Amino acids	Before fermentation	After fermentation
	(g/100 g of the fraction)	
Lysine	0.08	0.56
Histidine	0.08	0.036
Arginine	0.08	0.35
Threonine	0.14	0.84
Serine	0.88	1.99
Glutamic acid	1.07	1.77
Proline	0.72	1.97
Glycine	0.09	0.35
Alanine	1.04	3.61
Valine	0.57	2.60
Isoleucine	0.56	1.68
Leucine	0.45	4.75
Tyrosine	0.57	1.27
Phenylalanine	0.56	3.36

Pyrazines may be induced by fermentation. Indeed, Van Praag *et al.* [209] have observed the formation of tetramethylpyrazine in the presence of micro-organisms such as *'Bacillus subtilis'*.

Lipids and phospholipids. Cocoa lipid extract contains 98% neutral lipids (cocoa butter) and 1 to 2% polar lipids. The latter are composed approximately of 70% glycolipids and 30% phospholipids essentially composed of phosphatidyl choline [139]. The lipids are separated in the course of the cocoa powder manufacture and they are reincorporated when the chocolate is produced. They exert a determining influence on the physical properties of chocolate. Cocoa butter is composed of triglycerides. Under the action of a lipase, the following fatty acids are generated: palmitic (25.9%), stearic (34.6%), oleic (36.7%), and linoleic (2.8%) acids.

During storage of cocoa beans, under certain conditions, free acidity of 0.97 mg/g goes up to 2.45 mg/g. Accountable for this increase are the saturated C_{16} and C_{18} fatty acids. The presence of significant quantities of esters of aliphatic and aromatic acids in cocoa aroma is to be attributed to fermentation [151].

4.2.2 Precursors involved during roasting. Generally the compounds involved are amino acids, peptides, and sugars. A large number of heterocyclic compounds (about 100 including 70 pyrazines) and volatile compounds are obtained through their mutual interactions in the course of *Maillard* and *Strecker* reactions. They finally give rise to the overall cocoa aroma (310 volatile compounds [112,158, 178,216b]). Rohan & Stewart [157b] have shown that the destruction of reducing sugars requires a temperature lower than that needed for the total destruction of amino acids. Also, the destruction of α-amino acids at a given temperature varies widely according to the amino acid. For example, leucine, arginine, methionine, and lysine are only slightly degraded after heating at 100°C for an hour in the presence of glucose, whereas phenylalanine, threonine, glutamine, and valine are partly degraded. Reineccius *et al.* [149] have studied the influence of variety, fermentation, temperature, and roasting on aroma quality depending on pyrazine concentration. Trimethylethylpyrazine, which is an important flavour contributor, was found to be much more important in beans from Ecuador than in beans from Accra.

Aldehydes. Volatile aldehydes are generated from free amino acids by *Strecker* degradations. Thus, *L*-leucine gives 3-methylbutanal, and *L*-phenylalanine gives phenylethanal. These two aldehydes are certainly the precursors of an important flavour contributor in roasted cocoa, 5-methyl-2-phenyl-2-hexenal *via* an intermediary aldolisation product [209] (*see Chapter III*).

Bitter compounds. The pleasant bitterness of cocoa is due to diketopiperazines resulting from the interaction of cyclic dipeptides with theobromine [141]. These dipeptides are formed in trace amounts during cocoa roasting.

4.3 Coffee flavour precursors

The composition of roasted coffee not only varies according to the origin of green coffee but also the technique used, the degree of roasting, and roasting time. According to Lockhart [107] the composition of green coffee from various countries is as follows (*see Table I–13*).

Table I–13 – Green coffee content (%) according to various origins [107]

	Santos	Mocoa	El Salvador	Guatemala	Hawaii
Moisture	8.75	9.06	8.7	10.5	–
Ether extract	12.96	14.0	8.2	5.0	18.2
Nitrogen	–	–	2.3	1.8	2.6
Proteins	9.5	8.56	–	–	15.9
Crude fibre	20.7	22.46	21.3	23.1	13.8
Ash	4.41	4.20	3.6	3.2	3.6

Seed storage and humidity are responsible for modifications which lead to the emanation of off-flavours or a deterioration of aroma itself.

In the next table we indicate the proportions of the main green coffee components [107] (*see Table I–14*).

Table I–14 – Composition of green coffee [107]

	g/100 g		g/100 g
		Miscellaneous	
Water	8–12	Tannins	2
Oil (ether extract)	4–18	Coffee tannic acid	8–9
Unsaponifiable	0–2	Caffeic acid	1
Nitrogen	1.8–2.5	Pentosans	5
Proteins	9–15	Starch	5–13
Caffeine	0–2	Dextrin	0–0.85
Chlorogenic acid	2–8	Sucrose	5–10
Trigonelline	2.5–4.5	Reducing sugars	0–5
Mineral ash (mg/100 g)		Cellulose	10–20
Calcium	85–100	Hemicellulose	20
Phosphorous	130–165	Lignin	4
Iron	3–10	Vitamins (present in small amounts)	
Sodium	4		
Manganese	1–45		
Rubidium	traces		
Copper	traces		
Fluorine	traces		

In 1964, little was known about the formation mechanism of aromatic substances in roasted coffee. Today, the role of precursors in these reactions is far better appreciated, and this is mainly owing to analytical progress and the use of model systems. According to Russwurm's [159] findings the main chemical groups of precursors involved during coffee roasting are as follows [111,153]: (a) carbohydrates and sugars, (b) proteins and amino acids, (c) amines and tryptamines, (d) phenol acids, (e) trigonelline, (f) non-volatile acids such as citric acid.

During roasting, which constitutes the main stage, the temperature lies between 180° and 260°C. The duration may vary according to the roasting degree that is wanted : light, medium or dark.

The composition of the main coffee components before and after roasting reported in *Table I–15* [48] reveals the significant participation of proteins, sucrose, and chlorogenic acid during this stage.

Table I–15 – Composition of green and roasted coffee [48]

Constituents	Per cent, dry basis Green	Roasted
Hemicellulose	24.0	24.0
Cellulose	12.7	13.2
Lignin	5.6	5.8
Fat	11.4	11.9
Caffeine	1.2	1.3
Sucrose	7.3	0.3
Chlorogenic acid	7.6	3.5
Proteins	11.6	3.1
Trigonelline	1.1	0.7
Reducing sugars	0.7	0.5
Unknown	14.0	31.7

Carbohydrates and sugars. They are equivalent to 40 or 50% of the green coffee bean weight. These carbohydrates are highly polymerized water-soluble molecules whose components are mannose and galactose. Galactomannoses do not appear to be susceptible to roasting, and their hydrolysis is difficult to achieve [195]. Other polysaccharides have been mentioned by various authors [33,196].

During the roasting process, the amount of arabinose and galactose which result from thermally degradable polysaccharides, decreases, whereas the quantity of mannose increases proportionally [196]. Sucrose in green coffee accounts for 6 to 10% of its weight. This constituent is the most susceptible to heating, as shown in the following results [48] (*see Table I–16*).

Table I–16 – Effect of roasting on sucrose content in coffee [48]

Coffee	Colombia		Santos	
			Per cent, dry basis	
	% initial	(% loss)	% initial	(% loss)
Green	4.59	–	5.47	–
High roasted	0.45	(90.20)	0.68	(87.57)
Medium roasted	0.17	(96.30)	0.27	(95.06)
Dark roasted	0.06	(98.69)	0.10	(98.17)

The content of sugars in roasted coffee and in instant coffee extract is reported in *Table I–17* [100].

Table I–17 – Sugar content (% by weight) in roasted coffee (a) and instantly soluble coffee powder (b) extracts [100]

Sugars	(a)	(b)
Glucose	0–0.9	0–0.37
Fructose	0–0.9	0–0.5
Arabinose[c]	0–0.5	0.4–2.5
Galactose[c]	traces	0.2–0.9
Mannose[c]	–	0.1–1.0
Ribose	–	traces

(c) Arabinose, galactose, and mannose arise from hydrolysis of degradable polysaccharides.

In the course of the roasting process, the reducing sugars participate in the various steps of the *Maillard* reaction. It is worthy of note that they also undergo cyclodehydration reactions, in the course of which, furfural, along with its 5-hydroxymethyl derivative, are formed from pentoses and hexoses respectively. Apart from the two previous components, numerous furans are to be found in the sugar hydrolysates (*see Chapter III*).

Proteins, peptides and amino acids. The first thorough study of green coffee protein content dates back to 1952. Out of a total of fourteen amino acids separated, nine could be measured. The same amino acids can be found in roasted coffee. Subsequently, several studies were published on peptides and amino acids obtained from hydrolysis of the proteins of green and roasted coffee of various origins [48,194]. Out of 17 green coffee amino acids, 5 were also proved to be totally destroyed during roasting. They are: arginine, cysteine, lysine, serine, and threonine.

On the other hand, an increase can be observed in the quantities of glutamic acid, leucine, phenylalanine, and proline. The other amino acids, namely: alanine, asparagine, glycine, histidine, methionine, tyrosine, and valine remain unaltered. The structure of the amino acids decomposed during roasting, as well as the degree of roasting, have an influence on the formation of heterocyclic compounds present in coffee aroma. Consequently, cysteine leads to thiophenes and thiazoles, whereas hydroxyamino acids give rise to pyrazines. The composition of this aroma, which can be broken down into more than 650 constituents [210], 254 of which being heterocyclic compounds, formed the subject of advanced studies [215] *(see Chapter II)*.

Amines and tryptamines. The three polyamines: putrescine, spermine, and sperminidine which are present in green coffee, are either destroyed or transformed during the roasting process. Putrescine especially is converted into pyrrolidine [106]. Consequently, they are important aroma contributors. Besides, three *N*-alkoyl-5-hydroxytryptamines **19** (n = 18, 20, 22) have been identified in green coffee wax. These amines possess anti-oxidizing properties towards lard. They are also susceptible to temperature and to roasting time [51,226].

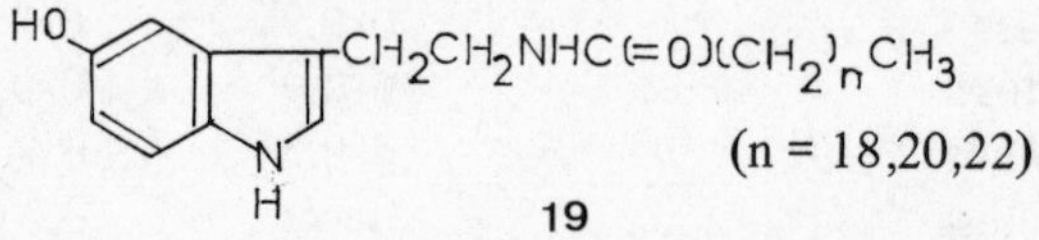

HO — $CH_2CH_2NHC(=O)(CH_2)_nCH_3$

(n = 18,20,22)

19

Lipids. Coffee lipids break down into glycerides, fatty and unsaponifiable acids. The saponifiable fraction contains linoleic acid (47%), oleic acid (8%), hexenedecenoic acid (1%), palmitic acid (32%), stearic acid (8%), behenic acid (5%), and higher-chain fatty acids. The unsaturated aldehydes: 2,4-alkadienal, 1,6-hexadienal, and (*E*)-2-nonenal, that play a prominent olfactory role in coffee aroma, are obtained from the oxidation of unsaturated fatty acids in C-18 [182]. (*E*)-2-Nonenal possesses a very low perception threshold (1 ppb).

Coffee oil also contains two diterpenic acids: kahweol **20** and cafestal **21**. Kaufmann & Schickel [93] studied the behaviour of lipids during the coffee roasting process.

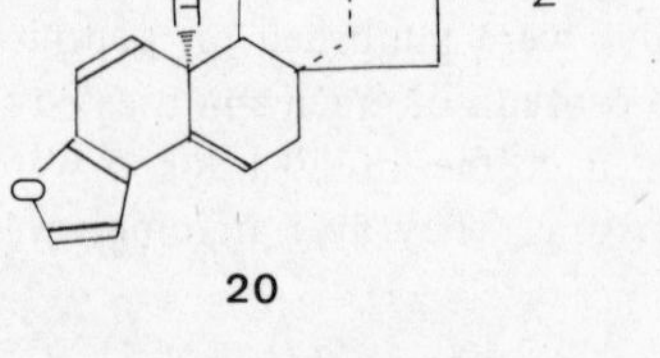

OH, CH_2OH, H

20

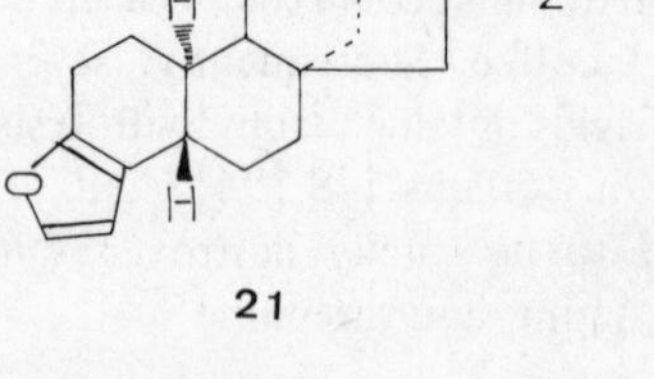

OH, CH_2OH, H, H

21

Phenol acids. Among the coffee phenol acids which account for 7% in weight, the most significant is chlorogenic acid **24**. It was isolated in 1908 and characterized in 1932 like 3-caffeoyl quinic acid. Since then other mono- and dicaffeoyl quinic acids have been identified in coffee.

These acids undergo considerable degradation in the course of the roasting process, notably chlorogenic acid [48].

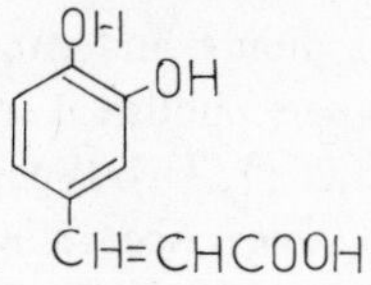

22 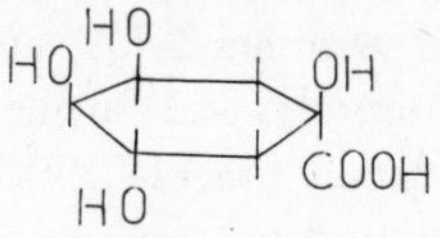 **23** 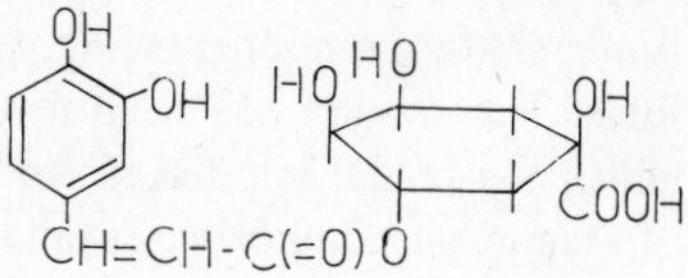 **24**

The acid fraction is also composed of ferulic, caffeic, and feruloyl quinic acids. *Para* coumaric acid is assumed to occur in *Robusta* coffee but to be absent in *Arabica* coffee [142]. The thermal degradation of quinic acid **23** at 240°C is a route to the following phenols: phenol, catechol, hydroquinone, pyrogallol. The degradation of caffeic acid **22** gives catechol (58%), 4-ethylcatechol, 4-vinylcatechol (0.8%), and 3,4-dihydroxy cinnamic aldehyde. These two acids are assumed to be the diphenol precursors found in roasted coffee aroma [205,206].

Trigonelline. This important coffee component **25** (1% in weight) is easily degraded by heating at 180–230°C for 15 minutes in a sealed tube. This reaction

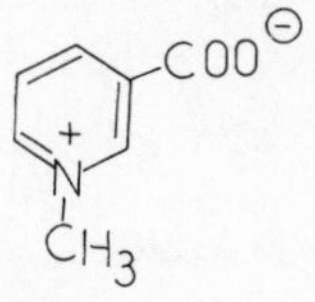

25

gives rise to a residue composed of pyridines (46%), pyrroles (3%), bicyclic compounds (29%: bipyridyls, piperidinylpyridine), and miscellaneous. Six pyridines (pyridine itself, its three methyl isomers, and 3-ethyl and 3-carbomethoxy pyridines) and three pyrroles (*N*-methyl-, *N*-methyl-2-formyl, and 5-methyl-2-formyl) have actually been identified in coffee aroma [214].

Non-volatile acids. Acidity plays an important role in the flavour acceptance of coffee beverages. During the roasting operation, citric **26** and malic **27** acids are partly degraded. Silicagel chromatographic analysis shows the formation of

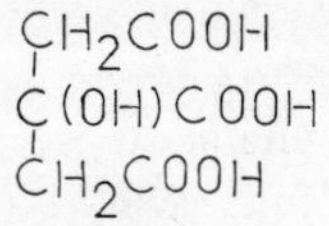

CH_2COOH
$C(OH)COOH$
CH_2COOH

26

$CHOHCOOH$
CH_2COOH

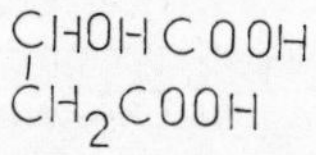

27

itaconic, *cis* citraconic, *trans* mesaconic, *cis* maleic, and *trans* fumaric acids [225]. These acids as well as phenolic acids and volatile acids contribute to the flavour of coffee beverages.

I–5 FRUITS AND VEGETABLES

Fruits and vegetables occupy a most prominent place in our food. In whatever way we prepare or consume them, their aroma is always obtained from the same fundamental components: i.e. proteins, α-amino acids, sugars, lipids, and vitamins. The subject has been reviewed recently, and further papers published to which the reader is referred for more details [45,119,130,165,201]. Two types of reaction are important aroma contributors in fruits and vegetables. According to whether the fresh or raw foodstuff is considered, the aroma results from enzymatic processes in which vegetable cells are involved. In return, in the foodstuff subjected to thermal treatment, non-enzymatic *Maillard* type browning reactions are responsible for the formation of aroma and in particular of heterocyclic compounds. With a view to illustrating these two routes, the three following examples have been chosen: tomato, asparagus and potato.

5.1 Tomato (*Lycopersicon esculentum*)

176 components were identified in tomato aroma in its various preparations (canned, juice, ketchup, purée). Although few heterocyclic compounds contribute to this aroma, one of them nevertheless is responsible for its peculiar flavour. This distinctive heterocyclic compound is 2-isobutylthiazole **28** [213]. Tomato composition is given in *Table I–18*.

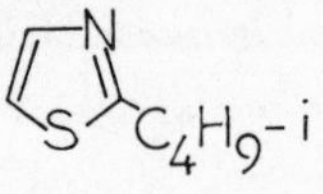

N, S, C_4H_9-i

28

Table I–18 – Tomato content [138]

	(g/100 g)
Water	93.4
Sugars	2.8
Nitrogen	0.14
Proteins	0.9
Fatty oil	traces
Carbohydrates	28
Vitamins[a)]	

a) Carotene: 600μg/100 g; vitamin C: 20 cg/100 g; vitamin E, folic acid, pantothenic acid.

Lipids. These are phospholipids, galactolipids, and triglycerides with a high linoleic and linolenic acid content. These unsaturated fatty acids are generated through the medium of acylhydrolases and phospholipase [45]. The tomato lipoxygenase converts [9,95] unsaturated fatty acids (C_{18}) into 9 and 13-hydroxyperoxides. The 13-isomer is assumed to undergo the cleavage with formation of hexanal in the case of linoleic acid and (*E*)-3-hexenal in the case of linolenic acid [211] (*Scheme 5*). In the case of cucumber, this enzymatic process is not the same. The cleavage originates from two hydroperoxides which lead mainly to aldehydes of C_9(53, 54, 55): 2-(*E*)-6-(*Z*)-nonadienal and 2-(*E*)-nonenal.

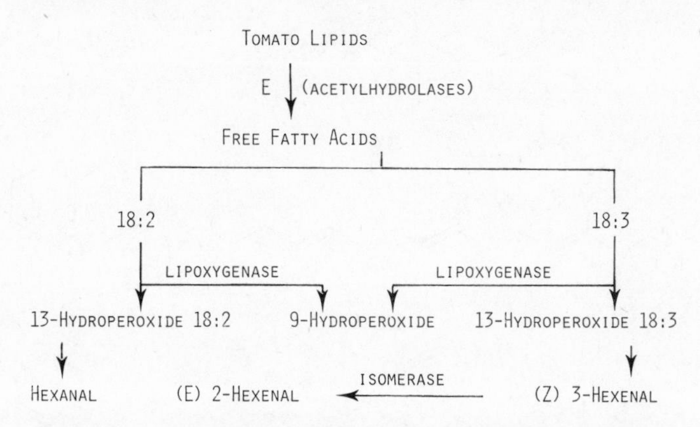

Scheme 5 – C_6 aldehyde formation in tomato [53,54,55]

Under the action of NAD^+ oxidoreductase these aldehydes are converted into the corresponding alcohols: hexenol, 3-hexenol and 2-(*E*)-hexenol [180, 189].

The method of fruit preparation greatly influences the enzyme systems. Thus, using ^{14}C-labelled linolenic acid and linoleic acid, it was shown that considerably less labelled carbon appeared in 3-hexenol when the tomatoes were blended rather than crushed. The reduction of 3-hexenol by crushed tomatoes at 250°C was shown to be faster than that of the 2-(*E*)-isomer.

Also, for the same type of fruit at various stages of its ripening, the biosynthetic aldehyde formation pathways might be different (case of mangoes and bananas) treated or not with ethylene [57,200].

α-Amino acids. Starting from ^{14}C-labelled amino acids it was proved that the carbonyl compounds present in aroma, might arise from their enzymatic degradation during the maturation process [232]. Thus, *L*-leucine yields 3-methylbutanal and 3-methylbutanol [232].

Carotenoids. The C_{40} carotenoid hydrocarbons of tomato (**32** to **35**) are assumed to be responsible for the formation of the three following ketones: 2,6-dimethyl-2,6-undecadiene-10-one **29**, 2,6-dimethyl-2,6,8-undecatriene-10-one **30** and 2,6,10-trimethyl-2,6,10-pentadecatriene-14-one **31** [95,213].

β-Ionone and 5,6-epoxy-ionone also identified from this aroma seem to be related to the thermal degradation of β-carotene [173].

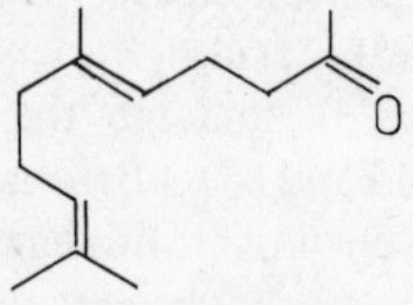

29

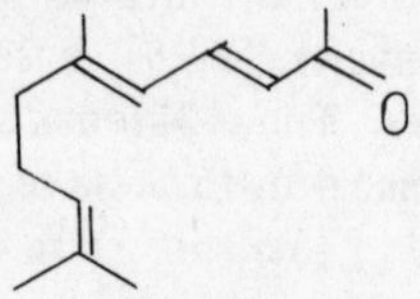

30

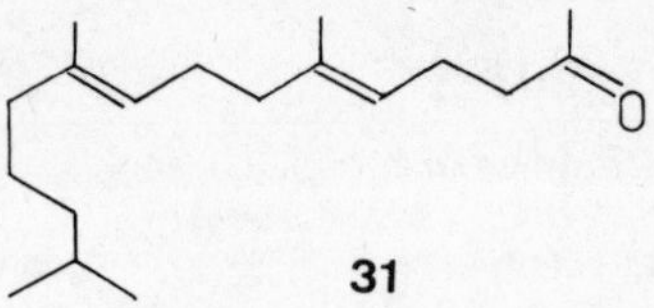

31

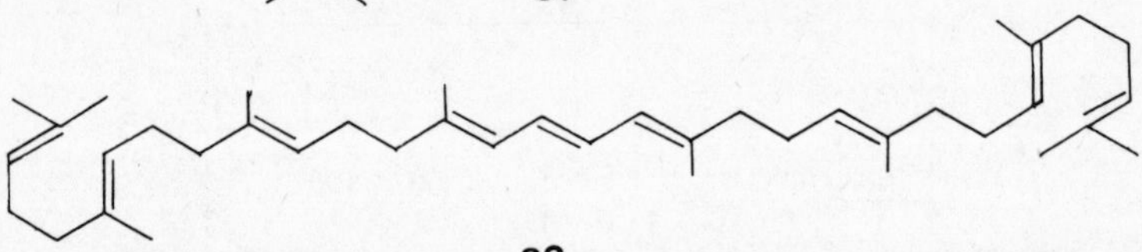

32

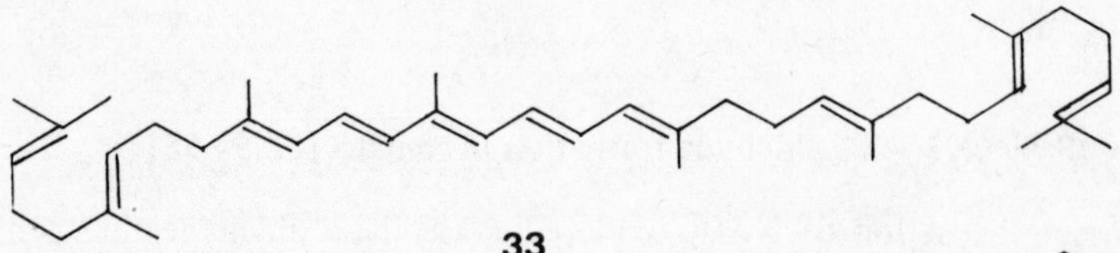

33

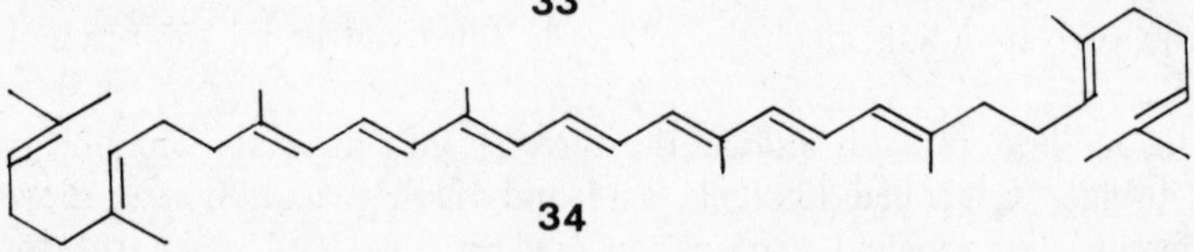

34

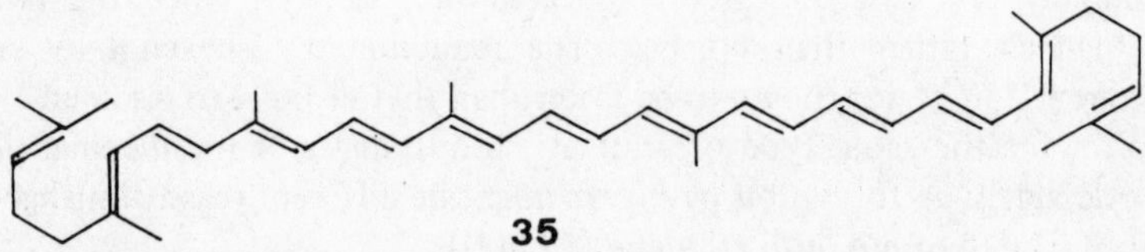

35

Fig. 3 – Tomato carotenoids and ketones [95,213].

***S*-Methylmethionine sulphonium salt.** It appears to be the precursor of dimethyl sulphide whose presence in tea aroma has already been mentioned. The proportion of this sulphur increases during storage and causes an alteration in aroma intensity and quality [62,63].

Miscellaneous. The tomato fruit acidity is due to a certain number of mineral and organic acids. Among the latter, the following can be mentioned: acetic, lactic, fumaric, malic, pyrolidone carboxylic, citric, and galacturonic.

Also, phenol acids and flavanones such as: chalconaringin, naringenin, naringenin -7-glucoside and the *meta* and *para* coumaric acids have been identified in tomato skin. These compounds appear during fruit ripening [77]. Sugars account for 2 to 3% of the fruit weight. They are responsible for the formation of 5-hydroxymethylfurfural in tomato juice and purée stored at 37°C [81].

Tomato possesses besides a high vitamin C (ascorbic acid) content (10 to 30 mg/100 g). Furthermore, tomato is easily degraded when stored or heated and gives rise to furans [192]. It can also be converted into dehydroascorbic acid which is light- and heat-sensitive. On account of their possessing α-diketone properties, these acids may react with amino acids in *Strecker* reactions [147].

5.2 Asparagus *(Asparagus officinals L.)*

The aroma precursors of asparagus are now well known especially from the findings of Tressl [203], Archimbault *et al.* [2a,2b]. They are essentially carboxylic acids, sugars, and amino acids as reported in *Table I–19* [35].

Table I–19 – Asparagus composition [35]

	g/100 g		mg/100 g
Water	92.9	Acids	
Proteins	2.1	Malic	100
Fats	0.2	Citric	100
Carbohydrates	4.1	Oxalic	5.2
Vitamins: A, B_1, B_2, B_6, nicotinic acid, pantothenic acid, vitamin C (33 mg/100 g).			

Carboxylic acids. Among the acids listed in *Table I–20*, the most abundant are citric and 4-hydroxycoumaric acids which by themselves account for 70% of the whole [28,50,124,127b].

Table I–20 – Carboxylic acids identified from *Asparagus officinalis L.* [2a,50,124,127b]

Acids		
Acetic	Glycolic	Pyrrolidone carboxylic
2,4-Dihydroxybenzoic	Glyceric	Quinic
2,4,6-Trihydroxybenzoic	Lactic	Shikimic
Caffeic	Maleic	Succinic
Citric	Malic	Vanillic
m-Coumaric	Malonic	(E) 3-Hydroxycoumaric
p-Coumaric	Oxalic	(E) 4-Hydroxycoumaric
Ferulic	Palmitic	(E) 4-Hydroxy-3-methoxycoumaric
Fumaric	Phthalic	3,4,5-Trihydroxybenzoic
Gallic	Protocatechic	3,4-Dihydroxycoumaric
α-Ketoglutaric		

Acids are quite significant flavour contributors to cooked asparagus. During the heating process phenolic, citric, and malic acids are likely to undergo degradation.

Sulphur-containing acids. Among the various sulphur acids identified from raw asparagus occurring only in trace amounts, the main representative acid is asparagusic acid (**36**; R = H) [202,203]. Its methyl and ethyl esters (**36**; R = CH_3, C_2H_5) are also present (*Fig. 4*).

The typical odour of asparagus is due to 1,2-dithiacyclopentene arising from the thermal degradation of asparagusic acid [202].

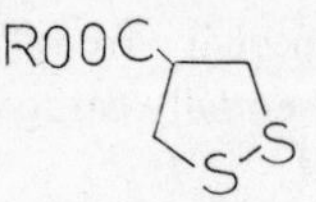

36 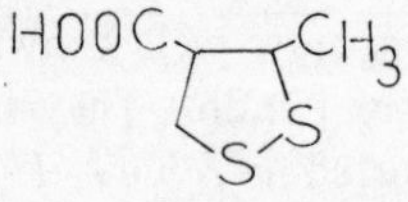 **37** 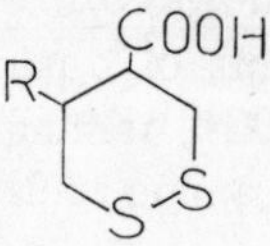 **38** (R=H, CH_3) **39**

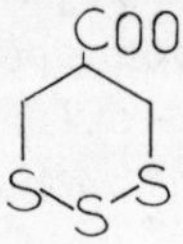

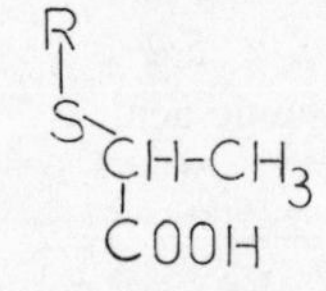

40 (R=H, CH_3) **41** (R=H, CH_3) **42**

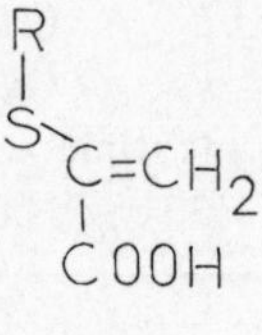

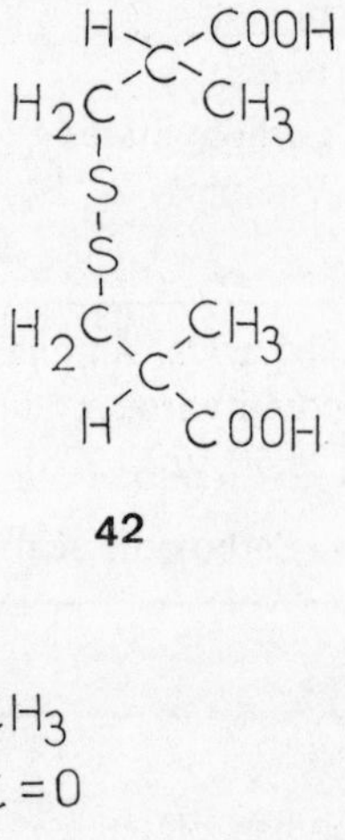

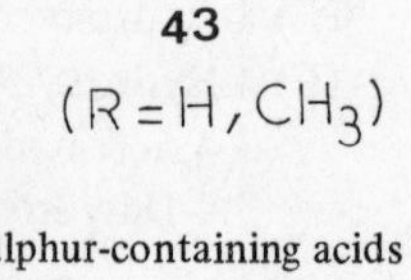

43 (R=H, CH_3) **44**

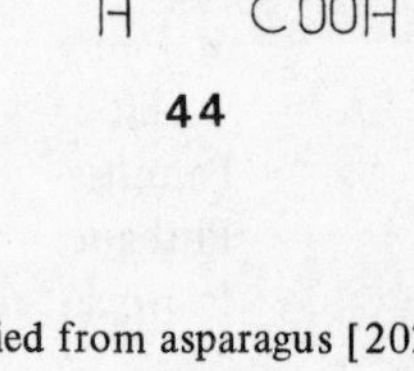

Fig. 4 – Sulphur-containing acids identified from asparagus [202,203].

Sugars. Among the various sugars occurring in asparagus (fructose, glucose, galactose, raffinose, rhamnose, sucrose, xylose), fructose and glucose by themselves account for more than 90% of the mixture [2c,127a].

Amino acids of raw and cooked asparagus. Various groups of research workers have studied the amino acid content in raw and cooked asparagus [58,183]. In the course of the cooking process the free amino acid content goes down by more than half. But this decrease varies according to the type of amino acids. The disappearance of amino acids goes hand in hand with the formation of the volatile compounds which contribute to aroma. The latter arise in the course of *Maillard* and *Strecker* reactions, taking place between sugars and amino acids. Among the various heterocyclic compounds that can be observed, 2-acetylthiazole contributes as well as asparagusic acid and 1,2-dithiocyclopentene to asparagus aroma [204]. 2-Phenylethanal results from the degradation of phenylalanine [204].

***S*-Methylmethionium salt**. Dimethyl sulphide, another significant odour constituent of asparagus, is produced by thermal degradation of the salt *S*-methylmethionium according to a process akin to the one described for tea. *S*-Methylmethionine is indeed a raw asparagus component.

Lipids. Under the action of certain enzymes (lipoxygenase, alcohol dehydrogenase. . .) unsaturated fatty acids of C_{18} are oxidized according to the mechanisms identical to those described for tomato. The oxidation of ^{14}C-labelled linoleic acid by an asparagus enzymatic homogenate, leads mainly to hexanal and pentanal (*Table I–21*).

Table I–21 – Enzymatic splitting of [μ-^{14}C] linoleic acid in the volatile homogenates[a] [204]

	Asparagus	Cucumber
Radioactivity in the extract (%)	9.3	5.0
Distribution of radioactivity among volatiles (%):		
Pentanal	9.5	0.5
Hexanal	57.5	12.6
2-Heptenol, hexanol	6.5	1.4
Pentanol	3.9	9.1
2-Nonenal	3.3	33.0
2,4-Decadienal	1.3	22.5
2-Nonenol	6.0	2.0
9-Oxononanoic acid	1.5	13.0

a) 12.5 μl of [μ-^{14}C] linoleic acid homogenised with 50 g of tissue for 5 min.

5.3 Potato (*Solanum tuberosum L.*)

Potato is perhaps the most important vegetable in the world. Native to the Andean highlands around Lake Titicaca in South America, it was introduced into Europe only in the 16th century and in France in the 18th century, mainly thanks to Parmentier's efforts.

This food can be prepared by a wide range of cooking methods (boiled, cooked, baked, fried etc. .) and presentation (dehydrated, frozen, flakes, mashed, puffed, chips etc. .). In every single case its aroma is different, and it is quite difficult to specify the role of precursors in the various stages of the processes taking place in the raw and processed vegetable [36,170].

The aroma of processed potatoes is indeed a mixture of products of various origins. Some of them result from enzymatic reactions taking place in the raw vegetable while others come from *Maillard* reaction. Thus, 3-isopropyl-2-methoxy-pyrazine found in raw potatoes to which it imparts a characteristic earth-like flavour [19b] could be formed *via* the enzymatic route. Methional, another characteristic odour contributor of this aroma, was identified from the volatile products of the essence obtained from steam distillation. It could be formed from *Strecker* degradation of methionine.

Comparatively few studies have been devoted to raw potato aroma [17], by contrast with dehydrated [32,64,169,190] or variously cooked potatoes [18, 19a,29,36,129,132]. From *Table I–22* it appears that raw potato is mainly composed of starch, proteins, and free amino acids, carbohydrates, organic acids, sugars, and minerals. But the influence of nucleotides on potato flavour is important.

Nucleotides. Along with amino acids, they are potato flavour contributors [188]. The precursor is RNA whose decomposition is important at 50°C (at pH 6) [15]. The nucleotide composition of two potato varieties is reported in *Table I–22* [187]. These are enzymatically decomposed into ribose-5′-phosphate. On heating the latter gives 5-methyl-3-(2*H*)-furanone [199].

Amino acids. Among all the current α-amino acids identified in boiled potatoes [187] asparagine, glutamine, glucosamine, and valine are predominant (≏ 70% of the mixture). A number of potato varieties have been studied [92]. Their amino acid composition changes appreciably from one variety to the other. These amino acids react with sugars and give rise to the numerous heterocyclic compounds (pyrazines, oxazoles, thiazoles, thiophenes) identified in cooked potato aroma. Aldehydes such as methional and 2-phenylethanal are formed respectively in the course of the *Strecker* degradation process from methionine and phenylalanine [17,73]. The significant formation of certain aldehydes such as 2-methyl- and 3-butanal is responsible for some off-flavours [169].

Sugars. Potato contains mono- and oligosaccharides: fructose, glucose, inositol, melibiose, and raffinose [190]. The sugar compositon for two potato varieties is indicated in *Table I–23*. The pyrazines induced by sugars interaction with amino acids are described in the next chapter. Among these 2-ethyl-3,6-dimethyl-pyrazine is a constituent of the aroma of chips [18].

Table I–22 – Composition of fresh potatoes [188]

Compounds	Average values (%)	Compounds	Average values (%)
Water	80	Sugars (glucose, fructose, sucrose)	0.5
Starch (amylose, amylopectose)	10.0		
Proteins	2.0	Lipids	0.2
Organic acids (citric, malic, succinic, fumaric)	1.5	Polyphenols	0.2
		Chlorogenic acid	
Minerals (K, Mg, Ca, P, Na, Mn, Fe)	1.0	Caffeic acid	
		Vitamins (ascorbic acid, B_1, B_2, B_6, nicotinic acid, and pantothenic acid)	0.02
Free amino acids (all current amino acids)	0.8		
Non-starch polysaccharides	0.7	Pigments	0.015
Hemicellulose		Anthocyanins	
Pectoses		Carotenoids	
Hexosans		Alkaloids	0.01
Pentosans		Solanine	
		Chaconine	

RNA nucleotides 0.01%	Varieties: Bintje (mg/100 g)	Ostaca (mg/100 g)
5′-AMP	3.0	2.45
5′-GMP	2.4	1.39
2′,3′-GMP	1.72	1.79
5′-UMP	2.14	1.78

Table I–23 – Sugar content in potato (g/100 g)

Sugars	Kennebeck	Russat-Burbank
Glucose	0.36–0.27	0.8 -3.0
Fructose	0.82–0.87	1.61–3.84
Sucrose	1.01–3.35	1.25–2.55

Lipids. During storage, potato lipids undergo enzymatic degradation which produce off-flavours. Among the degradation products of fatty acids, hydrocarbons, alkanals, and 2-alkenals are to be found [169].

C_4 to C_8-aldehydes, crotonic aldehyde, acetone [171], as well as 2-(*E*)-4-(*E*)-decadienal and 2-pentylfuran [17,19b] have been identified in fresh potato juice. These last two products along with hexenal result from enzymatic oxidation and auto-oxidation of linoleic acid. Like the case of soya and peas, their formation might be due to the action of lipoxygenase [45]. 2-Pentylfuran, an auto-oxidation product of linoleic acid, has also been identified in the aroma of soya bean oil [22]. Similarly, 1-octen-3-one and the corresponding alcohol arise from linoleic acid oxidation. The quality of the aroma of dehydrated granules is dependent on the amount of hexenal present. This aldehyde has been recommended as a criterion to assess the quality of aroma. Oxidation generated aldehydes may give rise to aldolization reactions. The formation of certain aldehydes (e.g. 2-phenyl-2-butenal, 2-phenyl-5-methyl-2-hexenal etc. .) in chips aroma is thus explained [19a].

Polyphenols. Hasegawa *et al.* [67] have studied the extent of the variations in chlorogenic acid content during protracted cold storage of potatoes. They were able to correlate the polyphenol content with the flavour (more especially bitterness and astringency) of several varieties and hybrid potatoes.

Vitamins. In decreasing order of importance the following vitamins can be found in potatoes: vitamin C (ascorbic acid), nicotinic acid, vitamin B_6 (pyridoxin), and vitamin B_1 (thiamin). These vitamins are thermally degraded and give rise to heterocyclic compounds so that it is amazing that they should have no influence on cooked potato flavour [123].

I–6 BREAD AND WHEAT FLOUR

Owing to the prominent place it occupies in man's food, a large number of studies have been devoted to the flavour of bread [14,27,38b,113,143]. The volatile substances identified in 1975 which constitute the aroma of white bread are the result of a whole series of biochemical and thermal processes which occur during bread-making and baking respectively. Moreover, bread is not a product resulting from just one raw material, and its making varies according to the countries considered. By this token, American or English-type breads are made from wheat flour, sodium chloride, yeast, and various additives, glycerides, milk, emulsifying agents (monoglycerides or lecithins), bromate, and potassium iodate. German bread is often based on rye meal. In France bread is made from a mixture of wheat flour, water, sodium chloride, and yeast to which small amounts of bean flour, ascorbic acid, and malt may be added. Under these conditions, the various types of bread-making lead to different organoleptic characters. This is clearly shown by qualitative and quantitative differences in aroma composition [38b]. In the course of the various bread-making stages, initial precursors undergo modifications leading to volatile products and new precursors.

6.1 Bread processing

a) Kneading. During the kneading step, the activity of several enzyme systems triggers off the formation of intermediate percursors. Thus, α-amino acids arise from proteins and peptides, sugars from oligosaccharides, and hexanal from linoleic acid originally present in wheat flour.

b) Fermentation. It affects amino acids and sugars and generates compounds of yeast intermediary metabolism. Volatile products yielded by normal dough fermentation are carbon dioxide, ethanol, carbonyl compounds, and volatile acids [88,164]. Various factors such as duration, concentration of salts, yeast type, and bacteria may alter the composition of these flavourants and have an effect on bread organoleptic qualities. For example, the bread made by the 'spongy dough' method tastes better than the one made by the 'dense dough' method. The role of microorganisms other than yeasts has been studied [155]. It has been shown that lactic bacteria, in particular, have flavour-enhancing effects [20,176].

c) Baking. Various reactions take place during the baking process and lead to the formation of aromatic substances. They are:

- *Maillard–Strecker* reactions between sugars and amino acids,
- esterification reactions,
- aldol condensation of aldehydes,
- decomposition of sulphur amino acids which give rise to hydrogen sulphide and methylmercaptan.

Brown pigments (or melanoidins) forming on the crust of bread during *Maillard* reaction are colour and bread flavour contributors. Optimal baking temperature conditions lie between 90° and 150°C.

d) Cooling and staling. As bread cools, volatile products concentrated on the crust diffuse into the crumb and induce a flavour alteration. In the case of stale bread, a decrease in carbonyl compounds is to be observed as a result of their interaction with starch [38b]. This hardening is easily reversed if the bread is reheated at 60°C or above. Freezing freshly baked bread to a temperature below the freezing point will retard the hardening rearrangement of water, starch, and proteins, so that immediately after thawing out the crumb of the bread will be soft. A few years ago, it was discovered that certain substances

Table I–24 – Wheat flour content [138]

	Flour (g/100g)	
	(100%)	(72%) (bread)
Water	14.0	14.5
Sugars	2.3	1.5
Starch	64.5	73.3
Nitrogen	2.26	1.98
Proteins	13.2	11.3
Carbohydrates	65.8	74.8
Lipids	2.0	1.2

called 'freshness preservers' retard the rate of hardening. These materials are emulsifying agents such as partially hydrolyzed fats, e.g. mono- and diglycerides. Wheat flour content [138] and ingredients used in bread processing are listed in *Tables I–24* and *I–25*, respectively.

Table I–25 – Ingredients for bread processing (Pre-ferment method) [89b]

Pre-ferment	(g)	Paste	(g)
Water (in sufficient amount to make one litre)	–	Shortening	21.0
		Sodium chloride	10.5
Sucrose	66.0	Yeast	7.0
Yeast	44.0		
Sodium chloride	11.0	Pre-ferment	320 ml
Ammonium acid phosphate	4.0	Water	128 ml
Calcium carbonate	1.0		
Potassium sulphate	1.0	Total weight	1228.5 g
Magnesium sulphate	0.5		
Potassium bromate	0.06		
Potassium iodate	0.02		

6.2 Precursors

Proteins, peptides and α-amino acids. As the flour is kneaded, proteins hydrolyze into amino acids, the latter then participating during the various stages of bread processing (in particular during baking) in *Maillard–Strecker* reactions. This evolution through the various processing stages has been studied [43a,43b]. Bread contains all essential amino acids: tryptophan, lysine, threonine, leucine, isoleucine, valine, phenylalanine, methionine. Carbonyl compounds formation in *Maillard–Strecker* reactions is influenced both by the nature of the amino acid and the sugar [164]. Thus, Kiely *et al.* [97] have shown that the speed of browning is influenced by sugar, whereas the smell depends on the amino acids. Model reactions between reducing sugars and leucine, histidine, or arginine reproduce the aroma of bread. The increase in carbonyl compounds obtained by adding amino acids during manufacture goes hand in hand with the deepening colour of the crust [164]. Carbonyl compounds generated from amino acids according to *Strecker* reaction [197] and those actually identified in the aroma of bread are more or less the same. However, the aroma contains more of them. Thus the formation of the following aldehydes: propanal, butanal, butanone, 2,3-butanedione, pentanal, isopentanal, and hexanal indicates a different origin. Also proteins participate in the formation of carbonyl compounds in bread crust [89]. During baking, they react with starch to generate numerous carbonyl compounds, but butanal and its 2- and 3-methyl derivatives are not included. The influence of a number of amino acids on aroma and bread crust colour has been studied [97,164]. Amino acids are introduced into bread during its making. Proline remains rather ineffective on crust colour. The quantity of amino acids formed by pyrolysis also has an influence on crust colour and aroma. A characteristic odour component of this aroma might arise from the action of dihydroxyacetone on proline. Indeed, when proline is added to the formula of bread, its aroma is strengthened. This observation was the starting-point of various findings. Among the model reactions studied, the following ones can be quoted: glycine-glucose, triose-sugars-isoleucine, proline-sugars etc. [89,113]. The characteristic component of this aroma could thus be identified as 2-acetyl-1,4,5,6-tetrahydropyridine [78]. It is to be prepared by heating proline with dihydroxyacetone or glycerol in the presence of sodium bisulphate [79]. Sulphur amino acids play a prominent role during the fermentation and baking processes, especially when they generate methylmercaptan [30].

Sugars present in flour, dough and bread. The sugars contributing to the flavour of bread have three origins:

– the sugars naturally present in the flour,
– the sugars formed during the fermentation step from oligosaccharides and polysaccharides,
– the flavour sweetening sugars that are added.

The usual sugars are to be found in flour: glucose, fructose, galactose, maltose, and sucrose. When stored for a long time exposed to the air, reducing sugars and amino acids participate in the *Maillard* reaction.

During fermentation (*Embden–Meyerhof* route), sugars are involved in a series of biochemical reactions resulting in the formation of organic acids, alcohols, carbonyl compounds, and esters. Among these products, ethanol, acetaldehyde, pyruvic acid, and carbon dioxide are predominant [88]. The most volatile of them are eliminated during baking, whereas the others take part in various reactions [197]. Sugars come again into play during the caramelization step and browning reactions. In the course of these reactions, the heterocyclic compounds and carbonyl compounds thus obtained constitute bread aroma. The brown pigments (melanoidins) appearing at the same time in bread crust are formed through polymerization of aldimines and ketimines [88]. The influence of sugar on aroma has been studied. Among the various sugars ($\simeq 20$) involved in model reactions with glycine or lysine, pentoses are the most reactive, and they are followed by hexoses and disaccharides. Arabinose, ribose, and xylose, which are particularly reactive, trigger off a deeper browning of the crust. The same effect can be observed with lactose and melobiose. The quantity of 5-hydroxymethylfurfural can be increased by adding sucrose to the initial bread-making formula.

Starch. Its particular texturation properties and mechanical qualities explain why we feel drawn towards bread. It has an effect on taste because the state of the starch determines the speed at which it will be attacked by salivary α-amylase. This reaction is responsible for the sweet flavour of bread crumb when it is chewed. Finally, it has an effect on the staling process which it is mainly responsible for [72]. Some authors concluded that hardening and other physical changes in bread on staling are caused by retro-gradation of branched molecules of the starch. Amylase is generally able to give rise to complexes with the various aroma and flavour components. These complexes can split up on heating. This explains why stale bread, when heated up, at last recovers its fresh bread taste [38b].

Lipids in wheat flour. While kneading and fermentation are in progress, lipids are oxidized by the action of lipoxygenase. If fats or vegetable oil are added to the bread formula, the aroma can be altered and another type of bread produced with a more supple crust and a softer crumb. During baking, odorous carbonyl compounds arise from oxidation of fats. The most prominent of them is hexanal which comes from linoleic acid oxidation due to lipoxygenase [38a].

Salts. They are used as taste-enhancing compounds in the bread formula. Bread flavour depends upon the nature of cation and also that of anion. Sodium chloride is used in most cases. It is generally admitted that it exerts an enhancing effect on the aroma and the flavour of aroma volatile compounds. It could also exert a role in the formation of intermediate precursors by interfering with enzymatic activity and/or microbial functions during fermentation [89].

Phenolic acids in wheat flour. The following acids have been identified: *p*-hydroxybenzoic, salicylic, vanillic, *p*-coumaric, *o*-coumaric, gentisic, ferulic, isoferulic, caffeic, protocatechic, sinapic, syringic, chlorogenic, and isochlorogenic. They possess their own astringency and organoleptic properties. During dough baking, some of them deteriorate and give simple phenols. Thus, in the case of rye crispbread, phenol, cresol, ethylphenol, 4-vinyl-2-methoxyphenol and 2-methoxyphenol have been identified [218].

I–7 MILK AND DAIRY PRODUCTS

Milk is one of the most remarkable products in the human diet. It can be used in various forms: dry whole milk, non-fat dry milk, spray dried whey, sterile concentrated, raw, or boiled etc. . . Besides, it goes into the composition of several dairy-based foods: cream, butter, yoghurt and cheese. As such, it has been the subject matter of thorough research work, and for further details the reader is referred to the various reviews concerning either milk or dairy products [40b, 80,84,116] and cheese [1,34].

7.1 Composition

Milk is composed of fats, proteins, sugars, mineral materials, vitamins, pigments, enzymes, and water. The composition of cow's milk is given in *Table I–26* [31].

The aroma constituents of milk (over 200 substances identified in 1975) have a double origin according to whether they be initially present in raw milk or formed in the course of various processing stages. In the latter case, they arise from the degradation of the main groups of precursors (fats, lactose, and proteins). The various processing stages of dairy products and cheese have been described by Dumont & Adda [40b]. Each group of precursors is the seat of two types of changes:

– biochemical changes with the participation of either enzymes and microbial flora contained in milk, or of bacteria and exogenous secondary microbial flora in the case of yoghurt and cheese,
– chemical changes due to oxygen (auto-oxidation) or to heat.

We shall not enlarge upon the changes cheese has to undergo, because the latter are very intricate and specific to each sort of cheese (Camembert, Cheddar, Gouda, Gruyère, Emmenthal, Parmesan, Pont l'Evêque, and Roquefort).

Table I–26 – Pasteurized composition of whole cow's milk [31]

Component	(g/100 g)
Water	87.2–88.5
Fats total	3.70
Polyunsaturated fatty acids	0.1
Cholesterol	0.01
Proteins	2–6
Carbohydrates (lactose)	4.90
Non-lipid-containing solids	9.1
Total solids	12.8
Ash	0.7–0.72

Vitamins	(mg)	Elements	(mg)
A	140 iu[a)]	Na	75
B_1	0.04	K	139
B_2	0.15	Ca	133
B_6	0.05	Mg	13
Nicotinic acid	0.07	Mn	0.002
Pantothenic acid	0.33	Fe	0.04
C	1	Cu	0.01
E	0.06	P	88
B_{12}	0.0006	S	29
Biotin	0.002	Cl	105
D	0.5–4 iu		
Folic acid	0.0001		

a) 1 iu Vitamin A = 0.0006 mg of β-carotene.

7.2 Precursors

Fats. They occur in milk in globule form being 2.4 to 7.2 μm in diameter. These are liquid lipids in a stable emulsion formed at the temperature of the animal's body. The average cow's milk fat composition is given in *Table I–27* [102]. The fatty acids in the glycerides are mainly oleic and myristic acids accompanied by palmitic, lauric, capric, caproic, and caprylic acids (*Table I–28*) [116].

Table I–27 – Fat content of cow's milk [102]

Compounds	Range of occurrence (%)
Triglycerides of fatty acids	97.0-98.0
Diglycerides	0.25–0.48
Monoglycerides	0.015–0.038
Ketoacidglycerides	0.85–1.28
Aldehydoglycerides	0.011-0.015
Glyceryl ethers	0.011-0.023
Free fatty acids	0.10–0.44
Phospholipids	0.2–1.0
Cerebrosides	0.013–0.066
Sterols	0.12–0.4
Compounds	**Range of occurrence (ppm)**
Squalene	70
Carotenoids	7-9
Vitamin A	6–9
Vitamin D	0.0085-0.021
Vitamin E	24
Vitamin K	1
Carbonyl compounds (neutral free)	0.1-0.8

The occurrence of a wide variety of keto- and hydroxy acids and of many other acids (more than 142) has been reported [85]. According to Jensen *et al.* [71] milk also contains small amounts of unsaturated fatty acids other than oleic acid. They are mainly C_{10} to C_{24} acids with one, two or three double bonds. Some acids (C_{20} and C_{22}) contain four or five double bonds [71]. Milk lipids are responsible for aroma production following enzymatic and chemical reactions. For example, triglyceride lipolysis (under the action of bacterial lipases) generates milk fatty acids. Lipid degradation induces carbonyl compound formation. The

Table I–28 – Fatty acids in sample of milk fat for cows fed normal rations [116].

Acids	Range	Average (%)
Butyric (C_4)	2.4–4.23	2.93
Caproic (C_6)	1.29–2.4	1.90
Caprylic (C_8)	0.13–1.04	0.79
Capric (C_{10})	1.19–2.01	1.57
Lauric (C_{12})	4.53–7.6	5.84
Myristic (C_{14})	15.56–22.62	19.78
Palmitic (C_{16})	5.78–29.0	15.17
Stearic (C_{18})	7.80–20.37	14.91
Oleic (C_{18} : 1)	25.27–40.31	31.90

latter may either result from free β-keto acid decarboxylation or from β-fatty acid oxidation. The condensation in carbon atoms of the methylketones thus formed increases in proportion to the number of processing stages. In milk, the two methylketones (C_{13} and C_{17}) are to be found, and six methylketones from C_5 to C_{15} are present in butter [3,70,172].

Among other characteristic milk and butter aroma carbonyl compounds, one must mention oct-1-en-3-one **45** (accompanied by the corresponding alcohol), (*E*)-2-(*Z*)-6-nonadienal **46**, (*E*)-6-nonenal **47**, and (*Z*)-4-heptenal **48**. The first one, that is responsible for the metallic aroma of butter, and the second result from linoleic acid and linolenic acid auto-oxidation respectively (*see Scheme 6*).

The presence of pent-1-en-3-one and of pent-1-en-3-al is accounted for by a shift in the linolenic acid double bond. C_5 to C_{16} saturated aldehydes, C_6 to C_{11} alkenals, and C_8 to C_{12} -2,4-alkadienals have been identified among the auto-oxidation products of milk fatty acids. In butter, (*Z*)-4-heptanal possessing the characteristic odour of cream is formed from (*E*)-11-(*Z*)-15-octadecadienoic acid, as shown in *Scheme 6* [90]. In 'Camembert', the presence of oct-1-en-3-ol which possesses a characteristic mushroom odour might be due to the cleavage in the unconjugated 10-hydroperoxide linolenate system. The latter is a specific metabolite of *Penicillium caseicolum* strains [122]. γ- and δ-Lactones also play an important role in milk and butter flavours (*see Chapters II and III*). In butter, lactone precursors are γ- and δ-hydroxy acids obtained from the polar glyceride fraction of milk fat. These δ-hydroxy acids easily cyclize on heating [44] in the presence of water.

In butter oil, up to a certain level, increasing temperature and time of heating result in higher lactone levels. C_8 to C_{14} lactones are also to be found in various types of cheese (Blue, Cheddar, Swiss). In that case, they arise from saturated fatty acids contained in the mammary gland. In milk fats such as in bovine milk fat, the complete range of saturated γ- and δ-lactones has been identified. δ-Lactones, especially δ-12 and δ-14, predominate. Finally, milk and dairy product aromas contain considerable quantities of free fatty acids and short-chain alcohols. The latter are converted into esters under the action of esterases from lactic bacteria [40a,76].

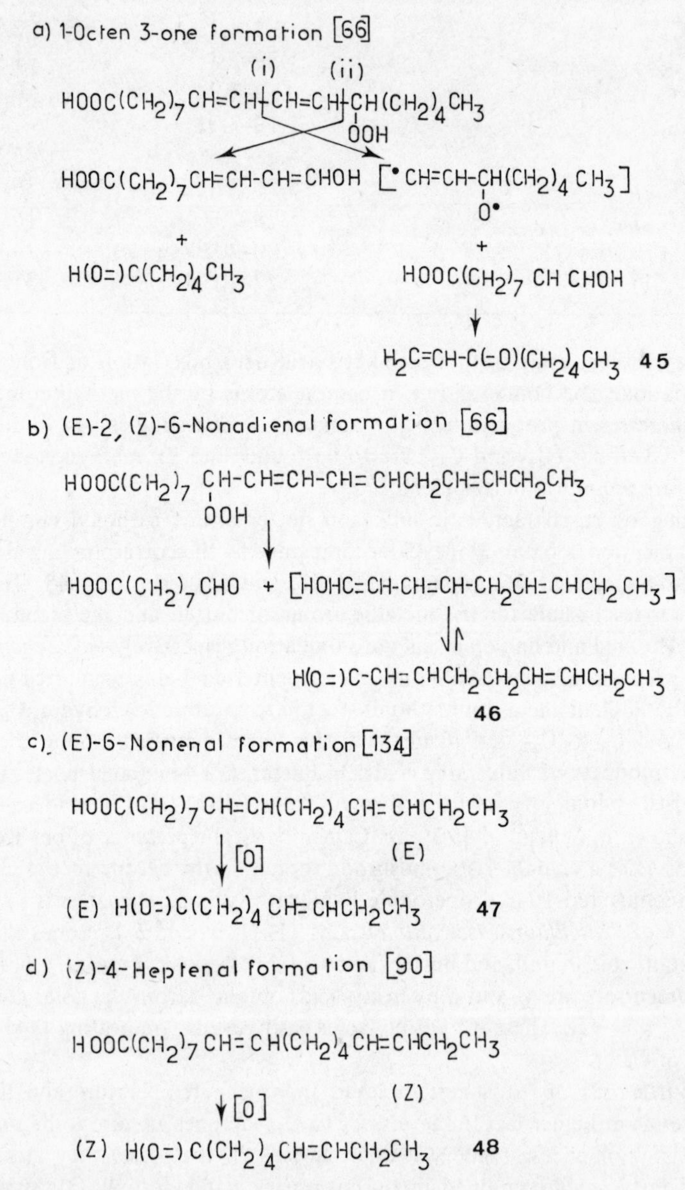

Scheme 6 – Carbonyl compound formation from fatty acids

Proteins and amino acids. Owing to the numerous highly nutritional proteins it contains, milk occupies an outstanding place in human food. The most important milk protein is casein whose amino acid composition is given in *Table I–29* [83, 84]. In fact, casein is a mixture of α- and β-caseins separated by precipitation at pH 3.5 at a temperature of 2°C for the first and pH 4.9 for the second.

Table I–29 – Amino acid content of casein [83,84]

	(g/100 g)
Nitrogen	15.63
Phosphorus	0.86
Sulphur	0.80
Amino-nitrogen	0.93
Amido-nitrogen	1.6

Amino acids	(g/100 g)		(g/100 g)
Glycine	0.93	Tryptophan	1.7
Alanine	3.0	Arginine	4.1
Valine	7.2	Histidine	3.1
Leucine	9.2	Lysine	8.2
Isoleucine	6.1	Aspartic acid	7.1
Proline	11.3	Glutamic acid	22.4
Phenylaline	5.0	Serine	6.3
Cysteine	0.34	Threonine	4.9
Methionine	2.8	Tyrosine	6.3

Other milk proteins are lactalbumin,lactoglobulin, euglobulin (milk cream), and pseudoglobulin. Under the action of milk protease, amino acids are generated. During the various stages of UHT sterilized milk production (ultra high temperature), they are going to react in *Maillard* reactions. It has been shown that the lactose-casein model system generates *Amadori* compounds. The latter split up into carbonyl compounds (e.g. diacetyl), or after undergoing dehydration give furans. Aldehydes such as phenylacetaldehyde are obtained from the *Strecker* reaction. When heated, lactose reacts with milk proteins and gives rise to lactulosyl-lysine. The latter may also appear when milk powder is stored under bad conditions [80]. *Ortho* aminoacetophenone is present in concentrated milk to whose particular flavour it contributes. It may arise from alkaline degradation or oxidation of tryptophan [3,146]. Hydrogen sulphide and mercaptans certainly spring from β-lactoglobulin which contains active sulphydryl groups [96]. As in the case of many other foodstuffs, methionine accounts for the formation of methanethiol, dimethyl sulphide, and methional. The routes in the formation of milk flavour from proteins can be collected in *Scheme 7* [40b].

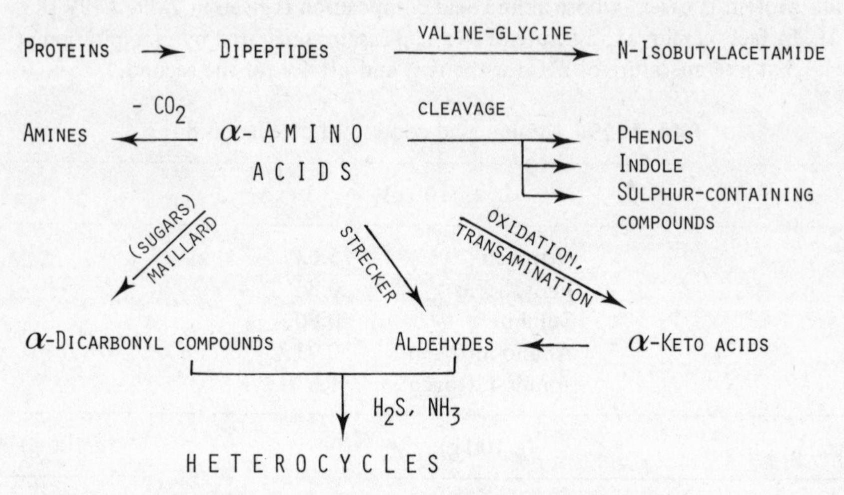

Scheme 7 – Milk flavour formation from proteins [40b]

Lactose. Practically the only sugar occurring in milk, lactose is composed of glucose and galactose. In the free state, these two compounds occur in milk only in very small amounts. In cow's milk, lactose averages 4.8%. It is absolutely essential for the making of fermented dairy products such as yoghurt and cheese. Under the action of lactic ferments, hydrolysis of lactose produces lactic acid. Biochemical degradations subsequently convert the latter into acetic and propionic acids. There are also some species of bacteria known as heterofermentative bacteria which yield ethanol during the fermentation process.

Miscellaneous. Vitamins, C (ascorbic acid) **49**, B_6 (pyridoxine) **50**, and B_1 (thiamin) **51**, are susceptible to thermal processes and degrade into heterocyclic compounds. Vitamin B_6 is assumed to participate in *Maillard*-like reactions [11, 60]. Other compounds such as aliphatic acids C_8, C_{10}, C_{12} and C_{16}, γ-lactone of

49 **50** **51**

C_{12}, γ-lactones of C_{10} and C_{12}, *p*-cresol, 4-ethylphenol, indole, vanillin, methyl- and ethyltosylates, *p*-hydroxyacetophenone can be found free or combined (glucuronides and sulphates) [11]. Phenolic compounds might arise from the lignin ingested by the animal when feeding. Other compounds such as benzylmethyl sulphide, benzylisocyanate, benzylcyanide, indole, and skatole might come from animal's feed. We have already indicated the element content in cow's milk in *Table I–26*. *N*-isobutylacetamide, which possesses a bitter taste, has recently been identified both in mould and surface ripened cheese [40]. Among the various salts accounting for 0.9% of milk, the most abundant are chlorides, phosphates, and citrates. In milk, the latter can be converted at pH 6.6 into acetoin and diacetyl. Finally, enzymes represent the last important group of milk components. They participate and interfere during biochemical conversion processes and become ineffective at a high temperature. The principal ones along with their respective functions are collected in *Table I–30* [116].

Table I–30 – Milk enzymes and their function as catalysts [116]

Enzymes	Function as catalysts and sensitivity
Lipase	Hydrolysis of fats to glycerol and fatty acids. Bitterness of dairy products will be due to this enzyme.
Arylesterase,	Hydrolysis of fats.
Cholinesterase	Acetylcholine destruction.
Alkaline phosphatase	Destroyed during pasteurizing.
Acid phosphatase	Unstable to light but resistant to heat.
Xanthine oxidase	Aldehyde oxidation.
Lactoperoxidase	Oxygen transfer from peroxides, especially H_2O_2 to other products.
Protease	Hydrolysis of peptide bonds of proteins.
Catalase	H_2O_2 decomposition in water.
Aldolase	Cleavage of fructose-1,6-dihydroxyacetoin phosphate and phosphoglyceraldehyde.

As a reminder, let us mention the presence of many mono- and polysulphur compounds (≙ 36) in ripened cheese flavour. They are responsible for its garlic note [1].

I–8 ALCOHOLIC BEVERAGES: WINE

Owing to its importance in many countries and particularly in France, wine is a product which has given rise to much research [91,173,210]. The aroma components of grape and wine consist of several hundred chemical compounds. Most of them are present in concentrations between 10^{-4} and 10^{-9} g/l. The total content of all aromatic substances in wine amounts to 0.8–1.2 g/l (e.g. 1% of the ethanol content). Fusel alcohols, acids, and fatty acid esters quantitatively form the greater part of aroma but also, carbonyl, phenolic, sulphur and nitrogen compounds as well as lactones, acetals, hydrocarbons, sugars and some other unclassified compounds rank among aroma compounds (*see Table I–31*).

Table I–31 – Wine composition [35]

Component	(g/100 g)	Component	(mg/l)
Alcohols	8.8–12.5	Acids	
Carbohydrates	0.2–8.0	Tartaric	163–234
Glycerol	0.8–2.6	Lactic	71–248
Polyphenolic	0.2–0.6	Succinic	90–130
compounds & tannins		Malic	0–280
		Citric	6–58
		Aromatic	56–136

Vitamins $\simeq$ 0.2 mg
(B_1, B_2, B_6, nicotinic acid, pantothenic acid, folic acid)
Elements 90–175 mg
(mainly, K, Mg, Ca, Na, P, S)[a)]

a) Lithium is also present in wine at the average level of 8 μg/l.

The quality of wine is determined by a balance of all these components as assessed by sensory analysis. Approximately, the same compounds occur in the aroma of other alcoholic beverages (beer, whisky, cognac, rum, vodka . . .). This is probably owing to the similar yeast and fermentation conditions which determine the most important aroma components in these beverages.

Alcohols and esters. Alcohols formed during fermentation, e.g. 2-methyl-1-propanol, 3-methyl-1-butanol, 2-methyl-1-butanol, 1-hexanol and 2-phenylethanol, account for 50% of all aromatic substances ($\triangleq$ 500 mg/l). They are present in a slightly higher concentration than their organoleptic perception threshold, and they are responsible for the sweet taste of wine [284]. The average concentration of fusel alcohols in other alcoholic beverages is 1.5 g/l in cognac, about 1 g/l in whiskies, and nearly 0.6 g/l in rums.

Esters are also produced by yeast during alcoholic fermentation. Among more than 50 esters reported in wine aroma, three of them are of more frequent occurrence: ethyl acetate (22 to 190 mg/l), ethyl lactate (9 to 17 mg/l) and ethyl-*n*-octanoate (1 to 6 mg/l). Ethyl acetate plays an important role; it contributes to give wine its hard character. The content of higher alcohols and esters in wine is more affected by the fermentation conditions than by the kind of yeast. The *Sacch. cerevisiae* is the dominant species in all cases of wine fermentation.

Acids [10,114,115,117,160,161]. High acidity level in wine (pH range from 2.8 to 3.8) is linked to a relatively strong acid found in grapes: tartaric acid. Lactic bacteria develop under anaerobic conditions in wine, transforming tartaric acid and other constituents (sugars, glycerol) into acetic acid. Glutamine is the main source of 2-pyrrolidone-5-carboxylic acid in the concentrated juice of the *Concord* variety grape. The hardness of red wines depends upon the level of acetic acid and ethyl acetate. When present in excessive amounts, they reduce the organoleptic qualities of wine.

Terpenes [6,223]. The natural aroma of some grape varieties results from a mixture of 8 to 10 terpene compounds (1 to 3 mg/l) including linalool and its oxides, geraniol, nerol, α-terpineol, and trienol along with diols and triols derivatives from linalool [223].

Phenolic compounds and tannins [39,186]. These substances are the main contributors to the bitter taste and odour of wines. The total concentration of all phenols ranges from 2000 to 6000 mg/l. The natural phenols pre-existing in grapes may be modified by the enzyme and exposure incident to crushing and preparation for fermentation. Microorganisms such as yeasts may also increase or modify wine phenols. Other reactions associated with their oxidation during processing normally produce taste and odour changes as well as colour changes.

The wine phenols may be classified into two large categories: a) non-flavonoids, b) flavonoids and tannins.

a) *Non-flavonoids* [89]. (≙ 200 mg of gallic acid equivalent (GAE) per litre of wine). Among non-flavonoids isolated from commercial red wines were the following phenols: *m*-cresol, 4-ethylphenol, 4-vinylphenol, 4-ethylguaiacol, tyrosol, *p*-cresol, guaiacol, 4-vinylguaiacol, isoeugenol, vanillin, phenol, salicylic acid and its methyl and ethyl esters, and vanillic acid. Some of these phenols can arise from thermal degradation of ferulic, caffeic, and *p*-coumaric acids. They appear to be near or below their individual flavour thresholds in most wines. Additional phenols of similar types have been found in distilled beverages aged in oak barrels. Non-flavonoids in typical wines are largely accounted for by cinnamic and benzoic acid derivatives.

b) *Flavonoids and tannins* [186]. They represent about 1200 mg GAE/l. Tannins are formed by the condensation of between 2 and 10 single-flavour molecules. They are an important element in the suppleness of red wine. The polymerization level affects the tannin capacity to combine with the protein. This capacity governs all their properties. Thus, polyhydroxyphenols with a molecular weight of the order of 600 (e.g. catechins) are considered the minimum for appreciable astringency. Astringency generally increases with the increasing degree of polymerization. Anthocyanins, mineral salts, and polysaccharides can intervene in the structure of tannins, leading to complexes of molecular weight ranging from 1000 to 5000.

Among flavonols reported in wine were aglycones, quercetin, rutin, hesperidin, naringin (a flavanone glycoside). The major flavan-3-ols of wine are (+)-catechin, (–)-epicatechin, gallocatechin, and gallate derivatives.

Sugars and α-amino acids [37,109,126]. Sugars present in wine are: α- and β-arabinose, α- and β-rhamnose, α-xylose, α- and β-mannose, fructose, α- and β-galactose, sorbose, α- and β-glucose, trehalose.

Some furan derivatives (e.g. furfural, 5-methylfurfural, 2-acetylfuran, 2-methylfuran etc.) found in wine aroma may arise from their degradation. The etherification of *Maillard* formed alcohols such as furfuryl alcohol, leads to 2-ethoxymethylfuran. Similarly, ethyl 2-furoate arises from esterification of 2-furoic acid. Certain lactones such as 4,5-dihydro-5-ethoxy-2-(3*H*)-furanone may arise from glutamic acid [126b].

Among α-amino acids contained in wine (≙ 1.8 mg/l) five predominate. In decreasing order of importance they are: proline, alanine, glutamic acid, glutamine, γ-aminobutyric acid, aspartic acid, asparagine. Methionine is the precursor of methional found in 'Cabernet-Sauvignon' grapes [126a].

The presence of alkyl substituted pyrazines in some alcoholic beverages such as rum may be explained by a *Maillard* reaction between amino acids and sugars during ageing. 2-Methoxy-3-isobutylpyrazine has been indicated in an aroma fraction characteristic of 'Cabernet-Sauvignon' grapes [5].

Acids derived from sugars and polyols. Wine contains free or combined uronic, galacturonic, and glucuronic acids in variable amounts as well as free gluconic and mucic acids. 2,3-Butane-diol and polyalcohols are also components of wine (*see Table I–32*).

Table I–32 – Average values of acids derived from sugars and polyalcohols in red and white wines [37]

Acids derived from sugars	(mg/l)	Polyalcohols	(mg/l)
Mucic acid	40	2,3-Butanediol	288
Gluconic acid	340	Erythrytol	79
Glucuronic acid	37	Arabitol	46
Galacturonic acid	620	Mannitol	100
		Sorbitol	45
		Scyllo-inositol	46
		Myo-inositol	354

I-9 REFERENCES

[1] Adda, J., S. Roger & J. P. Dumont, *Advances in the Knowledge of Cheese*, in Ref. 24, 65 (1978).

[2] Archimbault, Ph., R. Fellous & A. Puill, *Parf. Cosm. Arômes*, **32**, 91 (1980a); *ibid*, **33**, 85 (1980b).

[3] Arnold, R. G. & R. C. Lindsay, *J. Dairy Sci.*, **52**, 1097 (1969).

[4] Bajaj, K. L., *Riv. Ital. E P P O S*, **61**, 123 (1979).

[5] Bayonove, C., P. Cordonnier & P. Dubois, *Compt. Rend. Acad. Sci., Paris,* **281**, 75 (1975).

[6] Bayonove, C., H. Richard & P. Cordonnier, *Compt. Rend. Acad. Sci., Paris,* **243**, 549 (1976).

[7] Benterud, A., *Physical, Chemical and Biological Changes in Food Caused by Thermal Processing* (T. Hoyen & O. Kvale, Eds.); Applied Sci. Publ. London, 187 (1977).

[8] Blanc, M., *Lebensm.-Wiss. Technol.*, **5**, 95 (1972).

[9] Bonnet, J. L. & T. Crouzet, *J. Food Sci.*, **42**, 625 (1977).

[10] Bourzeix, M., J. Guitraud & F. Champagnol, *J. Chromatog.*, **50**, 83 (1970).

[11] Brewington, C. R., O. W. Parks & D. P. Schwarz, *J. Agric. Food Chem.*, **21**, 38 (1973); *ibid*, **22**, 293 (1974).

[12] Brown, D. F., *J. Amer. Peanut Res. Educ. Assoc.*, **3**, 208 (1971).

[13] Brown, D. F., J. V. Senn & F. G. Dollear, *J. Agric. Food Chem.*, **21**, 463 (1973).

[14] Bure, J., *Ann. Nutr. Aliment.*, **19**, A-371 (1965).

[15] Buri, R. & J. Solms, *Naturwiss.*, **58**, 56 (1971).;

[16] Buttery, R. G. & R. M. Seifert, *J. Agric. Food Chem.*, **16**, 1053 (1968).

[17] Buttery, R. G., R. M. Seifert & L. C. Ling, *J. Agric. Food Chem.*, **18**, 538 (1970).

[18] Buttery, R. G., R. M. Seifert, D. G. Guadagni & L. C. Ling, *J. Agric. Food Chem.*, **19**, 969 (1971).

[19] Buttery, R. G., *J. Agric. Food Chem.*, **21**, 31 (1973a); R. G. Buttery & L. C. Ling, *ibid*, **21**, 745 (1973b).

[20] Carlin, G. T., *Proc. Amer. Soc. Bakery Eng.*, **1968**, 56.

[21] Cerbulis, J., *Arch. Biochem. Biophys.*, **49**, 442 (1954); *ibid*, **51**, 406 (1956).

[22] Chang, S. S., *J. Amer. Oil Chemists' Soc.*, **56**, 908A (1979).

[23] Charalambous, G. & I. Katz, Eds., *Phenolic, Sulfur and Nitrogen Compounds in Food Flavors*; ACS Symposium Series No. **26** (1976).

[24] Charalambous, G. & G. E. Inglett, Eds., *Flavour of Foods and Beverages*; Academic Press, New York (1978).

[25] Co, H. & G. W. Sanderson, *J. Food Sci.*, **35**, 160 (1970).

[26] Cobb, W. Y., *J. Food Sci.*, **36**, 538 (1971).

[27] Coffman, J. R., *Baker's Dig.*, **46**, 39 (1972).

[28] Coggon, P., L. Ramanozyk, Jr. & G. W. Sanderson, *J. Agric. Food Chem.*, **25**, 278 (1977).

[29] Coleman, E. C. & C. T. Ho, *J. Agric. Food Chem.*, **28**, 66 (1980).

[30] Collyer, D. M., *J. Sci. Food Agric.*, **17**, 440 (1966).

[31] Corbin, E. A. & E. O. Whittier, *Composition of Milk* in *Fundamentals of Dairy Chemistry* (B. H. Webb & A. H. Johnson, Eds.); Avi. Publ. Co, Westport (1965).

[32] Cording, J. Jr. & J. F. Sullivan, *Food Engineering*, **45**, 95 (1973).

[33] Courtois, J. E., R. Percheron & J. C. Glomoud, *Café, Cacao, Thé*, **7**, 231 (1963).

[34] Day, E. A., *Cheese Flavor*, in Ref. 175, 331 (1967).

[35] Deatherage, F. E., Ed., *Food for Life*, Plenum Press, New York, London, 84 (1979).
[36] Deck, R. E., J. Pokorny & S. S. Chang, *J. Food Sci.*, **38**, 345 (1973).
[37] De Smet, P., P. A. P. Liddle, B. Cresto & A. Bossard, *Ann. Fals. Exp. Chim.*, **72**, 633, 781 (1979).
[38] Drapon, R. & Y. Beaux, *Compt. Rend. Acad. Sci., Paris,* **286**, 2598 (1969a); Drapon, R., *La France et ses Parfums,* **13**, 239 (1971b).
[39] Dubois, P., G. Brule & M. Ilic, *Ann. Technol. Agric.*, **20**, 131 (1971).
[40] Dumont, J. P. & J. Adda, *J. Agric. Food Chem.*, **26**, 364 (1978a); *Idem, Flavour Formation in Dairy Products*, in Ref. 103, 245 (1979b).
[41] Dwivedi, B. K. & R. G. Arnold, *J. Food Sci.*, **37**, 887 (1972).
[42] Dwivedi, B. K., *Meat Flavor* in *Fenaroli's Handbook of Flavors Ingredients*, CRC Press, 2nd ed. Vol. **II**, 822 (1975).
[43] Dash El, A. A. & J. A. Johnson, *Cereal Sci. Today,* **12**, 282, 286 (1968a); Dash El, A. A., *Baker's Dig.*, **45**, 36 (1971b).
[44] Erickson, S., *Milchwiss.*, **31**, 549 (1976).
[45] Erickson, C. E., *Review of Biosynthesis of Volatiles in Fruits and Vegetables since 1975*, in Ref. 103, 159 (1979).
[46] Eyton, W. B., *The Flavour Ind.*, **1972**, 23.
[47] Feenstra, W. H. & P. W. Meijboom, *J. Amer. Oil Chemists' Soc.*, **48**, 684 (1971).
[48] Feldman, J. R., W. S. Ryder & J. J. Kung, *J. Agric. Food Chem.*, **17**, 733 (1969).
[49] Flament, I., *Parf. Cosm. Arômes,* **35**, 75 (1980).
[50] Flanzy, C. & P. Benard, *Ann. Technol. Agric.*, **19**, 53 (1970).
[51] Folstar, P., *Agric. Res. Reports*, No. **854**, PUDOC, Wageningen (1976).
[52] François, R., *Revue des Fabricants de Confiserie, Chocolaterie, Confiturerie et Biscuiterie,* **1980**, 28.
[53] Galliard, T., D. R. Phillips & J. Reynolds, *Biochem. Biophys. Acta,* 1976, 181.
[54] Galliard, T., J. A. Matthew, A. J. Wright & M. J. Fishwick, *J. Sci. Food Agric.*, **28**, 863 (1977).
[55] Galliard, T. & J. A. Matthew, *Phytochemistry,* **16**, 339 (1977).
[56] Garnero, J., *Riv. Ital. E P P O S,* **62**, 253 (1980).
[57] Gholap, A. S. & C. Bandyopadhyay, *J. Amer. Oil. Chemists' Soc.*, **52**, 514 (1975).
[58] Giannone, L., G. Pezzani & M. Campanini, *Ind. Conserve (Parme),* **43**, 105 (1968).
[59] Gonzalez, J. G., P. Coggon & G. W. Sanderson, *J. Food Sci.*, **37**, 797 (1972).
[60] Gregory, J. F. & J. R. Kirk, *J. Nutrition,* **108**, 1192 (1978).
[61] Griffiths, L. A., *Biochem. J.*, **70**, 120 (1958).

[62] Guadagni, D. G., J. C. Miers & D. Venstrom, *Food Technol. Champaign.*, **22**, 1003 (1968); *Idem, J. Food Sci.*, **34**, 630 (1969).

[63] Guadagni, D. G. & J. C. Miers, *Food Technol. Champaign.*, **23**, 375 (1969).

[64] Guadagni, D. G., R. G. Buttery, R. M. Seifert & D. W. Venstrom, *J. Food Sci.*, **36**, 383 (1971).

[65] Hamm, R., *Changes of Muscle Proteins During the Heating of Meat* in *Physical Chemical and Biological Changes in Food Caused by Thermal Processing* (T. Hoyem & O. Kvale, Eds.); Applied Sci. Publ., London, 101 (1977).

[66] Hammond, F. C. & F. D. Hill, *J. Amer. Oil. Chemists' Soc.*, **41**, 180 (1961).

[67] Hasegawa, S., R. M. Johnson & W. A. Gould, *J. Agric. Food Chem.*, **14**, 165 (1966).

[68] Hatanaka, A. & T. Harada, *Phytochemistry*, **12**, 234 (1973).

[69] Hatanaka, A., T. Kahwara, J. Sekiya & Y. Kido, *Phytochemistry*, **16**, 1828 (1977).

[70] Hawrysh, Z. J. & C. M. Stine, *J. Food Sci.*, **38**, 7 (1973).

[71] Herb, S. F., P. Magidman, F. E. Luddy & R. W. Reimenschneider, *J. Amer. Oil. Chemists' Soc.*, **39**, 142 (1962).

[72] Herz, K. O., *Food Technol.*, **19**, 90 (1965).

[73] Ho, C. T. & E. C. Coleman, *J. Food Sci.*, **45**, 1094 (1980).

[74] Hodge, J. E., F. D. Mills & B. E. Fisher, *Cereal Sci. Today*, **17**, 34 (1972).

[75] Holaday, C. A. & J. L. Pearson, *J. Food Sci.*, **39**, 1206 (1974).

[76] Hosono, A., J. A. Elliot & W. A. Mac Gugan, *J. Dairy Sci.*, **57**, 535 (1974).

[77] Hunt, G. M. & E. A. Baker, *Phytochemistry*, **19**, 1415 (1980).

[78] Hunter, I. R., M. K. Walden, W. H. Mac Fadden & J. W. Pence, *Cereal Sci. Today*, **11**, 493 (1966).

[79] Hunter, I. R., M. K. Walden, J. R. Scherer & R. E. Lundin, *Cereal Chem.*, **46**, 189 (1969).

[80] Hurrell, R. F., *Interaction of Food Components During Processing* in *Food and Health: Science and Technology* (G. R. Birch & K. J. Parker, Eds.); Applied Sci. Publ., 369 (1980).

[81] Jacorzynski, B. & W. Iwanska, *Rocz. Panstw. Zakl. Hig.*, **21**, 65 (1970).

[82] Jarboe, J. K. & A. F. Mabrouk, *J. Agric. Food Chem.*, **22**, 787 (1974).

[83] Jenness, R., B. L. Larson, T. L. Mac Meekin, A. M. Swanson, C. H. Whitnah & R. M. Whitney, *J. Dairy Sci.*, **39**, 536 (1956).

[84] Jenness, R. & S. Patton, *Principles of Dairy Chemistry* (Chapman & Hall, London): J. Wiley & Sons Inc., New York (1959).

[85] Jensen, R. G., J. G. Quinn, D. L. Carpenter & J. Sampugna, *J. Dairy Sci.*, **50**, 119 (1967).

[86] Johnson, B. R., G. R. Waller & A. L. Burlingame, *J. Agric. Food Chem.*, **19**, 1020 (1971).

[87] Johnson, B. R., G. R. Waller & R. L. Foltz, *J. Agric. Food Chem.*, **19**, 1025 (1971).

[88] Johnson, J. A., L. Rooney & A. Salem, *Chemistry of Bread Flavor* in *Flavor Chemistry*; ACS Symposium Series No. **56**, 153 (1966).

[89] Johnson, J. A. & A. A. El Dash, *J. Agric. Food Chem.*, **17**, 740 (1969a); L. P. Carroll, B. S. Miller & J. A. Johnson, *Cereal Chem.*, **33**, 303 (1956b).

[90] De Jong, K., *Fette, Seifen, Anstrichm.*, **69**, 277 (1967).

[91] Kahn, J. H., *Compounds Identified in Whisky, Wine and Beer: a Tabulation, J. Assoc. Off. Anal. Chem.*, **52**, 1166 (1969).

[92] Kaldy, M. S. & P. Markakis, *J. Food Sci.*, **37**, 375 (1972).

[93] Kaufmann, H. P. & R. Schickel, *Fette, Seifen, Anstrichm.*, **65**, 1012 (1963).

[94] Kaufmann, H. P. & A. K. Sen Gupta, *Chem. Ber.*, **96**, 2489 (1963).

[95] Kazeniac, S. J. & R. M. Hall, *J. Food Sci.*, **35**, 519 (1970).

[96] Keenan, T. W. & R. C. Lindsay, *J. Dairy Sci.*, **51**, 112 (1968).

[97] Kiely, R. J., A. C. Nowlin & J. H. Moriarty, *Cereal Sci. Today*, **5**, 273 (1960).

[98] Kinlin, T. E., R. Muralidhara, A. O. Pittet, A. Sanderson & T. P. Walradt, *J. Agric. Food Chem.*, **20**, 1021 (1972).

[99] Koehler, P. E., *Formation of Alkylpyrazine Compounds and their Role in the Flavour of Roasted Food:* Ph. D. Thesis, Oklahoma State University, Stillwater.

[100] Kroplien, U., *J. Agric. Food Chem.*, **22**, 110 (1974).

[101] Kuninaka, A., in *The Chemistry and Physiology of Flavors* (H. W. Schulz, E. A. Day & L. M. Libbey, Eds.); Symposium of Foods, The Avi. Publ. Co., Westport, 515–535 (1967).

[102] Kurtz, F. E., *The Lipids of Milk: Composition and Properties* in *Fundamentals of Dairy Chemistry*; The Avi. Publ. Co., Westport (1965).

[103] Land, D. G. & H. E. Nursten, Eds., *Applied Sci. Publ.*, London (1979).

[104] Lawrie, R. A., *Flavour Ind.*, **1**, 591 (1970).

[105] Lien, Y. C. & W. W. Nawar, *J. Food Sci.*, **39**, 911, 914 (1974).

[106] Lijinski, W. & S. S. Epstein, *Nature* (London), **225**, 21 (1970).

[107] Lockhart, E. E., *Coffee & Tea Ind.*, **1957**, 71.

[108] Luc, V. & R. Dominique, *Adv. Chromatog. Proc. Int. Symp.*, **6**, 69 (1970).

[109] Lhuguenot, J. C., L. Ude, E. Dywarski & C. Baron, *Ann. Fals. Exp. Chim.*, **72**, 275 (1979).

[110] Mabrouk, A. F., in *Phenolic, Sulfur and Nitrogen Compounds in Food Flavors* (G. Charalambous & I. Katz, Eds.); ACS Symposium Series No. **26**, Amer. Chem. Soc., 146, 159 (1976).

[111] Mabrouk, A. F., *Flavor of Browning Reaction Products* in *Food Taste Chemistry* (J. C. Boudreau, Ed.); ACS Symposium Series, 115, 205 (1979).

[112] Maga, J. A., *CRC Crit. Rev. Food Technol.*, **4**, 39 (1973); *idem, J. Agric. Food Chem.*, **21**, 22 (1973).
[113] Maga, J. A., *Bread Flavor* in *Fenaroli's Handbook of Flavor Ingredients*; 2nd Edition, CRC Press, Vol. **II**, 539 (1975).
[114] Markakis, P. & A. Amon., *Food Technol.*, **23**, 11, 131 (1969).
[115] Markakis, P. & G. Kallifidas, *Amer. J. Enol. Viticult.*, **22**, 135 (1971).
[116] Marth, E. H., R. V. Hussons, L. F. Cremers, J. H. Guth, L. D. Hilker, H. W. Jackson, O. J. Krett, E. C. Stimpson, R. A. Sullivan & L. Tumerman, *Milk Products* in *Encyclopedia of Chemical Technology*, **13**, 506 (1967).
[117] Martin, G. E., J. G. Sullo & R. Schoeneman, *J. Agric. Food Chem.*, **19**, 985 (1971).
[118] Mason, M. E., J. A. Newell, B. R. Johnson, P. E. Koehler & G. R. Waller, *J. Agric. Food Chem.*, **17**, 728 (1962).
[119] MacLeod, A. J., *Volatile Flavour Compounds of the Cruciferae* in *The Biology and Chemistry of the Cruciferae* (J. G. Vaughan, A. J. MacLeod & B. M. C. Jones, Eds.); Academic Press, London (1979).
[120] Millin, D. J., D. J. Crispin & D. Swaine, *J. Agric. Food Chem.*, **17**, 717 (1969).
[121] Mohr, W., H. Röhrle & T. Severin, *Fette, Seifen, Anstrichm.*, **73**, 515 (1971).
[122] Mollard, R., D. Cahagnier, B. Poisson & R. Dapron, *Ann. Technol. Agric. INRA.*, **25**, 29 (1976).
[123] Mondy, N. I., C. Metcalf & P. L. Plaisted, *J. Food Sci.*, **36**, 459 (1971).
[124] Morot-Gaudry, J. F., H. Z. Nicol & E. Jolivet, *J. Chromatog.*, **87**, 430 (1973).
[125] Mueggler-Chavan, F., R. Viani, J. Bricout, J. P. Marion, H. Mechtler, D. Reymond & R. H. Egli, *Helv. Chim. Acta,* **52**, 549 (1969).
[126] Muller, C. J., R. E. Kepner & A. D. Webb, *Amer. J. Enol. Viticult.*, **22**, 156 (1971a); *Idem, J. Agric. Food Chem.*, **20**, 193 (1972b).
[127] Nakamura, H., K. Watanabe & J. Mitzutani, *Nippon Nogie Kagaku Kaishi.*, **48** (4), 275 (1974a); *Idem, Nippon Nogei Kagaku Kaishi.*, **49** (12), 665 (1975).
[128] Newell, J. A., M. E. J. Mason & R. S. Matlock, *J. Agric. Food Chem.*, **15**, 767 (1967).
[129] Nursten, H. E. & M. R. Sheen, *J. Sci. Food Agric.*, **25**, 643 (1974).
[130] Nursten, H. E., *Flavor Chemistry of Fruits and Vegetables* in *Agricultural and Food Chemistry, Past, Present, Future* (R. Teranishi, Ed.); The Avi. Publ. Co., Westport, 333 (1978).
[131] Ostovar, K. & P. G. Keeney, *J. Food Sci.*, **38**, 611 (1973).
[132] Pareless, S. R. & S. S. Chang, *J. Agric. Food Chem.*, **22**, 339 (1974).
[133] Park, R. J., J. D. Armitt & W. Stark, *J. Dairy Res.*, **36**, 37 (1969).
[134] Parks, O. W., N. P. Wong, C. A. Allen & D. P. Schwartz, *J. Dairy Sci.*, **52** (7), 953 (1969).

[135] Parliment, T. H., W. Clinton & R. Scarpellino, *J. Agric. Food Chem.*, **21**, 485 (1973).
[136] Pattee, H. E., J. A. Singleton, E. B. Johns & E. B. Mullin, *J. Agric. Food Chem.*, **18**, 353 (1970).
[137] Pattee, H. E. & J. A. Singleton, *J. Amer. Peanut Res. Educ. Assoc.*, **2**, 143 (1970).
[138] Paul, A. A. & D. A. T. Southgate, *The Composition of Foods* (McCance & Widdowson's, Eds.); 4th edition (1978).
[139] Pearsons, J. G., P. G. Keeney & S. Patton, *J. Food Sci.*, **34**, 497 (1969).
[140] Pepper, F. H. & A. M. Pearson, *J. Agric. Food Chem.*, **22**, 49 (1974).
[141] Pickenhagen, W., P. Dietrich, B. Keil, J. Polonski, F. Nouaille & E. Lederer, *Helv. Chim. Acta*, **58**, 1078 (1975).
[142] Pictet, G. & H. Brandenberger, *J. Chromatog.*, **4**, 396 (1960).
[143] Pomeranz, Y. & J. A. Shellenberger, *Bread Flavor* in *Bread Science and Technology*; Avi. Publ. Co., Westport, 180 (1971).
[144] Purr, A., H. Moreiner & R. Springer, *Rev. Intern. Chocolat.*, **20** (5), 94 (1965).
[145] Quesnel, V. C., *J. Sci. Food Agric.*, **56**, 596 (1965).
[146] Ramshaw, E. H. & E. A. Dunstone, *J. Dairy Res.*, **36**, 215 (1969).
[147] Ranganna, S. & L. Setty, *J. Agric. Food Chem.*, **16**, 529 (1968).
[148] Reineccius, G. A., D. A. Andersen, T. E. Kavanagh & P. G. Keeney, *J. Agric. Food Chem.*, **20**, 199 (1972).
[149] Reineccius, G. A., P. G. Keeney & W. Weissberger, *J. Agric. Food Chem.*, **20**, 202 (1972).
[150] Renold, W., R. Näf-Muller, U. Keller, B. Wilhalm & G. Ohloff, *Helv. Chim. Acta*, **57**, 1301 (1974).
[151] Reymond, D., *Chem. Tech.*, **7**, 664 (1977).
[152] Reymond, D. & D. Rostagno, *Flavor Aspects of Chocolate* in *Flavor of Foods and Beverages, Chemistry and Technology* (G. Charalambous & G. E. Inglett, Eds.); Academic Press, New York, 75 (1978).
[153] Reymond, D., *Flavor Chemistry of Tea, Cocoa and Coffee* in Ref. 130, 315 (1978).
[154] Ribereau-Gayon, P., J. N. Boidron & A. Terrior, *J. Agric. Food Chem.*, **23**, 1042 (1975).
[155] Robinson, R. J., T. H. Lord, J. A. Johnson & B. S. Miller, *Cereal Chem.*, **35**, 295 (1958).
[156] Rohan, T. A. & M. Connell, *J. Food Sci.*, **29**, 460 (1964a); *ibid*, **29**, 456 (1964b).
[157] Rohan, T. A. & T. Stewart, *J. Food Sci.*, **32**, 399 (1967a); *ibid*, **32**, 625 (1967b).
[158] Rohan, T. A., *Gordian.*, **1**, 5, 53, 111 (1970).
[159] Russwurm, H., *Fractionation and Analysis of Aroma Precursors in green Coffee, Colloquium on Coffee Chemistry*, Amsterdam, ASIC, Paris, 462 (1969).

[160] Ryan, J. J., *J. Assoc. Offic. Anal. Chem.*, **55**, 1104 (1972).
[161] Ryan, J. J. & J. A. Dupont, *J. Agric. Food Chem.*, **21**, 45 (1973).
[162] Saijo, R. & T. Takeo, *Agric. Biol. Chem.*, **34**, 227 (1970).
[163] Saito, S. & F. Kano, *Tokyo Kagyo Daigaku Nogaku Shaho.*, **10** (2), 32 (1964).
[164] Salem, A., L. W. Rooney & J. A. Johnson, *Cereal Chem.*, **44**, 576 (1967).
[165] Salunkhe, D. K., *CRC Crit. Rev. Food Technol.*, **8**, 161 (1977).
[166] Sanderson, H. C. & J. G. Gonzalez, *J. Food Sci.*, **36**, 231 (1971).
[167] Sanderson, G. W. & H. N. Graham, *J. Agric. Food Chem.*, **21**, 576 (1973).
[168] Sanderson, G. W., A. S. Ranadive, L. S. Eisenberg, F. J. Farrell, R. Simons, C. H. Man & P. Coggon, *Contribution of Phenolic Compounds to the Taste of Tea* in Ref. 23, 14 (1976).
[169] Sapers, G. M., *J. Agric. Food Chem.*, **23**, 1027 (1975).
[170] Sapers, G. M., O. Panasiuk, F. B. Talley, S. F. Osman & R. L. Shaw, *J. Food Sci.*, **37**, 579 (1972).
[171] Schörmüller, J. & J. Weber, *Z. Lebensm.-Unters. Forsch.*, **130**, 158 (1966).
[172] Schörmüller, J., M. Walther & W. Wachs, *Z. Lebensm.-Unters. Forsch.*, **139**, 273 (1969).
[173] Schreier, P., *Flavor Composition of Wines: A Review, CRC Critical Reviews in Food Science and Nutrition,* **11**, 59 (1978).
[174] Schreier, P., F. Drawert & B. Bhiwapurkar, *Mikrobiol. Technol. Lebensm.*, **6**, 90 (1979).
[175] Schulz, H. W., E. A. Day & L. M. Libbey, Eds., *Cheese Flavor* in *Chemistry and Physiology of Flavors*, Avi. Publ. Co., Westport (1967).
[176] Schuldt, E. E. Jr., *Baker's GDig.*, **41** (5) 90 (1967).
[177] Schutte, L. & E. B. Koenders, *J. Agric. Food Chem.*, **20**, 181 (1972).
[178] Schutte, L., *CRC Crit. Rev. Food Technol.*, **4**, 457 (1974).
[179] Schwimmer, S. & M. Fierman, *Flavour Ind.*, 137 (1972).
[180] Seck, S. & J. Crouzet, *Phytochemistry,* **12**, 2925 (1973).
[181] Sekiya, J., T. Kajiwara & A. Hatanaka, *Phytochemistry,* **16**, 1043 (1977).
[182] Sessa, D. J., *J. Agric. Food Chem.*, **27**, 235 (1979).
[183] Sheldon, R. M., R. C. Lindsay & L. M. Libbey, *J. Food Sci.,* **37** , 313 (1972).
[184] Shipe, W. E., *Enzymatic Modification of Milk Flavor* in *Enzymes in Food and Beverage Processing* (L. Ory & A. J. St.-Angela, Eds.); ACS Symposium Series No. **47**, 57 (1977).
[185] Shu, C. K. & G. R. Waller, *J. Food Sci.*, **36**, 579 (1971).
[186] Singleton, V. L. & A. C. Noble, *Wine Flavor and Phenolic Substances* in Ref. 23, 47 (1976).
[187] Solms, J., in *Gustation and Olfaction* (G. Ohloff & A. F. Thomas, Eds.); Academic Press, London, New York, 92 (1971).
[188] Solms, J. & R. Wyler, *Taste Components of Potatoes* in *Food Taste Chemistry*, Amer. Chem. Sty., 175 (1979).

[189] Stone, E. J., R. M. Hall & S. J. Kazeniac, *J. Food Sci.*, **40**, 1138 (1975).

[190] Sullivan, J. F., R. P. Konstance, M. J. Calhoun, F. B. Talley, J. Cording & O. Panasiuk, *J. Food Sci.*, **39**, 58 (1974).

[191] Tanaka, T. N. Saito, A. Okuhara & T. Yokotsuka, *Nippon Nogei Kagaku Kaishi*, **43**, 171, 263 (1969).

[192] Tatum, J. H., P. E. Shaw & R. E. Berry, *J. Agric. Food Chem.*, **17**, 38 (1969).

[193] Teranishi, R., Ed., *Agricultural and Food Chemistry, Past, Present, Future*; Avi. Publ. Co., Westport (1978).

[194] Thaler, H. & R. Gaigl, *Z. Lebensm.-Unters. Forsch.*, **118**, 22 (1962); *ibid*, **119**, 10 (1962); *ibid*, **120**, 357 (1963); *ibid*, **120**, 449 (1963).

[195] Thaler, H., *Café, Cacao, Thé*, **7**, 3, 240 (1963).

[196] Thaler, H. & W. Arneth, *3rd Int. Colloq. on Coffee*, Trieste, 127 (1967).

[197] Thomas, B. & M. Rothe, *Baker's Dig.*, **34**, 4, 50, 53 (1960).

[198] Tonsbeek, C. H. T., A. J. Plancken & T. Van De Weerdhof, *J. Agric. Food Chem.*, **16**, 1016 (1968).

[199] Tonsbeek, C. H. T., E. B. Koenders, A. S. M. Van Der Zijden & J. A. Losekoot, *J. Agric. Food Chem.*, **17**, 397 (1969).

[200] Tressl, R. & F. Drawert, *J. Agric. Food Chem.*, **21**, 560 (1973).

[201] Tressl, R., M. Holzer & M. Apetz, in *Aroma Research* (H. Maarse & P. J. Groenen, Eds.); PUDOC, Wagenigen, 41 (1975).

[202] Tressl, R., T. Kossa, R. Renner & H. Köppler, *Z. Lebensm.-Unters. Forsch.*, **162**, 123, 366 (1976).

[203] Tressl, R., M. Holzer & M. Apetz, *J. Agric. Food Chem.*, **25**, 455 (1977).

[204] Tressl, R., D. Bahri, M. Holzer & T. Kossa, *J. Agric. Food Chem.*, **25**, 459 (1977).

[205] Tressl, R., D. Bahri, A. Jensen & H. Köppler, *Z. Lebensm.-Unters. Forsch.*, **167**, 111 (1978).

[206] Tressl, R., K. G. Grünewald, R. Silvar & D. Bahri, *Chemical Formation of Flavour Substances* in *Progress in Flavour Research* (G. D. Land & H. E. Nursten, Eds.); Applied Sci. Publ., London, 197 (1979).

[207] Van Den Ouweland, G. A. M., H. Osman & H. G. Peer, *Challenges in Meat Flavor Research* in Ref. 130, 292 (1978).

[208] Van Der Wal, B., K. D. Kettenes, J. Stoffelsma & A. T. J. Semper, *J. Agric. Food Chem.*, **19**, 276 (1971).

[209] Van Praag, H., S. Stein & M. S. Tibbetts, *J. Agric. Food Chem.*, **16**, 1005 (1968).

[210] Van Straten, S., F. De Vrijer & J. C. De Beauveser, *List of Volatile Compounds in Food*, 3rd Ed. Suppl. 1-8, Central Institute for Nutrition and Food Research, TNO, Zeist (1977-1980).

[211] Veldink, G. A., J. F. G. Vliegenthart & J. Boldingh, *Prog. Chem. Fats Lipids*, **15**, 139 (1977).

[212] Vernin, G., *Ind. Aliment. Agric.*, **5**, 433 (1980); *idem, Parf. Cosm. Aromes.*, **31**, 77 (1980).

[213] Viani, R., J. Bricout, J. P. Marion, F. Müggler-Chavan, D. Reymond & R. H. Egli, *Helv. Chim. Acta*, **52**, 887 (1969).

[214] Viani, R., R. H. Egli & I. Horman, *J. Food Sci.*, **39**, 1216 (1974).

[215] Vitzthum, O. G. & P. Werkhoff, *Z. Lebensm.-Unters. Forsch.*, **156**, 300 (1974).

[216] Vitzthum, O. G., P. Werkhoff & P. Hubert, *J. Agric. Food Chem.*, **23**, 999 (1975a); *idem, J. Food Sci.*, **40**, 911 (1975b).

[217] Vitzthum, O. G., *Chemistry and Technology of Coffee* in *Kaffee und Koffein* (O. Eichler, Ed.); Springer Verlag, Berlin (1975).

[218] Von Sydow, E. & K. Anjou, *Lebensm.-Wiss. Technol.*, **2**, 15 (1969).

[219] Walradt, J. P., A. O. Pittet, T. E. Kinlin, R. Muralidhara & A. Sanderson, *J. Agric. Food Chem.*, **19**, 972 (1971).

[220] Wasserman, A. E. & A. M. Spinelli, *J. Agric. Food Chem.*, **20**, 171 (1972).

[221] Weinges, K. & O. Muller, *Chem. Ztg.*, **96**, 612 (1972).

[222] Wickremasinghe, R. L., *Phytochemistry*, **13**, 2057 (1974).

[223] Williams, P. J., C. R. Strauss & B. Wilson, *Phytochemistry*, **19**, 1137 (1980).

[224] Wolfrom, M. L., M. L. Laver & D. L. Patin, *J. Org. Chem.*, **26**, 4533 (1961).

[225] Woodman, J. S., A. Giddey & R. H. Egli, *3rd Int. Colloq. on Coffee Chemistry*, Trieste, ASIC, Paris (1967).

[226] Wurziger, J., *Ann. Fals. Exp. Chim.*, **66**, 1 (1973).

[227] Yamanishi, T., *Nippon Nogei Kagaku Kaishi.*, **49**, 9, R-1 – R-9 (1975).

[228] Yamazaki, A., I. Kumashiro & T. Takenishi, *Chem. Pharm. Bull.*, **17**, 1128 (1969).

[229] Yamazaki, A., I. Kumashiro & T. Takenishi, *Chem. Pharm. Bull.*, **16**, 338 (1968).

[230] Yanagawa, H., T. Kato, Y. Kitahara, N. Takahashi & Y. Kato, *Tetrahedron Letters*, **25**, 2549 (1972).

[231] Young, C. T., *J. Food Sci.*, **38**, 123 (1973a); *idem*, **45**, 1086 (1980b).

[232] Yu, M. H., D. K. Salunkhe & L. E. Olson, *Plant Cell. Physiol.*, **9**, 633 (1968).

CHAPTER II

Heterocyclic Aroma Compounds in Foods: Occurrence and Organoleptic Properties

Gaston VERNIN and Genevieve VERNIN, University of Aix-Marseilles, France

II–1 INTRODUCTION

The overall flavour of food results not only from the traditional classes of compounds such as carbohydrates, lipids, alcohols, phenols and carbonyl compounds but also from the various heterocyclic compounds, a newly identified class of flavouring substances. They are present in flavours in such minute traces that before the advent of gas-liquid chromatography, especially in combination with mass spectrometry, they were impossible to detect.

They have been isolated and identified from the following food systems:

- Meat, fats, and poultry products,
- Vegetables, mushrooms, and allium species,
- Fruits (juices, syrups. . .),
- Nut and oilseed products,
- Cereals and related products (bread. . .),
- Milk and dairy products (butter, cheese. . .),
- Tobacco and tobacco smoke,
- Alcoholic and non-alcoholic beverages,
- Essential oils,
- Fish (caviar, shrimp paste. . .),
- Miscellaneous foodstuffs.

These substances contain one or more heteroatoms (O, S and/or N) in five, six and more membered rings or in fused ring systems. Among them: pyrazines, furans and lactones, pyrroles, thiazoles, pyridines and oxazoles play an important role. Pyrimidines are less widely distributed, and no isothiazole or isoxazole have yet been found in foods. Among fused heterocyclic systems, cyclopentapyrazines and benzothiazoles are the most abundant. Some of these compounds possess an extraordinarily high odour strength resulting from their low detection threshold (down to 0.002 ppb) and contribute to certain flavour types in which they constitute character impact compounds (top-notes). For these reasons, they also are ideal compounding ingredients in food chemistry as is evident in the patent literature [152].

Table II–1 – Recent books and reviews dealing with food flavours

Books	References
"Chemistry and Physiology of Flavors"	177
"Volatile Compounds in Foods"	195
"Fenaroli's Handbook of Flavor Ingredients"	52
"Food Flavoring Processes"	152
"Phenolic, Sulfur and Nitrogen Compounds in Food Flavors"	29
"Agricultural Food Chemistry, Past, Present, Future"	200
"Effect of Heating on Foodstuffs"	161
"Flavor of Foods and Beverages"	30
"Progress in Flavour Research"	95
"Food Flavors" (I. D. Morton & J. A. MacLeod, Eds.), Elsevier Publ. Co. (1981)	
"Fragrances & Flavors. Recent Developments" (Chemical Technology)	
Review No. 156 (Ed. S. Torrey); Noyes Data Corporation, 1980	
"Maillard Reactions in Food, Chemical, Physiological and Technological Aspects" (C. Eriksson, Ed.), Pergamon Press, Oxford	in press
Reviews	
"Processed Food and Flavors"	31
"Chemistry and Flavour"	9
"Recent Developments in Studies of the Maillard Reaction"	135b
"The Role of Heteroatomic Substances in the Aroma Compounds of Foodstuffs"	137b
"Heterocyclics in Flavour Chemistry"	57
"Les Hétérocycles dans les Arômes Alimentaires"	212a
"La Chimie des Arômes: Les Hétérocycles"	212b
"Les Hétérocycles dans la réaction de Maillard: Nouvelles Substances Aromatisantes"	212c
"Recent Development in the Field of Naturally-Occurring Aroma Compounds"	137a
"The Role of Heterocyclic Compounds in Foods"	
– Pyrazines	105
– Thiazoles	106
– Thiophenes	107
– Lactones	108
– Oxazoles and Oxazolines	109
– Furans	111a
– Pyrroles	111b
– Pyridines	212d
"Chemical Formation of Flavour Substances"	210

The purpose of this chapter is to update the enumeration of the heterocyclic compounds now known to occur in food flavours or patented as flavourings, and to describe their organoleptic properties. For more information, the reader is referred to recently published books and reviews dealing with food flavours (*see Table II–1*).

Heterocyclic compounds are classified according to the ring size, the number of heteroatoms (disregarding the substituents), and the degree of oxidation of the ring itself. Condensed bicyclic components are classified in the same way as the monocyclic systems. In tables, substituents are listed according to the following hierarchy: H > $-CH_3$ > $-CH_2R$ > $-CHR_2$ > $-CR_3$ > –CH= > aryl > heteroaryl > –C≡ > $-NH_2$ > –NHR > $-NR_2$ > –N= > –OH > –OR > –SH > –SR. In these tables, food systems from which heterocyclic compounds have been identified are indicated by italic numbers (*1* - *142*) to avoid repetitions. For the corresponding reference, the reader is referred to the text or to the review given in caption.

II–2 FIVE-MEMBERED HETEROCYCLIC SYSTEMS CONTAINING ONE, TWO, OR MORE HETEROATOMS

2.1 Furans and reduced systems

2.1.1 Structure and occurrence

Formed mainly by thermal degradation of carbohydrates and ascorbic acid and from sugar-amino acids interaction during food processing, furans are therefore present in nearly all food aromas [111] (*see Table II–2, II–3 and II–4*).

Table II–2 Furan and 2-substituted derivatives[a)]

1

1; R	Name	Occurrence[b)]
H	Furan	*18,66,71,72,73,83,94, 104,113,127,137*
CH_3	2-Methylfuran	*3,4,9,12,18,66,69,71,73, 83,94,104,106,127,137*
C_2H_5	2-Ethylfuran	*3,4,12,18,25,66,71,73, 94,104,127*
n-C_3H_7	2-*n*-Propylfuran	*4,18,66,71,73,104,127, 131,136*
n-C_4H_9	2-*n*-Butylfuran	*4,12,18,46,71,73,82,94, 104*

Table II–2 (*continued*)

1; R	Name	Occurrence
n-C_5H_{11}	2-*n*-Pentylfuran	*3,4,9,12,17,19,21,25,28, 46,54,66,67,69,71,73,77, 78,79,82,83,85,94,104, 105,121*
C_6H_{13}	Methylpentylfuran	*19*
n-C_6H_{13}	2-*n*-Hexylfuran	*4,9,18,66,73,94,121,127*
n-C_7H_{15}	2-*n*-Heptylfuran	*9,18,19,66*
C_8H_{17}	Methylheptylfuran	*19,121*
n-C_8H_{17}	2-*n*-Octylfuran	*9,19,66*
$(CH_2)_3$ O	2,2′-Difurfurylmethane	*104*
i-C_5H_{11}	2-Isopentylfuran	*105*
$(CH_2)_2$ O CH_3	(5′-Methyl-2′-furyl)-2-furfurylmethane	*104*
CH_2CH_2CHO	3-(2′-Furyl)-propanal	*104*
$CH_2CH_2COCH_3$	4-(2′-Furyl)-butan-2-one	*104*
$CH_2CH(CH_3)_2$	2-Isobutylfuran	*104*
CH_2COCH_3	Furfuryl methyl ketone (2-Acetonylfuran)	*12,67,72,74,83,85,92, 104,133*
$CH_2COC_2H_5$	Furfuryl ethyl ketone (1-(2′-Furyl)butan-2-one	*44,104*
CH_2 O	2,2′-Difurylmethane	*12,92,104*
CH_2 O CH_3	(5′-Methyl-2′-furyl)-2-furylmethane	*12,79,92,104*
CH_2-N	1-Furfurylpyrrole	*66,67,79,83,104*
CH_2-N CHO	1-Furfuryl-2-formyl-pyrrole	*79,83,85*
CH_2OH	Furfuryl alcohol	*4,10,12,25,38,44,56,66, 67,74,76,79,81,83,85, 102,104,105,108,113*
CH_2OCH_3	Furfuryl methyl ether	*104*
$CH_2OC_2H_5$	Furfuryl ethyl ether	*12,83,109*

Table II–2 (*continued*)

1; R	Name	Occurrence
CH_2OCH_2-(2-furyl)	Difurfuryl ether	*12,83,85,92,104*
CH_2OCH_2-(5-methyl-2-furyl) CH_3	(5′-Methyl-2′-furfuryl) furfuryl ether	*92,104*
$CH_2OC(=O)H$	Furfuryl formate	*12,83,85,92,104*
$CH_2OC(=O)CH_3$	Furfuryl acetate	*12,66,67,76,83,85,91, 92,104*
$CH_2OC(=O)C_2H_5$	Furfuryl propionate	*12,92,104*
$CH_2OC(=O)C_3H_7$-*n*	Furfuryl butyrate	*12,89,92,104*
$CH_2OC(=O)CH(CH_3)C_2H_5$	Furfuryl-2′-methylbutyrate	*104*
$CH_2OC(=O)C_4H_9$-*n*	Furfuryl valerate (or pentanoate)	*12,89*
$CH_2OC(=O)C_5H_{11}$-*n*	Furfuryl hexanoate	*12*
CH_2SH	Furfuryl mercaptan	*104*
CH_2SCH_3	Furfurylmethyl sulphide	*12,83,104*
CH_2SCH_2-(2-furyl)	Difurfuryl sulphide	*104*
CH_2S-SCH_3	Furfurylmethyl disulphide	*12,83*
$CH(CH_3)_2$	2-Isopropylfuran	*104*
CHOHCO-(2-furyl)	Furoin	*104*
$CH{=}CH_2$	2-Vinylfuran	*66,71,104*
$CH{=}CHCH_3$	2-Propenylfuran	*73,94,104,127*
$C(CH_3){=}CHCH_3$	2-Isobutenylfuran	*4,104*
	2-(2′-Methylpropenylfuran)	
$CH{=}CHCH_2CH_3$	1-Butenylfuran	*71*
$CH{=}CH{-}CHO$	3-(2′-Furyl)-propen-1-al	*104*
$C(CH_3){=}CH{-}CHO$	α-Methylfuranacrolein	*67*
$CH{=}CHCOCH_3$	4-(2′-Furyl)-3-buten-2-one	*104*
CHO	Furfural	*3,4,12,25,38,46,47,50, 53,55,56,62,64,66,67, 68,69,72,74,76,79,81, 83,84,85,87,89,91,92, 93,102,104,105,106,108, 109,113,133,136*
$COCH_3$	Furyl methyl ketone (or 2-Acetylfuran)	*3,4,12,38,50,56,64,66, 67,76,79,81,82,83,89,92, 102,104,105,106,107,113*

Table II–2 (*continued*)

1; R	Name	Occurrence
COC_2H_5	Furyl ethyl ketone (or 2-Propionylfuran)	*12,25,83,85,92,102,104*
$COCH_2OH$	2-Hydroxyacetylfuran	*56,92*
$COCOCH_3$	1-(2′-Furyl)-propane-1,2-dione	*83,85,92,104*
$COCOC_2H_5$	1-(2′-Furyl)-butane-1,2-dione	*85,104*
COCO–(2-furyl)	Furil	*104*
COOH	Furoic acid (or 2-Furan carboxylic acid)	*25,72,83,106*
$COOCH_3$	Methyl furoate	*66,67,102,104*
$COOC_2H_5$	Ethyl furoate	*12,67,102,104,109*
$COOC_3H_7$-*n*	*n*-Propyl furoate	*104*
$COOC_4H_9$-*i*	Isobutyl furoate	*104*
$COOC_5H_{11}$-*i*	Isopentyl furoate	*104*
$COOCH_2CH(CH_3)C_2H_5$	2-Methylbutyl furoate	*104*
$C(=O)SCH_3$	Methylthio furoate	*104*
$C(=O)SC_2H_5$	Ethylthio furoate	*104*
2-furyl (O)	2,2′-*bis*-Furyl	*92*
pyrazinyl (N, N)	2-(2′Furyl)-pyrazine	*12,65,66,70,83,102,104, 105,113b*
methylpyrazinyl (N, N, CH_3)	2-(2′-Furyl)-5 (or -6) methylpyrazine	*12,83,102,105*
3-methylpyrazinyl (H_3C, N, N)	2-(2′-Furyl)-3-methyl-pyrazine	*65,70*
SCH_3	2-Methylthiofuran	*3*

a) Data compiled from Ref. [108].

b) Italic numbers are corresponding to the following aromas: *3* boiled beef; *4* canned beef; *9* roasted beef; *12* cooked pork liver; *17* chicken broth; *18* cooked chicken and chicken fats; *19* roasted turkey; *21* cabbage, broccoli and cauliflower; *25* asparagus; *28* potatoes; *38 Boletus edulis*; *46* American cranberry; *47* Arctic bramble; *50* canned mango; *53* cloudberry; *54* grapes; *55* maple syrup; *56* orange (juice or syrup); *62* raspberry; *64* tamarind; *66* roasted filberts;

67 roasted peanuts; *68* roasted pecans; *69* macadamia; *70* roasted sesame seed; *71* rapeseed protein; *72* hydrolysed soy protein; *73* soy protein isolate; *74* deep fat fried soy bean; *76* soy sauce; *77* soy bean; *78* corn oil and frozen corn; *79* popcorn; *81* roasted barley; *82* rice; *83* bread; *84* bread flavour; *85* rye crisp-bread; *89* stale non-fat dry milk; *91* butter; *92* dry whey powder; *102* cocoa; *104* coffee; *105* tea; *106* Jamaican rum; *108* beer and related products; *109* wine; *113a* wood smoke; *121* tuna oil; *127* fish concentrate; *131* virgin oil; *133* grape leaf; *136* cod.

The parent compound was identified in more than ten food systems including: cooked chicken [133], roasted filberts [90], rapeseed protein [163b], soy protein [163a], bread [123], sodium caseinate [163a], coffee [67a], wood smoke, fish protein concentrate [46], and caramel. 2-Methylfuran is one of the components of roasted and ground coffee [67a,193a,194a,217]. Among the higher homologues, 2-*n*-pentylfuran is the most widely distributed. It is described as having beany, grassy, liquorice, green, pungent, sweet, and fruity odour [52b,148a]. 2-Alkyl-furans found in canned beef [164] were associated with sensory terms such as wort-like, hay-like, musty, cooked meat, and retort off-flavour. Some furfuryl ketones were found in coffee [193a,217], leek [176], and roasted peanuts [221] aromas. Furfuryl alcohol not only occurred in the heat treatment of cultured dairy products [36] but also in a large variety of aromas. Furfuryl mercaptan has been identified as a characteristic component of coffee flavour [217]. 2-[(Methyldithio)-methyl]-furan identified in coffee [67a,193a,194a,217] and in fresh white bread [125] as well as furfuryldisulphide is described as the sole character impact compound of bread. Furfural was found naturally occurring in a large number of food flavours and essential oils. It was found to be present in meats [28,128b,149], beets [145], asparagus [208], artichoke [23], mushrooms [24], peach [191], orange juice [174], chestnut syrup [112], roasted green tea [237], rum [99], and blue cheese to mention only a few foodstuffs. Its formation in tomato juice during storage seems to be related to the degradation of ascorbic acid [199b]. The 5-methyl derivative, which has a bitter almond-like aroma similar to that of benzaldehyde and 2-methylpentenal, was found in the same food systems as furfural [24,63, 90,183,202,208,218a,221]. 5-Hydroxymethylfurfural results from the heat treatment employed in preparing various sorts of dry and sterile concentrated milk [143]. It was also present in tomatoes where its concentration increases during storage. 2-Acetylfuran is among the aroma constituents of sticks of liquorice [63] and that of chestnut syrup [112]. Its 3-hydroxy derivative or isomaltol has been identified in caramelization products of carbohydrates (*see Chapter III*). Some alkylfuroates and alkylthiofuroates were present in coffee aroma [193a,194a,217]. 2-(2′Furyl)-pyrazine, another interesting mono-substituted furan, was found in pressure-cooked pork liver [128b], roasted filberts [90], sesame seed [114], cocoa [216b], coffee [64], tea [237] and cigarette smoke [132].

2,5-Disubstituted furans **2** are more widely distributed in food aromas than their 2,3-, 2,4- and 3,4-isomers (*see Table II–3*). The most numerous are 5-methyl derivatives bearing an alkyl, acetonyl, hydroxymethyl, alkoxy, vinyl, propenyl, formyl, acetyl, propionyl, 2-pyrazinyl or methylthio groups in the 2-position. They have mainly been characterized in coffee [67a,193a,194,217], oilseed products [114,163b,226a], sodium caseinate [163a], dry whey [53a], popcorn [220]. . . 1-(5′-Methyl-2′-furyl)-1,2-propanedione was observed in bread [32], rye crispbread [196a], and coffee [67a,194a]. 5-Methyl-2-methylthiofuran was present in malt [210], and coffee [194], and 2-(5′methyl-2′-furyl)-5 (or 6) methylpyrazine in cocoa [216b].

Table II–3 – 2,5-Disubstituted furans[a)]

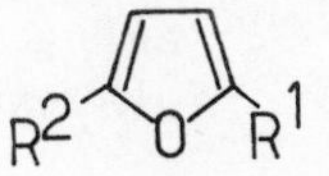

R^2 O R^1

2

2; R^1	R^2	Name	Occurrence
CH_3	CH_3	2,5-Dimethylfuran	*4,20,66,71,73, 82,94*
C_2H_5	CH_3	2-Ethyl-5-methylfuran	*4,71,73,94,127*
n-C_3H_7	CH_3	2-*n*-Propyl-5-methylfuran	*71,73,94*
n-C_4H_9	CH_3	2-*n*-Butyl-5-methylfuran	not reported
n-C_5H_{11}	CH_3	2-*n*-Pentyl-5-methylfuran	*73*
$CH_2CH_2COCH_3$	CH_3	1-(5′-Methyl-2′-furyl)-3-butanone	*104*
CH_2COCH_3	CH_3	1-(5′-Methyl-2′-furyl)-2-propanone	*104*
$CH_2COC_2H_5$	CH_3	1-(5′-Methyl-2′-furyl)-2-butanone	*104*
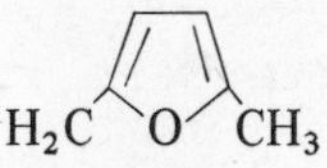 H_2C O CH_3 (furan ring)	CH_3	*Bis*-(5′-Methyl-2′-furyl)-methane	*79,92,104*
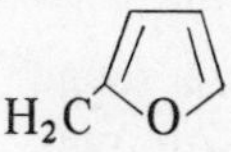 H_2C O (furan ring)	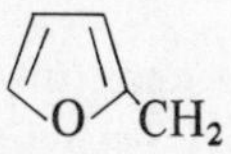 O CH_2 (furan ring)	2-(2′-Furfuryl)-5-(2′-furfuryl)-furan	*92*
CH_2SCH_3	CH_3	5-Methyl-2-methylfurfuryl sulphide	*104*
i-C_3H_7	CH_3	5-Methyl-2-isopropylfuran	*104*
$CH{=}CH_2$	CH_3	5-Methyl-2-vinylfuran	*104*
$CH{=}CH{-}CH_3$	CH_3	5-Methyl-2-propenylfuran	*104*

Table II–3 (*continued*)

2; R^1	R^2	Name	Occurrence
CHO	CH_3	5-Methylfurfural	*3,4,12,38,50,53, 56,64,65,67,70, 72,74,79,83,85, 92,93,102,105, 106,108,109, 113,133*
CHO	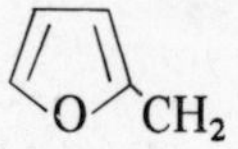 O CH_2	5-(2′-Furfuryl)-furfural	*92*
CHO	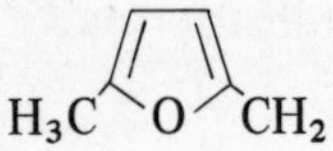 H_3C O CH_2	5-(5′-Methyl-2′-furfuryl)-furfural	*92*
CHO	$HOCH_2$	5-Hydroxymethylfurfural	*55,56,79,83,92, 104,113*
$COCH_3$	CH_3	Methyl 5-methylfuryl ketone (or 5-Methyl-2-acetylfuran)	*3,4,38,66,67,72, 79,85,92,102, 104*
$COCH_3$	C_2H_5CO	2-Acetyl-5-propionylfuran	*72*
COC_2H_5	CH_3	5-Methyl-2-propionylfuran	*104*
$COCOCH_3$	CH_3	1-(5′-Methyl-2′-furyl)-propane-1,2-dione	*83,85,104*
CN	CH_3	5-Methyl-2-furyl nitrile	*104*
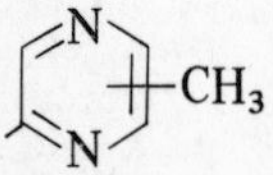 N N $-CH_3$	CH_3	2-(5′-Methyl-2′-furyl)-5 (or -6) methylpyrazine	*102*
SCH_3	CH_3	5-Methyl-2-methylthiofuran	*104,138*

a) See the caption for Table II–2; *20* eggs; *65* roasted almonds; *93* casein; *94* sodium caseinate; *138* malt.

Relatively few 3-substituted and polysubstituted furans **3** were found (*see Table II–4*). Three compounds (2,3-dimethyl-5-ethyl-; 5-ethyl-2-formyl, and 3,4-dimethyl-5-ethyl- furans) were identified in the volatile fraction of commercial aromas of cigarettes [165]. 3-Methyl-, 3-phenyl-, and 2-vinylfurans were observed in coffee [193a,217]. 3-Furan carboxaldehyde and 3-furan carboxylic acid are constituents of bread [124a] and rum [99] aromas, respectively. 2-*n*-Pentyl-3-methylfuran was found in virgin olive oil [50]. Furan composition and contents of arctic bramble [83], rum [99], canned beef [154b], unheated and heated rapeseed protein [163b], and soy protein [163a], fish protein [163a], as well as the influence of heating time on furan levels in maple syrup have been reported in the literature [111].

Table II–4 – Different substituted furans[a)]

R3 R2 R4 O R1

3

3; R^1	R^2	R^3	R^4	Name	Occurrence[a)]
H	CH_3	H	H	3-Methylfuran	*4,71,73,94,104, 127*
H	C_6H_5	H	H	3-Phenylfuran	*12,66,67,102, 104*
H	CHO	H	H	3-Furan carboxaldehyde	*83*
H	COOH	H	H	3-Furan carboxylic acid	*106*
n-C_5H_{11}	CH_3	H	H	2-*n*-Pentyl-3-methylfuran	*131*
$COCH_3$	OH	H	H	2-Acetyl-3-hydroxyfuran (isomaltol)	*83*
$CH{=}CH_2$	CH_3	H	H	2-Vinyl-3-methylfuran	*104*
$CH{=}CH_2$	H	CH_3	H	2-Vinyl-4-methylfuran	*104*
CH_3	CH_3	H	CH_3	2,3,5-Trimethylfuran	*104*
CH_3	CH_3	H	C_2H_5	2,3-Dimethyl-5-ethylfuran	*113b*
$CH{=}CH_2$	CH_3	H	CH_3	2-Vinyl-3,5-dimethylfuran	*104*
N, N, $-CH_3$	CH_3	H	CH_3	(3′ or (4′),5′Dimethyl-2-furyl-5 (or -6)-methyl-pyrazine	*102*
$CH{=}CH_2$	H	CH_3	CH_3	2-Vinyl-4,5-dimethylfuran	*104*

a) See the caption for Tables II-2 and II–3.

Monoterpenoid furan derivatives occur in a limited number of flavour-yielding plant species [137a]. Perillene **4** was isolated from the essential oil of the leaves of *Perilla citriodora makino* as well as from *Lasius fuliginosus latr.* Rose furan **5** was a constituent of the rose oil. Several sesquiterpene furans were isolated from various plant species and tobacco [137a], and can be used for flavouring tobacco.

The furan saturated ring (*see Fig. 1*) appears in some terpenoids such as *cis* and *trans* linalool oxides **6**. These compounds have been found in numerous essential oils as important aroma flavouring substances [137a]. They were also identified in tea [237], coffee [64,193a,194a], cocoa [114], tomato [213b], citrus fruits [88], apricot [36], arctic bramble [162], cloudberry [76], cranberry [1], mango [77c], passion fruit [232], grapes [173], and wine [172, 173]. Other related structures such as **7**, **8** and **9** have been isolated from coffee [193], passion fruit [232] and citrus fruits [88], respectively. The diastereoisomeric lilac alcohols **10** and the corresponding aldehydes should probably be considered as potential flavour components [137a].

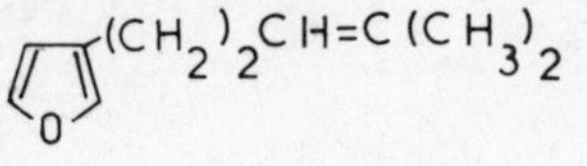

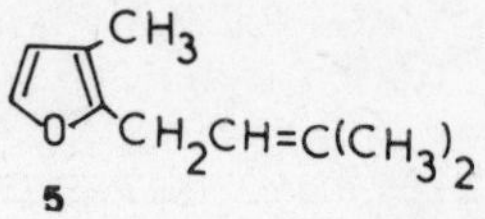

$(CH_2)_2CH{=}C(CH_3)_2$

CH_3 $CH_2CH{=}C(CH_3)_2$

5

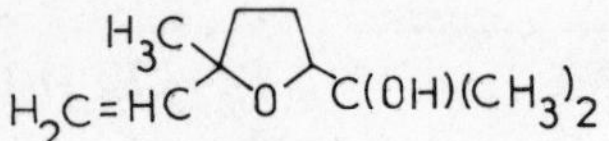

H_3C $H_2C{=}HC$ O $C(OH)(CH_3)_2$

6

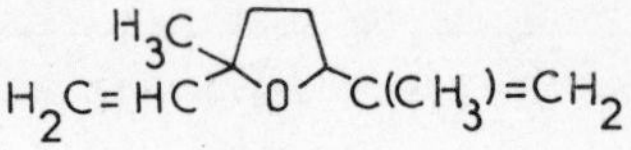

H_3C $H_2C{=}HC$ O $C(CH_3){=}CH_2$

7

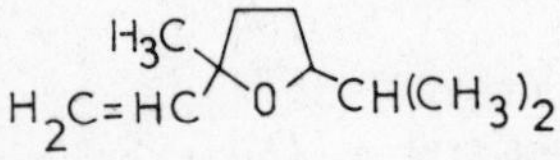

H_3C $H_2C{=}HC$ O $CH(CH_3)_2$

8

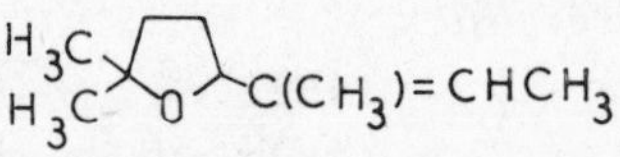

H_3C H_3C O $C(CH_3){=}CHCH_3$

9

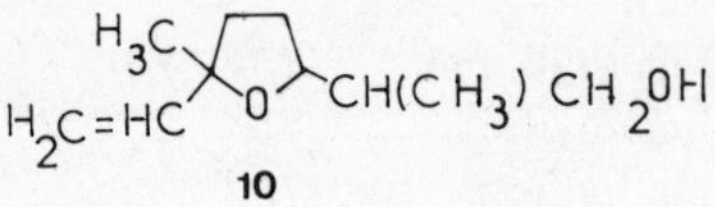

H_3C $H_2C{=}HC$ O $CH(CH_3)\,CH_2OH$

10

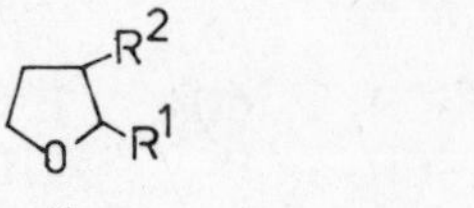

R^2 O R^1

11

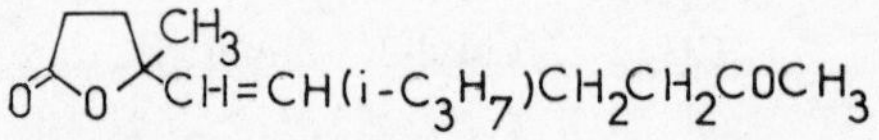

O O CH_3 $CH{=}CH(i\text{-}C_3H_7)CH_2CH_2COCH_3$

12

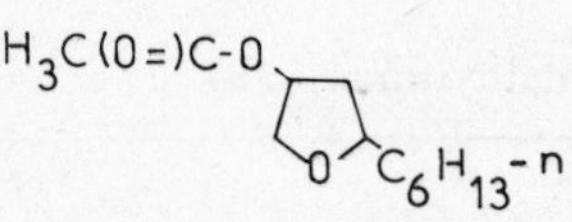

$H_3C(O{=})C\text{-}O$ O $C_6H_{13}\text{-}n$

13

H_3C CH_3 O CH_3 CH_2

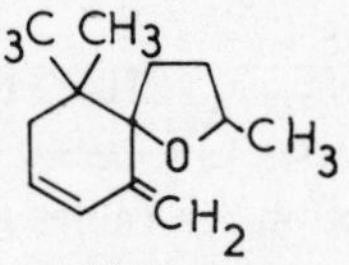

14

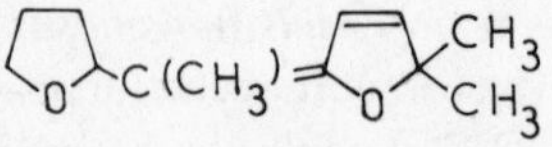

O $C(CH_3){=}$ O CH_3 CH_3

15

O $C_3H_7\text{-}n$

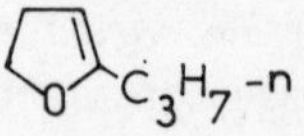

16

Fig. 1 – Terpenoid- and saturated furan derivatives in foods, essential oils and tobacco.

Some tetrahydrofurans **11** (R^1 = H, CH_3, $COCH_3$, R^2 = H) were identified in coffee [64,213a], soy protein isolate [163a], rapeseed protein [163b], and cooked chicken [133]. 2-Acetonyl-3-isopropyltetrahydrofuran is an unusual component of the 'Burley' tobacco flavour [39]. It is considered as well as **12** to be a biodegradation product of thunberganoids. Methyl 2-*n*-hexyltetrahydrofuranyl-4 acetate **13** was also an interesting odorous compound which has the same utilization as certain furan derivatives [71]. It has a sweet and fruity aroma and the taste of peach and apricot. Vitispirane **14** and davana ether **15** were isolated from vanilla bean extracts and from *Artemisia pallens wall.* [137a], respectively. Finally, 2-*n*-propyl-4,5-dihydrofuran **16** was identified in virgin olive oil aroma [50].

2.1.2 Sensory properties and patents

As it is quite apparent from the data reported in *Table II–5*, furans possess a wide range of sensory properties [49,52b,75].

Table II–5 – Flavour characteristics of various furans and tetrahydrofurans [49,52b,75]

Compounds	Odour description
Furan	Sickly, nasty smelling, ethereal
2-Methylfuran	Sickly ethereal
2-Ethylfuran	Powerful, burnt, sweet, coffee-like
2-*n*-Pentylfuran	Beany,grassy, liquorice, green, pungent, sweet fruity
2-Furfurylfuran	Caramellic
2,2′-Difurylethylene	Fatty
Furfuryl alcohol	Bitter taste, coconut, burnt potato
Ethyl 2-furan propionate	Fruity, aromatic, pungent
Isobutyl 2-furan propionate	Camomile
Isoamyl 2-furan propionate	Fruity, winy, brandy-like
Isoamyl 2-furan butyrate	Sweet,green,floral
Isoamyl 2-furan butyrate	Sweet, buttery, fruity
2-Furylmethane thiol	Coffee-like, unpleasant
Difurfuryl sulphide	Toasted
Furfuryl disulphide	Roasted, powerful, repulsive, unpleasant
2-Methyl-3 (-5 or -6)-furfuryl-thiopyrazine	Roasted, coffee-like
Furfural	Rotten, sweet, caramel-like flavour, sharp, penetrating odour, sweet, bread-like, sweet taste

Table II–5 (*continued*)

Compounds	Odour description
2-Acetylfuran	Powerful, balsamic, sweet, burning, sweetish taste, tobacco-like, cinnamic
2-Ethylfuryl ketone	Sweet
2-Furylhydroxymethyl ketone	Burning, sweetish taste
Furil	Mild, sweet
2-Furoic acid	Stinging aroma, sour taste
Methylfuroate	Pleasant, fruity, fungi, tobacco
α-Furyl cyanide	Sweet, pungent
2,5-Dimethylfuran	Pungent
5-Methylfurfural	Sweet, caramel-like flavour, sharp, grape sweet spicy, warm odour
5-Hydroxymethylfurfural	Bitter, astringent taste
5-Methyl-2-furanthiol	Burnt, sulphury
5-Methyl-2-methylthiofuran	Fruity, sulphury
2,5-Dimethyl-3-furanthiol	Strong meaty taste, roasted meat aroma
Bis-(2,5-dimethyl-3-furyl) sulphide	Bloody aroma, boiled meat taste
Bis-(2,5-dimethyl-3-furyl) disulphide	Meaty
2-Methyl-3-furanthiol	Roasted meat aroma
Bis-(2-methyl-3-furyl) trisulphide	Brothy
Bis-(2-methyl-3-furyl) tetrasulphide	Braised beef aroma and taste
2-Methyl-4-furanthiol	Green, meaty, herbaceous
Bis-(5-methyl-2-furyl) disulphide	Chemical, rubbery
Cis and *trans* linalool oxides	Rosy aroma
2,5-Diethyltetrahydrofuran	Sweet herbaceous, caramellic odour
Tetrahydrofurfuryl acetate	Faint, fruity, ethereal
Tetrahydrofurfuryl alcohol	Faint, warm, oily
Tetrahydrofurfuryl butyrate	Sweet, apricot, pineapple
Tetrahydrofurfuryl cinnamate	Sweet, balmy, vinous
2-Hexyl-4-acetoxytetrahydrofuran	Sweet, floral, fruity
3-Mercapto-2-methyl-4,5-dihydro-(2*H*)-furan	Roasted meat

They are mainly associated with a caramel-like, sweet, fruity, nutty, meaty, and burnt odour impression. However, few furan taste [181] and odour thresholds [164] have been reported. Owing to their olfactory properties, several compounds of this kind have proved to be important flavouring chemicals in the food industry (*see Chapter VIII*). The following 2-substituted furans **1** (R = CHO, $COCH_3$, CH_2OH, $CH_2OC(=O)R$, CH_2COR, $CH_2C(=O)OR$, $(CH_2)_3C_6H_5$, $CH{=}CHC(=O)OC_3H_7$-*n*, $OC(=O)CH_3$, SH) were found to have fruity aromas with the mild flavour of caramel (burnt sugar) when they are added in small amounts to non-alcoholic beverages, ice-creams, ice . . . High doses cause a bitter taste [147]. *N*-Alkylidene furfurylamines **17** (R = *i*-C_3H_7, *i*-C_4H_9, *i*-C_5H_{11}, *i*-C_6H_{13}) have chocolate or cocoa-like flavour [167]. 3-Acetyl-2,5-dimethylfuran **18** (R^1 = R^2 = CH_3) was useful in imitation nut flavours, especially the hazelnut and walnut flavour, and for milk, chocolate, coffee, and caramel flavours [129]. A mixture of sulphur derivatives of furan including alkylfuran-3-thiols **19** (R^1 = CH_3, R^2 = H) and *bis*-(alkyl-3-furyl) sulphides **20**; (n = 1, R^1 = H, CH_3, R^2 = H) are recommended in flavour formulations for meaty and roasted aroma and flavour in a soup base [47,49]. Isoamyl-(2-methyl-3-furyl) sulphide [21] enhanced the nutty taste of meat flavoured foods. This compound may be used in conjunction with other known flavour adjuvants such as 4-methyl-5-β-hydroxyethylthiazole [78a]. 2-Methyl-3-mercaptotetrahydro-3-furyl-2′-methyl-4,4′-dihydro-3′-furyl sulphide **22** (R = H, CH_3) possesses meat-like aroma [219].

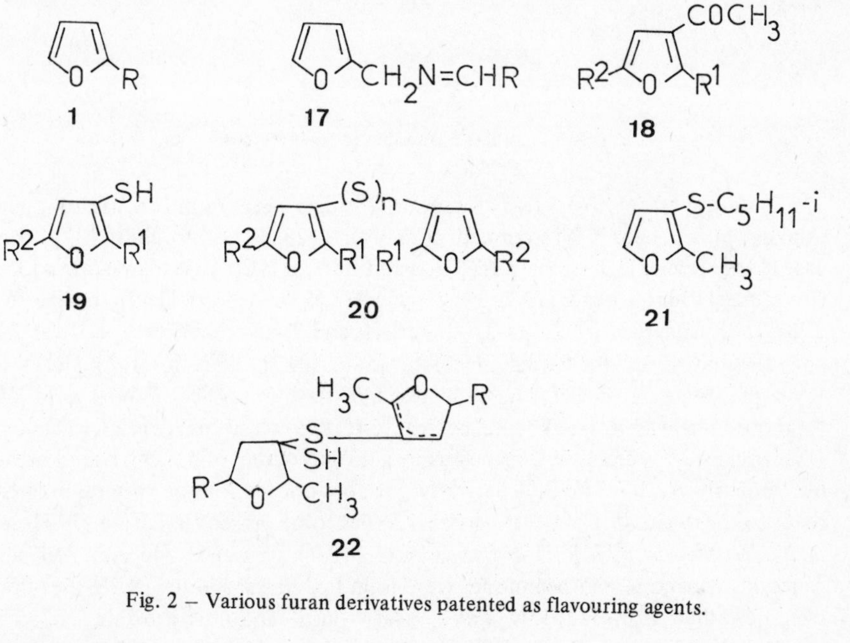

Fig. 2 – Various furan derivatives patented as flavouring agents.

2.2 Furanones

2.2.1 3-(2H)-Furanones

Saturated and unsaturated furanones **23–26** are widely distributed in food aromas [137b] (*see Fig. 3*).

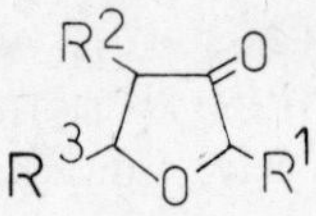

23

a) $R^1 = CH_3$, $R^2 = R^3 = H$
b) $R^1 = R^3 = CH_3$, $R^2 = H$
c) $R^1 = R^2 = R^3 = CH_3$

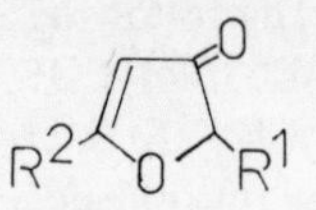

24

a) $R^1 = CH_3$, $R^2 = H$
b) $R^1 = R^2 = CH_3$
c) $R^1 = \underline{n}\text{-}C_6H_{13}$, $R^2 = CH_3$
d) $R^1 = \underline{n}\text{-}C_8H_{17}$, $R^2 = CH_3$

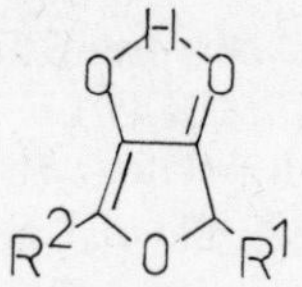

25

a) $R^1 = R^2 = CH_3$
b) $R^1 = H$, $R^2 = CH_3$
c) $R^1 = C_2H_5$, $R^2 = CH_3$
d) $R^1 = CH_2OH$, $R^2 = CH_3$
e) $R^1 - R^2 = C_2H_5$

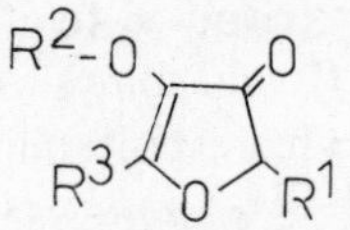

26

a) $R^1 = R^2 = R^3 = CH_3$
b) $R^1 = R^3 = CH_3$, $R^2 = CH_3CO$

Fig. 3 – 3-(2*H*)-Furanones in food aromas.

2-Methyl-tetrahydro-3-(2*H*)-furanone **23a** has been found in the following aromas: boiled beef [74], cooked pork liver [128b], potato chips [37], tomatoes [16], dried mushrooms [202], roasted filberts [90], macadamia nuts [34], soy sauce [134a], bread [123], whey powder [53a], cocoa [160], coffee [67, 193a,217], and rum [99]. Its 2,5-dimethyl- and 2,4,5-trimethyl derivatives **23b** and **23c** have been identified in coffee [64,67,217]. The thiols **23** (R^1 = H, R^2 = SH, R^3 = H or CH_3) have a meat-like flavour [140c]. 2-Methyl-3-(2*H*)-furanone **24a** was isolated from coffee [64] and white bread [123,124]. 2,5-Dimethyl-3-(2*H*)-furanone **24b** possesses a strong odour of freshly baked bread, but similarly to the 2-methyl derivative, it does not seem to be very reminiscent of this aroma [124b,124c]. It was also found to be present in coffee [125] and in mango flavour [77c]. The two 2-alkyl-3-(2*H*)-furanones **24c** and **24d**, also called norcepanone and cepanone, were found in allium species [8,36,65]. Both were described as having fatty, waxy, meaty, burnt, and musty aroma [65]. The most interesting compound of this series is the 2,5-dimethyl-4-hydroxy-3-(2*H*)-furanone **25a** also known as furaneol [136]. It has been identified in beef broth [204,205], freshly cooked bread, roasted almonds [197b], roasted filberts [180], popcorn [220], coffee [67], wood smoke, pineapple [189], strawberries, and in some berries in *Genus Rubus*, especially cloudberry and hybrids between raspberry and arctic bramble [162]. In canned orange juice [199] it imparts an unnatural flavour and taste if the concentration exceeds its threshold level (T = 0.04 ppb in water).

According to Hodge [75], the caramel-like flavour of this compound and analogous substances may be due to an association by hydrogen bond between the 4-hydroxy group and the adjacent carbonyl groups, the molecule having a planar configuration. The *O*-alkyl derivatives **26** have less powerful organoleptic properties. The monomethyl derivative **25b** is a flavour component of soy sauce [134b] and is considered to be an important aroma contributor. It has also been isolated from beef broth [204], arctic bramble [83], and mango [77c]. It possesses a caramel-like cooked beef aroma and a roasted chicory root-like flavour. The ethyl derivative **25c** which exists in two tautomeric forms have been characterized in molasses and in soy sauce [134a]. Its flavour was described as sweet and reminiscent of shortcake and cooked fruits. The 5-hydroxymethyl-furanone **25d** has a charred sugar flavour. 2,5-Dialkyl-4-hydroxy-3(2*H*)-furanones **25** in mixture with *L*-proline are used to generate highly acceptable honey and slightly browned flavours [154]. The compounds **25c** and **25d** are also effective as flavouring agents for products containing meat or meat simulating ingredients [152]. The 3-(2*H*-)-furanones **25a** and **25b** are also used as precursors or reactants in meat flavour compositions. In another process they are reacted with hydrogen sulphide or substances capable of liberating hydrogen sulphide to produce an artificial flavouring resembling roasted or fried meats [140b].

The trimethylfuranone **26a** has been isolated from the strawberry, canned mango [77c], and arctic bramble [137b] aromas. Its threshold value in water is 0.03 ppm, and its use is recommended in alcoholic beverages. Odourless precursors such as formate or acetate **26** (R^2 = CHO or CH_3CO) can also be used. The acetate was found in coffee [137b]. The *n*-amyl ether of furaneol **26** ($R^1 = R^2 = R^3 = CH_3$, $R^2 = n\text{-}C_5H_{11}$) has jasmine-like odour properties whereas the *n*-butyl ether has a weaker and unspecific floral odour [137b].

*2.2.2 4,5-Dihydro-2-(3*H*)-furanones (or γ-lactones) and 2-(5*H*)-Furanones*

From the comprehensive reviews of Maga [108] and Ohloff [137a] it appears that nearly all major food classes contain γ-lactones or butanolides. γ-Butyrolactone and its γ-alkyl derivatives having an even number of carbon atoms are the most frequently found in aromas, but as the number of carbon atoms increases, they occur less frequently (*see Table II–6*).

The following γ-lactones **27** (R = CH_3, C_2H_5, . . . $n\text{-}C_6H_{13}$) have been found in cooked asparagus flavour [208]. γ-Butyrolactone **27a** is described as having faint, sweet aromatic slightly buttery odour [108], and its γ-methyl derivative **27b** has hay, tobacco, and faint aromatic notes. γ-Caprolactone **27c** is also an aroma compound of the pineapple flavour [189]. It possesses a warm, powerful, herbaceous, sweet odour and a tobacco, sweet coumarin-caramel taste. The higher homologous **27e** and **27f** give a strong aroma of coconut, and the *n*-octyl

Table II–6 – 5-Substituted-4,5-dihydro-2-(3*H*)-furanones (or γ-lactones) [108,137a,209]

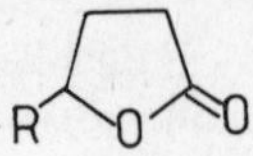

27

27; R =	Name	Occurrence[a)]
(a) H	4,5-Dihydro-2-(3*H*)-furanone (γ-butyrolactone)	*3,9,10,35,38, 56,60,66,67, 79,83,85,102, 104,105,109, 120a*
(b) CH_3	5-Methyl. . . (γ-valerolactone)	*3,11,38,57,58, 66,67,77,85, 101,102,104, 105,108,109, 111*
(c) C_2H_5	5-Ethyl. . . (γ-hexalactone)	*10,11,25,35, 48,58,60,61, 63,66,77,83, 102,105,108, 108c*
(d) n-C_3H_7	5-*n*-Propyl. . . (γ-heptalactone)	*10,11,25,58, 66,67,105, 108,108c,143*
(e) n-C_4H_9	5-*n*-Butyl. . . (γ-octalactone)	*10,11,17,25, 35,48,58,60, 61,63,66,77, 91,105,108, 108b*
(f) n-C_5H_{11}	5-*n*-Pentyl. . . (γ-nonalactone)	*9,10,11,25, 35,58,61,66, 77,77a,85,91, 102,105,108, 108b,108c*
(g) n-C_6H_{13}	5-*n*-Hexyl. . . (γ-decalactone)	*10,11,13,18, 20b,48,56,58, 61,63,77,91, 96,105,108, 108b,108c,120a*

Table II–6 (*continued*)

27; R =	Name	Occurrence
(h) *n*-C_7H_{13}	5-*n*-Heptyl. . . (γ-undecalactone)	*10,11,77,86, 91,108c*
(i) *n*-C_8H_{17}	5-*n*-Octyl. . . (γ-dodecalactone)	*10,11,13,18, 20b,48,58,63, 77,86,88,91, 96,101*
(j) *n*-C_9H_{19}	5-*n*-Nonyl. . . (γ-tridecalactone)	*91*
(k) *n*-$C_{10}H_{21}$	5-*n*-Decyl. . . (γ-tetradeca-lactone)	*10,11,13,18, 20b,91,105*
(l) *n*-$C_{11}H_{23}$	5-*n*-Undecyl. . . (γ-pentadeca-lactone)	*10,91*
(m) *n*-$C_{12}H_{25}$	5-*n*-Dodecyl. . . (γ-hexadeca-lactone)	*91*
(n) (Z) *n*-$C_5H_{11}CH{=}CH(CH_2)_2$	5-*Cis*-3′-nonen.. . (4-hydroxy-*cis*-tridecen-7-enoic acid lactone)	*108c*
(o) (Z) *n*-$C_5H_{11}CH{=}CH{-}CH_2$	5-*Cis*-2′-dodecen. . . (4-hydroxy-dodec-*cis*-6-enoic acid lactone)	*13,91,108c*
(p) *i*-C_3H_7	5-Isopropyl. . . (4-hydroxy-5-methyl-hexanoic acid lactone)	*111,120a*
(q) $H_2C = C(CH_3)$	5-Isopropenyl. . . (4-hydroxy-5-methyl-hexen-5-enoic acid lactone)	*122*
(r) $CH_3CO(CH_2)_2CH(i\text{-}C_3H_7)$- -CH=CH	5-[(3′isopropyl)-hept-1′-ene-6′-one] . . .	*111*
(s) CH_3CHOH	5-(α-hydroxyethyl). . . (4,5-Dihydroxyhexanoic acid lactone)	*35,108b,109, 110*
(t) CH_3CO	5-Acetyl. . . (solerone)	*108b,109,110*
(u) C_2H_5O	5-Ethoxy. . .	*108b,109,110*
(v) $C_2H_5OC(=O)$	5-Carboethoxy. . .	*108b,109,110*

a) See also the caption for Table II–2; *10* beef fat; *11* pork fat; *13* mutton meat; *20b* horse fat; *35* tomatoes; *48* apricot; *58* peaches and pears; *60* pineapple; *61* plums; *63* strawberry; *91* butter; *108a* beer; *108b* fermentation products; *108c* culture media from *Sporolomyces odorus*; *110* sherry; *111* 'Burley' tobacco; *143* hibiscus oil.

group that of musky peach. The γ-decalactone **27g** found in the volatile fraction of orange juice [30], has peach-like taste as well as γ-undecalactone **27h**. Some taste threshold values of these lactones in water and oil have been reported [188]. The lactone **27o** which has been identified in butter, mutton meat, mushrooms, and in culture media from *Sporolomyces odorus* [137a] contributes significantly to the cooked lamb flavour [142]. 5-Methyl-5-hexen-4-olide **27q** is a component of the essential oil of hop, and the lactone **27r** was found in tobacco [39]. Some γ-lactones with a functional group in the γ-positions **27s–27v** were identified in alcoholic beverages such as sherry and wine [4]. The 5-acetyl derivative isolated from wine aroma possesses a typically wine-like odour similar to that of Pinot noir [4,171c]. Furthermore, the hydroxylactone **27s** was reported in tomatoes [179]. These lactones are formed by chemical and enzymatic reactions and are flavour-contributing components in numerous fermentation products [209].

Among the α-alkyl substituted γ-butyrolactones, the α-methyl derivative **28** (R = CH_3) was identified in dried mushrooms [202], non-fat dry milk [54], spray dried whey [53b], cocoa [114], and coffee [222].

Polysubstituted γ-lactones also occur in food aromas (*see Fig. 4*).

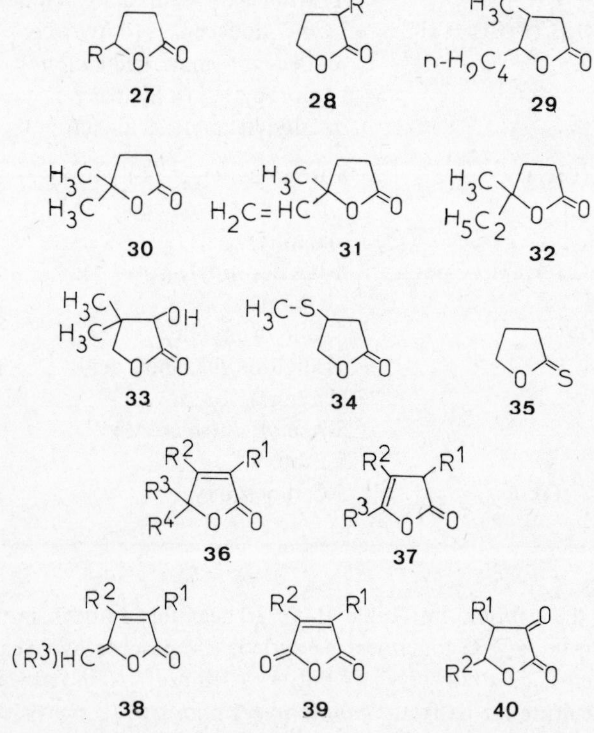

Fig. 4 – Saturated and unsaturated γ-lactones in food aromas.

Thus the γ-*n*-butyl lactone **29** known as 'whisky lactone' was found in alcoholic beverages such as wine, brandy, and fermentation products [209]. The 5,5-dimethyl-2-(3*H*)-furanone **30** which exists in two diastereoisomeric forms was identified in beer and in fermentation products [209]. The vinyl derivative **31** was found in lavender oil [203], wine [171c], black tea, and Japanese hop. Its hydrogenation product was identified from 'Burley' tobacco [39]. 4,5-Dihydro-3-hydroxy-4,4-dimethyl-2-(3*H*)-furanone **33** or pantolactone, which is a decomposition product of pantothenic acid, was characterized in a Spanish sherry [137a]. It is relatively abundant in the neutral fraction of roasted meat flavour [58]. 4,5-Dihydro-4-methylthio-2-(3*H*)-furanone **34** identified by thermolysis of precursors of roasted meat flavour, has a characteristic sulphur-like note of onions [58]. γ-Thiobutyrolactone **35** was found in dried mushrooms [202].

α,β-Unsaturated γ-lactones or 2-(5*H*)-furanones [36] (*see Table II–7*) were

Table II–7 – 2-(5*H*)-Furanones (or α, β-unsaturated γ-lactones [108,137a,209]

R2 R1 R3 R4 O O

36

36; R^1	R^2	R^3	R^4	Occurrence[a)]
H	H	H	H	*66,83,104,108,111,113a,139*
H	H	CH_3	H	*54,67,74,113a,140*
H	H	C_2H_5	H	*62,108,111*
H	H	n-C_5H_{11}	H	*108*
H	H	n-C_8H_{17}	H	*90*
H	CH_3	H	H	*43,111*
H	i-C_3H_7	H	H	*111*
CH_3	H	H	H	*43,111,113a*
H	H	CH_3	CH_3	*108,108b,111,120,122*
H	CH_3	CH_3	H	*104,113a*
CH_3	CH_3	H	H	*104,113a*
CH_3	H	CH_3	H	*43,111,113a*
CH_3	CH_3	CH_3	H	*104,113a*
CH_3	CH_3	n-C_5H_{11}	H	*105,111,125b*
OH	CH_3	CH_3	H	*108,108b,110b*
OH	CH_3	C_2H_5	H	*72*

a) See also the caption for Table II–2; *43* roasted onions; *90* milk caramel; *108b* fermentation products; *110b* sake; *111* 'Burley' tobacco; *120* lavender oil; *122* hop oil; *125b* peppermint oil; *139* saffron; *140* burnt sugar.

principally found in the following food aromas [108,137a,209] : fried onions, grapes, raspberry, roasted filberts and peanuts, soy protein hydrolysates, soy beans, white bread, milk, caramel, roasted coffee, Ceylon tea, beer, sake, hop, burnt sugar, saffron, lavender oil and Japanese peppermint. 5-*n*-Pentyl-2-(5*H*)-furanone **36** ($R^1 = R^2 = R^3 = H$, $R^4 = n\text{-}C_5H_{11}$) identified in beer [209], imparts deep fat fried aroma and flavour to foods [27a]. The 5-hydroxy furanone **36**; ($R^1 = R^2 = H$, $R^3 = OH$, $R^4 = n\text{-}C_5H_{11}$) enhances potato chips aroma, and its 4,5-dimethyl derivative **36** ($R^1 = H$, $R^2 = R^4 = CH_3$, $R^3 = OH$) imparts burnt aroma to aged sake.

2.3 Thiophenes and thiophenones

Thiophene derivatives **41**, **42** are well known compounds in several areas of the chemistry industry (drugs, dyes, pesticides. . .), but only in recent years has their potential importance in the flavour chemistry been assessed [107]. They were identified in a wide range of food systems in which they significantly contribute to the sensory properties [233] (*see Table II–8, II–9*), but they are not yet as numerous as their furan homolgues.

Table II–8 – 2-Substituted thiophenes [107]

S R

41

41; R =	Name	Occurrence[a)]	Sensory Properties [233]
H	Thiophene	*5,6,73,93,104, 105,127*	Benzene-like
CH_3	2-Methylthiophene	*5,6,18,44, 3,93,127*	Heated onion, sulphury, roasted, bitter, green
C_2H_5	2-Ethylthiophene	*6,73,127*	Styrene-like
$n\text{-}C_3H_7$	2-*n*-Propylthiophene	*73,127*	Chemical-like
$n\text{-}C_4H_9$	2-*n*-Butylthiophene	*6,73*	
$n\text{-}C_5H_{11}$	2-*n*-Pentylthiophene	*5,73,79*	
$n\text{-}C_8H_{17}$	2-*n*-Octylthiophene	*6*	
$n\text{-}C_{14}H_{29}$	2-*n*-Tetradecylthiophene	*6*	
CH_2COCH_3	1-(2′-Thienyl)-propanone	*6*	
CH_2OH	2-Hydroxymethylthiophene	*6,25,104*	
CH=CHCHO	2-Thiophenacrolein	*6,12*	
CHO	5-Formylthiophene	*5,6,12,25,38, 66,82,83,104*	Benzaldehyde-like

$COCH_3$	2-Acetylthiophene	*6,12,25,67, 104*	Mustard-like, onion-like,malty, roasted
$COCH_2CH_3$	2-Propionylthiophene	*104*	Cream, caramel
$COCOCH_3$	1-(2′-Thienyl)1,2-propanedione	*104*	Praline-like, woody, coffee
$COOCH_3$	2-Carbomethoxythiophene or (methylthiophene-2-carboxylate)	*104*	Burnt, aromatic, grape
$COOC_2H_5$	Ethylthiophene-2-carboxylate	–	Grape,winy, burnt
$COOC_3H_7$-*n*	*n*-Propylthiophene-2-carboxy-late	–	Burnt
$COOC_4H_9$-*n*	*n*-Butylthiophene-2-carboxy-late	–	Apricot,oily, burnt
$COOC_5H_{11}$-*i*	Isopentylthiophene-2-carboxylate	–	Oily, burnt, sulphury, toasted
C(=O)OCH_2-(2-furyl)	Furfurylthiophene-2-carboxylate	–	Earthy
CH_2OC(=OH)	Formyloxymethyl-2-thiophene	*104*	Mustard, sweet
$CH_2OC(=O)CH_3$	Acetoxymethyl-2-thiophene	*104*	Acetate-like

a) Data compiled from Ref. [107] ; *3* boiled beef; *5* canned beef stew; *6* cooked beef; *12* cooked pork liver; *18* cooked chicken; *25* asparagus· *38 Boletus edulis*; *44* leek; *66* roasted filberts; *67* roasted peanuts; *73* soy protein isolate *79* pop-corn; *82* rice; *83* bread; *104* coffee.

Thiophenes most frequently found are thiophene itself, its 2-alkylated, 2- and 3-acylated derivatives. Concentrations of thiophene and its 2-methyl derivative found in canned beef [148a] are influenced by variations in processing such as temperature, heating time, and various additives. They are described as being sickly pungent and green-sweet, respectively.

Among 46 sulphur-containing compounds twenty thiophenes were identified in pressure-cooked lean ground beef [228]. Mussinan & Walradt [128] found ten thiophenes from pressure-cooked pork liver among 179 compounds. Thiophene and 2-methylthiophene were also identified in fried chicken and cooked chicken meat [133], respectively. Three 2-substituted- **41** (R = CH_2OH, CHO, $COCH_3$) and one 2,3-disubstituted thiophene **42** (R^1 = CHO, R^2 = CH_3, R^3 = R^4 = H) were found in cooked asparagus [208]. 2-Formylthiophene and its 2,3-dimethyl derivative were identified from *Boletus edulis* [202]. Boelens *et al.* [8] in their analyses of onion oil identified 2,3-, 2,4-, and 3,4-dimethylthiophene

Table II–9 – Mono-, di-, and trisubstituted thiophenes [107]

42

42; R^1	R^2	R^3	R^4	Name	Occurrence[a)]	Sensory properties [233]
H	CH_3	H	H	3-Methylthiophene	*5,37,44,93,127*	Fatty, winy
H	$CH=CH_2$	H	H	3-Vinylthiophene	*104*	Hydrocarbon-like
H	CHO	H	H	3-Formylthiophene	*6,12,82*	
H	$COCH_3$	H	H	3-Acetylthiophene	*6,12,83,104*	
H	$COCOCH_3$	H	H	1-(3′-Thienyl)-1,2-propanedione	*104*	
CH_3	CH_3	H	H	2,3-Dimethylthiophene	*5*	
CHO	CH_3	H	H	2-Formyl-3-methylthiophene	*12,25*	Saffron, camphor-like
CHO	H	H	n-C_3H_7	2-Formyl-5-*n*-propylthiophene	–	Caramel
$COCH_3$	CH_3	H	H	2-Acetyl-3-methylthiophene	*104*	Methylbenzoate, honey-like, nutty, starchy
CH_3	H	CH_3	H	2,4-Dimethylthiophene	*43*	Fried onion
CH_3	H	C_2H_5	H	2-Methyl-4-ethylthiophene	*104*	
$COCH_3$	H	CH_3	H	2-Acetyl-4-methylthiophene	*104*	C_{11} alcohol
CH_3	H	H	CH_3	2,5-Dimethylthiophene	*5,43,45*	Green-like
CH_3	H	H	C_2H_5	2-Methyl-5-ethylthiophene	–	Diphenyloxide-like
CH_3	H	H	n-C_3H_7	2-Methyl-5-propylthiophene	–	Onion-like

Table II–9 *(continued)*

42; R^1	R^2	R^3	R^4	Name	Occurrence[a)]	Sensory properties [233]
CH_3	H	H	n-C_8H_{16}	2-Methyl-5-octylthiophene	*131*	
C_2H_5	H	H	C_2H_5	2,5-Diethylthiophene	*131*	
C_2H_5	H	H	n-C_4H_9	2-Ethyl-5-*n*-butylthiophene	*79*	
C_2H_5	H	H	n-C_6H_{13}	2-Ethyl-5-*n*-hexylthiophene	*131*	
CH_2COCH_3	H	H	CH_3	2-Acetonyl-5-methylthiophene	*6*	
CHO	H	H	CH_3	2-Formyl-5-methylthiophene	*6,12,67,79*	Sherry-like, bitter almond
$COCH_3$	H	H	CH_3	2-Acetyl-5-methylthiophene	*6,12,104*	Phenylpropyl-alcohol
H	CH_3	CH_3	H	3,4-Dimethylthiophene	*43,44,45*	Fried onion
CHO	CH_3	CH_3	H	2-Formyl-3,4-dimethyl-thiophene	*38*	
CH_3	CHO	H	CH_3	2,5-Dimethyl-3-formyl-thiophene	*6,12*	

a) See also the caption for Table II–8; *37* mushrooms; *43* roasted onions; *45* shallots; *93* casein; *127* fish protein concentrate; *131* virgin olive oil.

isomers. The same thiophenes were found in shallots [38]. According to Galetto & Hoffman [65] the aroma of fried onion, initially attributed to these compounds, would be due to intermediate sulphides and disulphides. Threshold values of 2-methyl-, 3-methyl- and dimethylthiophene isomers ranged from 3 to 5 ppm [65]. 2-Formyl-, and 2-acetylthiophenes as well as their 5-methyl derivatives are the flavour components of roasted peanuts [221] and filberts [90]. As they worked on the volatile products associated with commercially available soy protein isolate, casein and fish protein concentrate, Qvist and Von Sydow [163a] identified seven thiophenes. The incorporation of fat and potato flour into protein decreases the amount and number of thiophenes during heating [163a]. Three thiophenes of which the *n*-pentyl derivative were found in popcorn [220] and two others in rye crispbread and white bread [124a]. 2-Formyl- and 3-formylthiophenes were recently identified in rice [211]. Among the fifteen mono- and disubstituted thiophenes reported in the coffee aroma [193a,194a], seven acylated derivatives have been identified. Cazenave *et al.* [26] have also reported on the presence of thiophene in the black tea aroma.

3,4-Dimethyl-2,5-dihydro-2-thiophenone **43** (*see Fig. 5*), was found in onion oil [8] and leek, and 4,5-dihydro-2-(3*H*)-thiophenone **44** or thiobutyrolactone

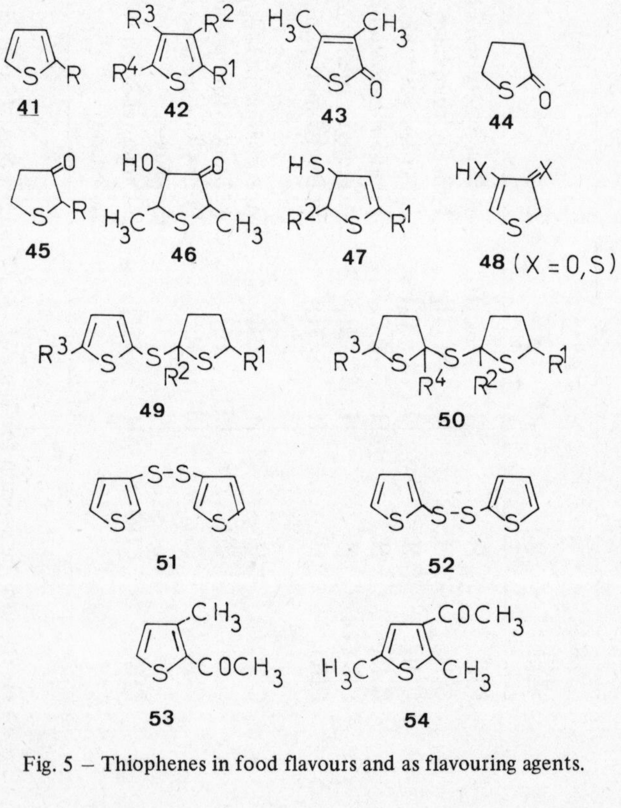

Fig. 5 – Thiophenes in food flavours and as flavouring agents.

in coffee [217] and mushroom flavours [202]. 4,5-Dihydro-3-(2*H*)-thiophenone **45** (R = H) and its 2-methyl derivative were identified in roasted peanuts [221], roasted filberts [90], coffee [194], cooked beef [228], pressure-cooked beef [128b], and wine [171b]. This latter compound constitutes the major portion of the volatile sulphur-containing compounds in beer [171c,172]. These compounds were described as having an onion, garlic taste and a green burnt coffee taste, respectively [233].

2,5-Dimethyl-4-hydroxy-3-(2*H*)-thiophenone **46**, which is an artificial product isosteric with furaneol [140a], is described as a flavouring agent having a roasted and burnt meat aroma [89]. Mercaptothiophenes **47** result from the reaction of hydrogen sulphide with 4-hydroxy-5-methyl (or 4,5-dimethyl)-3-(2*H*)-furanone [140a]. They were found as well as the compound **48** to be important enhancers of aroma of roasted meats and various foodstuffs. 2,2′-Dithienyl sulphides **49** and **50** are also recommended to improve the meaty flavour of foodstuffs [151]. Likewise *bis*-2,2′- or 3,3′-dithienyl sulphides **51** and **52** are capable of providing a meat-like flavour, e.g. roasted meat, having a slight onion or garlic character with fatty green notes, and may be used in combination with other edible materials to impart a meaty or roasted meat organoleptic impression to foods [85b]. As a result of their pronounced roasted meat character the *bis*-thienylsulphides are adapted for use in the preparation of spices, nut flavours, e.g. peanuts and walnuts, as well as meat and gravies having in addition a mushroom flavour note. Acylated thiophenes are also interesting as flavouring ingredients. Thus 2-acetyl-3-methylthiophene **53** was described as imparting a honey-like flavour to a syrup base. According to another patent, it possesses a characteristic coffee-like aroma [159]. 3-Acetyl-2,5-dimethylthiophene **54** is recommended to improve the aroma and taste of foodstuffs, tobacco, and perfumes [78c]. It is also useful in imitation of nut flavours. However, it appears that thiophenes have limited applications in food flavour formulations, because their sensory properties may indeed be quite potent, but not necessarily desirable.

2.4 Pyrroles and reduced systems

Beside pyrazines, furans, thiazoles, and thiophenes, pyrroles are among the most widespread heterocyclic compounds in food flavours with about fifty compounds (*see Table II–10, II–11 and II–12*).

Twenty-five pyrroles were identified in tobacco and/or tobacco smoke [132,170], twenty-three in coffee [194a], fourteen in roasted peanuts [221], ten in roasted filberts [90], and seven in popcorn [220]. Few pyrroles were also identified in canned beef stew [149], cooked beef [98a], shallow fried beef [160,225f], cooked pork liver [128b], cooked asparagus [208], *Boletus edulis* [202], leek [175], bread [32], rye crispbread [196a], roasted cocoa [218], roasted green tea [237], and liquorice sticks [62].

N-Substituted pyrroles **55** (*see Table II–10*), 2-substituted pyrroles **56** (*see Table II–11*) and 2-formylpyrrole derivatives **57** (*see Table II–12*) were the most widely distributed pyrrole compounds in food flavours. They impart a burnt and earthy characteristic note.

Table II–10 – *N*-Substituted pyrroles

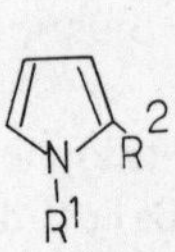

N, R^2, R^1

55

55; R^1	R^2	Name	Occurrence [a)]
CH_3	H	1-Methylpyrrole	*66,67,83,104, 113b*
C_2H_5	H	1-Ethylpyrrole	*105,113b*
n-C_3H_7	H	1-*n*-Propylpyrrole	*113b*
n-C_4H_9	H	1-*n*-Butylpyrrole	*113b*
i-C_5H_{11}	H	1-*n*-Isopentylpyrrole	*113b*
i-C_4H_9	H	1-Isobutylpyrrole	*113b*
$CH_2CH(CH_3)C_2H_5$	H	1-(2′-Methylbutyl)-pyrrole	*104*
H_2C– 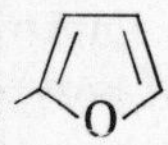 (furan ring, O)	H	1-Furfurylpyrrole	*25,66,67,79, 83,104,113b*
CHO	H	1-Formylpyrrole	*67,79*
$COCH_3$	H	1-Acetylpyrrole	*66,67,79*
CH_3	CHO	1-Methyl-2-formylpyrrole	*25,66,67,92, 102,104,113b*
CH_3	$COCH_3$	1-Methyl-2-acetylpyrrole	*7,25,104,105*
CH_3	COC_2H_5	1-Methyl-2-propionylpyrrole	*105*
CH_3	C≡N	1-Methyl-2-cyanopyrrole	*113b*
C_2H_5	$COCH_3$	1-Ethyl-2-acetylpyrrole	*105*
n-C_5H_{11}	$COCH_3$	1-*n*-Pentyl-2-acetylpyrrole	*102*
H_2C– 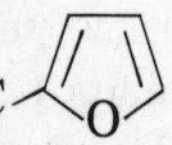 (furan ring, O)	$COCH_3$	1-Furfuryl-2-acetylpyrrole	*104*
CHO	CH_3	1-Formyl-2-methylpyrrole	*67*

a) *7* shallow fried beef [225f]; *25* asparagus [208]; *66* roasted filberts [90]; *67* roasted peanuts [221]; *83* bread [124a]; *79* popcorn [220]; *92* dry whey [53a]; *102* cocoa [114,218]; *104* coffee [67a]; *105* tea [237]; *113b* tobacco smoke [170].

Table II–11 – 2-Substituted pyrroles

56

56; R	Name	Occurrence[a)]
H	Pyrrole	*6,67,83,104,113b*
CH_3	2-Methylpyrrole	*6,67*
C_2H_5	2-Ethylpyrrole	not reported
n-C_3H_7	2-*n*-Propylpyrrole	not reported
n-C_4H_9	2-*n*-Butylpyrrole	not reported
n-C_5H_{11}	2-*n*-Pentylpyrrole	*66*
i-C_4H_9	2-Isobutylpyrrole	*66*
CHO	2-Formylpyrrole	*25,38,44,66,67,83, 102,104,105,113b*
$COCH_3$	2-Acetylpyrrole	*25,38,66,67,83,92, 102,104,105,113b, 135*
$COCH_2CH_3$	2-Propionylpyrrole	*38,66,67,104,105, 113b*
$COCOC_2H_5$	1-(2′-pyrrolyl)-1,2- butanedione	*104*
COOH	2-Pyrrole carboxylic acid	*25*
$COOCH_3$	2-Methoxycarbonylpyrrole	*102,113b*
C≡N	Pyrrole-2-carbonitrile	*113b*

a) See also Table II–10; *6* cooked beef [98a]; *38 Boletus edulis* [202]; *44* leek [175]; *83* bread [124a]; *135* sticks of liquorice [62].

2-Acetylpyrrole, one of the important odorous components of liquorice sticks [62] was found in a great variety of foods. Its *N*-methyl derivative was characterized in shallow fried beef [160,225f], cooked asparagus [208], coffee [67a], and green tea [237]. 2-Formylpyrrole was found in 16 natural products, and its *N*-methyl- and *N*-ethyl derivatives were also frequently found. The latter compound is a typical component of coffee with a burnt roasted flavour. Variously substituted pyrroles are claimed to have excellent odour characteristics useful in perfumes and flavours [157]. *N*-Substituted pyrroles have applications for nutty and roasted flavours.

1-Pyrroline **58** imparts a significantly enhanced butter flavour added to manufactured margarine [131a]. 2-Pyrroline **59** is a flavour component of numerous foods [3].

Table II–12 – 2-Formylpyrrole derivatives

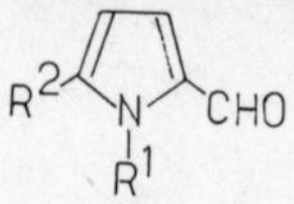

57

57; R^1	R^2	Name	Occurrence[a)]
H	H	2-Formylpyrrole	(*See Table II–11*)
CH_3	H	1-Methyl-2-formylpyrrole	(*See Table II–10*)
C_2H_5	H	1-Ethyl-2-formylpyrrole	*38,67,79,92,102,104,105*
n-C_5H_{11}	H	1-*n*-Pentyl-2-formyl-pyrrole	*102*
$CH_2CH(CH_3)C_2H_5$	H	1-(2′-Methylbutyl)-2-formylpyrrole	*104*
n-C_7H_{11}	H	1-*n*-Heptyl-2-formyl-pyrrole	*38*
i-C_5H_{11}	H	1-Isopentyl-2-formyl-pyrrole	*38,79,104,113b*
$CH_2CH_2C_6H_5$	H	1-(2′-Phenylethyl)-2-formylpyrrole	*38*
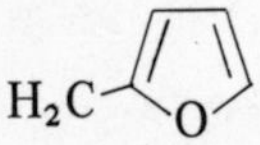 H_2C–(2-furyl)	H	1-Furfuryl-2-formyl-pyrrole	*67,79*
$CH(C_3H_7$-*i*$)COOH$	H	2-(2′-Formyl-1′-pyrrolyl)-3-methyl-butanoic acid	*113b*
$CH(C_4H_9$-*s*$)COOH$	H	2-(2′-Formyl-1′-pyrrolyl)-3-methyl-pentanoic acid	*113b*
$CH(C_4H_9$-*i*$)COOH$	H	2-(2′-Formyl-1′-pyrrolyl)-4-methyl-pentanoic acid	*113b*
H	CH_3	2-Formyl-5-methylpyrrole	*67,79,92,104,113b*
CH_3	CH_3	2-Formyl-1,5-dimethyl-pyrrole	*104*
$CH_2CH_2C_6H_5$	CH_3	1-(2′-Phenylethyl)-2-formyl-5-methylpyrrole	*38*

a) See the caption for Tables II–10 and II–11.

3-Pyrroline **60**, its *N*-methyl- and *C*-alkyl derivatives as well as several substituted pyrrolidines **61** were identified in tobacco [132,175]. Pyrrolidine **61** ($R^1 = R^2 = R^3 = H$) was found in fish products [82], caviar, cheese, beer, wine, and tobacco [137b]. According to Ohloff & Flament [137b] its particularly powerful and characteristic amine smell accounts for the fishy and fermented

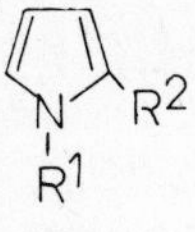

Fig. 6 – Pyrroles and reduced systems in food flavours.

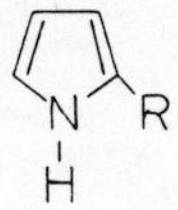

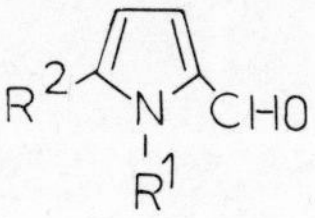

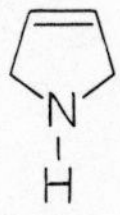

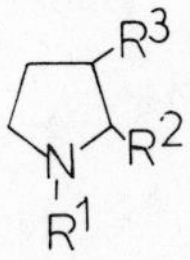

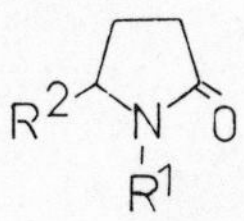

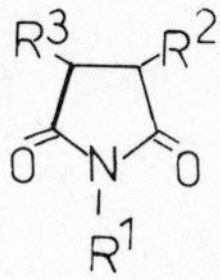

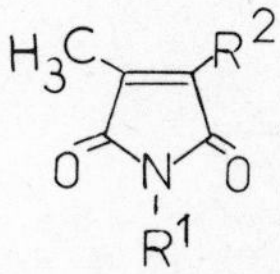

notes of these foodstuffs. 2-Acetyl-1-methylpyrrolidine was found in bread and has its typical flavour [77a]. It has been pointed out that the presence of *N*-nitrosopyrrolidine and its 3-hydroxy derivative, along with other carcinogenic aliphatic amines, are present in the aroma of fried bacon and some cooked meats [46]. α-Pyrrolidone **62** (R^1 = R^2 = H) was identified as a volatile flavour component of cooked rice [239]. It was also found as well as its *N*-methyl derivative in roasted filberts [90], and in tobacco [103]. The higher homologues **62** (R^1 = C_2H_5, C_3H_7) were characterized in cooked pork and in soya protein hydrolysate, respectively [137b]. Pyrrolidone carboxylic acid **62** (R^1 = H, R^2 = COOH) was isolated from tomatoes. It is a well-known compound which is in equilibrium with glutamic acid in an aqueous medium. In a mixture with succinic acid, glutamic acid, and 5′-ribonucleotides, meaty characteristics and flavour notes of fish were observed [240]. Succinimide **63** (R^1 = R^2 = R^3 = H) and its *N*-methyl-, 2-ethylidene-, 3-methyl-, 2-ethyl-3-methyl-, 2-ethyl-*N*-methyl derivatives were reported in tobacco [132,170]. *N*-Ethylsuccinimide was identified in the black tea aroma [73]. Unsaturated compounds or maleimides **64** have also been found in tobacco [39,103] and in meat aroma [58].

Recent studies suggested that pyrroles arise mainly *via Maillard* or *Strecker* reactions between sugars (or α-dicarbonyl compounds) and amino acids (proline, hydroxyproline) during processing [41,103] (*see Chapter III*).

2.5 Dioxolanes, dithiolanes, trithiolanes, imidazoles, and pyrazoles

Five-membered heterocyclic compounds with two or more identical heteroatoms in their rings are not common in food flavours (*see Fig. 7*).

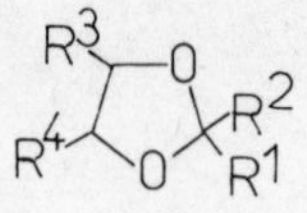

Fig. 7 – Two or more identical heteroatoms containing five-membered heterocyclic systems.

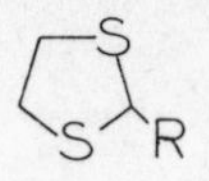

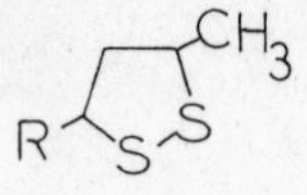

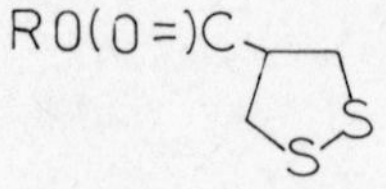

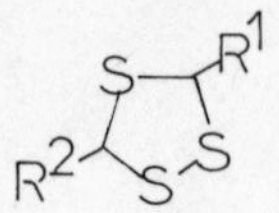

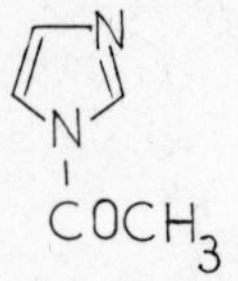

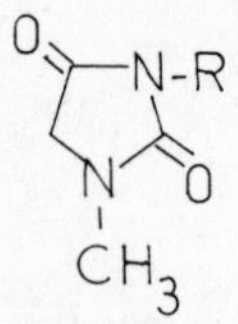

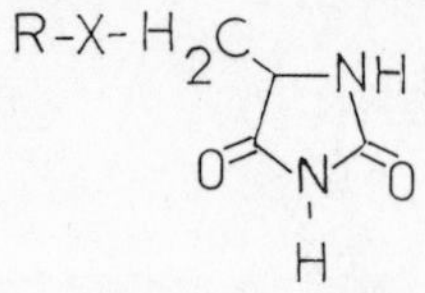

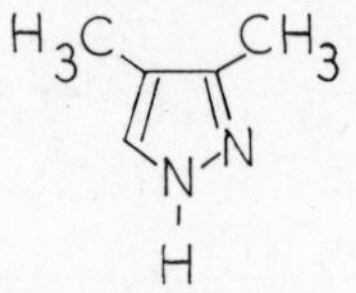

1,3-Dioxolanes **65** were principally found in fruits and wines [57]. Thus 2,4,5-trimethyl-1,3-dioxolane is present in wine [4], sherry, grape and cranberries [1]. 2,4-Dimethyl-5-ethyl- and 2,2,4-trimethyl-1,3-dioxolanes were characterized in wine and tomatoes, respectively [57]. These compounds are ketals resulting from condensation of acetaldehyde and acetone with glycols. They have an agreeable fruity-note. 2-Substituted 4-methyl-1,3-dioxolanes were recently reported as solvent interaction products in some commercial beef flavourings and alkyl- or alkenyl substituted 2,2-dialkyl-4-pentenalethylene acetals were recommended as perfumes.

1,3-Dithiolanes **66** (R = H, CH_3) and 1,2-dithiolanes **67** (R = H, CH_3) have an allium-like flavour of onion or garlic with metallic background notes. They are especially suitable for stewed vegetable-type flavours [229]. Methyl- and ethyl-1,2-dithiolane-4-carboxylate **68** (R = CH_3, C_2H_5) were found in cooked asparagus [208]. 1,2-Dithiacyclopentene **69** was also found in asparagus. It possesses its characteristic odour. 1,2,4-Trithiolane **70** (R^1 = R^2 = H) is a component of mushrooms and red algae [137b] flavours. The diastereomeric 3,5-dimethyl-1,2,4-trithiolanes **70** (R^1 = R^2 = CH_3) were identified in the flavour of various foodstuffs including [137b]: boiled beef, pork meat, potato, mushrooms, dry red beans, roasted filberts, toasted cheese, and grape leaves. Flament *et al.* [58], in addition to 3,5-dimethyltrithiolane, identified two higher homologue both in *cis* and *trans* forms **70** (R^1 = C_2H_5, *i*-C_3H_7, R^2 = CH_3) in a commercial beef extract. The *cis* 3,5-diphenyltrithiolane **70** ($R^1 = R^2 = C_6H_5$) has been identified in Guinea hen weed [57].

So far very few imidazole compounds have been found in flavours, probably because of their low volatility. However, they contribute to the flavourous properties of fish muscle [82] and of several food systems. The presence of 1-acetylimidazole **71** in 'Burley' tobacco was reported [39].

1,3-Dimethylhydantoin **72** ($R = CH_3$) and 1-methyl-3-ethylhydantoin **72** ($R = C_2H_5$) were identified for the first time in an aroma of commercial beef extract [58]. They probably are formed by degradation of uric acid. Imidazoline -2,4-diones **73** (X = S or SO, R = alkyl or alkenyl groups) are garlic flavouring materials.

Only one pyrazole **74** has been reported in food flavours [170].

2.6 Oxazoles, Oxazolines and oxazolidines

Relatively few oxazoles **75**, oxazolines **76** and **77** and oxazolidines **78** have been found in food flavours [109].

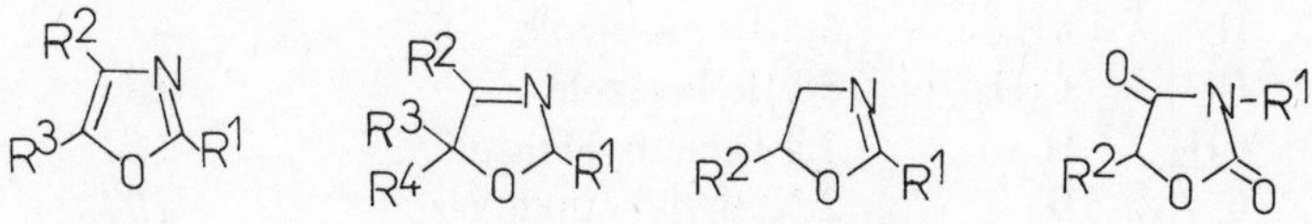

R2 R3 N O R1 **75**   R2 R3 R4 N O R1 **76**   R2 N O R1 **77**   O N-R1 R2 O O **78**

Fig. 8 – Oxazoles, oxazolines and oxazolidines in food flavours.

However, about thirty substituted oxazoles with alkyl- and/or acetyl groups in the 2-, 4-, and 5-positions were found in coffee [193,215a], cocoa [216b] and meat aromas [27b,28,128b,130,149]. They are listed in *Table II–13*. Among them four derivatives (2,5-dimethyl-, 4,5-dimethyl-, trimethyl- and 2-*n*-propyl-5-methyloxazoles) were found in the volatile basic fraction of roasted cocoa [216b]. 5-Acetyl-2-methyloxazole was identified in coffee flavour [194a]. Except for these two foodstuffs and the heated meat system where 2,4,5-trimethyloxazole was identified [128b,130], no other oxazole and no isoxazoles were reported. The flavour description of 2,4-dimethyloxazole is nutty and sweet, similar to that of 2,4,5-trimethyloxazole which has also a green note [130]. The threshold value of the trimethyloxazole in water is 5 ppb. In mixture with diacetyl, this compound imparts an earthy potato-like and mushroom-like flavour to foods [7]. 2,4-Dimethyl-, 2,4-dimethyl-5-ethyl-, 2,5-dimethyl-4-ethyl-, and 2,4,5-trimethyl-3-oxazolines **76** were found in cooked beef aroma, and it is apparent that they contribute to the overall flavour properties of this aroma [129,130]. The first report of a 3-oxazoline in meat was made by Chang *et al.* [27]. They isolated and identified 2,4,5-trimethyl-3-oxazoline in boiled beef. This compound was described as having a "characteristic boiled beef aroma".

Table II–13 – Oxazoles [109]

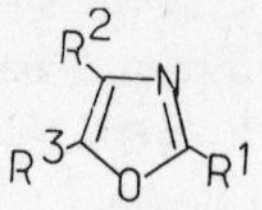

75

75; R^1	R^2	R^3	Name	Occurrence[a)]
H	H	H	Oxazole	–
CH_3	H	H	2-Methyloxazole	–
C_2H_5	H	H	2-Ethyloxazole	*104*
n-C_3H_7	H	H	2-*n*-Propyloxazole	–
$COCH_3$	H	H	2-Acetyloxazole	–
C_6H_5	H	H	2-Phenyloxazole	*104*
H	CH_3	H	4-Methyloxazole	–
H	C_2H_5	H	4-Ethyloxazole	*104*
H	H	CH_3	5-Methyloxazole	–
H	H	C_2H_5	5-Ethyloxazole	*104*
CH_3	CH_3	H	2,4-Dimethyloxazole	*104*
CH_3	C_2H_5	H	2-Methyl-4-ethyloxazole	*104*
C_2H_5	CH_3	H	2-Ethyl-4-methyloxazole	*104*
CH_3	H	CH_3	2,5-Dimethyloxazole	*102,104*
CH_3	H	C_2H_5	2-Methyl-5-ethyloxazole	*104*
C_2H_5	H	CH_3	2-Ethyl-5-methyloxazole	*104*
n-C_3H_7	H	CH_3	2-*n*-Propyl-5-methyloxazole	*102,104*
CH_3	H	$COCH_3$	2-Methyl-5-acetyloxazole	*104*
H	CH_3	CH_3	4,5-Dimethyloxazole	*102,104*
H	CH_3	C_2H_5	4-Methyl-5-ethyloxazole	*104*
H	C_2H_5	CH_3	4-Ethyl-5-methyloxazole	*104*
CH_3	CH_3	CH_3	2,4,5-Trimethyloxazole	*3,5,12,102,104*
CH_3	CH_3	C_2H_5	2,4-Dimethyl-5-ethyloxazole	*104*
CH_3	C_2H_5	CH_3	2,5-Dimethyl-4-ethyloxazole	*104*
n-C_3H_7	CH_3	CH_3	2-*n*-Propyl-4,5-dimethyloxazole	*104*
CH_3	CH_3	$COCH_3$	2,4-Dimethyl-5-acetyloxazole	*104*

a) *3* boiled beef [27b]; *5* canned beef stew [149]; *12* cooked pork liver [128b]; *102* cocoa [216b]; *104* coffee [215b].

The occurrence of the trimethyloxazoline in beef was later confirmed [74]. Olfactory properties and threshold values (T) of some oxazolines are reported in *Table II–14*.

Table II–14 – Olfactory properties and threshold values (T) of some 3-oxazolines [73,130]

76

76; R^1	R^2	R^3	R^4	Name	T(ppm)	Flavour description
H	H	H	H	3-Oxazoline	–	–
CH_3	H	CH_3	H	2,5-Dimethyl-3-oxazoline	1.0	Nutty, vegetable
CH_3	CH_3	CH_3	H	2,4,5-Trimethyl-3-oxazoline	1.0	Woody, musty, green
CH_3	C_2H_5	CH_3	H	2,5-Dimethyl-4-ethyl-3-oxazoline	0.5	Nutty, sweet, green, woody
CH_3	CH_3	C_2H_5	H	2,4-Dimethyl-5-ethyl-3-oxazoline	1.0	Nutty, sweet, vegetable
C_2H_5	C_2H_5	C_2H_5	H	2,4,5-Triethyl-3-oxazoline	–	Fresh carrots
i-C_3H_7	CH_3	CH_3	H	2-Isopropyl-4,5-dimethyl-3-oxazoline	–	Cocoa
i-C_3H_7	C_2H_5	C_2H_5	H	2-Isopropyl-4,5-diethyl-3-oxazoline	–	Cocoa
i-C_3H_7	n-C_3H_7	n-C_3H_7	H	2-Isopropyl-4,5-dipropyl-3-oxazoline	–	Banana
i-C_3H_7	CH_3	CH_3	CH_3	2-Isopropyl-4,5,5-trimethyl-3-oxazoline	–	Rum
i-C_4H_9	CH_3	CH_3	CH_3	2-Isobutyl-4,5,5-trimethyl-3-oxazoline	–	Earthy (fungal), butter, green leaf
t-C_4H_9	CH_3	CH_3	CH_3	2-Tertiobutyl-4,5,5-trimethyl-3-oxazoline	–	Cool minty

Various pathways for oxazoles and oxazolines formation were discussed [215] (*see Chapter III*). They are formed either by *Strecker* degradation of α-amino ketones [184a] from furfural, hydrogen sulphide, and ammonia or by the thermal interaction of ammonia, acetaldehyde, and acetoin [130].

The utility of 3-oxazolines as flavouring agents is exemplified by the patent literature. For instance, Hetzel & Torres [73] recommended the use of a series of 2,4,5-alkyl substituted 3-oxazolines to give cocoa, banana, cool minty, melon, butter, rum and carrot-like aromas. Thus 2-isopropyl-4,5-diethyl-3-oxazoline (1 to 50 ppm) enhanced a milk-like aroma and taste. The cocoa-like aroma of this compound is about five times more intense than that of the 4,5-dimethyl derivative.

2-Oxazolines **77** also have interesting olfactory properties [150]. 2-Ethyl-5-methyl-2-oxazoline possesses a melon fruity-like aroma and was useful in enhancing rum or caramel flavoured products such as rum punch beverages. 2-*n*-Propyl-5-methyl-2-oxazoline has a cool minty flavour and aroma, while the 2-*n*-pentyl-5-methyl-2-oxazoline has a fruity sweet flavour and aroma potentially useful in flavouring or enhancing flavours of a great variety of products. A 3,5-dimethyl-1,3-oxazolidine-2,4-dione **78** was identified in a meat aroma [58].

It is quite apparent that oxazoles and oxazolines will have further food flavouring applications in the near future. The identification of numerous heterocyclic compounds of this kind will be forthcoming.

2.7 Thiazoles and reduced systems

2.7.1 Structure and occurrence

Since 1967 thiazoles have been receiving increased attention because of their natural occurrence in various food systems [106] and their noteworthy organoleptic properties [157b]. In our laboratory, specializing in thiazole synthesis and reactivity studies, we could again and again appreciate the extraordinary potent smell of this unique class of compounds now essential in flavour chemistry.

As shown in *Table II–15*, thiazole derivatives **79** play an important role in the following flavours: meat products [100,128b,149,228], vegetables [17,19, 20,32,62a,87a,192,208,213b], passion fruit [232], roasted products [183,221], popcorn [220], rice bran [211], milk [2], cocoa [194], coffee [194a,215b], tea [216a], rum [236], whisky [236], and malt [210].

In most cases, thiazole itself was identified in these foodstuffs. 2-Isobutylthiazole is the most typical one of the 2-alkylthiazoles. It was found by Viani *et al.* [213b] in a volatile extract of raw tomatoes. It is described as having a strong green odour similar to that of tomato leaf. The pure compound in water had a spoiled wine-like highly horseradish type flavour, and its threshold value ranged from 1.3 to 2 ppb [87a,192]. According to Stevens [192] the relative ratio of this compound in tomatoes changes with location and variety. It represents an exceptional case of a sulphur-containing flavour component formed biologically in the intact plant. 2-Acetylthiazole, another interesting thiazole derivative, was found in ground beef [228], canned beef [149], boiled potatoes [19], baked potatoes [32], cooked asparagus [208], dry red beans [20], and rice bran [211]. 2-Acetyl- and 2-propionyl-4-methylthiazole were components of the coffee flavour [194a,215b]. 2,4-Dialkylthiazoles were principally detected in ground beef [228] and coffee [215b], and possess more interesting olfactory properties than their 2,5-disubstituted isomers. 4-Methyl-5-vinylthiazole was characterized from roasted filberts [90], cocoa extracts [194], passion fruit [232], and malt [210]. Two 5-β-hydroxyethylthiazole derivatives (4-methyl- and 2,4-dimethyl) were found in malt [210] and in meat flavour [228], respectively. Three 2,4,5-trialkylated thiazoles (**79**; $R^1 = i\text{-}C_3H_7, i\text{-}C_4H_9, R^2 = R^3 = CH_3$; $R^1 = i\text{-}C_3H_7, R^2 = CH_3, R^3 = C_2H_5$) were observed in potato products [19].

Table II–15 – Thiazoles: occurrence [106] and sensory properties [157a]

79

79; R^1	R^2	R^3	Name	Occurrence[a)]	Flavour description[b)]
H	H	H	Thiazole	*2,12,66,67,79, 82,104,106,107, 108*	Pyridine-like
CH_3	H	H	2-Methylthiazole	*2,82,104*	Green vegetable
C_2H_5	H	H	2-Ethylthiazole	*104*	Green, nutty
n-C_3H_7	H	H	2-*n*-Propylthiazole	–	Green, herby, nutty
n-C_4H_9	H	H	2-*n*-Butylthiazole	–	Raw, green, herby
n-C_5H_{11}	H	H	2-*n*-Pentylthiazole	–	–
i-C_4H_9	H	H	2-Isobutylthiazole	*35*	Green tomato leaves, wine (2 ppb)
i-C_3H_7	H	H	2-Isopropylthiazole	–	Green vegetable
$COCH_3$	H	H	2-Acetylthiazole	*2,5,12,25,26, 27,34,82*	Nutty, cereal, popcorn
(2-furyl)	H	H	2-Furylthiazole	–	Nutty, pungent
OCH_3	H	H	2-Methoxythiazole	–	Sweet, roasted, phenolic
OC_2H_5	H	H	2-Ethoxythiazole	–	Phenolic, burnt, nutty

Table II–15 (*continued*)

79;R^1	R^2	R^3	Name	Occurrence[a)]	Flavour description[b)]
OC_4H_9-*n*	H	H	2-Butoxythiazole	—	Green vegetable
H	CH_3	H	4-Methylthiazole	*2,12,25,67,104*	Green, nutty
H	C_2H_5	H	4-Ethylthiazole	*104*	
H	CH_3CO	H	4-Acetylthiazole	—	Nutty, cereal
H	H	CH_3	5-Methylthiazole	*25,105*	—
H	H	C_2H_5	5-Ethylthiazole	*104*	—
H	H	CH_3CO	5-Acetylthiazole	—	Roasted meaty, onion
H	H	C_2H_5O	5-Ethoxythiazole	—	Cooked onion
CH_3	CH_3	H	2,4-Dimethylthiazole	*2,70,88,104,105*	Meaty, cocoa-like (0,1 ppm)
CH_3	C_2H_5	H	2-Methyl-4-ethylthiazole	*2*	—
C_2H_5	CH_3	H	2-Ethyl-4-methylthiazole	*104*	—
C_2H_5	C_2H_5	H	2,4-Diethylthiazole	*104*	—
$COCH_3$	CH_3	H	2-Acetyl-4-methylthiazole	*104*	—
COC_2H_5	CH_3	H	2-Propionyl-4-methylthiazole	*104*	—
CH_3	H	CH_3	2,5-Dimethylthiazole	*104,105*	—
CH_3	H	C_2H_5	2-Methyl-5-ethylthiazole	*104,105*	—
CH_3	H	CH_3CO	2-Methyl-5-acetylthiazole	—	Coffee, sulphury
H	CH_3	CH_3	4,5-Dimethylthiazole	*104*	Roasted, nutty, green
H	CH_3	C_2H_5	4-Methyl-5-ethylthiazole	*2,104*	Nutty, green, earthy
H	CH_3	$HOCH_2CH_2$	4-Methyl-5-(β-hydroxyethyl)thiazole	*138*	—
H	CH_3	$CH_2{=}CH$	4-Methyl-5-vinylthiazole	*57,66,102,138*	—
H	CH_3	CH_3CO	4-Methyl-5-acetylthiazole	—	Roasted, nutty, sulphury

H	*i*-C_4H_9	C_2H_5	4-Isobutyl-5-ethylthiazole	–	Cucumber, green, potato
H	*n*-C_4H_9	*n*-C_3H_7	4-*n*-Butyl-5-*n*-propylthiazole	–	Bell pepper-like (0.003 ppb)
H	*i*-C_4H_9	CH_3O	4-Isobutyl-5-methoxythiazole	–	Green pepper
CH_3	CH_3	CH_3	2,4,5-Trimethylthiazole	*2,27,30,34,109*	Cocoa, nutty (0.05 ppm)
CH_3	CH_3	C_2H_5	2,4-Dimethyl-5-ethylthiazole	*27,104*	Nutty, roasted (2 ppb)
CH_3	CH_3	$CH_2{=}CH$	2,4-Dimethyl-5-vinylthiazole	*2*	–
CH_3	CH_3	CH_3CO	2,4-Dimethyl-5-acetylthiazole	–	Roasted, nutty, meaty
CH_3	C_2H_5	CH_3	2,5-Dimethyl-4-ethylthiazole	*27,29,104,105*	–
CH_3	*n*-C_3H_7	CH_3	2,5-Dimethyl-4-*n*-propylthiazole	–	–
CH_3	*n*-C_4H_9	CH_3	2,5-Dimethyl-4-*n*-butylthiazole	*29*	–
CH_3	CH_3CO	CH_3	2,5-Dimethyl-4-acetylthiazole	–	Roasted, meaty, sulphury
C_2H_5	CH_3	C_2H_5	2,5-Diethyl-4-methylthiazole	*29*	Meaty, green, nutty
C_2H_5	CH_3	CH_3	4,5-Dimethyl-2-ethylthiazole	*104*	–
i-C_4H_9	CH_3	CH_3	4,5-Dimethyl-2-isobutylthiazole	*27,30,34*	–
i-C_3H_7	CH_3	CH_3	4,5-Dimethyl-2-isopropylthiazole	*27,30,34*	–
i-C_3H_7	CH_3	C_2H_5	2-Isopropyl-4-methyl-5-ethylthiazole	*30,34*	–
CH_3	*i*-C_4H_9	CH_3O	2-Methyl-4-isobutyl-5-methoxythiazole	–	Green vegetable
CH_3	*i*-C_4H_9	C_2H_5O	2-Methyl-4-isobutyl-5-ethoxythiazole	–	Green, vegetable, onion

a) A dipropyl- and an ethyldimethylthiazole were tentatively identified in boiled potatoes [19]. Food systems: *2* ground beef [228]; *5* canned beef stew [149]; *12* cooked pork liver [128b]; *25* asparagus [208]; *26* cooked artichoke [23]; *27* dry red beans [20]; *29* baked potatoes [32]; *30* boiled potatoes [19]; *34* freeze-dried potato peel [19]; *66* roasted filberts [183]; *67* roasted peanuts [221]; *70* roasted sesame seed [91]; *79* popcorn [220]; *82* rice [211]; *88* sterilized milk concentrate [2]; *104* coffee [215b]; *105* tea [216a]; *106* Jamaican rum [236]; *107* Scotch whisky [236]; *108* beer [72]; *109* wine [216a]; *138* malt [210].

b) Some threshold values in brackets are given.

Boiled potatoes were found to contain a greater number of thiazole derivatives than more severely heat-treated products, namely potato chips. Therefore, it seems that the formation of thiazoles may be substantially influenced by temperature and also by the availability of free water. Significant variations in thiazole levels were observed among various lots of potatoes. Some 2,4,5-trisubstituted thiazoles were also found in the volatile flavour of baked potatoes [32] and among the volatile constituents of dry red beans [20]. In these food flavours thiazoles arise mainly by thermal degradation of reducing sugars in the presence of hydrogen sulphide and ammonia, or more frequently by degradation of cysteine and cystine alone or in the presence of sugars or α-diketones.

2.7.2 Sensory properties and patents

Owing to their specific flavours, a wide range of mono-, di- and trisubstituted thiazoles were patented by Pittet & Hruza [157]. The sensory properties of most thiazoles were described as being green, roasted, or nutty with some specific green vegetable notes. The lower 2-alkylthiazoles possessed green vegetable-like properties. When 2-isobutylthiazole was added to canned tomato, purée, or paste, a more intense tomato-like flavour resulted [87a]. In the case of tomato juice added levels of 20 to 50 ppb were found to be more acceptable, since at higher levels sensory notes described as rancid, medicinal, and metallic became apparent. 2-Alkylthiazoles were arranged in the order from those having greatest to least effect in tomato soup as follows [87b] : $i\text{-}C_4H_9 > s\text{-}C_4H_9 > s\text{-}C_5H_{11} > i\text{-}C_3H_7 > t\text{-}C_4H_9 > n\text{-}C_4H_9 > n\text{-}C_5H_{11} > n\text{-}C_3H_7$. From this study one can conclude that secondary alkyl groups are more interesting than those with a straight chain. Increased substitution (dialkyl- and trialkyl derivatives) added nutty, roasted, and meaty notes. 2,4,5-Trimethylthiazole has dark chocolate and light green flavour aroma notes. 2,4-Dimethyl-5-ethylthiazole has a *buchu* leaf oil odour and a meaty, liver flavour [157b]. 4,5-Dialkylthiazoles have a strong aroma of green pepper. Five compounds of this series were tested by Buttery *et al.* [21], and it was found that the more interesting is 4-*n*-butyl-5-*n*-propylthiazole which possesses the lower odour threshold value in water (0.003 ppb). The olfactory properties of 2-alkoxythiazoles were in between those of 2-alkoxypyridines and those of the corresponding pyrazines. Alkoxy substitution in the 5-position resulted in a more sulphury character than the same substitution in the 2-position. 2-Methyl-4-isobutyl-5-methoxythiazole has a pepper, onion aroma and a vegetable flavour. 4-Isobutyl-5-methoxythiazole has a vegetable soup 'minestrone' fragrance, and pepper, onion, celery, and green vegetable flavour characteristics.

Acetylthiazoles were also found to possess interesting olfactory properties and were patented [201,230]. 2- and 4-Acetylthiazoles possessed burnt, nutty, cereal and popcorn odours similar to the corresponding pyrazines and pyridines [157b]. However, 5-acetyl alkylthiazoles had more roasted, sulphury, and meaty odours. Thus the 4-methyl-5-acetylthiazole possesses an earthy peanut aroma and a sulphury roasted nut flavour with a bitter overtone, while the 2,4-dimethyl-5-acetylthiazole has a meaty sulphur fragrance and a boiled beef flavour. Pittet & Hruza [157] concluded that in the case of thiazoles substitution in the 2- or 4-position adjacent to the ring nitrogen resulted in more potent and characteristic flavours than substitution in the 5-position. They also described the olfactory properties of 1-(2′-thiazolyl)-propanone and those of 2-(2′-furyl)thiazole as being biscuit or nutty and pungent in character, respectively.

Most of the alcohol and carbonylthiazole derivatives **80** (R = CH_2OH, $CHOHCH_3$, COR′) patented as flavour compounds of coffee [233] were characterized as being green, fruity, woody, and burnt. The following 5-β-hydroxy-alkylthiazoles **81** (R^1 = CH_3, C_2H_5, *n*-C_4H_9, R^2 = H; R^1 = R^2 = CH_3; R^1 = $CH_3CH{=}CH(CH_2)_2$, R^2 = C_2H_5) and 4-ethyl-5-(γ-acetoxypropyl)-thiazole **82** specified for meat and chicken flavourings [85], and 4-methyl-5-vinylthiazole [70] have been patented.

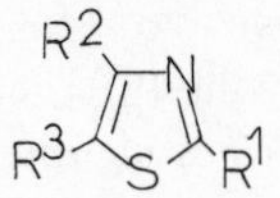

79 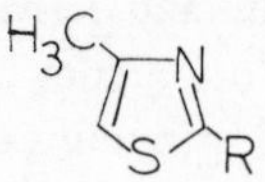 **80**

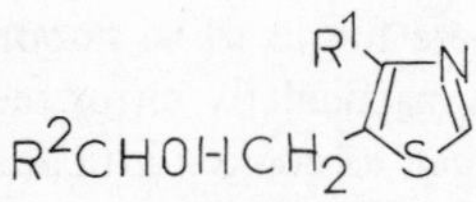

81 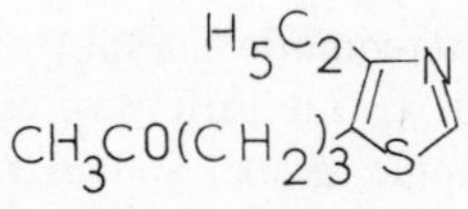 **82**

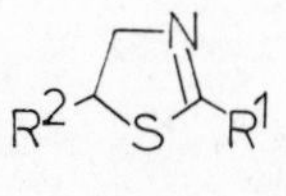

83 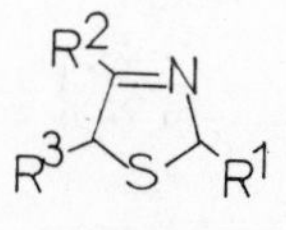 **84** 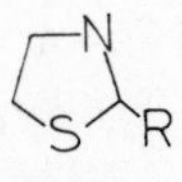 **85**

Fig. 9 – Thiazoles, thiazolines, and thiazolidines in food flavours.

Other uses of thiazoles as flavouring agents in perfumes and cosmetics were also described (**U.S. Pat.**, 3.678.064–18.7.1972; *ibid*, 3.778.518–11.12.1973).

Thiazolines **83**, **84** and thiazolidines **85** also possess interesting sensory properties, and some of them were patented.

2.4.5-Trimethyl- and 2-acetyl-2-thiazolines were reported to occur in dry red beans [20] and in beef broth [206], respectively. The latter compound was described as typical of freshly baked bread crust in character [157b]. 2-Acetyl-5-alkyl-2-thiazolines and 2-propionyl-2-thiazolines were found to be useful in the flavour of meat (stock cubes), fried and roasted foods [33]. 2,4-Dimethyl-3-thiazoline found in cooked beef aroma [130] has a nutty, roasted and vegetable aroma, while the corresponding oxazoline only is nutty and vegetable. Similarly, 2,4,5-trimethyl-3-thiazoline [130] is meaty, nutty, and onion-like, while its oxazoline analogue is woody, musty, and green. Some 3-thiazolines **84** ($R^1 = R^3 = CH_3$, $R^2 = n\text{-}C_3H_7$, $i\text{-}C_4H_9$) were recommended as flavouring ingredients in perfumes and Cologne [187].

Thiazolidines **85** ($R = i\text{-}C_3H_7$, $n\text{-}C_3H_7$, $n\text{-}C_4H_9$) mixed with pyrazines and 2-cyclohexanones impart important cereal, nut and meat-like flavour notes [56]. 2-Isopropylthiazolidine was useful to enhance chocolate-like and cocoa-like flavours. A similar result was obtained from equal amounts of 2-isobutyl-, 2-isopentyl-, and 2-isohexylthiazolidine.

II–3 SIX OR MORE MEMBERED HETEROCYCLIC SYSTEMS CONTAINING ONE, TWO, OR MORE HETEROATOMS

3.1 Pyrans, Thiapyrans and derivatives

The diastereomeric Rose oxides **86** of the (4R) series were identified in various essential oils: rose oil, geranium oil, *Ribes nigrum*, verbena oil, tea oil, in grapes and wines and in the secretion of the thoracic glands of *Aromia moschata L.* [137a]. 2-Alkyl-4-phenyl (2*H*, 1*H*) dihydropyrans **87** ($R = i\text{-}C_3H_7$) were used to enhance the aroma of certain products such as tobacco and canned orange juice [214]. The tetrahydropyran itself **88** was detected in *buchu* leaf oil and citrus oils. The diastereomeric linolool oxides **89** were found to be important aroma components of a wide variety of essential oils, particularly citrus oils, in fruits: apricot, pineapple, grape, and in vegetables such as cooked artichoke [23]. In black tea, they exert a favourable flavouring effect [67b]. 2-Acetonyltetrahydropyran derivative **90** isolated from 'Burley' tobacco is described as having a sweet, phenolic, and woody odour [169].

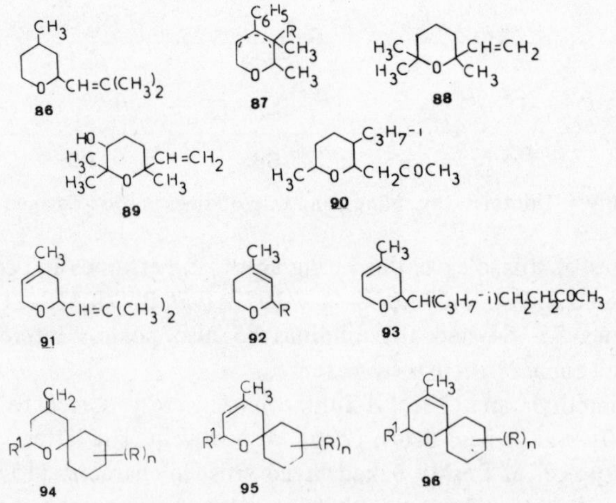

Fig. 10 – Tetrahydro- and dihydropyrans in foods.

Nerol oxide **91** was identified in Bulgarian rose oil, *verbena* oil, grapes [171b], and wines [173] where it is an important aroma component. Compounds such as 2-substituted-4-methyl-5,6-dihydro-(2*H*)-pyrans **92** (R = alkenyl-, alkyl-, alkoxy-, cyclohexyl-, and 2-hydroxy-3-butenyl groups) were used as flavouring ingredients in perfumes, foods and beverages [138]. The dihydropyran **93** was isolated from 'Burley' tobacco. Tetrahydropyranspirocyclohexanes **94, 95** and **96** in a mixture have been patented for augmenting, enhancing, or modifying the organoleptic properties of perfumes and Colognes [78b].

But the most important substances in this class are derived from α- and γ-pyrones and their reduced rings (*see Fig. 11*). 3,4,5,6-Tetrahydro-2-pyrones or δ-lactones **97** are closely related to the corresponding five-membered rings [108]. They are mainly found in meat products [100], animal fats [100,225a, 225b], fruits [52b,232], roasted filberts [90], vegetable oils, milk products [2, 104], black tea [26,238], beverages produced by alcoholic fermentation such as beer, rum, and whisky [209], and in culture media from *Sporobolomyces odorus* [209]. Finally, some essential oils such as tuberose oil contain δ-lactones (*see Table II–16*).

Table II–16 – Tetrahydro-2-pyrones (or δ-lactones) [108,137a]

97

97; R^1	R^2	R^3	Name	Occurrence[a)]
(a) H	H	H	Tetrahydro-2-pyrone (δ-valerolactone)	*10,66,67*
(b) H	H	CH_3	6-Methyl. . . (δ-hexalactone)	*10,77,86,91, 132*
(c) H	H	C_2H_5	6-Ethyl. . . (δ-heptalactone)	*10,63,77,86, 105*
(d) H	H	n-C_3H_7	6-*n*-Propyl. . . (δ-octalactone)	*10,48,60,63, 77,86,91,96, 101a,101c,132*
(e) H	H	n-C_4H_9	6-*n*-Butyl. . . (δ-nonalactone)	*10,11,12,77, 86*
(f) H	H	n-C_5H_{11}	6-*n*-Pentyl. . . (δ-decalactone)	*10,11,12,13, 18,20b,41,58, 62,77,86,87, 88,91,96,101c, 105,108,132*

Table II–16 (*continued*)

97;	R^1	R^2	R^3	Name	Occurrence[a)]
(g)	H	H	n-C_6H_{13}	6-*n*-Hexyl. . . (δ-undecalactone)	*10,77,91*
(h)	H	H	n-C_7H_{15}	6-*n*-Heptyl. . . (δ-dodecalactone)	*10,11,13,18, 20b,58,77,86, 87,88,91,96, 101a,101c,132*
(i)	H	H	n-C_8H_{17}	6-*n*-Octyl. . . (δ-tridecalactone)	*10,91*
(j)	H	H	n-C_9H_{19}	6-*n*-Nonyl. . . (δ-tetradecalactone)	*10,11,13,18, 20b,86,91,96, 101c*
(k)	H	H	n-$C_{10}H_{21}$	6-*n*-Decyl. . . (δ-pentadecalactone)	*10,91*
(l)	H	H	n-$C_{11}H_{23}$	6-*n*-Undecyl. . . (δ-hexadecalactone)	*10,11,13,86, 91,101c*
(m)	H	H	n-$C_{13}H_{27}$	6-*n*-Tridecyl. . . (δ-octadecalactone)	*86*
(n)	H	H	$C_2H_5CH{=}CH(CH_2)_3$	δ-Dodec-9-ene lactone (Z)-9-Dodecen-5-olide	*91*
(o)	H	H	n-$C_4H_9CH{=}CH(CH_2)_3$	δ-Tetradec-9-ene lactone (Z)-9-tetradecen-5-olide	*91*
(p)	H	H	$C_2H_5CH{=}CHCH_2$	(Z)-7-Decen-5-olide (Jasmine lactone)	*105,120b*
(q)	H	H	n-$C_5H_{11}CH{=}CH$	δ-Dodec-6-ene lactone (Z)-6-Dodecen-5-olide	*91*
(r)	H	H	CH_2OH	δ-6-Hydroxyhexalactone (6-hydroxy-5-hexanolide)	*35*
(s)	H	CH_3	CH_3	*Trans*-4-methyl-5-hydroxy-hexanoic acid lactone (5,6-dimethyl-tetrahydro-2-pyrone)	*86*
(t)	i-C_3H_7	H	H	3-Isopropyl-5-hydroxypenta-noic acid lactone (4-isopropyl-tetrahydro-2-pyrone)	*111*

a) *10* beef fats; *11* pork, chicken and sheep fats; *12* pork liver; *13* lamb meat; *18* chicken fats; *20b* horse fats; *35* tomatoes; *48* apricot; *57* passion fruit; *58* peach; *60* pineapple; *62* raspberry; *63* strawberry; *66* roasted filberts; *77* oxidized and hydrogenated soy bean oil; *86* evaporated milk and milk fats; *87* dry whole milk; *88* sterilized concentrated milk; *91* butter; *96* blue cheese; *101a* Swiss cheese; *101c* Cheddar cheese; *105* black tea; *108* culture media from *Sporobolomyces odorus*; *111* 'Burley' tobacco; *120b* essential oil of tuberose; *132* coconut oil.

As seen, δ-decalactone **97f** and δ-dodecalactone **97h** are among the most widely distributed. The first possesses a creamy, fruity, coconut-like, peach-like, and milk-like aroma [61], while δ-dodecalactone was described as possessing a powerful, fresh-fruit, oily odour and low levels of peach, pear, plum-like flavour [52b]. Three δ-lactones bearing an unsaturated chain in 6-position **97n, 97o, 97q** were identified in the butter flavour, while the jasmine lactone **97p** was found in black tea aroma [73], tuberose oil, and among fermentation products. 5,6-Dimethyl-pyranone **97s** was isolated from milk. The isopropyl derivative **97t** imparts a coconut-like aroma to tobacco.

R1 R2 R3 O O **97** R1 R2 O O **98** OH O O **99**

O OH R O O **100** R2 O R1 **101** R2 R1 S **102**

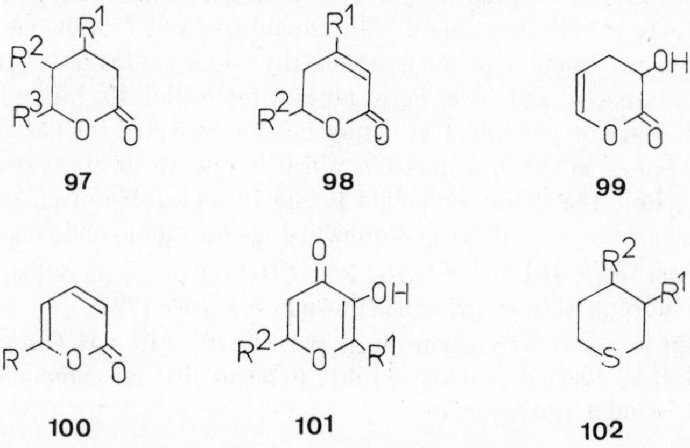

Fig. 11 – Saturated and unsaturated δ-lactones and tetrahydrothiapyrans.

Several α,β-unsaturated δ-lactones **98** were also found in food aromas [137a]. 3-Methyl-2-penten-5-olide **98** (R^1 = CH_3, R^2 = H) or anhydromevalactone and the isopropyl derivative **98** (R^1 = i-C_3H_7, R^2 = H) were identified in 'Burley' tobacco flavour [39]. The latter compound imparts a coconut-like aroma to tobacco [178]. Parasorbic acid **98** R^1 = H, R^2 = CH_3), a cytotoxic compound, is the main volatile component of *Sorbus aucuparia L*. The *n*-pentyl derivative **98** (R^1 = H, R^2 = n-C_5H_{11}) known as *Massoia* lactone which occurs naturally in the essential oil of *Massoia aromatica* bark was also found in tuberose oil, wine, and sugar molasses [137a]. Tuberolactone **98** (R^1 = H, R^2 = C_2H_5CH=$CHCH_2$) occurs also in the characteristic aroma of the tuberose essential oil.

3-Hydroxy-2-pyrone **99** was identified among the degradation products in off-flavoured orange juice [199c]. Two doubly unsaturated δ-lactones **100** (R = H, n-C_5H_{11}) were identified in peach flavour. The pentyl derivative has a powerful butter-like aroma. It was also found in mushrooms.

Several detailed reports [52b,101,158], exist on the characteristic organoleptic properties of α-pyrone derivatives. Their odour threshold values average about 0.1 ppm [188]. The most typical compound of γ-pyrones class is certainly the 2-methyl-3-hydroxy-4-(4*H*)-pyrone **101** (R^1 = CH_3, R^2 = H) or maltol, which is the character impact compound in malt. It was also identified as an important component in various processed foodstuffs such as caramel, coffee, roasted filberts, and cultured dairy products [137b]. Maltol is described by Hodge [75] as having a characteristic caramel-like aroma, like the burnt sugar of confectionery. Its taste in dilute solution gives burnt and fruity notes. The threshold value in water is 35 ppm [153]. Besides these pleasant olfactory properties, maltol possesses the peculiarity to inhibit the bitterness of some products and to exert a synergetical effect with sugar [10]. Owing to these sensory properties, maltol was used as a flavouring agent in foods [52]. In mixture with *L*-proline (100 to 150°C) it generates honey and slightly browned flavour notes [154a]. Reacting it with sulphur-free amino acids, Pittet & Seitz [156] obtained reaction products having the odour and/or taste properties of baked goods, cocoa, and sweet burnt notes. Ethylmaltol, the higher homologue of maltol, was not yet found in nature but was accepted for flavouring foodstuffs. It causes an odour impression which is four to six times stronger than that of maltol. The latter compound is said to be capable of masking natural bitter principles such as those of wormwood, gentian, hop, and cola [146]. Its 6-methyl derivative **101** ($R^1 = C_2H_5$, $R^2 = CH_3$) isolated from refinery molasses has an odour of preserved tang boiled down in soy sauce [79].

Except for the two tetrahydrothiapyrans **102** ($R^1 = R^2 = H, CH_3$) tentatively identified from roasted peanuts [115b], it seems that no compounds of this class were found in food aromas.

3.2 Pyridines and derivatives

Pyridine and its substituted derivatives **103** and **104** are principally distributed in cooked beef [92,98a,184b,225e], beetroots [145], cooked asparagus [208], dry red beans [20], potatoes [15,18], leek [175], roasted filberts [90], peanuts [221], barley [185], corn frozen [11], popcorn [220], rice bran [211], cheese [44], cocoa [216b], coffee [67a], tea [216a], beer [72], tobacco [132], and jasmine oil [207]. Pyridine itself is a flavour component of a wide variety of foodstuffs (*see Table II–17*).

Two 2-alkylpyridines **103** ($R^1 = CH_3$, n-C_3H_7, $R^2 = R^3 = H$) are flavour components of cooked beef [98a]. Some alkylpyridines were also found in beef by Kohler *et al.* [92] and Watanabe & Sato [225e]. 2-Ethylpyridine was characterized in cooked asparagus flavour [208]. In potato chips and potatoes, one notable component not found in baked potatoes but previously reported in potato chips was 2-acetylpyridine [18]. It is described as having a popcorn odour. Among seventeen components of cooked beetroots aroma, pyridine and its 4-methyl derivative account for 60% of volatile products [145]. Eight pyridines were identified in a rice bran aroma [211], and sixteen among fifty new volatiles of the black tea aroma. The five following pyridines: pyridine, dimethyl-, trimethyl-, 2-methyl-5-ethyl-, and 2-*n*-propyl-3-methoxypyridines were found in 'Gruyère de Comté' cheese [44]. The latter compound will be responsible for

Table II–17 – 2-, 3- and 4-Substituted pyridines

R^3, R^2, R^1 substituted pyridine (structure **103**)

103

03; R¹	R²	R³	Name	Occurrence[a)]
I	H	H	Pyridine	*22,25,29,44,67,78, 79,82,104,105,113b*
H_3	H	H	2-Methylpyridine	*6,67,82,113b,105*
$_2H_5$	H	H	2-Ethylpyridine	*25,105,113b*
-C_3H_7	H	H	2-*n*-Propylpyridine	*6*
-C_5H_{11}	H	H	2-*n*-Pentylpyridine	*66,67*
H=CH_2	H	H	2-Vinylpyridine	*113b*
OCH_3	H	H	2-Acetylpyridine	*29,66,67,105,108, 113b*
$_6H_5$	H	H	3-Phenylpyridine	*105,113b*
I	CH_3	H	3-Methylpyridine	*82,104,105,113b*
I	C_2H_5	H	3-Ethylpyridine	*105,113b,117*
I	*n*-C_3H_7	H	3-*n*-Propylpyridine	—
I	*n*-C_4H_9	H	3-*n*-Butylpyridine	*105*
I	$CH{=}CH_2$	H	3-Vinylpyridine	*113b,117*
I	$COCH_3$	H	3-Acetylpyridine	*66,113b,108*
I	COC_2H_5	H	3-Propionylpyridine	*113b*
I	COC_3H_7-*n*	H	3-Butyrylpyridine	*113b*
I	$COOCH_3$	H	Methyl nicotinate	*104,117*
I	$COOC_2H_5$	H	Ethyl nicotinate	*117*
I	C_6H_5	H	3-Phenylpyridine	*113b*
I	C≡N	H	3-Cyanopyridine	*113b*
I	CHO	H	3-Formylpyridine	*113b*
I	$NHCH_3$	H	3-Aminomethylpyridine	*113b*
I	OH	H	3-Hydroxypyridine	*81,113b*
I	OCH_3	H	3-Methoxypyridine	*105*
I	$OC(=O)C_2H_5$	H	3-Pyridylpropionate	*66,67*
I	H	CH_3	4-Methylpyridine	*22,82,113b*
I	H	C_2H_5	4-Ethylpyridine	—
I	H	$HC{=}CH_2$	4-Vinylpyridine	*105*

) *6* cooked beef; *22* carrot, parsnip, beetroot, beets; *25* asparagus; *29* baked potatoes; *4* leek; *66* roasted filberts; *67* roasted peanuts; *78* corn frozen; *79* popcorn; *81* roasted arley; *82* rice; *104* coffee; *105* tea; *108* beer; *113b* tobacco; *117* jasmine.

the off-flavour of this cheese. 3-Hydroxypyridine was found in roasted barley aroma [185]. In coffee, alkylpyridines are mainly formed by thermal degradation of trigonelline (*see Chapter III*). Four alkylated pyridines and ten alkyl substituted nicotinates were recently identified by the analysis of the basic fraction of the Chinese jasmine [207].

Among 42 mono- and polysubstituted pyridines **104**, tobacco alkaloids (nicotine, cotinine, nornicotine etc. . .) are the most numerous [170]. Some nornicotine derivatives have recently been patented in Japan as tobacco flavourings.

The di- and trisubstituted pyridines found in tobacco also contain alkyl [39], vinyl [132], phenyl [169], cyano [132], acetyl [132] and hydroxy substituents (*see Table II–18*).

Pyridine, piperidine, and their derivatives were reported to occur in fish products [82]. 2-Pyridinemethanethiol was recommended as a key ingredient in a synthetic meat flavour similar to roasted pork [51].

Sensory properties of some pyridines were described by Pittet & Hruza [157b]. 2-Alkoxypyridines were found to have unpleasant phenolic odours, whereas corresponding 2-alkoxypyrazines were sweet and nutty. Pyridines of similar structure to that of alkylthiazoles were also found to possess green odours. But insufficient numbers of di- and trialkylpyridines were evaluated to make significant conclusions regarding their flavouring properties.

Relatively few saturated pyridines were found in food flavours (*see Fig.12*). One of the most interesting products with a partially saturated skeleton is 2-acetyl-1,4,5,6-tetrahydropyridine **105** which has a typical bread aroma. Prepared from proline and dihydroxyacetone, it is effective as bread flavouring [177]. 1,2,5,6-Tetrahydropyridine **106**, *N*-alkyl-5,6-dihydro-2-pyridinones **107**, the following monosubstituted piperidines **108** (R = H, 2-, 3-, 4-alkyl-, 2-ethyl-, 3-methyl-, isopropyl), the dimethylpiperidines (2,3-, 2,4-, and 2,6-isomers), 2-piperidinone and its methyl derivative **109** (R = H, CH_3) were found in tobacco [170]. The *N*-nitrosopiperidine **110** found in cured meat, is formed during the

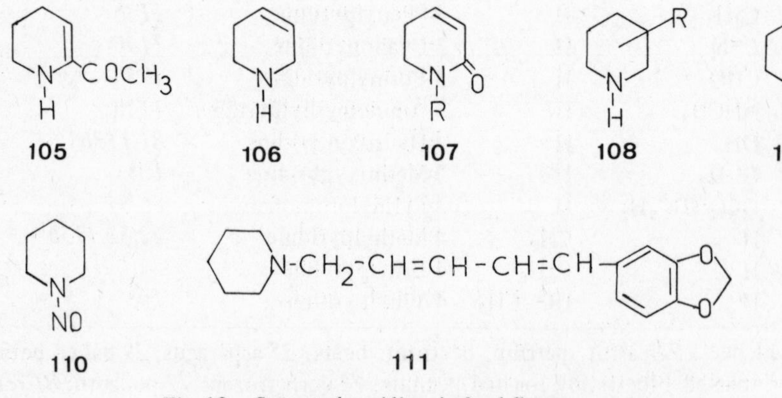

Fig. 12 – Saturated pyridines in food flavours.

Table II–18 – Polysubstituted pyridines

104

104; R^1	R^2	R^3	R^4	R^5	Name	Occurrence[a)]
CH_3	CH_3	H	H	H	2,3-Dimethylpyridine	*113b*
n-C_3H_7	OCH_3	H	H	H	2-*n*-Propyl-3-methoxy-pyridine	*97*
CH_3	H	CH_3	H	H	2,4-Dimethylpyridine	*82,113b*
CH_3	H	H	C_2H_5	H	2-Methyl-5-ethylpyridine	*27,97,105*
CH_3	H	H	i-C_3H_7	H	2-Methyl-5-isopropyl pyridine	*113b*
CH_3	H	H	CH_3CO	H	2-Methyl-5-acetylpyridine	*113b*
CH_3	H	H	C_6H_5	H	2-Methyl-5-phenylpyridine	*113b*
CH_3	H	H	$C{\equiv}N$	H	2-Methyl-5-cyanopyridine	*113b*
CH_3	H	H	H	CH_3	2,6-Dimethylpyridine	*82,105*
CH_3	H	H	H	C_2H_5	2-Methyl-6-ethylpyridine	*105*
H	CH_3	CH_3	H	H	3,4-Dimethylpyridine	*82,113b*
H	C_2H_5	CH_3	H	H	3-Ethyl-4-methylpyridine	*117*
H	$HC{=}CH_2$	CH_3	H	H	3-Vinyl-4-methylpyridine	*117*
H	$COOCH_3$	CH_3	H	H	Methyl 4-methylnicotinate	*117*
H	OH	H	H	CH_3	3-Hydroxy-6-methylpyridine	*113b*
H	CH_3	H	CH_3	H	3,5-Dimethylpyridine	*82,113b*
H	$COOCH_3$	H	C_2H_5	H	Methyl 5-ethylnicotinate	*117*
H	$COOC_2H_5$	H	C_2H_5	H	Ethyl 5-ethylnicotinate	*117*
H	$COOCH_3$	H	$HC{=}CH_2$	H	Methyl 5-vinylnicotinate	*117*
H	$COOC_2H_5$	H	$HC{=}CH_2$	H	Ethyl 5-vinylnicotinate	*117*
CH_3	H	CH_3	CH_3	H	2,4,5-Trimethylpyridine	*113b*
CH_3	H	CH_3	i-C_3H_7	H	2,4-Dimethyl-5-isopropyl-pyridine	*113b*
CH_3	H	CH_3	H	CH_3	2,4,6-Trimethylpyridine	*113b*
CH_3	CH_3	H	H	CH_3	2,3,6-Trimethylpyridine	*113b*
					Dimethylcyanopyridine	*113b*
H	$COOCH_3$	CH_3	C_2H_5	H	Methyl 4-methyl-5-ethyl-nicotinate	*117*
H	$COOCH_3$	CH_3	$HC{=}CH_2$	H	Methyl 4-methyl-5-vinyl-nicotinate	*117*
H	$COOC_2H_5$	CH_3	C_2H_5	H	Ethyl 4-methyl-5-ethyl-nicotinate	*117*

a) See the caption for Table II–17; *27* beans; *97* ‘Gruyere de Comté’.

cooking of meats [46]. Piperine **111** is an important component of *Piper nigrum L.* Its level in natural extracts of black pepper is 33.4% [83].

3.3 Dioxanes, oxathianes, dithianes, dithiins, trithianes, oxadithianes, and dioxathianes

Several six-membered heterocyclic compounds containing two or three identical or different O- and/or S- atoms in their rings were found in *Allium* species (*see Fig. 13*) and in model reactions (*see Chapter III*). Moshonas & Shaw [122] identified 1,4-dioxane **112** in lemon. 3-Ethoxy-4-hydroxybenzaldehyde-2,2-dimethylpropanediol acetal **113** is recommended as a flavouring in tobacco and bakery products [94]. The cyclic acetals **114** and **115** have woody odours and improve known perfume compositions [97]. A mixture of diastereomeric 2-methyl-4-propyl-1,3-oxathianes **116** was identified in a flavour concentrate of the yellow passion fruit [234]. Both products have a strong and natural fruity odour with a green and slightly burnt note, particularly the *cis* diastereomer.

Several di- and trisulphur-containing heterocyclic compounds were identified within the last few years as flavour components of numerous foods [120,171b]. They were found especially in cooked beef and boiled meat aromas [13,187]. 1,2-Dithiane **117** was found in a model reaction between cysteine and xylose and among the pyrolysis products of cysteine [96c]. 3-Vinyl-3,4-dihydro-1,2-dithiin **118** and the *homo* conjugated isomer **119** were found in the sulphur-containing flavour fraction of cooked asparagus [208]. 3-(E)-Propenyl-4-methyl-(4*H*)-1,2-dithiin **120** were isolated from the reaction of hydrogen sulphide on 2-butenal [5]. These compounds are described as being strongly odorous. 1,3-Dithiins **121** and **122** were detected in a synthetic onion aroma obtained by reacting aliphatic aldehydes with hydrogen sulphide and ammonia [96a].

3-Methyl-1,4-thiazane-5-carboxylic acid *S*-oxide **123** or cycloalliin is a garlic sulphoxide amino acid isolated from onion [25]. *s*-Trithiane **124** (R = H) or trithioformaldehyde, its 2,4,6-trimethyl- **124** (R = CH_3) and 2,2,4,4,6,6-hexamethyl derivatives **125** were identified in meat aroma [228]. The parent compound **124** (R = H) was also found in chicken flavour [117]. 1,2,4-Trithiane and its 3-methyl derivative **126** (R = H and CH_3) are reported among the degradation products of cysteine and cystine [96c]. The latter compound was recently found in a commercial beef extract [58]; added to foods it imparts a roasted or meat-like aroma [59]. Various alkyl substituted 1,3,5-dithiazines **127**, perhydro-1,3,5-thiadiazine **128**, triethyl-1,3,5-oxadithiane **129** (R = C_2H_5), 1,3,5-dioxathianes **130** (R = CH_3, C_2H_5) were reported in model reactions (*see Chapter III*). Some of these products possess a typical flavour reminiscent of onion and meat.

3.4 Pyrazines

Several reviews [105,135,137b] have drawn attention to this important class of aroma compounds. Tetramethylpyrazine and a higher homologue were first isolated in 1879 from fusel oil obtained from molasses of sugar beet. In 1928, Reichstein & Staudinger reported finding pyrazine, its methyl-, 2,5-dimethyl-, and 2,6-dimethylpyrazines and higher homologues in coffee. Some years later,

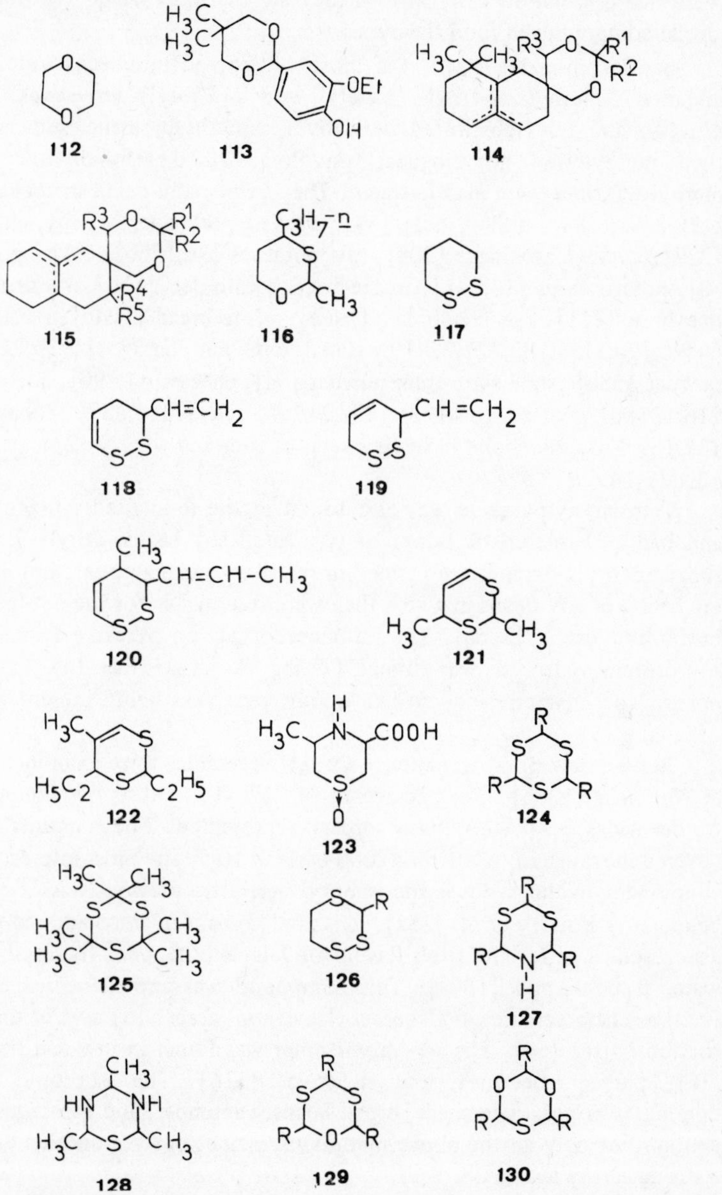

112 113 114

115 116 117

118 119

120 121

122 123 124

125 126 127

128 129 130

Fig. 13 – Six-membered heterocyclic aroma compounds containing two or three identical or different O-, S- and/or N-atoms.

some of these compounds were reported in a chocolate extract. New pyrazines **131** have been found in more than fifty natural or processed, cooked, roasted, or fermented products of vegetable or animal origin [137b] (*see Table II–19*). With nearly a hundred derivatives, they are the most widely distributed heterocyclic compounds in food flavours.

Besides pyrazine itself, the most widespread monosubstituted pyrazines included 2-methyl-, 2-ethyl-, 2-acetyl- and 2-(2′furyl) derivatives. Among the di-, tri-, and tetrasubstituted derivatives, dimethylpyrazine isomers as well as their methylethyl homologues, trimethyl-, 2,5-dimethyl-3-ethyl-, and tetramethylpyrazines were also frequent. They principally occur in cooked or boiled beef [98a,128a], grilled meat [55], cooked pork liver [128b], chicken broth [227], cooked asparagus [208], raw potatoes [36,126b], baked potatoes [18, 83], potato chips [18,141], freeze-dried potato peel [116], mushrooms [202], rice bran [211], rye crispbread [196a], white bread [124b], roasted nut products [90,115a,183,221,224], roasted oats and barley [113b,223b], whey powder [53b], stale skim milk powder [54], cheese [44,190], cocoa [114,160, 216b,218b], coffee [64,193a,194a,215b], tea [216a,237], tobacco [39,41, 132,169,170], alcoholic beverages such as rum and whisky [236], and essential oils [12,14,45].

Tetramethylpyrazine was also found in the fermented soy products [93], and had a characteristic odour of fermented soy beans. Ethyl-, 2,5-dimethyl-, and trimethylpyrazines are formed in relatively large amounts during the deep-fat frying of soy beans in which they were responsible for the nut-like or peanut butter-like aroma [226a]. The influence of pH on pyrazine formation during soy protein hydrolysis was studied [113b]. At an acid pH, the mixture did not possess the characteristic nut-like odour, pyrazines being present in the form of salts.

Besides raw potatoes, twenty-six raw vegetables were examined by Murray & Whitfield [126b]. The occurrence of 2-alkyl-3-methoxypyrazines was found to be widespread with many orders represented. The amounts determined covered the range from >1 to 20 000 parts in 10^{13}. The most interesting odorous component in this series is the isobutyl derivative previously isolated from bell peppers by Buttery *et al.* [15a]. It is predominant in green and red peppers and also responsible for the fresh flavour of Jalapeno peppers. Its threshold value in water is 0.002 ppb [180a]. This compound was also identified in an aroma fraction characteristic of 'Cabernet Sauvignon' grapes [6] and in the flavour of roasted coffee [64]. The *sec*-butyl isomer was found in raw and boiled carrots [35], parsnip, beetroot, and silverbeet [126]. The isopropyl homologue dominates in peas, pea shells, broad beans, cucumber, and asparagus. Beetroots contain not only all the above methoxypyrazines but yet another odorant with a strong earthy aroma.

Burrel *et al.* [14] reported finding three alkylpyrazines and five alkoxypyrazines in galbanum oil. 2-Methoxy-3-*sec*-butylpyrazines and 2-methoxy-3-

Table II–19 – Pyrazines [105]

131; R¹	R²	R³	R⁴	Name	Occurrence[a)]
H	H	H	H	Pyrazine	*6,25,66,67,81, 85,92,104,108, 113b*
CH_3	H	H	H	2-Methylpyrazine	*3,6,25,29,35,44, 66,67,68,79,81, 82,83,85,92,93, 102,104,105,106, 107,108,113b*
C_2H_5	H	H	H	2-Ethylpyrazine	*3,6,25,29,66,67, 83,85,104,105, 113b*
n-C_3H_7	H	H	H	2-*n*-Propylpyrazine	*66,67,85,104*
i-C_5H_{11}	H	H	H	2-Isopentylpyrazine	*29*
CH_2COCH_3	H	H	H	2-Acetonylpyrazine	*6*
i-C_3H_7	H	H	H	2-Isopropylpyrazine	*66,67,102*
$CH{=}CH_2$	H	H	H	2-Vinylpyrazine	*6,66,67,85,104*
Trans- -$HC{=}CH(CH_3)$	H	H	H	2-(*Trans*-1-propenyl) pyrazine	*6,104,113b*

Table II–19 (*continued*)

131; R^1	R^2	R^3	R^4	Name	Occurrence[a)]
$COCH_3$	H	H	H	2-Acetylpyrazine	*6,12,65,66,67, 70,113b*
O	H	H	H	2-(2′-Furyl)pyrazine	*12,65,66,102, 104,105,113b*
OCH_3	H	H	H	2-Methoxypyrazine	*33*
CH_3	CH_3	H	H	2,3-Dimethylpyrazine	*6,12,25,29,38, 66,67,68,70,81, 82,83,85,92,102, 104,105,106,107, 108,113*
CH_3	C_2H_5	H	H	2-Methyl-3-ethylpyrazine	*29,66,67,72,81, 85,104,105*
CH_3	n-C_3H_7	H	H	2-Methyl-3-*n*-propylpyrazine	—
CH_3	n-C_4H_9	H	H	2-Methyl-3-*n*-butylpyrazine	*29,70*
CH_3	i-C_4H_9	H	H	2-Methyl-3-isobutylpyrazine	*29,104*
CH_3	OCH_3	H	H	2-Methyl-3-methoxypyrazine	*104*
C_2H_5	C_2H_5	H	H	2,3-Diethylpyrazine	*29,66*
C_2H_5	OCH_3	H	H	2-Ethyl-3-methoxypyrazine	*33*
n-C_3H_7	OCH_3	H	H	2-*n*-Propyl-3-methoxypyrazine	—
n-C_4H_9	OCH_3	H	H	2-*n*-Butyl-3-methoxypyrazine	*115*
s-C_4H_9	OCH_3	H	H	2-*sec*-Butyl-3-methoxypyrazine	*23*
i-C_3H_7	OCH_3	H	H	2-Isopropyl-3-methoxypyrazine	*23,29*
$COCH_3$	CH_3	H	H	2-Acetyl-3-methylpyrazine	*83,113b*

$COCH_3$	C_2H_5	H	H	2-Acetyl-3-ethylpyrazine	*3,12,65,67,102*
$COCH_3$	OCH_3	H	H	2-Acetyl-3-methoxypyrazine	*70*
i-C_4H_9	OCH_3	H	H	2-Isobutyl-3-methoxypyrazine	*23,36,54*
CH_3	H	CH_3	H	2,5-Dimethylpyrazine	*6,25,29,38,67, 68,70,72,81,82, 83,85,92,97,102, 104,105,106,107, 108,113b*
CH_3	H	C_2H_5	H	2-Methyl-5-ethylpyrazine	*3,6,25,29,38,67, 68,81,102,108*
CH_3	H	i-C_3H_7	H	2-Methyl-5-isopropylpyrazine	*104,105,106, 107,113b*
CH_3	H	n-C_3H_7	H	2-Methyl-5-*n*-propylpyrazine	*85*
CH_3	H	n-C_5H_{11}	H	2-Methyl-5-*n*-pentylpyrazine	*66*
CH_3	H	i-C_4H_9	H	2-Methyl-5-isobutylpyrazine	*29*
CH_3	H	$CH_2{=}CH$	H	2-Methyl-5-vinylpyrazine	*25,29,66,67,104*
CH_3	H	*trans*– -CH_3-$CH{=}CH$	H	2-Methyl-5-(*trans*-1-propenyl) pyrazine	*104*
CH_3	H	O	H	2-Methyl-5-(2-furyl) pyrazine	*105,113b*
C_2H_5	H	C_2H_5	H	2,5-Diethylpyrazine	*66,102,104, 105,113b*
$COCH_3$	H	CH_3	H	2-Acetyl-5-methylpyrazine	*6,66,67*
$COCH_3$	H	C_2H_5	H	2-Acetyl-5-ethylpyrazine	*6*
$COCH_3$	H	OCH_3	H	2-Acetyl-5-methoxypyrazine	*65*
$COCH_3$	H	$COCH_3$	H	2,5-Diacetylpyrazine	*65*

Table II–19 (*continued*)

131; R^1	R^2	R^3	R^4	Name	Occurrence[a)]
CH_3	H	H	CH_3	2,6-Dimethylpyrazine	*3,6,12,25,29, 35,38,44,67, 68,70,72,82,92, 102,104,105, 106,107,108, 113b*
CH_3	H	H	C_2H_5	2-Methyl-6-ethylpyrazine	*3,6,12,25,29,66, 67,68,85,92,102, 105,106,107,108, 113b*
CH_3	H	H	*n*-C_4H_9	2-Methyl-6-*n*-butylpyrazine	*29*
CH_3	H	H	$C_2H_5CH(CH_3)CH_2$	2-Methyl-6(2′-methylbutyl)-pyrazine	*102*
CH_3	H	H	*i*-C_5H_{11}	2-Methyl-6-isopentylpyrazine	*102*
CH_3	H	H	CH_2=CH	2-Methyl-6-vinylpyrazine	*6,66,67,85,104, 106,113b*
CH_3	H	H	*trans*-CH_3–CH=CH	2-Methyl-6-(*trans*-1-propenyl) pyrazine	*104*
CH_3	H	H		2-Methyl-6-(2′-furyl) pyrazine	*113b*
C_2H_5	H	H	C_2H_5	2,6-Diethylpyrazine	*6,66,104,105, 113b*
C_2H_5	H	H	*n*-C_3H_7	2-Ethyl-6-*n*-propylpyrazine	*29*

$COCH_3$	H	H	C_2H_5	2-Acetyl-6-ethylpyrazine	*67,113b*
$COCH_3$	H	H	$CH_2{=}CH{-}CH_2$	2-Acetyl-6-allylpyrazine	*65*
Undetermined				Dimethylpyrazine	*101b*
Undetermined				Methylethylpyrazine	*66,101b*
Undetermined				Methylpropylpyrazine	*6,113b*
Undetermined				Methylisopropylpyrazine	*113b*
Undetermined				Methylisobutylpyrazine	*3*
Undetermined				Methylpentylpyrazine	*29*
Undetermined				Ethylvinylpyrazine	*29*
Undetermined				Acetylethylpyrazine	*66*
CH_3	CH_3	CH_3	H	Trimethylpyrazine	*3,6,12,25,29,38, 66,67,68,70,74, 81,85,92,97,101b, 102,105,106,107, 108,113b*
CH_3	CH_3	C_2H_5	H	2,3-Dimethyl-5-ethylpyrazine	*3,6,66,67,97,102, 104,105,108,115*
CH_3	CH_3	n-C_4H_9	H	2,3-Dimethyl-5-*n*-butylpyrazine	*29*
CH_3	CH_3		H	2,3-Dimethyl-5-(2′-furyl)-pyrazine	*113b*
C_2H_5	C_2H_5	CH_3	H	2,3-Diethyl-5-methylpyrazine	*6,29,66,67*
CH_3	C_2H_5	CH_3	H	2,5-Dimethyl-3-ethylpyrazine	*3,29,38,66,67, 68,70,74,81,85, 102,105,106, 107,108,113b*
CH_3	n-C_3H_7	CH_3	H	2,5-Dimethyl-3-*n*-propyl-pyrazine	not reported

Table II–19 (*continued*)

131;R^1	R^2	R^3	R^4	Name	Occurrence[a)]
CH_3	i-C_4H_9	CH_3	H	2,5-Dimethyl-3-isobutylpyrazine	*29,102*
CH_3	$CH_2CH(CH_3)C_2H_5$	CH_3	H	2,5-Dimethyl-3-(2′-methylbutyl) pyrazine	*102*
CH_3	i-C_5H_{11}	CH_3	H	2,5-Dimethyl-3-*i*-pentylpyrazine	*102*
C_2H_5	CH_3	C_2H_5	H	2,5-Diethyl-3-methylpyrazine	*6,29,66,105*
$COCH_3$	CH_3	CH_3	H	2-Acetyl-3,5-dimethylpyrazine	*104*
OCH_3	i-C_3H_7	CH_3	H	2-Methoxy-3-isopropyl-5-methylpyrazine	*115*
OCH_3	i-C_3H_7	CH_3O	H	2,5-Dimethoxy-3-isopropyl-pyrazine	*115*
OCH_3	$(CH_3)_2C(OH)$	CH_3	H	2-Methoxy-3-(α-hydroxyiso-propyl)-5-methylpyrazine	*115*
OCH_3	$COCH_3$	CH_3	H	2-Methoxy-3-acetyl-5-methyl-pyrazine	*115*
CH_3	C_2H_5	H	CH_3	2,6-Dimethyl-3-ethylpyrazine	*25,29,70,81,102, 108,113b*
CH_3	i-C_5H_{11}	H	CH_3	2,6-Dimethyl-3-*i*-pentylpyrazine	*102*
C_2H_5	CH_3	H	C_2H_5	2,6-Diethyl-3-methylpyrazine	*29,66,70,105,115*
C_2H_5	C_2H_5	CH_3	H	2,3-Diethyl-5-methylpyrazine	*70*
C_2H_5	C_2H_5	H	C_2H_5	Triethylpyrazine	*6,66*
	Undetermined			Dimethylethylpyrazine	*101b,113b*
	Undetermined			Dimethylpropylpyrazine	*29*
	Undetermined			Dimethylbutylpyrazine	*29*

	Undetermined			Dimethylisobutylpyrazine	*29,66*
	Undetermined			Dimethylisopentylpyrazine	*29*
	Undetermined			Dimethylvinylpyrazine	*29*
	Undetermined			Diethylpropylpyrazine	*29*
	Undetermined			Methylethylisobutylpyrazine	*29*
CH_3	CH_3	CH_3	CH_3	Tetramethylpyrazine	*6,38,66,67,75, 82,97,101b,102, 105,106,107, 113b,142*
CH_3	CH_3	CH_3	C_2H_5	Trimethylethylpyrazine	*29,38,97,101b, 102,104*
CH_3	C_2H_5	CH_3	C_2H_5	2,5-Dimethyl-3,6-diethyl-pyrazine	*66,102*
	Undetermined			Dimethylethylvinylpyrazine	*29*
OCH_3	*i*-C_3H_7	CH_3	CH_3O	2,6-Dimethoxy-3-isopropyl-5-methylpyrazine	*115*

a) See the caption for Table II–2; *6* cooked beef; *7* shallow fried beef; *10* beef fats; *22* carrot, parsnip, beetroot, silver-beet; *23* green peas; *29* baked potatoes; *30* boiled potatoes; *32* dehydrated potatoes; *33* freeze-dried potato peel; *34* potato chips; *35* tomatoes; *36* bell-, green and red peppers; *44* leek; *65* roasted almonds; *75* fermented soy beans; *80* roasted oats; *93* casein; *97* 'Gruyère de Comté'; *101b* Emmental cheese; *103* chocolate; *107* whisky; *111–113* tobacco; *115* galbanum oil; *116* Petitgrain Paraguay; *123* molasses fusel oil; *142 Glycyrrhiza glabra*.

isobutylpyrazine were previously reported in this oil by Bramwell *et al.* [12]. Owing to their extremely low odour thresholds, alkoxypyrazines contribute to the characteristic aroma associated with galbanum oil. Duprey & James [45] reported the presence of 2-methoxy-3-isopropyl-, 2-methoxy-3-isobutyl-, and 2-methoxy-3-isopropyl-5-methylpyrazines in the oil of Petit Grain. The latter compound was also found in absinth, rosemary, clary sage, angelica root, coriander seed, carrot seed, and parsley seed [17].

Six alkylpyrazines were reported in 'Emmental cheese' [190] and 2-ethyl-3-methoxypyrazine occurs as an off-flavour in 'Gruyère de Comté' cheese [44]. Methyl-, 2,5-, and 2,6-dimethyl- and trimethylpyrazines are predominant in rum and whisky [236], but only 2-methyl-6-vinylpyrazine was found in rum.

2-Acetylpyrazine and its 3-, 5- and 6-methyl-, ethyl-, and dimethyl homologues were identified mainly in roasted products including peanuts [221], filberts [90], almonds [197b], sesame seed [113c], popcorn [220], coffee [225a]... They possess a characteristic roasted note reminiscent of popcorn [168]. Surprisingly, no acetylpyrazines were reported in roasted pecans [224].

2-(2′-Furyl)-pyrazine, identified for the first time in coffee flavour [64] seems to be widely distributed in cigarette smoke [132] and roasted foodstuffs [90,113c,128a,197a,198,216b,237].

Organoleptic properties and threshold values of some mono- and polysubstituted pyrazines were reported by several authors [15a,32,36,92b,144a,157b, 159,180]. But some published odour descriptions remain vague. For example, terms like nutty, roasted, green, earthy, vegetable, fruity, were frequently used. By comparing the odour description of variously substituted pyrazines, Takken *et al.* [198] showed that methoxyalkylpyrazines have green odours, while tetrasubstituted derivatives have a more bell pepper-like character than the corresponding trisubstituted ones. Replacement of a methoxy group in 2-methoxy-3-isobutylpyrazine by a methyl group has a small influence on the odour character. However, a change to methylthio causes sharp alteration of the odour properties, and threshold values were higher than that of the corresponding 2-methoxy derivatives. As in other series, it was pointed out that the most potent odorant was found when both bulky substituents are on the same side of the molecule [92b,180].

Pyrazines represent 4% of aroma compounds identical to natural substances used as flavouring agents, and patent literature clearly shows their importance as flavour constituents or perfume components [152].

Thus pyrazinium salts can be incorporated into the food to be flavoured in the same manner as in the free base form. For example, the hydrochloric acid salt of 2-methoxy-3-ethylpyrazine added to foods imparts a potato flavour [69b]. 2-Acetylpyrazine and its 5- or 6-methyl derivatives can be used to flavour or season snack foods such as potato or corn chips, puffed products and crackers, cereals, and processed cereals such as popcorn, wheat and flour [168]. 2-Acetyl-3-ethylpyrazine with an odour and taste reminiscent of slightly roasted potatoes is mentioned as seasoning or flavouring [119]. Addition of certain pyrazines (2,5-dimethyl-, methyl-) is effective for peanut butter and fried chicken flavour [131b]. 2-Ethyl-3,6- (or 3,5-) dimethylpyrazines were recommended in non-alcoholic beverages, ice-cream, ices, candy, gelatin, and pudding [92b]. Tetramethylpyrazine is a flavouring compound of bread, biscuits, roasted meats, cocoa, chocolate, coffee, tobacco... By mixing it with carbonyl compounds such as vanillin, or aliphatic aldehydes, it imparts a specific aroma to cocoa, chocolate, and coffee [121]. A mixture of pyrazines (tetramethyl-, 2-ethyl-3,6-dimethyl-, 2-ethoxy-3-methyl-, and 2-methyl-5- (or -6)-methoxy) improved instant coffee flavour described as an earthy roasted nut [159]. A mixture of aldehydes (2,4-nonadienal), pyrazines (2,5-dimethyl-3-ethyl-, tetramethyl-) and pyroligneous acid, is added to vegetable oil to give the flavour and aroma of lard. The flavour of roasted hazelnuts, peanuts, or almonds can be produced by the incorporation of certain pyrazines (2-ethoxy-3-methyl-, 2-methyl-3-methoxy-, 2-methyl-5 (or -6) methoxy) into ice-cream, milk, puddings, and fondants [231]. Certain pyrazines (2,6-dimethyl-, methyl-, acetyl-, 2,5-dimethyl- and methyl-pyrazinoate) blended with sulphides, phenols, and aldehydes give desirable chocolate flavours and aromas [131c]. The green flavours of 2-methoxy-3-alkylpyrazines were found to add desirable coffee notes [144b,144c]. 2-Ethoxy-6-methylpyrazine exhibits strong odour and flavour of fresh pineapple. It can be incorporated into liquid or solid carrier for addition to an unlimited variety of systems where a pineapple flavour is desired. 2-Acetonylpyrazine used in combinations with thiazolidines and/or cyclohexanones, imparts a toasted and burnt note to foodstuffs [230].

Several classes of pyrazines (alkyl-, alkadienyl-, ethers, alcohols, polyols, carbonyl and sulphur derived) were recommended either in various perfume compositions or as tobacco flavourings. But by far the most extensive pyrazine flavour related patents were granted to Winter *et al.* [230,231,233] showing their exceptional organoleptic properties.

3.5 Miscellaneous: cyclic polysulphides and macrocyclic lactones

A series of cyclic polysulphides **132** to **136** (*see Fig. 14*) was identified in reaction models, mushrooms and/or red algae [137b].

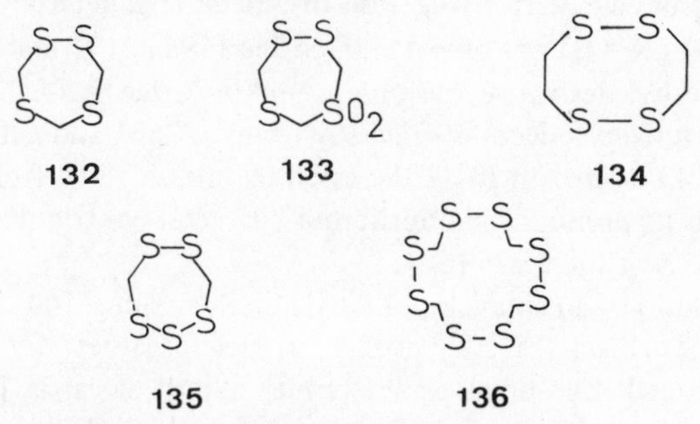

Fig. 14 – Cyclic polysulphides in food flavours.

1.2.4.6-Tetrathiepane **132** and its 4,4-dioxide **133** were characterized in the red algae *Chondria california*. The compound **132** was also found in mushrooms and in the reaction products of furfural with hydrogen sulphide and ammonia [184a]. 1,2,5,6-Tetrathiocane **134** was isolated from a steam-distillate of 'Chenin blanc' grape leaves. 1,2,3,5,6-Pentathiepane or lenthionine **135** was first identified in an edible mushroom highly appreciated in Asian countries. It possesses the typical mushroom odour and taste. Lenthionine **135** was also found in red algae, as well as the twelve-membered heterocyclic compound **136** containing eight sulphur atoms.

Macrocyclic lactones **137** to **140** (*see Fig. 15*) occur in some essential oils [137a]. The mushroom-like odour of the angelica root oil is attributed to the γ-lactone **137b** (n = 12). Three additional γ-lactones **137** (n = 10 and 14), **138** were also found in this oil. The lactone **139** was reported in Galbanum oil and ambrettolide **140** in ambrette seed oil. This latter compound is recommended for flavouring ice-cream, candy, pudding, and chewing-gum.

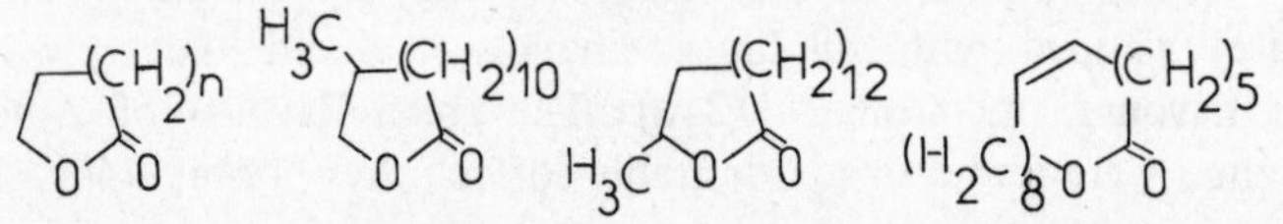

$(CH_2)_n$ H_3C $(CH_2)_{10}$ $(CH_2)_{12}$ H_3C $(CH_2)_5$ $(H_2C)_8$

137 **138** **139** **140**

(a): n = 10, (b): n = 12, (c): n = 14

Fig. 15 – Macrocyclic lactones in food flavours.

II–4 FUSED RING HETEROCYCLIC COMPOUNDS

More than 65 fused ring *N*-heterocyclic compounds were reported in tobacco and/or tobacco smoke [170], but bicyclic fused heterocyclic compounds derived from five- and six-membered heterocyclic systems also occur in some food systems and possess characteristic flavours.

4.1 Fused heterocyclic derivatives from five-membered heterocyclic compounds

Benzofuran **141** (R = H) was reported in coffee [194a,217] and in whey powder [53a]. Its 2-methyl derivative was only found in coffee. 2,3-Dihydrobenzofuran **142** was characterized in deep fat-fried soy beans [226b] and coffee [194a,217]. Menthofuran **143** is present in all the essential oils obtained from *Mentha piperata L.* in which its pronounced odoriferous character contributes to the pleasant flavour properties of these oils [68].

2,2,6-Trimethyl-7-oxabicyclo [4.3.0] non-9-en-8-one **144** or dihydroactinidiolide is present in black tea, cranberries, currants, tomatoes, *Actinidia polygama*, tobacco, crispbread, rice bran, passion fruit, and oil of cassia [108,137a]. The corresponding 4-keto derivative **145** was also found in tobacco [39]. These two

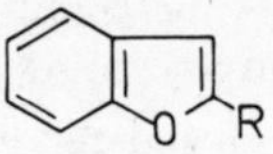

O R
141

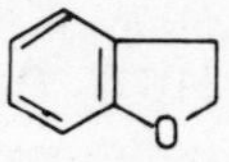

O
142

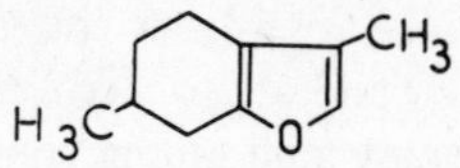

CH_3 H_3C O
143

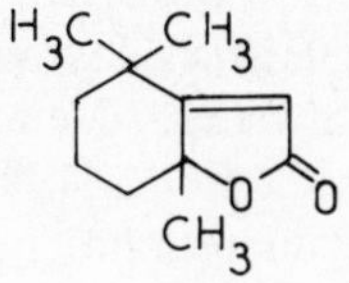

H_3C CH_3 O O CH_3
144

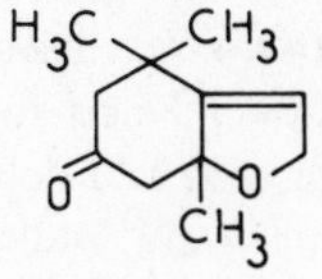

H_3C CH_3 O O CH_3
145

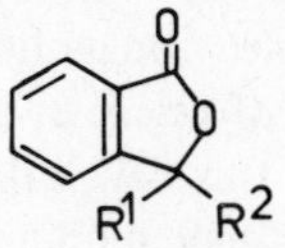

O O R^1 R^2
146

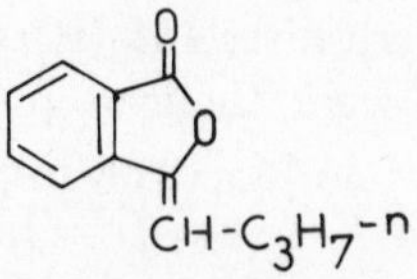

O O $CH{-}C_3H_7{-}n$
147

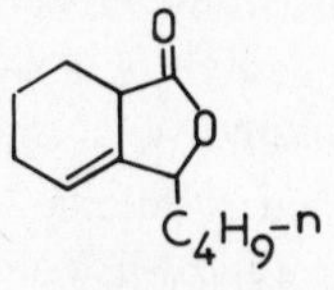

O O $C_4H_9{-}n$
148

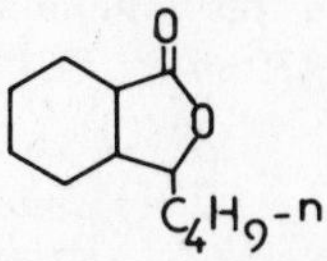

O O $C_4H_9{-}n$
149

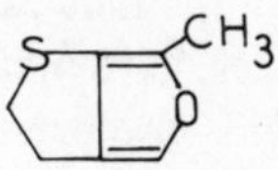

S CH_3 O
150

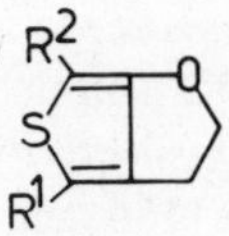

R^2 O S R^1
151

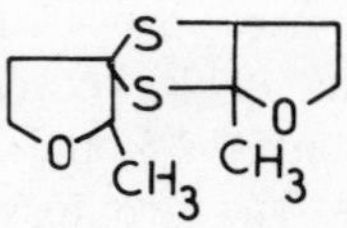

S S O O CH_3 CH_3
152

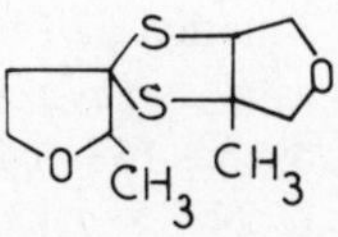

S S O O CH_3 CH_3
153

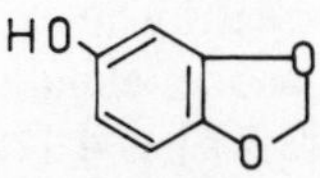

HO O O
154

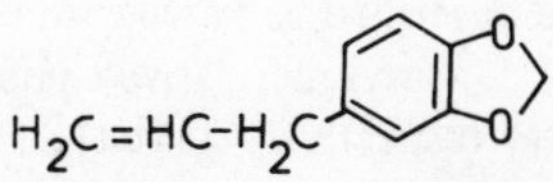

$H_2C{=}HC{-}H_2C$ O O
155

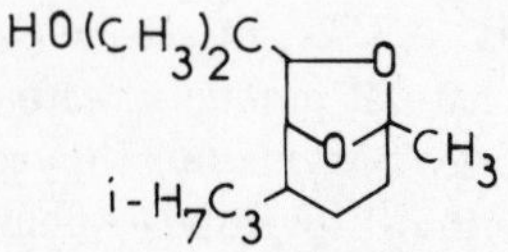

$HO(CH_3)_2C$ O O CH_3 $i{-}H_7C_3$
156

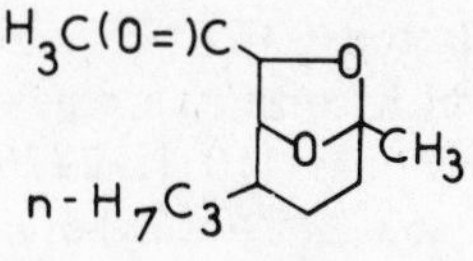

$H_3C(O{=})C$ O O CH_3 $n{-}H_7C_3$
157

Fig. 16 – Fused heterocyclic derivatives from furan and dioxolane.

products arise from enzymatic oxydation of β-carotene or its metabolites (β-ionone) and are tobacco metabolites [169]. The compound **145** is described as a pleasant mild aroma compound. Bicyclic γ-lactones or 3-alkyl- and alkylidene phthalides **146** and **147** occur in spice plants of certain *Umbelliferae* species. They constitute outstanding aroma components [68]. 3-Butylphthalide **146** (R^1 = H, R^2 = n-C_4H_9) and 3-butylidenephthalide **147** were identified in the essential oils of celery and lovage [81,136], as well as their more or less saturated derivatives, sedanolides **148** and **149** which have the characteristic odour and flavour of celery at a dose weaker than 1 part for 10^7 parts of water. 2-Methyl-3-oxa-8-thiabicyclo [3.3.0]-1,4-octadiene **150** isolated from coffee aroma [194a] has a violent sulphur odour in a pure state, while in a highly dilute solution, it develops a pleasant roasted or smoky odour. It is useful to improve or modify flavours or flavouring compositions such as nut, pecan, pistachio, candy, chocolate, cocoa, coffee, burnt sugar, roasted cereal, meat, and spice flavours [66]. When this compound was replaced by 2-oxa-7-thiabicyclo [3.3.0]-5,8-octadiene **151** (R^1 = R^2 = H) and its 8-methyl- or 6,8-dimethyl derivatives **151** (R^1 = H, CH_3, R^2 = CH_3), toasted, roasted, and smoky notes were obtained [66b]. 2′,3-(and 2′-4)-Dimethyl-3′,4-(and 3′5)dioxa-2,8-dithiabicyclo [3.3.0] octanspiro-cyclopentane **152** and **153** have a meat-like aroma [219].

Piperonal **154** is an odorous component of several essential oils [68]. 1-Allyl-3,4-methylene dioxybenzene **155** or safrole was identified in cinnamon, cocoa, mace, pepper, nutmeg, and anis seed [57]. Related compounds such as myristin, apiole, and dillapiole identified principally from spices, have methoxy groups in 5-, 2,5- and 2,3-positions, respectively. This skeleton is also present in piperin. The two bicyclic dioxolans **156** and **157** were reported in 'Burley' tobacco [40].

Benzothiophene **158** (R = H) was identified in peas [186] and coffee [194a,222]. Its 6-methyl derivative was found in milk caramel. Benzothiophene was described as having rubbery and earthy notes, while its 2-methyl-, 2-formyl- and 2-acetyl derivatives possess phenolic, almond-caramel, and earthy-roasted notes respectively [196a]. Thieno [3,2-b] and [2,3-b] thiophenes **159** and **160** were found to be present in coffee aroma [194a]. Thieno [3,2-b] and [3,4-b] thiophenes **159** and **161** were isolated from a model reaction between ribose and a cysteine/cystine mixture (*see Chapter III*).

Pyrrole lactones **162** (R^1 = H, CHO, R^2 = CH_3, i-C_3H_7, n-C_4H_9, i-C_4H_9, $C_6H_5CH_2$) which constitute a new class of natural products, were found to be tobacco flavourings [103]. 2-(2′-Formyl-5-hydroxymethyl-1′-pyrrolyl) acetic acid lactone and 2-(2′-formyl-5′-hydroxymethyl-1′-pyrrolyl)-3-phenylpropionic acid lactone were identified in flue cured tobacco [41]. These formyl derivatives improve the taste and/or smell of tobacco products by giving them a spicy peppery flavour. Bicyclic compounds **163** (R = H, CH_3) and **164** (R^1 = $COCH_3$ or CHO; R^2 = H or CH_3) were isolated from a proline/glucose model reaction (*see Chapter III*).

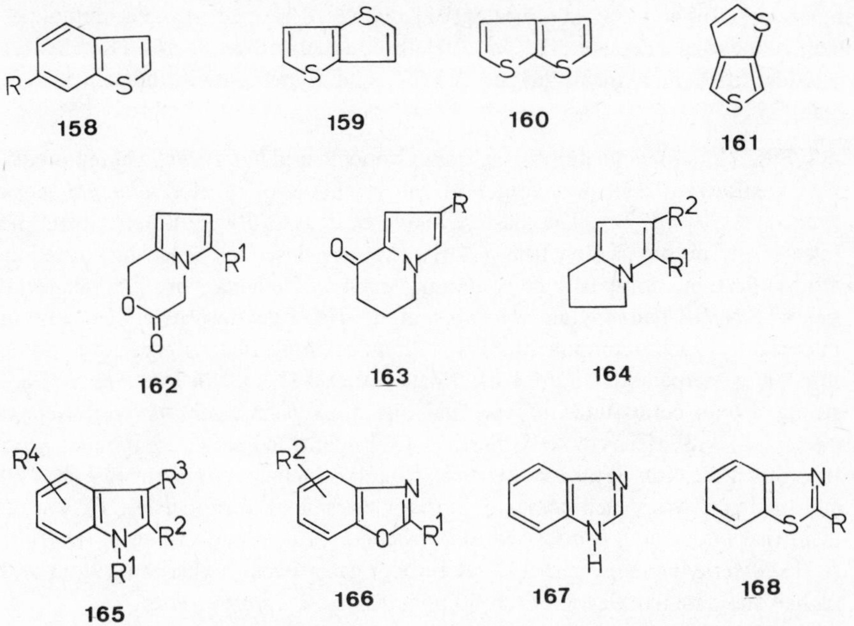

Fig. 17 – Fused heterocyclic derivatives from thiophene, pyrrole, oxazole, imidazole and thiazole.

The importance of indole **165** ($R^1 = R^2 = R^3 = R^4 = H$) and skatole **165** ($R^1 = R^2 = R^4 = H$, $R^3 = CH_3$) in the perfume industry is well known. Indole occurs naturally in numerous essential oils: jasmine, neroli, tuberose, ylang-ylang, *Celtis reticulosa* [68]. It possesses a jasmine and faeces-like odour. It was identified in the aroma compounds of some foodstuffs such as cauliflower, rice bran, roasted filberts. . . Different *N*- and *C*- alkyl substituted indoles and dimethyl indole isomers were reported in tobacco and/or tobacco smoke [170]. Skatole is the most flavourful compound with this skeleton. It was isolated from fish, tea, shrimp paste, meat, dry milk, rice bran, *Celtis reticulosa*, and civet.

Benzoxazole **166** ($R^1 = R^2 = H$) was identified in the black tea aroma [216a] and its 2-methyl derivative was found in meat flavour [137b] and in roasted coffee [215a]. Some dimethylbenzoxazoles **166** ($R^1 = CH_3$, R^2 = 4-, 5- and 6-CH_3) were also present in coffee aroma [215a]. Benzimidazole **167** was identified in tobacco [170]. Benzothiazole **168** (R = H) is certainly the most widespread aroma compound of this series. It has been identified in the following food systems: canned beef stew [149], cooked pork liver [235], broccoli [22], cooked asparagus [208], potatoes [15b], leek [175], cranberry [1], roasted filberts [90], roasted peanuts [221], whole-fat soy bean milk [254], butter [61], casein, Swiss cheese, cocoa [218], black tea [216a], beer, tobacco [39], and coconut meal [101]. It was described as being quinoline-like, rubbery in character [157b]. 2-Methylthiobenzothiazole **168** (R = SCH_3) was identified in a meat flavour [137b], and 2-methylbenzothiazole in rice bran [211].

4.2 Fused heterocyclic derivatives from six-membered heterocyclic compounds

Cyclopentapyran **169** was found in the essential oil of *Actinidia polygama Franch. & Sav.* [137a]. The diastereomeric edulans **170** were isolated from the juice of the purple passion fruit [126]. They are described as having a rose-like aroma. Perfume compositions containing 2,3,4,4a,5,6-hexahydro-2,5,5,8a-tetra-methyl-8a-(*H*)-1-benzopyran were patented. The oxo derivative **171** was an important tobacco component [39] with an oriental tobacco type note. Some fused monoterpenoid lactones of iridane series **171** to **176** (*see Fig. 18*) are strong aroma compounds of essential oils of *Nepeta cataria L.* and *Nepeta mussini L.* [137a]. Demole & Berthet [39] identified some compounds with the edulan skeleton in tobacco flavour. This class includes coumarin **177** (R = H) and its derivatives which possess important physiological properties. They were identified in cocoa, green tea, whisky, raspberry, cassis, and various citrus fruits [88]. 6-Methylbenzopyrone **177** (R = CH_3) has a sweet herbaceous odour and fig-like and date-like flavour. Over 50 ppm it possesses a bitter taste.

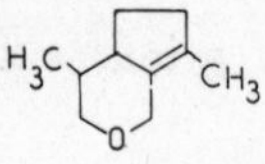
169

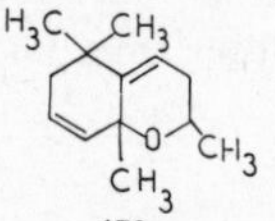
170

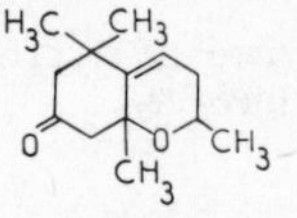
171

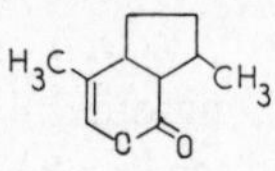
172

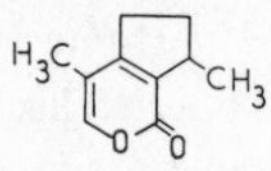
173

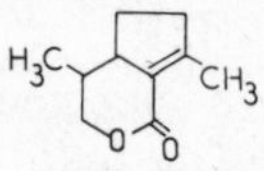
174

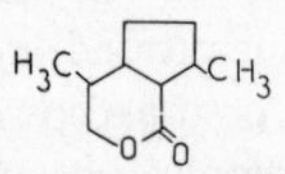
175

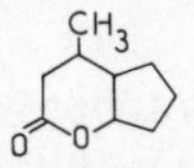
176

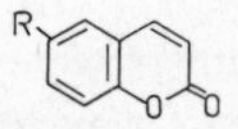
177

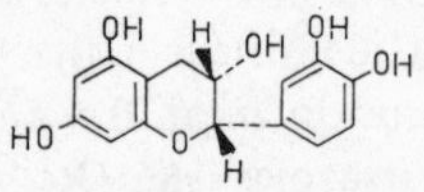
178

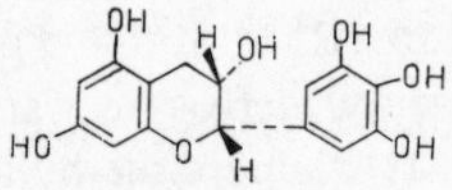
179

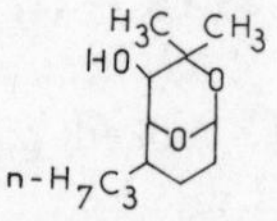
180

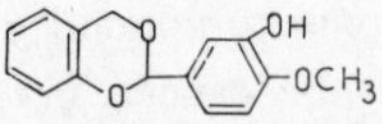
181

Fig. 18. – Fused heterocyclic derivatives from pyrans and dioxanes.

Flavanols of tea such as (-) epicatechin **178** and (-) epigallocatechin **179** which by oxidative polymerisation give theaflavin, exhibit benzopyran skeletons.

The fused 1,3-dioxane **180** was identified in 'Burley' tobacco, and the compound **181** is described as having a very sweet flavour similar to that of phyllodulcine.

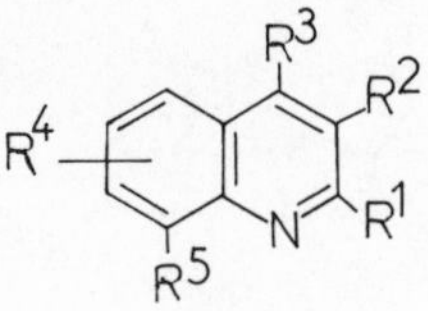

182 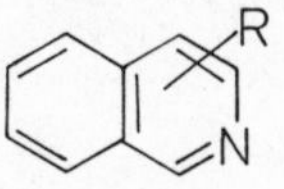 183 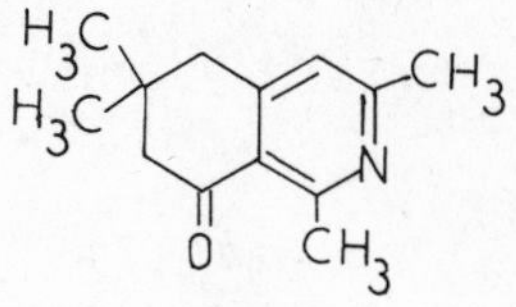 184

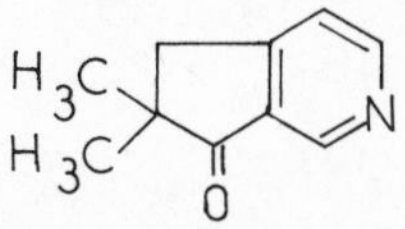

185 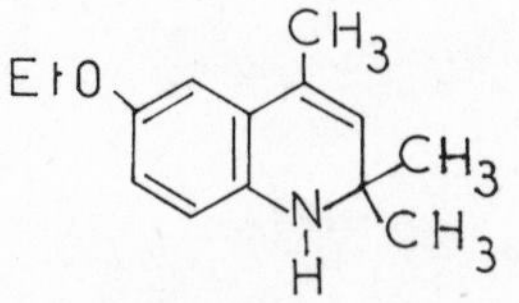 186 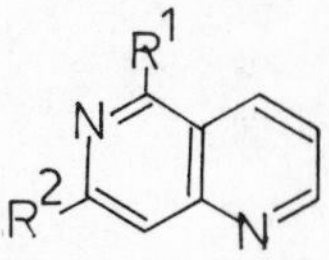 187

Fig. 19 – Fused heterocyclic derivatives from pyridine.

Quinoline **182** ($R^1 = R^2 = R^3 = R^4 = H$), its methyl- and dimethyl derivatives were found in the following aromas: tea [216a], rice bran [211], tobacco [39, 132], whisky, and fusel oils [57].

Isoquinoline **183** (R = H) was also reported in tobacco smoke [132]. Some quinoline and isoquinoline derivatives possess reminiscent benzaldehyde-like and anise-like odours. They have the same use as furans, principally in non-alcoholic beverages [71]. 1,3,6,6-Tetramethyl-5,6,7,8-tetrahydro-8-isoquinolinone **184** and 3,6,6-trimethyl-5,6-dihydro-7-pyridinone **185** were identified among the components of 'Burley' tobacco. 6-Ethoxy-2,2,4-trimethyl-1,2-dihydroquinoline **186** decreases the amount of nitrosamines in cooked meats. Some nitrogen-containing bicyclic compounds and their use as aroma ingredients were patented by Flament [60]. Thus 1,6-naphthyridine **187** ($R^1 = R^2 = H$) found in a meat aroma and its 5-methyl derivative **187** ($R^1 = CH_3$, $R^2 = H$) had a burnt meat taste. When R^1 and R^2 are lower alkyl groups, the corresponding compounds have flavours resembling food.

6,7-Dihydro-(5*H*)-cyclopenta (b)pyrazine **188** ($R^1 = R^2 = R^3 = R^4 = H$) and its 2-methyl, 2-ethyl-, 5-methyl-, 2,3-dimethyl- 2, (or 3) 5-dimethyl-, 2,3,6-trimethyl-5-hydroxy- and 2-acetyl-3-hydroxy-5-methyl derivatives were mainly identified in cooked or roasted products such as beef [127], pork liver [128b], asparagus [208], potatoes [32], almonds [197a], filberts [90], peanuts [221], sesame seed [91,113c], cocoa [216b], coffee [217], green tea [237] and beer [72]. Other bicyclic pyrazine derivatives **189** to **196** were also reported in food flavours or as flavouring products. Some of them have roasted, grilled, burnt, and animal notes interesting for the aromatization of meat and tobacco products [137b].

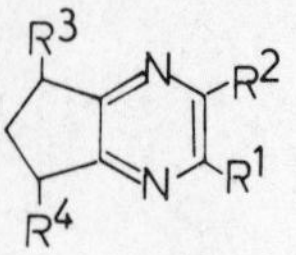

188

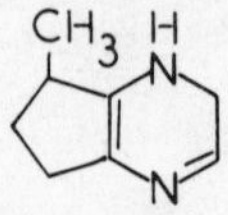

189

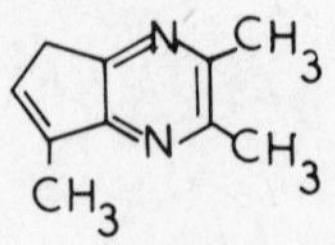

190

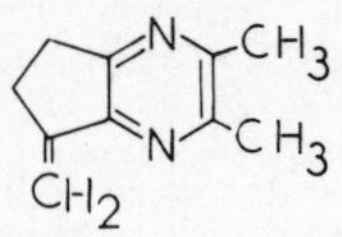

191

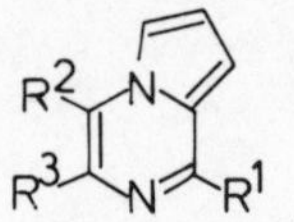

192

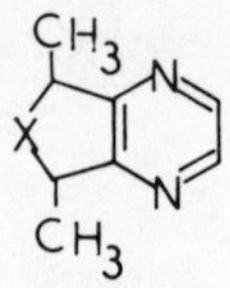

193

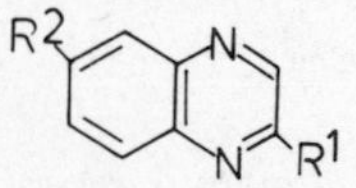

194

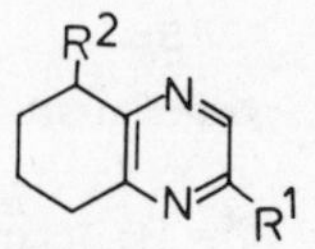

195

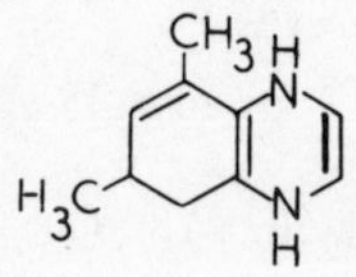

196

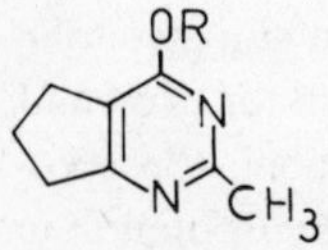

197

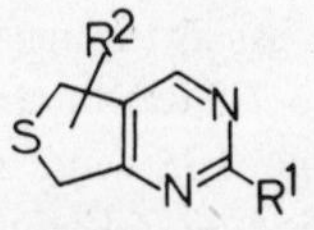

198

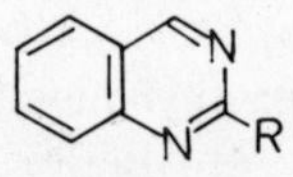

199

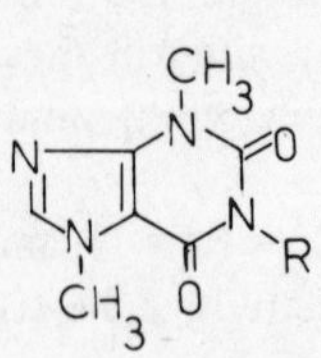

200

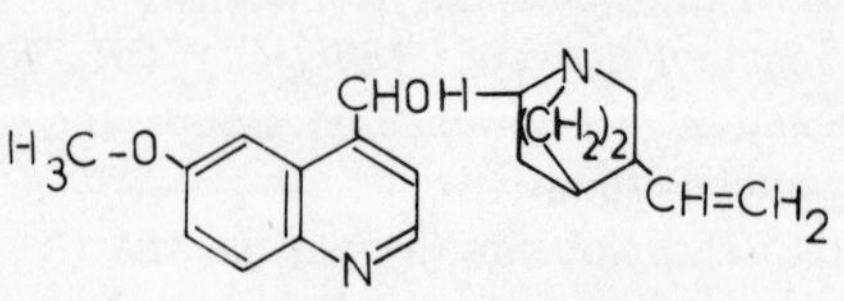

201

Fig. 20 – Fused bicyclic pyrazines and pyrimidines in food flavours.

Relatively few pyrimidine derivatives were found in food flavours. Among non-fused ring derivatives, 4,6-dimethylpyrimidine was identified among the basic compounds of meat aroma [58], and 4-acetyl-2-methylpyrimidine possesses a very interesting roasted note. This latter compound is detectable at a dose of 0.5 ppm [58]. 6,7-Dihydro-(5*H*)-2-methylcyclopenta [d] pyrimidines **197** (R = CH_3, *i*-C_4H_9) impart meaty and nut-like flavours to food products [48]. Dihydrothienopyrimidines **198** with one or more substituted alkyl group on the sulphur ring were patented. Thus, 2-methyl-5,7-dihydrothieno [3,4-d] pyrimidine, in aqueous solution at 1 ppb, has a sweet roasted nut, popcorn character with a light astringency [152]. At 0.03 parts per million in water, it has a popcorn, roasted peanut flavour with a peanut skins-like astringency. At 10^{-6} ppm in chicken broth it accentuates the root vegetables flavour notes.

A chocolate-like flavour note material is prepared by mixing 65 parts of the thienopyrimidine **198** (R^1 = R^2 = H) with 35 parts of 2,3,5,6-tetramethylpyrazine. 2-Methylquinazoline **199** (R = CH_3) was found in dried mushrooms [202]. A chemical widely consumed in coffee and tea is the stimulant caffeine **200** (R = CH_3), a natural constituent of coffee beans and tea leaves. The caffeine content of tea (8.6%) is considerably higher than that of coffee. It is used in non-alcoholic beverages [71] as well as quinine **201**. Theobromine **200** (R = H) and caffeine impart the bitter taste and stimulating effect of cocoa.

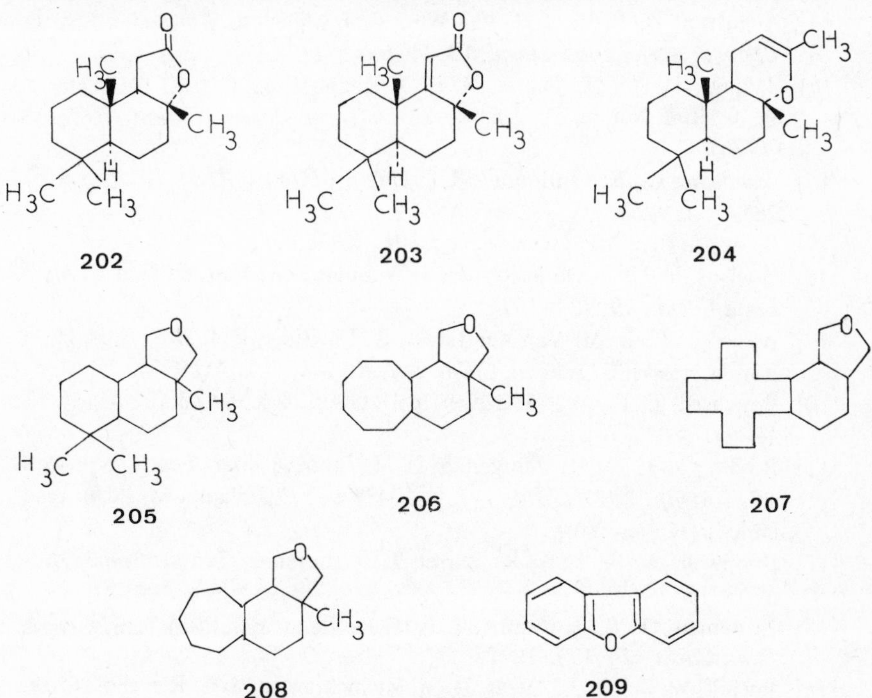

Fig. 21 – Polycyclic oxygen-containing heterocyclic systems.

4.3 Polyfused heterocyclic compounds

Owing to their low volatility, polycyclic fused heterocyclic systems are not very frequent in food aromas. Fused furans such as *Nor*-ambreinolide **202**, its dehydro derivative **203**, and sclareol oxide **204** considered to be the breakdown flavours from labdane diterpenoids, were exclusively found in oriental type tobaccos. They impart a pleasant cedar-like odour [137a].

Compounds **205** to **208** were described as having amber type odours. Dibenzofuran **209** was identified in cranberry flavour [1]. Some tricyclic pyrazines were patented for nut flavourings [155]. But by far the greatest number of polyfused heterocyclic systems were found in tobacco [170]. Among them, the most prevalent ones are carbazole, its methyl-, dimethyl- and *N*-alkyl derivatives. Other polyfused ring compounds such as acridans, dibenzo-carbazoles, and acridines, benzoquinolines etc... were also identified. These compounds were found to be carcinogenic in experimental animals.

II–5 REFERENCES

[1] Anjou, K. & E. Von Sydow, *Acta Chem. Scand.*, **21**, 945, 2076 (1967a); *ibid.*, **23**, 109 (1969b).

[2] Arnold, R. G., L. M. Libbey & E. A. Day, *J. Food Sci.*, **31**, 566 (1966).

[3] Askar, A. & H. J. Bielig, *Alimenta,* **15**, 3 (1976).

[4] Augustyn, O. P. H., C. J. Van Wijk, C. J. Miller, R. E. Kepner & A. D. Webb, *J. Agric. Food Chem.*, **19**, 1128 (1971).

[5] Badings, H. T., H. Maarse, R. C. J. Kleipool, A. C. Tas, R. Neeter & M. C. Ten Noever De Brauw, *Z. Lebensm.-Unters. Forsch.*, **161**, 53 (1976).

[6] Bayanove, C., R. Cordonnier & P. Dubois, *Compt. Rend. Acad. Sci.*, D, **281**, 75 (1975).

[7] Bentz, A. P. & J. F. Mezzino, *U.S. Pat.*, 3,666,494.

[8] Boelens, M., P. J. De Valois, H. J. Wobben & A. Van der Gen, *J. Agric. Food Chem.*, **19**, 984 (1971).

[9] Boelens, M., L. M. Van der Linde, R. De Rijke, P. J. De Valois, H. M. Van Dort & H. C. Takken, *Chem. Soc. Review*, **7**, 167 (1978).

[10] Bonchard, E. F., C. P. Hetzel & R. D. Olson, *U.S. Pat.*, 3,409,441 (5.11. 1968).

[11] Boyko, A. L., M. E. Morgan & L. M. Libbey, *Anal. Foods Beverages, Proc. Symp. 1977 (Publ. 1978)*, 57–58 (G. Charalambous, Ed.); Academic Press, New York.

[12] Bramwell, A. F., J. W. K. Burrel & G. Riezebos, *Tetrahedron Letters,* **37**, 3215 (1969).

[13] Brinkman, H. W., H. Copier, J. J. M. De Leuw & S. Boen Tjan, *J. Agric. Food Chem.*, **20**, 177 (1972).

[14] Burrell, W. K., R. A. Lucas, D. M. Michalkiewiczs & G. Riezebos, *Chem. Ind., (London)*, **1970**, 1409.

[15] Buttery, R. G., R. M. Seifert, D. G. Guadagni & L. C. Ling, *J. Agric. Food Chem.*, **17**, 1322 (1969a); *ibid.*, **18**, 538 (1970b).
[16] Buttery, R. G., R. M. Seifert, D. G. Guadagni & L. C. Ling, *J. Agric. Food Chem.*, **19**, 524 (1971).
[17] Buttery, R. G., R. M. Seifert, D. G. Guadagni & L. C. Ling, *J. Agric. Food Chem.*, **19**, 969 (1971).
[18] Buttery, R. G., D. G. Guadagni & L. C. Ling, *J. Sci. Food Agric.*, **24**, 1125 (1973).
[19] Buttery, R. G. & L. C. Ling, *J. Agric. Food Chem.*, **22**, 912 (1974).
[20] Buttery, R. G., R. M. Seifert & L. C. Ling, *J. Agric. Food Chem.*, **23**, 516 (1975).
[21] Buttery, R. G., D. G. Guadagni & R. E. Lundin, *J. Agric. Food Chem.*, **24**, 1 (1976).
[22] Buttery, R. G., D. G. Guadagni, L. C. Ling, R. M. Seifert & W. Lipton, *J. Agric. Food Chem.*, **24**, 829 (1976).
[23] Buttery, R. G., D. G. Guadagni & L. C. Ling, *J. Agric. Food Chem.*, **26**, 791 (1978).
[24] Card, A. & C. Avise, *Ann. Technol. Agric.*, **26**, 287 (1977).
[25] Carson, J. F., *See Ref.* 177, p. 390.
[26] Cazenave, P., I. Horman, F. Müggler-Chavan & R. Viani, *Helv. Chim. Acta*, **57**, 206 (1974); *ibid.*, p. 209.
[27] Chang, S. S., C. Hirai, B. R. Reddy, K. O. Herz, A. Kato & G. Sipma, *Chem. Ind. (London)*, **1968**, 1639.
[28] Chang, S. S. & W. A. May, *U.S. Pat.*, 3,767,427(23.10.1976a); Chang, S. S. *et al.*, *J. Agric. Food Chem.*, **25**, 450 (1977b); *ibid.*, *J. Food Sci.*, **42**, 298 (1977c).
[29] Charalambous, G. & I. Katz, Eds., *Phenolic, Sulfur and Nitrogen compounds in food flavors, ACS Symposium series* No. **26**, Washington (1976).
[30] Charalambous, G. & G. E. Inglett, Eds., *Flavor of Foods and Beverages; Chem. Technol. (Proc. conf.)*, Academic Press, New York (1978).
[31] Chida, N., *Shokukin to Kagaku.*, **20**, 96–101 (1978).
[32] Coleman, E. C. & Chi Tang Ho, *J. Agric. Food Chem.*, **28**, 66 (1980).
[33] Coppier, H. & C. H. T. Tonsbeek, *U.S. Pat.*, 3,678,064 (18.7.1972); *idem*, 3,778,518 (11.12.1973).
[34] Crain, W. O. & C. S. Tang, *J. Food Sci.*, **40**, 207 (1975).
[35] Cronin, D. A. & P. Stanton, *J. Sci. Food Agric.*, **27**, 145 (1976).
[36] Deck, R. E. & S. S. Chang, *Chem. Ind. (London)*, **1965**, 1343.
[37] Deck, R. E., J. Pokorny & S. S. Chang, *J. Food Sci.*, **38**, 345 (1973).
[38] Dembele, S. & P. Dubois, *Ann. Technol. Agric.*, **22**, 121 (1973).
[39] Demole, E. & D. Berthet, *Helv. Chim. Acta*, **55**, 1866 (1972).
[40] Demole, E. & C. Demole, *Helv. Chim. Acta*, **58**, 523 , 1867 (1975).

[41] Dickerson, J. P., D. L. Roberts, C. W. Miller, R. A. Lloyd & C. E. Rix, *Tobacco*, **20**, 71 (1976).

[42] Dietrich, P. & E. Sundt, *Swiss Pat.*, 483,415 (30.1.1970a); *Idem, U.S. Pat.*, 3,884,247 (20.9.1975b).

[43] Dubois, P., *Sci. Techn.*, **47**, 29 (1978).

[44] Dumont, J. P., S. Roger & J. Adda, *Le Lait*, **55**, 479 (1975); *ibid.*, **56**, 18 (1976).

[45] Duprey, R. J. H. & J. F. James, *Amer. Perf. Cosm.*, **86**, 53 (1971).

[46] Eisenbrand, G., C. Janzowski & R. Preussmann, *Proc. Intern. Symposium Nitrite Meat Products; 2nd*, 1976, 155 (Publ. 1977) (B. Tinbergen & B. Krol, Eds.), **1977**.

[47] Evers, W. J., *U.S. Pat.*, 3,666,495 (30.5.1972).

[48] Evers, W. J. & M. H. Vock, *U.S. Pat.*, 3,843,804 (22.10.1974).

[49] Evers, W. J., H. H. Heinsohn, B. J. Mayers & A. Sanderson, *See Ref.* 29, p. 184.

[50] Fedeli, E., *Riv. Ital. Sost. Gras.*, **54**, 202 (1977).

[51] Feldman, J. R. & J. H. Berg, *U.S. Pat.*, 3,803,330 (9.4.1974).

[52] *Fenaroli's Handbook of Flavor Ingredients* (T. E. Furia & N. Bellanca, Eds.); Publ. The Chemical Rubber Co. (1971a); vol. 2, CRC Press, Cleveland (1975b).

[53] Ferretti, A. & V. P. Flanagan, *J. Dairy Sci.*, **54**, 1764, 1769 (1971a,b).

[54] Ferretti, A. & V. P. Flanagan, *J. Agric. Food Chem.*, **20**, 695 (1972).

[55] Flament, I. & G. Ohloff, *Helv. Chim. Acta*, **54**, *1911 (1971).*

[56] Flament, I., *U.S. Pat.*, 3,881,025 (29.4.1975); *Swiss Pat.*, 559,518 (31.1.1975).

[57] Flament, I., *Proc. Int. Symp. Aroma Research Zeist*, Pudoc, Wageningen (1975), pp. 221–237.

[58] Flament, I., B. Willhalm & G. Ohloff, *See Ref.* 30, p. 15.

[59] Flament, I., *Ger. Pat.*, 2,852,783 (7.5.1979).

[60] Flament, I., *Eur. Pat. Appl.*, 4,352 (03.10.1979).

[61] Forss, D. A., *Progress in the Chemistry of Fats and other Lipids*, Vol. XIII, Part 4, 177 (R. T. Holmann, Ed.), Oxford, Pergamon Press (1972).

[62] Frattini, C., C. Buchi & G. Nano, *Chem. Ind. (London)*, **59**, 522 (1977).

[63] Frattini, C., C. Buchi, C. Barettini & G. Nano, *J. Agric. Food Chem.*, **25**, 1238 (1977).

[64] Friedel, P., V. Krampl, T. Radford, J. A. Renner, F. W. Shephard & M. A. Gianturco, *J. Agric. Food Chem.*, **19**, 530 (1971).

[65] Galetto, W. G. & P. G. Hoffman, *J. Agric. Food Chem.*, **24**, 854 (1976).

[66] Gautschi, F., B. Willhalm & G. Buchi, *U.S. Pat.*, 3,652,593 (28.03.1972); *ibid.*, 3,753,738 (23.08.1973).

[67] Gianturco, M. A., *See Ref.* 177, p. 431 (1967a); Gianturco, M. A., R. E. Giggers & B. H. Ridling, *J. Agric. Food Chem.*, **22**, 758 (1974b).

[68] Gildemeister, E. & Fr. Hoffmann, *Die ätherischen Oele, 4 Aufl.*, Bd. IIId, S-622 (W. Treibs & D. Markel, Eds.), Berlin, Akademie-Verlag (1966).
[69] Guadagni, D. G., R. G. Buttery, R. M. Seifert & D. W. Venstrom, *J. Food Sci.*, **36**, 363 (1971a); *Idem, U.S. Pat.*, 3,772,039 (13.11.1973b).
[70] Gubler, B. & K. Jaeggi, *Fr. Dem.*, 2,198,948 (5.4.1974).
[71] Hall, R. L. & B. L. Oser, *Food Technol.*, **19**, 253 (1965).
[72] Harding, R. J., H. E. Nursten & J. J. Wren, *J. Sci. Food Agric.*, **28**, 225 (1977).
[73] Hetzel, D. S. & A. Torres, *U.S. Pat.*, 3,769,293 (30.10.1973); *ibid.*, 3,892,879 (1.7.1975).
[74] Hirai, C., K. O. Herz, J. Pokorny & S. S. Chang, *J. Food Sci.*, **38**, 393 (1973).
[75] Hodge, J. E., *See Ref.* 177, p. 465.
[76] Honkanen, E. & T. Pyysalo, *Z. Lebensm.-Unters. Forsch.*, **160**, 393 (1976).
[77] Hunter, I. R. & M. K. Walden, *U.S. Pat.*, 3,425,840 (4.2.1969a); *ibid.*, 3,620,771 (16.11.1971b); Hunter, C. L. K. *et al.*, *J. Food Sci.*, **39**, 900 (1974c).
[78] I.F.F. Inc., N.L. 551,150 (28.09.1979a); *idem, U.S. Pat.*, 4,192,782 (11.03.1980b); *idem, Brit. Pat.*, 1,393,189.
[79] Ito, H., *Agric. Biol. Chem.*, **40**, 827 (1976).
[80] Jennings, W. G., *See Ref.* 177, p. 419; *idem, 4th Symp. Foods,* **1967**, 419; Oregon State University.
[81] Johnson, A. E., H. E. Nursten & A. A. Williams, *Chem. Ind. (London)*, **1971**, 1212.
[82] Jones, N. R., *See Ref.* 177, p. 267.
[83] Kallio, J., *J. Food Sci.*, **41**, 555 (1976a); *ibid.*, **41**, 563 (1976b).
[84] Kastrzeva, E. & K. Karwaska, *Pr. Inst. Lab. Badaw Przem. Spozyw*, **27**, 93-102 (1977).
[85] Katz, I., R. A. Wilson & C. Giacino, *U.S. Pat.*, 3,681,088 (1.8.1972a); *ibid.*, 3,706,577 (19.12.1972b).
[86] Katz, I., R. A. Wilson, W. J. Evers, M. H. Vock & G. W. Verhoeve, *U.S. Pat.*, 3,726,692 (10.4.1973).
[87] Kazeniac, S. J. & R. M. Hall, *J. Food Sci.*, **35**, 519 (1970a); *idem, U.S. Pat.*, 3,660,112 (2.5.1972b).
[88] Kefford, J. F. & B. V. Chandler, *Advances in Food Research, Supplement 2*, p. 106, Academic Press (1970).
[89] King, B., E. Demole & A. F. Thomas, *Dos*, 2,214,540 (5.10.1972); *idem, Swiss Pat.*, 539,631.
[90] Kinlin, T. E., R. Muralidhara, A. O. Pittet, A. Sanderson & J. P. Walradt, *J. Agric. Food Chem.*, **20**, 1021 (1972).
[91] Kinoshita, S. & T. Yamanishi, *Nippon Nogei Kagaku Kaishi*, **47**, 737 (1973).

[92] Koelher, P. E., M. E. Sason & J. A. Newell, *J. Agric. Food Chem.*, **17**, 393 (1969a); *ibid., J. Food Sci.*, **36**, 816 (1971b).
[93] Kosuge, T., H. Kamiya & T. Adachi, *Yakugaku Zasshi*, **82**, 190 (1962).
[94] Kulka, K., F. Mild & F. Fischetti, *Ger. offen.*, 2,836,208 (06.3.1980).
[95] Land, D. G. & H. E. Nursten, Eds., *Progress in Flavour Research*, Applied Science Publishers Ltd., London (1979).
[96] Ledl, F., *Z. Lebensm.-Unters. Forsch.*, **157**, 38, (1975a); *ibid.*, **157**, 229, (1975b); *ibid.*, **161**, 125 (1976c).
[97] Lenselink, W., *Brit. U.K. Pat.*, 2,018,251 (17.10.1979).
[98] MacLeod, G. & B. M. Coppock, *J. Agric. Food Chem.*, **24**, 835 (1976a); *ibid.*, **25**, 113 (1977b).
[99] Liebich, H. M., W. A. Koenig & E. Bayer, *J. Chromatog. Sci.*, **8**, 527 (1970).
[100] Liebich, H. M., D. R. Douglas, A. Zlatkis, F. Mueggler-Chavan & A. Donzel, *J. Agric. Food Chem.*, **20**, 96 (1972).
[101] Lin, F. M. & W. F. Wilkens, *J. Food Sci.*, **35**, 538 (1970).
[102] Lindsay, R. C., *See Ref.* 177, p. 315.
[103] Lloyd, R. A., C. W. Miller, D. L. Roberts, J. A. Giles, J. P. Dickerson, N. H. Nelson, C. E. Rix & P. H. Ayers, *Tobacco*, **20**, 43 (1976).
[104] Loney, B. E. & R. Bassette, *J. Dairy Sci.*, **54**, 343 (1971).
[105] Maga, J. A. & C. E. Sizer, *CRC Crit. Rev. Food Technol.*, **4**, 39–115 (1973); *idem, J. Agric. Food Chem.*, **21**, 22 (1973).
[106] Maga, J. A., *See Ref.* 52, Vol. **I**, 228 (1975).
[107] Maga, J. A. & I. Katz, *CRC Crit. Rev. Food Sci. Nutr.*, **1975**, 241–270.
[108] Maga, J. A., *CRC Crit. Rev. Food Sci. Nutr.*, **8**, 1 (1976).
[109] Maga, J. A., *J. Agric. Food Chem.*, **26**, 1049 (1978).
[110] Maga, J. A. & C. E. Sizer, *Lebensm.-Wiss. Technol.*, **11**, 181 (1978).
[111] Maga, J. A., *CRC Crit. Rev. Food Sci. Nutr.*, **11**, 355 (1979a); *idem, J. Agric. Food Chem.*, **29**, 691 (1981b).
[112] Manabe, T., *Hiroshima Noggo Tanki Daigaku, Kenhgu Hokoku*, **5**, 433 (1977).
[113] Manley, C. H. & I. S. Fagerson, *J. Food Sci.*, **35**, 286 (1970a); *idem, J. Agric. Food Chem.*, **18**, 340 (1970b); Manley *et al.*, *J. Food Sci.*, **34**, 73 (1974).
[114] Marion, J. P., F. Mueggler-Chavan, R. Viani, J. Bricout, D. Reymond & R. H. Egli, *Helv. Chim. Acta*, **50**, 1509 (1967).
[115] Mason, M. E., B. Johnson & M. Hamming, *J. Agric. Food Chem.*, **14**, 454 (1966a); *ibid.*, **15**, 66 (1967b).
[116] Meigh, D. F., A. A. E. Filmer & R. Self, *Phytochemistry*, **12**, 987 (1973).
[117] Minor, L. J., A. M. Pearson, I. E. Dawson & B. S. Schweigert, *J. Food Sci.*, **30**, 686 (1965).
[118] Montedoro, G., M. Bertuccioli & F. Anichini, *See Ref.* 30, pp. 247–281.

[119] Mookherjee, B. D., M. G. Beets & A. O. Pittet, *Dos.*, 2,117,926 (14.4. 1971a); Mookherjee *et al.*, *U.S. Pat.*, 3,684,809 (15.8.1972b); *ibid.*, 3,711,482 (16.1.1973d); *ibid.*, 3,767,428 (23.10.1973e).

[120] Morita, K., *Yuki Gosei Kagaku Kyokaishi*, **35**, 375 (1977).

[121] Moroe, T., S. Hattori, A. Komatsu, A. Saito & S. Miraki, *U.S. Pat.*, 3,459,556 (5.8.1969).

[122] Moshonas, M. G. & P. E. Shaw, *J. Agric. Food Chem.*, **20**, 1029 (1972).

[123] Mulders, E. J., H. Maarse & C. Weurman, *Z. Lebensm.-Unters. Forsch.*, **150**, 68 (1972).

[124] Mulders, E. J., M. C. Ten Noever De Brauw & Van Straten, *Z. Lebensm.-Unters. Forsch.*, **150**, 306 (1973a); Mulders, E. J., *ibid.*, **151**, 310 (1973b); *ibid.*, **152**, 193 (1973c).

[125] Mulders, E. J., R. J. C. Kleipool & M. C. Ten Noever De Brauw, *Chem. Ind. (London)*, **1976**, 613.

[126] Murray, K. E., J. Shipton & F. B. Whitfield, *Austral. J. Chem.*, **25**, 1971 (1972a); *idem*, *J. Sci. Food Agric.*, **26**, 973 (1975b).

[127] Mussinan, C. J., R. A. Wilson & I. Katz, *J. Agric. Food Chem.*, **21**, 871–873 (1973).

[128] Mussinan, C. J. & J. P. Walradt, *J. Agric. Food Chem.*, **22**, 827 (1974a); *ibid.*, **22**, 539 (1974b).

[129] Mussinan, C. J., M. Vock, E. J. Shuter & A. D. Quinn, *U.S. Pat.*, 3,855,051 (20.5.1975).

[130] Mussinan, C. J., R. A. Wilson, I. Katz, A. Hruza & M. Vock, *See Ref.* 29, p. 133.

[131] Nakel, G. M., *U.S. Pat.*, 3,336,138 (15.8.1967a); *ibid.*, 3,579,353 (18.5. 1971b); *ibid.*, 3,619,210 (9.11.1971c).

[132] Neurath, G. B., *Recent Advances in Knowledge of Chemical Composition of Tobacco Smoke* in *The Chemistry of Tobacco and Tobacco Smoke* (I. Schmeltz, Ed.), Plenum Press, New York (1972).

[133] Nonaka, M., D. R. Black, E. L. Pippen, *J. Agric. Food Chem.*, **15**, 713 (1967).

[134] Numomura, N., M. Sasaki, Y. Asao & T. Yokotsuka, *Agric. Biol. Chem.*, **40**, 485, 491 (1976a); *ibid.*, **43**, 1361 (1979b).

[135] Nursten, H. E., *See Ref.* 200, p. 335a; *idem*, *Food Chem.*, **6**, 263 (1981b).

[136] Ohloff, G., *Fortschritte der Chemischen Forschung* (A. Davison, M. J. S. Dewar, K. Hafner, Eds.), Vol. 12, No. **2**, 185–251, Berlin, Heidelberg, New York, Springer (1969).

[137] Ohloff, G., *Fortsch. Chem. Org. Naturst.*, **35**, pp. 431–527 (W. Herz, H. Grisebach & G. W. Kirby, Eds.), Wien, Springer Verlag (1978a); Ohloff, G. & I. Flament, *ibid.*, **36**, 231–283 (1979b).

[138] Ohloff, G., N. Giersch & E. Demole, *Swiss Pat.*, 604560 (1978).

[139] Okumura, J., T. Yamato, I. Yajima, *Japan Kokai Tokkyo Koho*, 7,946, 857 (13.4.1979).

[140] Ouweland Van den, G. A. M. & H. G. Peer, *Ger. Pat.*, 1,932,800 (8.1.1970a); *idem, U.S. Pat.*, 3,697,295 (10.10.1972b); *idem, J. Agric. Food Chem.*, **23**, 501 (1975c).
[141] Pareless, S. R. & S. S. Chang, *J. Agric. Food Chem.*, **22**, 339 (1974).
[142] Park, R. J., K. E. Murray & G. Stanley, *Chem. Ind. (London)*, 380 (1974).
[143] Parks, O. W., *See Ref.* 177, p. 296.
[144] Parliment, T. H. & M. F. Epstein, *J. Agric. Food Chem.*, **21**, 714 (1973a); Parliment *et al., U.S. Pat.*, 3,767,425 (23.10.1973b); *ibid.*, 3,824,421 (14.7.1974c).
[145] Parliment, T. H., M. G. Kolor & Y. Maing, *J. Food Sci.*, **42**, 1592 (1977).
[146] Partipharm, A. G., *Swiss Pat.*, 441,543 (31.10.1969).
[147] Peer, H. G., G. A. M. Van den Ouweland & C. N. De Groot, *Rec. Trav. Chim. Pays-Bas*, **87**, 1011–1017 (1968).
[148] Persson, T. & E. Von Sydow, *J. Food Sci.*, **38**, 377 (1973a); *ibid.*, **39**, 406 (1974b); *ibid.*, **39**, 537 (1974c).
[149] Peterson, R. J., H. J. Izzo, E. Jungermann & S. S. Chang, *J. Food Sci.*, **40**, 948 (1975).
[150] Pfizer Inc., *U.S. Pat.*, 3,769,293 (1970).
[151] P. F. W. Beheer, *Ger. Pat.*, 2,024,696.
[152] Pintauro, N., Ed., *Food Flavoring Processes*, Noyes Data Corporation, Park Ridge, N.J. (1976).
[153] Pittet, A. O., P. Rittersbacher & R. Muralidhara, *J. Agric. Food Chem.*, **18**, 929 (1970).
[154] Pittet, A. O. & A. W. Seitz, *U.S. Pat.*, 3,687,692 (29.8.1972a); Pittet *et al., U.S. Pat.*, 3,705,158 (5.12.1972b); *ibid.*, 3,773,525 (20.11.1973c).
[155] Pittet, A. O. & R. Muralidhara, *U.S. Pat.*, 3,754,934 (28.8.1973).
[156] Pittet, A. O. & E. W. Seitz, *U.S. Pat.*, 3,782,973 (1.1.1974).
[157] Pittet, A. O. & D. E. Hruza, *U.S. Pat.*, 3,769,040 (30.10.1973a); *Ger. Pat.*, 2,312,996; *idem, J. Agric. Food Chem.*, **22**, 264 (1974b).
[158] Pittet, A. O. & E. M. Klaiber, *J. Agric. Food Chem.*, **23**, 1189 (1975).
[159] Polaks Frutal Works, *Brit. Pat.*, 1,247,829; *ibid.*, 1,248,380 (29.9.1971).
[160] Praag Van, M., H. S. Stein & M. S. Tibbetts, *J. Agric. Food Chem.*, **16**, 1005 (1968).
[161] Priestley, R. J., Ed., *Effects of Heating on Foodstuffs*, Applied Sci., London (1979).
[162] Pyysalo, T., *Z. Lebensm.-Unters. Forsch.*, **162**, 263 (1976).
[163] Qvist, I. H. & E. Von Sydow, *J. Agric. Food Chem.*, **22**, 1077 (1974a); *ibid.*, **24**, 437 (1976b).
[164] Qvist, I. H., E. Von Sydow & C. A. Akesson, *Lebensm.-Wiss. Technol.*, **9**, 311, (1976).
[165] Radecki, A., J. Grzybrowski, J. Halkiewicz & H. Lamparezyk, *Acta Aliment. Pol.*, **3**, 203 (1977).

[166] Reymond, D., *See Ref.* 200, p. 315.

[167] Rizzi, G. P., *U.S. Pat.*, 3,617,310 (2.11.1971a); *ibid.*, 3,625,710 (7.12. 1971b).

[168] Roberts, D. L., *U.S. Pat.*, 3,402,051 (17.9.1968).

[169] Roberts, D. L. & W. A. Rhode, *Tobacco*, **16**, 107 (1972).

[170] Schmeltz, I. & D. Hoffmann, *Chem. Rev.*, **77**, 295 (1977).

[171] Schreier, P. & F. Drawert, *Z. Lebensm.-Unters. Forsch.*, **154**, 273 (1974a); *idem, Geruch. Geschmakstoffe. Int. Symp.*, 1974 (Publ. 1975b), (F. Drawert, Ed.); *idem, Chem. Mikrobiol. Technol. Lebensm.*, **3**, 154 (1974c).

[172] Schreier, P., F. Drawert, Z. Kergnyi & A. Junker, *Z. Levensm.-Unters. Forsch.*, **161**, 249 (1976).

[173] Schreier, P., F. Drawert & A. Junker, *J. Agric. Food Chem.*, **24**, 331 (1976).

[174] Schreier, P., F. Drawert, A. Junker & W. Mick, *Z. Lebensm.-Unters. Forsch.*, **164**, 188 (1977).

[175] Schreyen, L., P. Dirinck, F. Van Wassenhove & N. Schamp, *J. Agric. Food Chem.*, **24**, 336 (1976).

[176] Schreyen, L., P. Dirinck, F. Van Wassenhove & N. Schamp, *J. Agric. Food Chem.*, **24**, 1147 (1976).

[177] Schultz, H. W., E. A. Day & L. M. Libbey, Eds., *The Chemistry and Physiology of Flavors*, Westport, Connecticut, The AVI Publ. Co. Inc. (1967).

[178] Schumacher, J. N., W. A. Rhode & D. L. Roberts, *U.S. Pat.*, 3,380,457 (30.4.1968).

[179] Seck, S. & J. Crouzet, *Phytochemistry*, **12**, 2925 (1973).

[180] Seifert, R. M., R. G. Buttery, D. G. Guadagni, D. R. Black & J. G. Harris, *J. Agric. Food Chem.*, **18**, 246 (1970a); *ibid.*, **20**, 135 (1972b).

[181] Shaw, P. E., J. H. Tatum, T. J. Kew, C. J. Wagner & R. E. Berry, *J. Agric. Food Chem.*, **18**, 343 (1970).

[182] Shaw, P. E. & R. E. Berry, *J. Food Sci.*, **41**, 711 (1976a); *idem, J. Agric. Food Chem.*, **25**, 641 (1977b).

[183] Sheldon, R. M., R. C. Lindsay & L. M. Libbey, *J. Food Sci.*, **37**, 313 (1972).

[184] Shibamoto, T., *J. Agric. Food Chem.*, **25**, 206 (1977a); *ibid.*, **28**, 237–243 (1980b).

[185] Shimizu, Y., S. Matsuto, Y. Mizunuma & I. Okada, *Eiyo to Shokuryo*, **23**, 276 (1970a); *idem, Agric. Biol. Chem.*, **34**, 845 (1970b).

[186] Shipton, J., F. B. Whitfield & J. H. Last, *J. Agric. Food Chem.*, **17**, 1113 (1969).

[187] Shuster, E. J., D. A. Withycombe, B. D. Mookherjee & C. S. Mussinan, *U.S. Pat.*, 4,040,987 (9.8.1977).

[188] Siek, T. J., J. A. Albin, L. A. Sather & R. C. Lindsay, *J. Dairy Sci.*, **54**, 1 (1971).
[189] Silverstein, R. M., *See Ref.* 177, p. 450.
[190] Sloot, D. & H. Hofman, *J. Agric. Food Chem.*, **23**, 358 (1975).
[191] Spencer, M. D., R. M. Pangborn & W. G. Walter, *J. Agric. Food Chem.*, **73**, 725 (1978).
[192] Stevens, M. A., *J. Amer. Soc. Hort. Sci.*, **95**, 9 (1970).
[193] Stoffelsma, J., G. Sipma, D. K. Kettenes & J. Pypker, *J. Agric. Food Chem.*, **16**, 1000 (1968a); *idem, Rec. Trav. Chim. Pays-Bas*, **87**, 241 (1968b).
[194] Stoll, M., M. Winter, F. Gautschi, I. Flament & B. Willhalm, *Helv. Chim. Acta*, **50**, 628 (1967a); Stoll *et al.*, *ibid.*, **50**, 2065 (1967b).
[195] Straten Van, S., F. De Vrijer, *Rapport N.R.R. 4030, List of Volatile Compounds in Food*, 3rd Centraal Institut voor Voedingsonderzoek Zeist, Netherlands, 3rd ed., June 1973, 4th ed., Suppl. **1**, (1977); **2** (1978).
[196] Sydow Von, E. & K. Anjou, *Lebensm.-Wiss. Technol.*, **2**, 15 (1969a); *ibid.*, **2**, 78 (1969b).
[197] Takei, Y., S. Shimada, S. Watanabe & T. Yamanishi, *Agric. Biol. Chem.*, **38**, 645 (1974a); *ibid.*, **38**, 2329 (1974b).
[198] Takken, H. J., L. M. Van Der Linde, M. Boelens & J. M. Van Dort, *J. Agric. Food Chem.*, **23**, 638 (1975).
[199] Tatum, J. H., P. E. Shaw & R. E. Berry, *J. Agric. Food Chem.*, **15**, 773 (1967a); *ibid.*, **17**, 38 (1969b); Tatum *et al.*, *J. Food Sci.*, **40**, 707 (1975c).
[200] Teranishi, R., Ed., *Agricultural and Food Chemistry: Past, Present, Future, Amer. Chem. Soc.*, Avi Publ. Co. Inc., Westport, Connecticut (1978).
[201] Theehandel, N. V., *Neth. Appl.* 6,812,899 (13.3.1969).
[202] Thomas, A. F., *J. Agric. Food Chem.*, **21**, 955 (1973).
[203] Timmer, R., R. Ter Heide, P. J. De Valois & H. J. Wobben, *J. Agric. Food Chem.*, **23**, 53 (1975).
[204] Tonsbeek, C. H. T., A. J. Plancken & T. Van De Weerdhof, *J. Agric. Food Chem.*, **16**, 1016 (1968).
[205] Tonsbeek, C. H. T., E. B. Koenders, A. S. M. Van Der Zijden & J. A. Losekoot, *J. Agric. Food Chem.*, **17**, 397 (1969).
[206] Tonsbeek, C. H. T., H. Copier & A. J. Plancken, *J. Agric. Food Chem.*, **19**, 1014 (1971).
[207] Toyoda, T., S. Muraki & T. Yoshida, *Agric. Biol. Chem.*, **42**, 1901 (1978).
[208] Tressl, R., D. Bahri, M. Holzer & T. Kassa, *J. Agric. Food Chem.*, **25**, 459 (1977).
[209] Tressl, R., M. Apetz, R. Arrieta & K. G. Grünewald, *See Ref.* 30, p. 145.

[210] Tressl, R., K. G. Grünewald, R. Silvar & D. Bahri, *See Ref.* 95, pp. 197–213.

[211] Tsugita, T., T. Kurata & M. Fujimaki, *Agric. Biol. Chem.*, **42**, 643 (1978).

[212] Vernin, G., *Parf. Cosm. Arômes.*, **27**, 77 (1979); *idem, Ind. Aliment. Agric.*, 433 (1980a); Vernin, G. & J. Metzger, *Bull. Soc. Chim. Belges*, **90**, 553 (1981b); Vernin, G., *Riv. Ital. EPPOS*, **53**, 2 (1981c); Vernin, G., Perf & Flav. (1982d) in press.

[213] Viani, R., F. Mueggler-Chavan, D. Reymond & R. H. Egli, *Helv. Chim. Acta*, **48**, 1809 (1965a); *idem*, **52**, 887 (1969b).

[214] Vinals, J. F., J. Kiwala, D. E. Hruza, J. B. Hall & M. H. Vock, *U.S. Pat.*, 4,070,491 (1978).

[215] Vitzthum, O. G. & P. Werkhoff, *Z. Lebensm.-Unters. Forsch.*, **156**, 300 (1974a); *idem, J. Food Sci.*, **39**, 1210 (1974b).

[216] Vitzthum, O. G., P. Werkhoff & P. Hubert, *J. Agric. Food Chem.*, **23**, 999 (1975a); *idem, J. Food Sci.*, **40**, 911 (1975b).

[217] Vitzthum, O. G. & P. Werkhoff, *Z. Lebensm.-Unters. Forsch.*, **160**, 277 (1976).

[218] Wal Van der, B., G. Sipma, D. K. Kettenes & A. Th. J. Semper, *Rec. Trav. Chim., Pays-Bas*, **87**, 238 (1968a); *idem, J. Agric. Food Chem.*, **19**, 276 (1971).

[219] Wal Van der, B., *U.S. Pat.*, 3,803,172 (9.4.1974).

[220] Walradt, J. P., R. C. Lindsay & L. M. Libbey, *J. Agric. Food Chem.*, **18**, 926 (1970).

[221] Walradt, J. P., A. O. Pittet, T. E. Kinlin, R. Muralidhara & A. Sanderson, *J. Agric. Food Chem.*, **19**, 972 (1971).

[222] Walter, W. & H. L. Weidemann, *Z. Ernahrungwiss.*, **9**, 123 (1968).

[223] Wang, P., H. Kato & M. Fujimaki, *Agric. Biol. Chem.*, **32**, 501 (1968a); *ibid.*, **33**, 1775 (1969b); *ibid.*, **34**, 561 (1970c).

[224] Wang, P. & G. V. Odell, *J. Agric. Food Chem.*, **20**, 206 (1972).

[225] Watanabe, K. & Y. Sato, *Agric. Biol. Chem.*, **32**, 191 (1968a); *ibid.*, **32**, 1318 (1968b); *ibid.*, **33**, 242 (1969c); *ibid.*, **35**, 756 (1971d); *idem, J. Agric Food Chem.*, **19**, 1017 (1971e); *idem*, **20**, 174 (1972f).

[226] Wilkens, W. F. & F. M. Lin, *J. Agric. Food Chem.*, **18**, 333 (1970a); *ibid.*, **18**, 337 (1970b).

[227] Wilson, R. A. & I. Katz, *J. Agric. Food Chem.*, **20**, 741 (1972).

[228] Wilson, R. A., C. J. Mussinan, I. Katz & A. Sanderson, *J. Agric. Food Chem.*, **21**, 873 (1973).

[229] Wilson, R. A., C. J. Mussinan, I. Katz, C. Giacino, A. Sanderson & E. J. Shuster, *U.S. Pat.*, 3,863,013 (28.1.1975).

[230] Winter, M., F. Gautschi, I. Flament & M. Stoll, *French Pat.*, 1,530,436 (28.6.1968).

[231] Winter, M., *U.S. Pat.*, 3,622,346 (23.11.1971).

[232] Winter, M. & R. Klöti, *Helv. Chim. Acta*, **55**, 1916 (1972).

[233] Winter, M., F. Gautschi, I. Flament, M. Stoll & I. M. Goldman, *U.S. Pat.*, 3,702,253 (7.11.1972a); *ibid.*, 3,900,581 (19.8.1975b); *ibid.*, 3,900,582 (19.8.1975c).

[234] Winter, M., A. Furrer, B. Willhalm & W. Thomen, *Helv. Chim. Acta,* **59**, 1613 (1976).

[235] Withycombe, D. A., J. P. Walradt & A. Hruza, *See Ref.* 29, p. 85.

[236] Wobben, H. J., R. Timmer, R. Heide & P. J. De Walms, *J. Food Sci.,* **36**, 464 (1971).

[237] Yamanishi, T., S. Shimojo, M. Ukita, K. Kawashima & Y. Nakatani, *Agric. Biol. Chem.,* **37**, 2147 (1973). *See Ref.* 30, pp. 305-328.

[238] Yamanishi, T., M. Kawatsu, T. Yokoyama & Y. Nakatani, *Agric. Biol. Chem.,* **37**, 1075 (1973).

[239] Yajima, I., T. Yanai & M. Nakamura, *Agric. Biol. Chem.,* **43**, 2425 (1979).

[240] Zijden Van der, A. S. M., *U.S. Pat.*, 3,578,465 (11.5.1971).

CHAPTER III

Mechanisms of Formation of Heterocyclic Compounds in Maillard and Pyrolysis Reactions

Gaston VERNIN, University of Aix-Marseilles, France, and
Cyril PÁRKÁNYI, The University of Texas, El Paso, USA.

III–1 INTRODUCTION

Besides flavourless melanoidins with a high molecular weight, heterocyclic aroma compounds are produced non-enzymatically during food processing such as, e.g., the roasting of coffee, cocoa, tea and nut products, cooking or frying of meat, baking of bread, potatoes and baked goods, boiling of vegetables, etc. . . In all cases, they are formed from the same precursors [114a,116b] which are the basic food components, i.e., reducing sugars, free amino acids or dipeptides, and triglycerides or their derivatives. Reactions induced by heating these precursors (*see Chapter I*) known as non-enzymatic browning or *Maillard* reactions were first described in 1912 [70].

The mechanism, control, and nutritional consequences of these important reactions were the subject of several reviews [3,8b,25b,29,30,46b,47,49,75,104, 106,122,130,132b,132c,136,137,138,143], and the chemistry of non-enzymatic browning is a rapidly expanding subject [97].

Other thermal reactions such as the pyrolytic degradation of the main components of foods including sugars, amino acids, dipeptides, vitamins, as well as the oxidative degradation of lipids, contribute to the formation of heterocyclic compounds responsible for the flavour of foodstuffs.

Phenolic substances, flavonoids and carotenoids are likely substrates for non-enzymatic browning reactions, and they can be the precursors of some aromas [101].

Finally, heterocyclic compounds may be also formed enzymatically, e.g., in vegetables (tomatoes, asparagus) and fruits (pineapple), during the ripening of cheese, and during the fermentation of alcoholic beverages [21].

In the present chapter the main steps of *Maillard* reactions will be briefly summarized, and the occurrence of heterocyclic compounds in model systems and their formation pathways will be discussed. As in Chapters II and IV, each group of heterocyclic compounds will be examined separately.

III–2 IMMEDIATE PRECURSORS OF HETEROCYCLIC COMPOUNDS IN FOODS. THE VARIOUS STEPS OF THE *MAILLARD* REACTIONS

The origin of heterocyclic compounds in food flavours lies in a series of complex reactions in which reducing sugars and amino acids play a dominant role. These reactions were first described by Hodge *et al.* [47,49,75] and are summarized in *Scheme 1*.

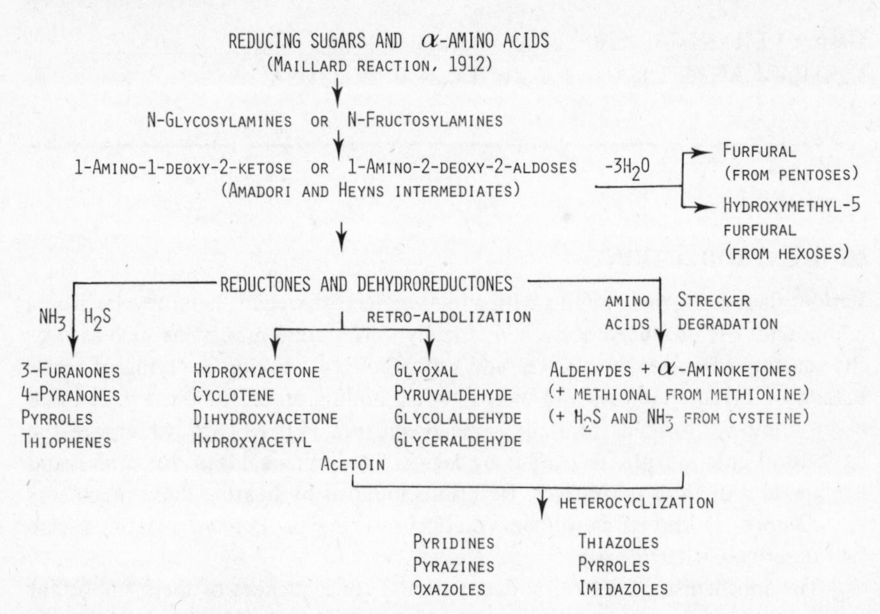

Scheme 1 – Formation of heterocyclic compounds in food products.

However, substances arising from other sources can also participate in the formation of heterocyclic compounds.

2.1 *N*-Glycosylamines

The first step in the amino acid-reducing sugar reaction is a sugar-amine condensation in equimolar ratios, leading to *N*-glycosylamines **2** [25] (*see Scheme 2*).

The condensation reaction involves the opening of the cyclic form of the sugar, nucleophilic addition of the amine to the carbonyl group and subsequent elimination of a molecule of water. The free compounds have not been prepared, but amorphous sodium salts [74] and amorphous and crystalline metal complexes have been described. In the case of primary amines a diketoseamine can also be formed [97a]. Arylamines such as *p*-chloroaniline undergo the *Maillard* reaction with glucose [8a]. In some model experiments utilizing *D*-glucose and amino acids, it was found that optimal conversion to rearrangement products takes places at a water content of 25–35% [90]. The effect of some inorganic ions on promoting or inhibiting browning reactions is known. Thus Hashiba [45] has shown that 1% of sodium or calcium chloride inhibited the browning of hydrolyzed vegetable protein or aqueous model systems containing sugars and amino acids. Ludwig [67] has observed an alteration of lysine during the production of β-lactoglobulin and during freeze-drying of a protein-sugar mixture. Other proteins are expected to behave analogously in the presence of reducing sugars.

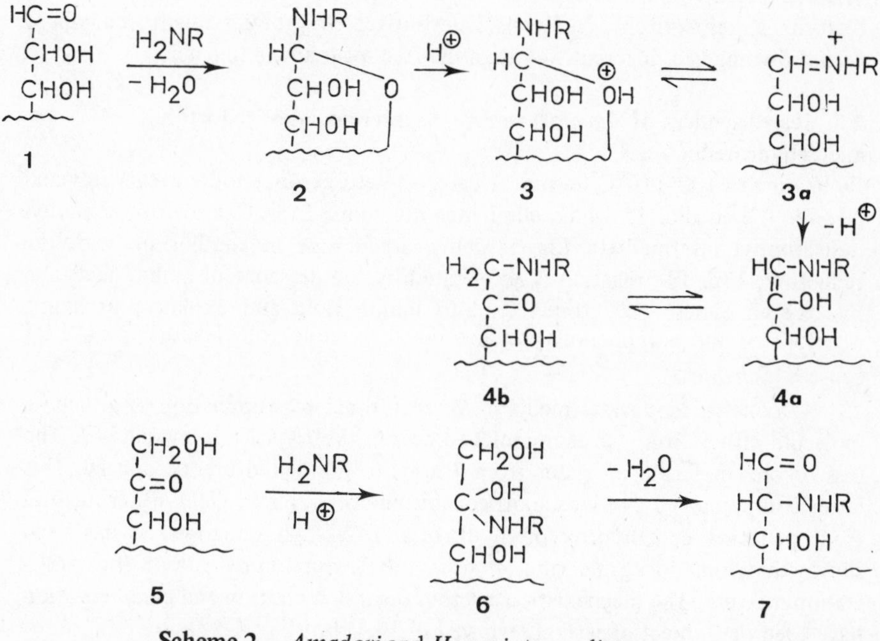

Scheme 2 – *Amadori* and *Heyns* intermediates.

2.2 *Amadori* and *Heyns* intermediates

The isomerization of *N*-substituted *N*-glucosylamines **2** to 1-amino-1-deoxy-2-ketoses **4** (*see Scheme 2*) [41,48,75] known as the *Amadori* rearrangement [4] occurs under the catalytic influence of the amino acid itself. It is readily detected because the reaction mixture develops strong reducing properties. It has been shown that blocking of the *Amadori* rearrangement of an *N*-substituted glycosylamine prevented the browning reaction which would have occurred [49]. The formation and stability of these intermediates in foods have been studied by Ciner-Doruk [17].

Similarly, 1-amino-2-deoxy-2-aldoses **7** or *Heyns* intermediates are formed from fructose [46] and as 2-ketoses they act as non-volatile flavour precursors in processed foods [75a,90]. Several methods of preparation of *Amadori* and *Heyns* intermediates were used [6a,6b,22,41,46,97], and their thermal degradation was studied [75,76,77]. Takeoka *et al.* [121] have isolated 1-amino-1-deoxy-2-ketoses (as silylated derivatives) arising from the reaction of *D*-glucose and *D*-ribose with *L*-tryptophan and *L*-phenylalanine, respectively.

Amadori and *Heyns* rearrangement products have also been isolated from natural products [90] including: apricots, beet molasses, cured tobacco leaves, liver, liquorice, tomato powder, black tea, green tea, roasted meat, etc. . . These essential precursors are heat labile and decompose *via* 1-deoxyosones and 3-deoxyosones, respectively. *N*-Glycosyl derivatives of proteins might decompose with browning in a different way than the free sugar and amino acid.

2.3 Rearrangement of *Amadori* and *Heyns* intermediates: reductones and dehydroreductones

By heating to 100–110°C 1-amino-1-deoxy-2-ketoses, **4b**, enolise irreversibly and give the 2,3-enediol **12** which eliminates the amine from C–1 to form a methyl α-dicarbonyl intermediate **13a** or dehydroreductone in equilibrium with the reductone **13b**. The reaction is accelerated by the presence of amino acids and by oxygen. These reductones tend to inhibit lipid and oxidative browning reactions of the polyphenol type and possibly reduce off-flavours from these sources (*Scheme 3*).

3-Deoxyhexosone intermediates **9** are formed by elimination of an amine molecule either from 1,2-enaminol **11** or from the *Heyns* intermediates **7**. The loss of one molecule of water from **9** leads to the dehydroreductone **10**. This latter compound **10** can lose another molecule of water to yield either furfural from pentoses or 5-hydroxymethylfurfural from 2-ketohexoses. It has been shown that some inorganic salts intensify the flavour and red colour of caramelization mixture. The mechanistic aspects of sugar dehydration and caramelization have been the subject of several reviews [29,30,49].

2.4 Retro-aldolization of rearranged *Amadori* and *Heyns* intermediates

The fragmentation of rearranged *Amadori* and *Heyns* intermediates leads to α-dicarbonyl compounds such as pyruvaldehyde **16**, diacetyl **21**, glyoxal **22**, hydroxyacetone **23** etc. The latter gives cyclotene **24** by self-condensation. Other reactive substances such as glycolaldehyde, glyceraldehyde and dihydroxyacetone are also formed [90]. Fructose gave much more pyruvaldehyde than did glucose with both glycine and β-alanine. The accepted mechanism for carbohydrate chain fission is a retro-aldolization reaction [49] (*see Scheme 4*).

2.5 Aldehydes

The main source of aldehydes in the *Maillard* reaction is the *Strecker* oxidative

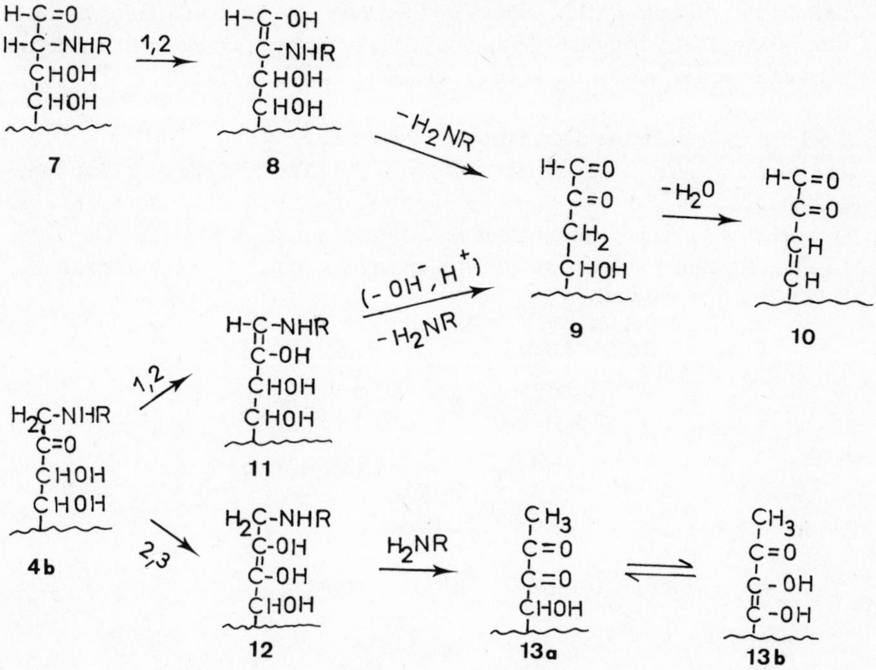

Scheme 3 – Rearrangement of *Amadori* and *Heyns* intermediates: reductones and dehydroreductones.

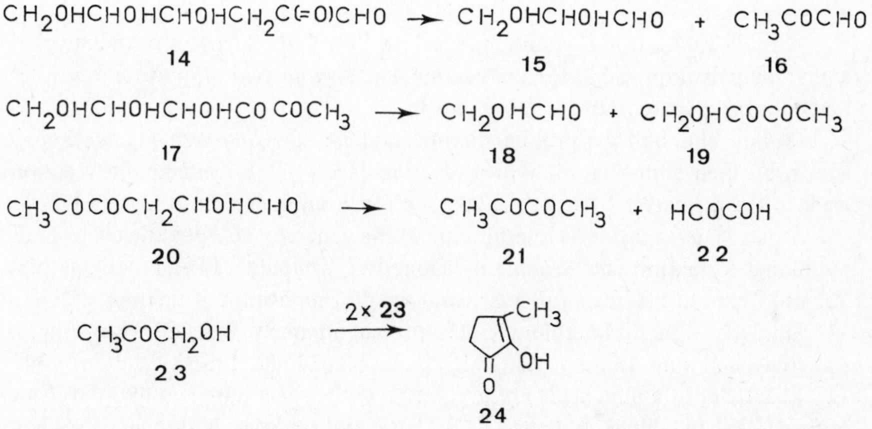

$$\underset{\mathbf{14}}{CH_2OHCHOHCHOHCH_2C(=O)CHO} \rightarrow \underset{\mathbf{15}}{CH_2OHCHOHCHO} + \underset{\mathbf{16}}{CH_3COCHO}$$

$$\underset{\mathbf{17}}{CH_2OHCHOHCHOHCOCOCH_3} \rightarrow \underset{\mathbf{18}}{CH_2OHCHO} + \underset{\mathbf{19}}{CH_2OHCOCOCH_3}$$

$$\underset{\mathbf{20}}{CH_3COCOCH_2CHOHCHO} \rightarrow \underset{\mathbf{21}}{CH_3COCOCH_3} + \underset{\mathbf{22}}{HCOCOH}$$

$$\underset{\mathbf{23}}{CH_3COCH_2OH} \xrightarrow{2\times\ \mathbf{23}} \mathbf{24}$$

Scheme 4 – Retro-aldolization of rearranged *Amadori* and *Heyns* intermediates.

degradation of amino acids. However, aldehydes can also be obtained by reactions such as the hydrolysis of *D*-glucose in the presence of amino acids, oxidative degradation of polyunsaturated fatty acids, and aldol condensation.

2.5.1 *Strecker* oxidative degradation of amino acids

The *Strecker* degradation of α-amino acids **25** in the presence of α-dicarbonyl compounds is undoubtedly one of the most important steps of *Maillard's* reaction (*see Scheme 5*). This reaction leads to the formation of aldehydes **29** (*see Table III-1*) and α-amino ketones **30** which play an important role in the formation of heterocyclic compounds.

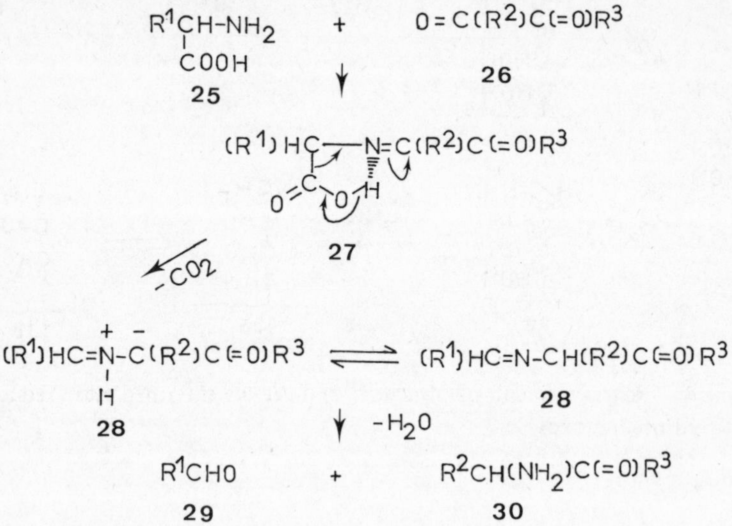

$R^1CH(NH_2)COOH$ (**25**) + $O=C(R^2)C(=O)R^3$ (**26**)

↓

$(R^1)HC(COOH)-N=C(R^2)C(=O)R^3$ (**27**)

↙ $-CO_2$

$(R^1)HC=\overset{+}{N}(H)-\overset{-}{C}(R^2)C(=O)R^3$ (**28**) ⇌ $(R^1)HC=N-CH(R^2)C(=O)R^3$ (**28**)

↓ $-H_2O$

R^1CHO (**29**) + $R^2CH(NH_2)C(=O)R^3$ (**30**)

Scheme 5 – *Strecker* degradation of amino acids.

With *D*-fructose or *D*-arabinohexosulose [94b], the *Strecker* degradation of amino acids is responsible for the caramel-like flavour of the mixture but other reactions participate in the reaction as well.

Proline and hydroxyproline do not undergo the *Strecker* degradation as shown by their abnormal ninhydrin reaction [131]. With cysteine, the reaction leads to the formation of acetaldehyde, hydrogen sulphide, and ammonia [79a].

Strecker degradation of methionine is the generally accepted mechanism of methional formation (*see Scheme 6*). Dimethyl disulphide **33** and methanethiol **32**, also found in this reaction, arise from the decomposition of methional **29i**.

Similarly *S*-methylmethionine **25j** present in many vegetables (asparagus, tomatoes) and milk decomposes to afford acrolein **31** and dimethyl sulphide **34**.

Reactivity of amino acids in the reaction with α-dicarbonyl compounds have been studied by Piloty & Baltes [92]. Basic and hydroxy amino acids reacted strongly, while the acidic nonpolar amino acids react less so. After heating of

Table III–1 Volatile *Strecker* aldehydes from α-amino acids and α-dicarbonyl compounds

Amino acids **25**; $R^1CH(NH_2)COOH$	Name of the amino acid	Aldehydes **29**; R^1CHO
R = (a) H	Glycine	Formaldehyde
(b) CH_3	α-Alanine	Acetaldehyde
(c) C_2H_5	α-Aminobutyric	Propionaldehyde
(d) i-C_3H_7	Valine	Isobutyraldehyde
(e) i-C_4H_9	Leucine	Isovaleraldehyde
(f) s-C_4H_9	Isoleucine	2-Methylbutanal
(g) $HOCH_2$	Serine	Glycolaldehyde
(h) CH_3CHOH	Threonine	Lactaldehyde
(i) $CH_3SCH_2CH_2$	Methionine	Methional
(j) $(CH_3)_2\overset{+}{S}CH_2CH_2$	*S*-Methylmethionine	Acrolein
(k) $HSCH_2$	Cysteine	Acetaldehyde (+H_2S)
(l) C_6H_5	Phenylglycine	Benzaldehyde
(m) $C_6H_5CH_2$	Phenylalanine	2-Phenylethanal
(n) p-$HOC_6H_4CH_2$	Tyrosine	2-(p-Hydroxyphenyl)-ethanal

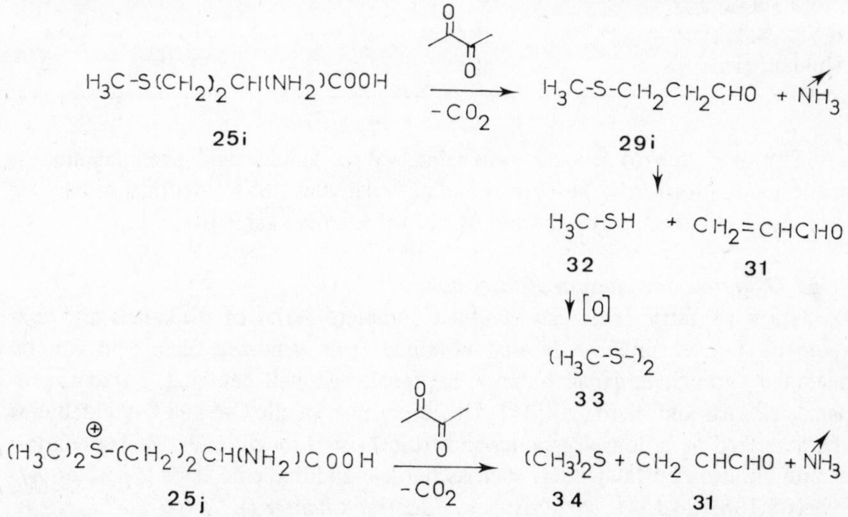

$$H_3C\text{-}S(CH_2)_2CH(NH_2)COOH \ (\mathbf{25i}) \xrightarrow[-CO_2]{\text{diacetyl}} H_3C\text{-}S\text{-}CH_2CH_2CHO \ (\mathbf{29i}) + NH_3\uparrow$$

$$\mathbf{29i} \rightarrow H_3C\text{-}SH \ (\mathbf{32}) + CH_2{=}CHCHO \ (\mathbf{31})$$

$$\mathbf{32} \xrightarrow{[O]} (H_3C\text{-}S\text{-})_2 \ (\mathbf{33})$$

$$(H_3C)_2\overset{\oplus}{S}\text{-}(CH_2)_2CH(NH_2)COOH \ (\mathbf{25j}) \xrightarrow[-CO_2]{\text{diacetyl}} (CH_3)_2S \ (\mathbf{34}) + CH_2\ CHCHO \ (\mathbf{31}) + NH_3\uparrow$$

Scheme 6 – *Strecker* degradation of methionine and its *S*-methyl derivative [106a].

glycine, α-alanine, β-alanine, phenylalanine, glycylalanine and ammonia with diacetyl in an aqueous medium, nitrogen-containing heterocyclic products (oxazoles, pyrazines, pyrroles, and pyridines) were identified.

According to Rizzi [100], α,β-unsaturated carbonyl compounds which may arise from lipid oxidation [35] react with phenylalanine in a non-enzymatic transamination analogously to the *Strecker* reaction and yield the corresponding amines. These amines may in turn react with other food constituents.

2.5.2. Hydrolysis of D-*glucose in the presence of amino acids*
Aldehydes obtained from hydrolysis of *D*-glucose in the presence of amino acids without sulphur are listed in *Table III–2* [102]. Acetone was also found in all reaction mixtures.

Table III–2 Aldehydes from *D*-glucose and α-amino acids without sulphur (95°, 12 h, pH 5.5) [102]

Compounds	Isoleucine	Leucine	Valine	Alanine
Formaldehyde	+	+	+	+
Acetaldehyde	+	+	+	+
Propionaldehyde	+	+	+	+
Isobutyraldehyde	+	+	+	+
n-Butyraldehyde	+	−	+	−
2-Methylbutanal	+	+	+	+
Isovalderaldehyde	+	+	+	+
n-Valeraldehyde	+	−	−	+
Unidentified	2	2	1	4

The oxidation of leucine, isoleucine, valine, alanine and phenylalanine in acidic medium affords the corresponding aldehydes [65]. Identical aldehydes have also been obtained by thermolysis of these amino acids [64,65].

2.5.3 Oxidative degradation of fatty acids
Oxidation of fatty acids can produce complete series of *n*-alkanals and *n*-2-alkenals. For example, *n*-decanal obtained from oxidized oleic acid can be degraded through nonanal, octanal, heptenal, hexenal, pentanal, butanal, propanal, ethanal and methanal [35]. However, most of the C-6 and C-9 aldehydes characterized in various vegetables and fruits arise through oxidative degradation of polyunsaturated fatty acids such as linoleic and linolenic acids [26,37,66,87, 89b,107,108,128,134], *via* hydroperoxides (*see Chapter I*).

It seems that plants contain several enzymes or enzyme systems of broad specificity that enable the plant to produce a large number of volatile compounds from a limited number of non-volatile precursors [26]. In most plant materials *trans* 2-hexenal rather than *cis* 3-hexenal is found. This is due to the presence of an isomerase which converts the *cis*-3 double bond into a *trans*-2 double bond. Polyunsaturated fatty acids yield malonaldehyde and other substances which react with amino acids, proteins, glycogen and other food constituents and are reactive browning intermediates. The thermal degradation of methyl ricinoleate yields heptanal and methyl undecenoate. The catalytic effects of heavy metal ions on lipid oxidations and subsequent browning reactions have been recognized for a long time.

2.5.4 Aldol condensation

The *Strecker* aldehydes and phenylacetaldehyde formed from the corresponding amino acids and phenylalanine, respectively, may undergo aldol-type reactions. The products obtained are 2-phenyl-2-alkenals **36**. Some of them were identified in cocoa [95], malt [131] and cooked asparagus [128]. Reduction of **36** leads to the corresponding alcohols **37** (*see Scheme 7*). Thus, 5-methyl-2-phenyl-2-hexenal **36**($R^1 = i\text{-}C_4H_9$) is presumably formed from the condensation of phenylacetaldehyde **29m** and isovaleraldehyde **29e** [71]. In cooked asparagus [128] 2-phenyl-2-butenal may arise from 2-phenylethanal and acetaldehyde. Aldol condensation of 2-oxobutyric acid **38**, an oxidation product of threonine, with *Strecker* aldehydes can lead to reactive compounds **39**.

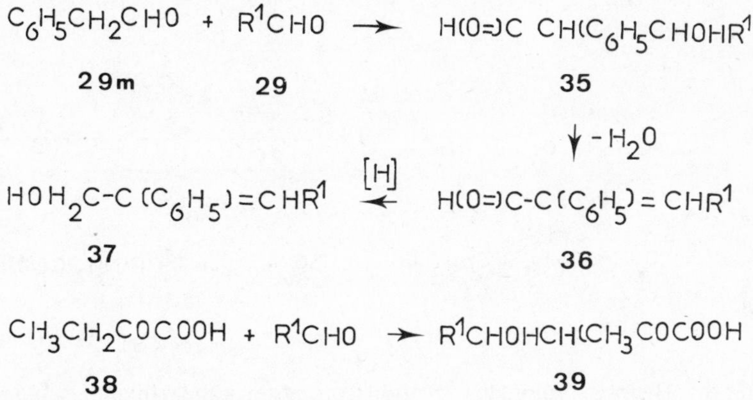

$$\underset{\mathbf{29m}}{C_6H_5CH_2CHO} + \underset{\mathbf{29}}{R^1CHO} \longrightarrow \underset{\mathbf{35}}{H(O{=})C\,CH(C_6H_5)CHOHR^1}$$

$$\downarrow -H_2O$$

$$\underset{\mathbf{37}}{HOH_2C{-}C(C_6H_5){=}CHR^1} \xleftarrow{[H]} \underset{\mathbf{36}}{H(O{=})C{-}C(C_6H_5){=}CHR^1}$$

$$\underset{\mathbf{38}}{CH_3CH_2COCOOH} + R^1CHO \longrightarrow \underset{\mathbf{39}}{R^1CHOHCH(CH_3)COCOOH}$$

Scheme 7 – Aldol condensation of aldehydes and 2-oxoacids.

2.5.5 Miscellaneous

The formation of carbonyl compounds from β-ketoacids and β-hydroxyacids has been described [131]. After the thermal oxidation of short chain 2- and 3-enoic acids, the corresponding epoxy acids, 4-hydroxy-2-enoic acids, 4-oxo-2-enoic acids, aldehydes and ketones, and saturated acids have been found [145]. All these compounds may participate in browning reactions to produce various types of flavours.

Enzymatic reactions and photo- and thermal oxidative reactions of carotenoids give several carbonyl compounds which act as precursors in the formation of some furan and pyran derivatives [88].

Catalysed addition of aldehydes to activated double bonds has been reported in a comprehensive review by Stetter [118] as being a new synthetic approach.

Reaction of lower alkanals with proteins has also been described [94a]. The reaction takes place within several days even under storage conditions. The primary reaction product was mainly a labile colourless Schiffs base further transformed into brown macromolecular products by secondary reactions analogous to aldolization. In model systems of alkanal and protein, and alkanal and cellulose, the changes of odour and colour were determined by sensory tests.

2.6 Hydrogen sulphide and alkyl sulphides

Besides α-dicarbonyl compounds, aldehydes, α-amino ketones and ammonia (or amines) arising in *Maillard* reactions, hydrogen sulphide is another important and highly reactive product. It is formed mainly during the *Strecker* degradation of cysteine with α-diketones [58]. The reaction involves oxidation of cysteine and a subsequent reduction as represented in *Scheme 8*. The formation of methanethiol from methionine and α-diketones can be explained in a similar manner.

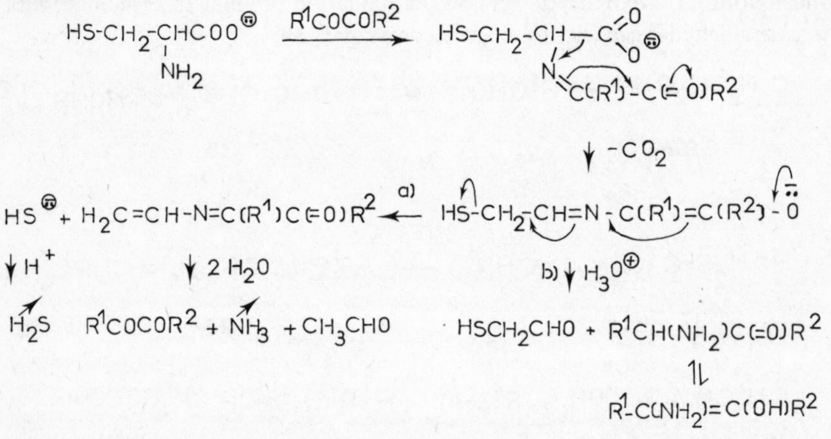

$$HS\text{-}CH_2\text{-}CH(NH_2)COO^{\ominus} \xrightarrow{R^1COCOR^2} HS\text{-}CH_2\text{-}CH(\text{-}N{=}C(R^1)\text{-}C({=}O)R^2)\text{-}COO^{\ominus}$$

$$\downarrow -CO_2$$

$$HS^{\ominus} + H_2C{=}CH\text{-}N{=}C(R^1)C({=}O)R^2 \xleftarrow{a)} HS\text{-}CH_2\text{-}CH{=}N\text{-}C(R^1){=}C(R^2)\text{-}\bar{O}^{\ominus}$$

$$\downarrow H^+ \quad\quad \downarrow 2\,H_2O \quad\quad\quad b) \downarrow H_3O^{\oplus}$$

$$H_2S \quad R^1COCOR^2 + NH_3 + CH_3CHO \quad\quad HSCH_2CHO + R^1CH(NH_2)C({=}O)R^2$$

$$\rightleftharpoons R^1\text{-}C(NH_2){=}C(OH)R^2$$

Scheme 8 – Hydrogen sulphide formation from cysteine and α-diketones [36,58].

Not only sulphur-containing amino acids [7,36,39,58,62,79b] but also thiamin [24b] and glutathione [39] contribute to the formation of hydrogen sulphide. Biotin and coenzyme A were also presumed to contribute to the formation of sulphur-containing heterocyclic compounds in food flavours.

1-Methylthioethanethiol **40** another important precursor is formed from acetaldehyde and methanethiol [106] (*see Scheme 9*).

$$H_3C\text{-}CHO + HSCH_3 \rightleftharpoons H_3C\text{-}CHOH\text{-}SCH_3$$

$$\downarrow HS^{\ominus}$$

$$H_3C\text{-}CH(SH)\text{-}SCH_3$$

40

Scheme 9 – 1-Methylthioethanethiol formation [106].

The first step occurs by specifically base-catalysed and generally acid-catalysed pathways. The second step involves nucleophilic displacement of OH^- by SH^- ions which are excellent nucleophiles.

III–3 OCCURRENCE AND FORMATION OF HETEROCYCLIC COMPOUNDS IN MODEL SYSTEMS

Model systems are of great help in the analysis of heterocyclic compounds in food flavours. They can be divided into the following categories:

– Reactions involving reducing sugars and amino acids.
– Thermal degradation of the *Amadori* and *Heyns* rearrangement products.
– Thermal degradation of sugars.
– Pyrolysis of α-amino acids and dipeptides.
– Pyrolysis of vitamins and related substances.
– Reactions of α-dicarbonyl compounds and aldehydes in the presence of either α-amino acids or ammonia and/or hydrogen sulphide.
– Other miscellaneous model systems.

The above reactions can lead to the following furans, thiophenes, pyrroles, oxazoles, imidazoles, thiazoles, pyrans, pyridines, pyrazines, and cyclic sulphides and polysulphides.

3.1 Furans and furanones

2-, and 2,5-disubstituted furans **41** and **42** listed in *Table III–3* and *III–4* were identified in more than thirty model systems. They were mainly found in products arising from sugar-amino acids interaction [31,55c,61,84,90,105,112, 116b] from the degradation of *Amadori* and *Heyns* intermediates [75,90], from the thermal degradation of sugars [29,46c,46d,141] and vitamins [23,24, 123,135], and from the reactions of sugars with hydrogen sulphide and/or ammonia [114a,114c,149].

Table III–3 2-Substituted furans

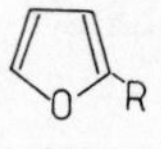

O R

41

41; R	Name	Model systems[a)]
H	Furan	*28,44*
CH_3	2-Methylfuran	*28,35,44,60,75*
C_2H_5	2-Ethylfuran	*28,44*
n-C_3H_7	2-*n*-Propylfuran	*44*
n-C_4H_9	2-*n*-Butylfuran	–
n-C_5H_{11}	2-*n*-Pentylfuran	*85*

Table III–3 (*continued*)

41;R	Name	Model systems[a)]
$CH_2CH_2COCH_3$	1-(2′-Furyl)-3-butanone	*28*
$CH_2CH_2SCH_3$	Methyl-2-ethylfuryl sulphide	*76,80*
$CH_2C(CH_3){=}CH_2$	2-Methyl-3-(2′-furyl)-2-propene	*85*
Cis and *trans*- $CH_2\text{-}CH{=}CH\text{-}C_2H_5$	2-(2′-Pentenyl)furans	*85*
CH_2COCH_3	2-Furfuryl methyl ketone	*28,75,76*
$CH_2COC_2H_5$	2-Furfuryl ethyl ketone	*28,75*
H_2C–(2-furyl)	2,2′-Difurylmethane	*61,80*
CH_2NH_2	Furfuryl amine	*67*
CH_2OH	Furfuryl alcohol	*4,22,23,24,28,29,34, 35,37,48,61,62,74,75, 76,80*
CH_2OCH_3	Furfuryl methyl ether	*75*
$CH_2\text{–}O\text{–}CH_2$–(2-furyl)	Difurfuryl ether	*28,29*
$CH_2OC(=O)H$	Furfuryl formate	*28,29*
$CH_2OC(=O)CH_3$	Furfuryl acetate	*28,29*
$CH_2OC(=O)CH{=}CH_2$	Furfuryl acrylate	*28*
CH_2SH	Furfuryl mercaptan	*18,75,76,80*
CH_2SCH_3	Furfurylmethyl sulphide	*75,76*
CH_2SCH_2–(2-furyl)	Difurfuryl sulphide	*80*
i-C_3H_7	2-Isopropylfuran	*44*
$CH{=}CH_2$	2-Vinylfuran	*44*
Cis and *trans*- $CH{=}CHCH_3$	*cis* and *trans* 2-Propenylfurans	*44*
$CH{=}CH$–(2-furyl)	1,2-(2,2′-difuryl)ethylene	*80*
CHO	Furfural	*3,4,5,9,10,14,22,23, 24,25–28,34,35,37,44, 61,62,75,76,77*

Table III–3 (*continued*)

41;R	Name	Model systems[a)]
$COCH_3$	2-Acetylfuran	*3,4,5,9,10,14,22,23, 24,25–28,34,35,37,44, 61,62,75,76,77,78,80*
COC_2H_5	2-Propionylfuran	*28,75,76*
COC_3H_7-*n*	2-*n*-Butyrylfuran	*24,37*
$COCH_2$–(2-furyl)	Deoxyfuroin	*61,62,80*
$COCH_2OH$	2-Hydroxyacetylfuran	*22,23,61,62*
COCHOH–(2-furyl)	Furoin	*61,62*
COCO–(2-furyl)	Furil	*44,61,62,80*
COOH	2-Furoic acid	*28,35,44,61,62,65*
$C(=O)SCH_3$	Methyl thiofuroate	*75,76*
(2-furyl)	2,2′-Difuryl	*28*
CN	2-Cyanofuran	*80*
$HNCH_2$–(2-furyl)	*N*-Furfuryl-2-aminofuran	*80*

a) *3* glucose-cysteine; *4* glucose-cystine; *5* glucose-glycine; *9* glucose-*DL*-alanine; *10* glucose-valine; *14* glucose-lysine; *18* xylose-cysteine; *19* xylose-cystine; *22* fructose-alanine; *23* fructose-γ-aminobutyric acid; *24* fructose-glycine; *25* lactose-glycine; *26* lactose-lysine; *27* lactose-valine; *28* lactose-casein; *29* D-lactose-*N*-α-formyl-*L*-lysine; *34* RP's glucose-glycine; *35* RP's glucose-theanine; *37* RP's fructose-glycine; *44* glucose; *48* base catalysed fructose; *61* ascorbic acid; *62* *L*-dehydroascorbic acid; *65a* cysteine-pyruvaldehyde; *65b* cystine-pyruvaldehyde; *67* α-amino acids-furfural; *71* glucose-NH_4OH(5M); *74* glucose-AcOH-CH_3NH_2; *75* glucose-H_2S; *76* glucose-H_2S-NH_3; *77* *L*-rhamnose-ammonia; *78* *L*-rhamnose-H_2S-NH_3; *80* furfural-H_2S-NH_3; *85* unsaturated fatty acids.
RP: reaction products.

Table III–4 2,5-Disubstituted furans

(structure 42: furan ring with R^2 at C-5 and R^1 at C-2)

42; R^1	R^2	Name	Model systems[a)]
CH_3	CH_3	2,5-Dimethylfuran	*44,75*
CH_3	C_2H_5	2-Methyl-5-ethylfuran	*4,44*
CH_3	n-C_3H_7	2-Methyl-5-*n*-propylfuran	*44*
CH_3	5-methylfurfuryl (H_3C–furan–CH_2)	2-Methyl-(5′-methylfurfuryl)-furan	*28,29*
CH_3	pyrrol-1-yl-CH_2 (N-CH_2)	1-(5′-methyl-2′-furfuryl)-pyrrole	*9*
CH_3	$HOCH_2$	2-Methyl-5-hydroxymethyl-furan	*5,9,10,14,24,34, 36,37,48,74*
CH_3	$CH_3C(=O)OCH_2$	2-Methyl-5-furfuryl acetate	*74*
CH_3	i-C_3H_7	2-Methyl-5-isopropylfuran	*44*
CH_3	$CH_2=CH$	2-Methyl-5-vinylfuran	*44*
CH_3	*cis* and *trans*- CH_3–$CH=CH$	*cis* and *trans*-2-methyl-5-propenyl-furans	*44*
CH_3	CH_3S	2-Methyl-5-methylthiofuran	*19,60*
C_2H_5	C_2H_5	2,5-Diethylfuran	*44*
H_2C–furyl	furyl–CH_2	2,5-Difurfurylfuran	*28*
CHO	CH_3	5-Methylfurfural	*4,5,10,14,23, 25–28,29,31, 34–37,44,76*
CHO	C_2H_3	5-Ethylfural	*71*(tentative)
CHO	furyl–CH_2	5-Furfurylfurfural	*28*
CHO	H_3C–furan–CH_2	5-(5′-Methylfurfuryl)-furfural	*28*
CHO	CH_2OH	5-Hydroxymethylfurfural	*22,23,28,29,35,74*
$COCH_3$	CH_3	2-Acetyl-5-methylfuran	*3,9,28,44,75*
COC_2H_5	CH_3	5-Methyl-2-propionylfuran	*28*

a) See the caption for Table III–3.

Among these reactions considerable attention was paid to the thermal degradation of *D*-glucose and carbohydrates [29,30,46c,49], and more than thirty furan derivatives were identified as the products. The major volatile products resulting from the pyrolysis of starch, cellulose, sucrose were essentially identical as the products obtained from glucose, as determined by gas chromatography. The relative amounts of furans generally increase with temperature [46d].

Several mechanisms have been proposed to explain the formation of furfural and 5-hydroxymethylfurfural from pentoses, and hexoses, respectively [30]. These two compounds were formed from dehydroreductones **10a** according to *Scheme 10*. The corresponding alcohols **45** formed by reduction have been reported in a mixture resulting from the base catalysed degradation of fructose [111a].

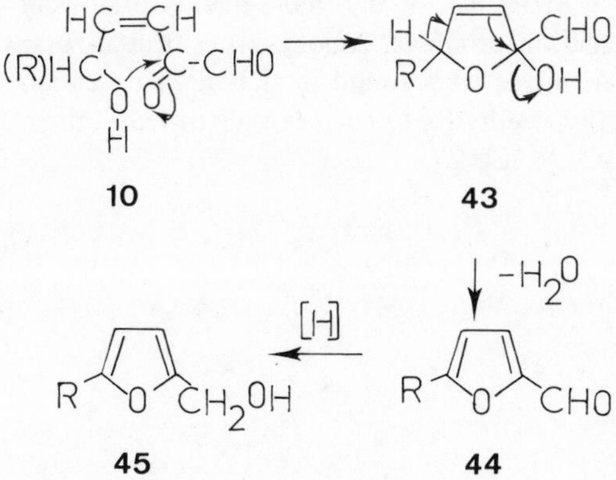

10 → **43** → ($-H_2O$) **44** → ([H]) **45**

Scheme 10 – Furfural formation from dehydroreductones.

Furfural is an important precursor of furan derivatives and heterocyclic flavour compounds obtained from food and model systems. Thus Shibamoto & Russell [114c] investigating the volatiles produced by reacting furfural, hydrogen sulphide and ammonia, identified ten furan derivatives, the major portions of which were the products of a reaction of the aldehyde group of furfural (e.g., difurfurylethylene, furil, deoxyfuroin).

The reaction of furfural with *L*-valine gives two aldimines from which furfurylamine and isobutylamine are formed by hydrolytic cleavage [99b]. Furfural may also react with compounds containing an active methylene group to give coloured substances [62d]. 2-Acetylfuran was identified in several model systems (*see Table III–3*). Its 3-hydroxy derivative or isomaltol is formed from 1-deoxyhexosone [13b], an *Amadori* rearranged intermediate (*see Scheme 3*). Pyrolysis of 1-deoxy-1-piperidino *D*-fructose affords 2-acetyl-3-piperidinofuran and its 4,5-dihydro derivative [75a]. By allowing 1-deoxy-1-*L*-prolino-*D*-fructose to decompose under vacuum, first at 140°C and then at 240°C, Mills & Hodge [77] obtained 2-pyrrolino substituted furans at higher temperatures. Low-temperature degradation of the *Amadori* compound yielded 2,5-dimethyl-2,4-dihydroxy-3(2*H*)-furanone and 2,5-dimethyl-4-hydroxy-3(2*H*)-furanone among other products. 2-*n*-Pentylfuran which significantly contributes to the beany and grassy flavour of the soy bean oil, is formed from linoleate through autoxidation [14,117]. In kilned malts it may be formed by thermal degradation of trihydroxy-3-octadecenoic acids. Chang *et al.* [117] assume that the linolenate undergoes the same autoxidation reactions as the linoleate to produce *cis* and *trans*-2-(1′-pentenyl)-furans and *cis*- and *trans*-2-(2′-pentenyl)-furans.

Ten furans have been reported to be the major thermal decomposition products of ascorbic acid **46** and *L*-dehydroascorbic acid **47** [135] (*see Scheme 11*).

Furfuryl alcohol, 2-acetylfuran, 2-furfurylfuran, furil **49** and deoxyfuroin **50** were also observed in furfural-hydrogen sulphide-ammonia model system [114c]. 2-Methylfuran and 2-methylthio-5-methylfuran were reported [23,24] as thiamin degradation products. Among 3-substituted furans, only 3-methyl-, and 3-formyl derivatives were found in the thermal degradation of *D*-glucose [46c]. 2,5-Dimethyl-3-ethylfuran has been identified in the reaction of cysteine with pyruvaldehyde [55c].

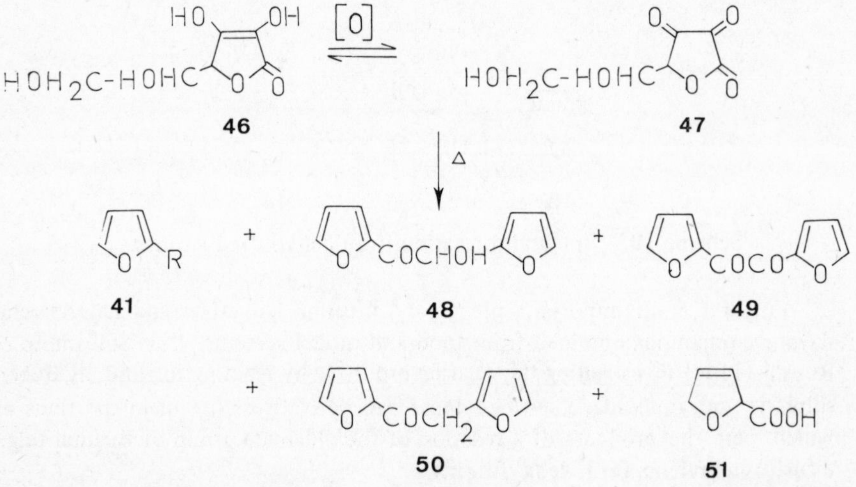

HO OH [O] O O
HOH_2C-HOHC O O HOH_2C-HOHC O O
46 **47**
Δ
O R + O COCHOH O + O COCO O
41 **48** **49**
+ O $COCH_2$ O + O COOH
50 **51**

Scheme 11 – Furans from the thermal degradation of ascorbic acid and *L*-dehydroascorbic acids.

4-Hydroxy-5-methyl-3-(2*H*)-furanone **52**, a compound with an active methylene group, has been identified as the degradation product of *L*-dehydroascorbic acid [135] and in the thermal degradation of *D*-xylose [109]. According to Tonsbeek *et al.* [126], it may also be formed in meat by enzymatic hydrolysis of ribonucleotides *via* a ribose-5′-phosphate intermediate (*see Scheme 12*). It reacts with hydrogen sulphide to afford mercaptofurans **53** to **57** which are responsible for meat aromas.

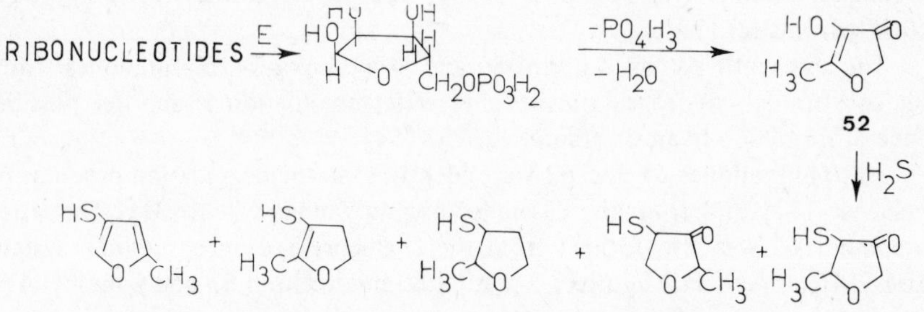

RIBONUCLEOTIDES —E→ (ribose-5′-phosphate, $CH_2OPO_3H_2$) —($-PO_4H_3$, $-H_2O$)→ **52**

52 —H_2S→ **53** + **54** + **55** + **56** + **57**

57 **56** **55** **54** **53**

Scheme 12 – Mercaptofuran derivatives from 3-(2*H*)-furanone.

Condensation of aldehydes [89] such as furfural, 2-thiophene carboxaldehyde or 1-methyl-2-pyrrole carboxaldehyde **58** (X = O, S, N-CH_3) with the furanone **52** yielded coloured substances **59** [62c] (*see Scheme 13*).

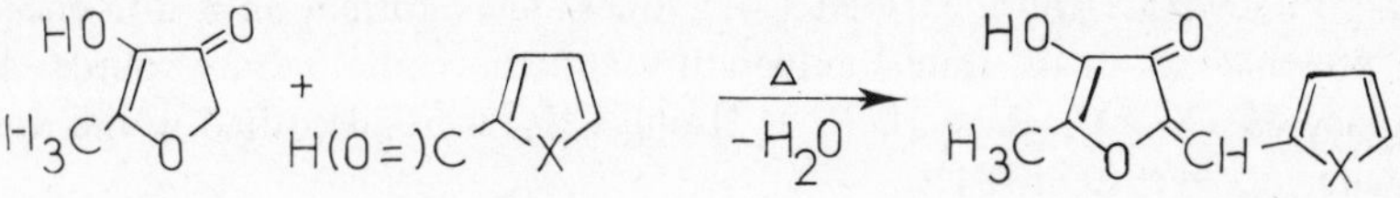

52 + H(O=)C–(X-ring) **58** —(Δ, $-H_2O$)→ **59**

Scheme 13 – Condensation of 4-hydroxy-5-methyl-3-(2*H*)-furanone with heterocyclic aldehydes.

The two 3-(2*H*)-furanones **60b** (R = CH_3 and CH_2OH), identified in the base-catalysed degradation of fructose [111a] were formed from 1-deoxyhexosone intermediates **13b** according to *Scheme 14*.

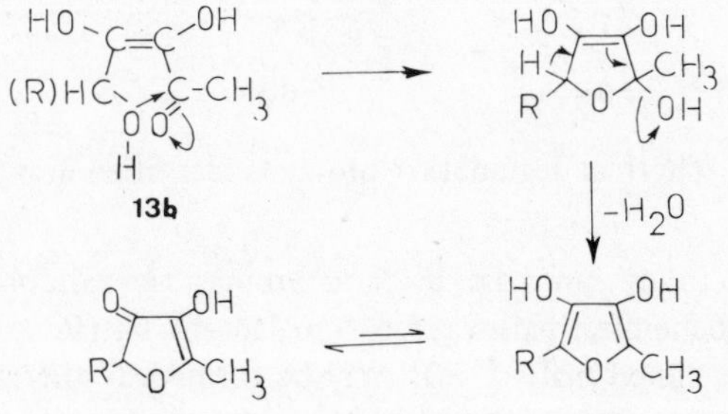

13b → (cyclic hemiketal) —$-H_2O$→ **60a** ⇌ **60b**

Scheme 14 – Formation of 3-(2*H*)-furanones from 1-deoxyhexosones

Furaneol **60b** (R = CH_3) was characterized in several model systems including: pyrolysis of *D*-glucose alone [29,46c,52] or in the presence of acetic acid and methylamine [53], pyrolysis of 1-deoxy-1-piperidino-*D*-fructose [75], roasting of alanine with rhamnose [112], heating of proline with 5-hydroxy-5,6-dihydromaltol [77].

The biosynthesis of 4-hydroxy and 4-methoxy-3-(2*H*)-furanones from fructose through hydrogen transfer and water elimination is another possible mode of formation in arctic bramble berries [54].

3-(2*H*)-Furanones **61** and **62** were identified in the degradation products of *D*-glucose [52] and from the casein-lactose browning system [31]. The latter compound **62** was also formed from the *D*-glucose-hydrogen sulphide system [104,114], as well as 3-hydroxy-2-methyltetrahydrofuran **63** and 2-methyl-4,5-dihydrofuran **64** (*see Scheme 15*).

γ-Butyrolactone **65** (R^1 = R^2 = H, X = O) was identified in the following model systems: lactose-casein [31], base-catalysed degradation of fructose [111a], dry degradation of RP's derived from glucose-theanine [90], *N*-formyl-*L*-lysine-*D*-lactose [31c] and thermal degradation of ascorbic acid [111b]. The β-hydroxy derivative **64** (R^1 = OH, R^2 = H, X = O) was found from *N*-formyl-*L*-lysine-*D*-glucose system [31c]. Ishizu *et al.* [51] let *D*-xylose react with calcium hydroxide and identified 13 lactones. Their formation is influenced by the presence of alkali, time, temperature and the carbohydrate source. Thiolactones **65** (X = O, R^1 = H, R^2 = H and CH_3) were identified in the xylose-cysteine model system [84].

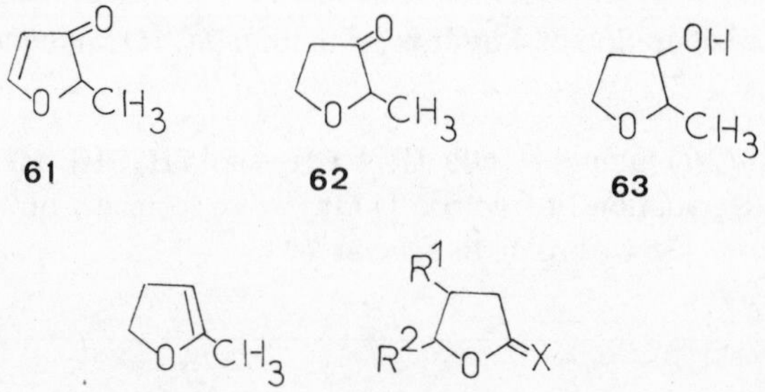

Scheme 15 – Thermal degradation products identified in model systems.

γ-Alkyl-γ-lactones can arise in food aromas *via* various routes including chemical and biochemical pathways. γ-Nonalactone **64** (R^1 = H, R^2 = n-C_5H_{11}, X = O) found in kilned malts [130] may be formed by thermal degradation of trihydroxy-3-octadecenoic acids [130,131].

According to Watanabe & Sato [144b] the thermal oxidation of C-12 saturated fatty acids in fats (*see Scheme 16*) affords γ- and δ-lactones **66** or **67**

(R = CH_3, C_2H_5, *n*-C_3H_7, *n*-C_4H_9, *n*-C_5H_{11}, *n*-C_6H_{13}, *n*-C_7H_{15}). γ-Hexalactone can be formed from peroxidized and hydrogenated linoleic acid [144a].

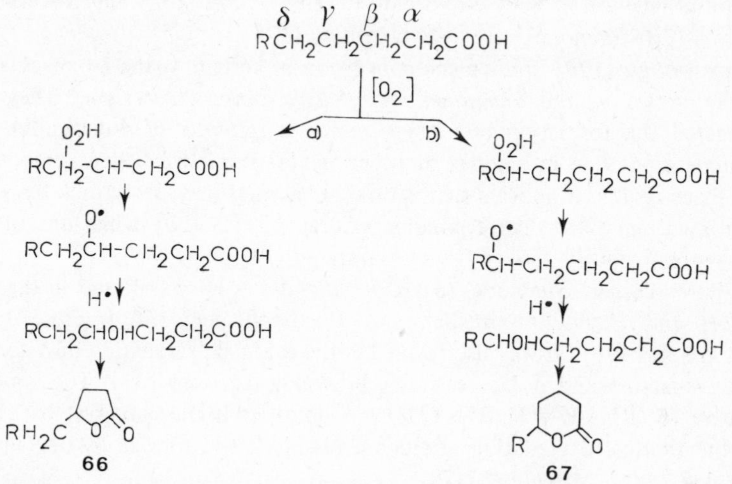

δ γ β α
$RCH_2CH_2CH_2CH_2COOH$
$[O_2]$
a) b)

a) RCH_2–CH(O_2H)–CH_2COOH → RCH_2CH(O•)–CH_2CH_2COOH →(H•) $RCH_2CHOHCH_2CH_2COOH$ → 66

b) RCH(O_2H)–$CH_2CH_2CH_2COOH$ → RCH(O•)–$CH_2CH_2CH_2COOH$ →(H•) $RCHOHCH_2CH_2CH_2COOH$ → 67

66 **67**

Scheme 16 – γ and δ-lactones from thermal oxidation of saturated fatty acids [144b].

Enzymatic reduction of oxo-fatty acids could lead to hydroxy fatty acids which are the primary lactone precursors in many fermentation products and foods [81b,124,129,148]. γ-Butyrolactone **65a** may be formed during yeast fermentation *via* 2-oxoglutaric acid **69** arising from glutamic acid **68** (*see Scheme 17*).

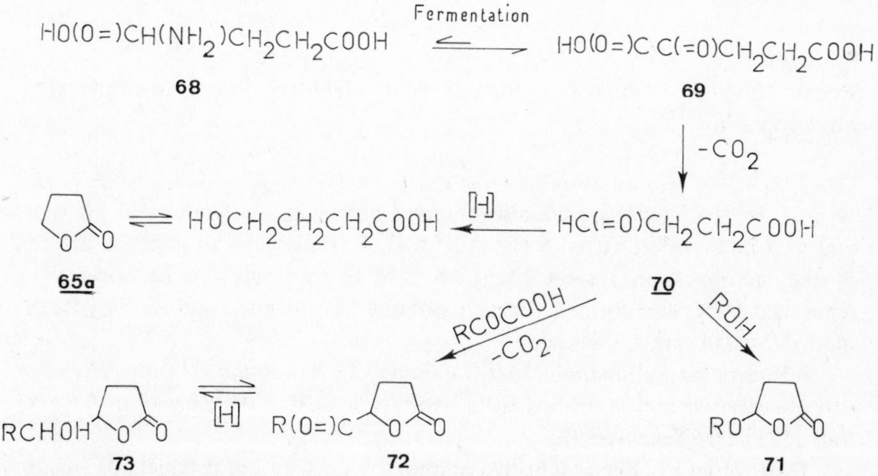

$HO(O=)CH(NH_2)CH_2CH_2COOH$ (**68**) ⇌ (Fermentation) $HO(O=)C\,C(=O)CH_2CH_2COOH$ (**69**)

69 → ($-CO_2$) $HC(=O)CH_2CH_2COOH$ (**70**) → ([H]) $HOCH_2CH_2CH_2COOH$ ⇌ **65a**

70 → ($RCOCOOH$, $-CO_2$) $R(O=)C$-lactone (**72**) ⇌ ([H]) $RCHOH$-lactone (**73**)

70 → (ROH) RO-lactone (**71**)

65a **73** **72** **71**

Scheme 17 – Formation of γ-butyrolactone and its γ-alkoxy- and γ-acyl derivatives from oxoacids [81b,129].

Degradation of the compound **69** leads to 4-oxobutyric acid **70** which is supposed to be a precursor in the formation of alkoxy- **71** and acyllactones **72** by aldol condensation with 2-oxoacids [81]. Only 5-ethoxy- and 5-acetyl-4,5-dihydro-2-(3*H*)-furanones have been identified [125].

Tressl *et al.* [129] studied the conversion of certain 4- and 5-oxoacids into optically active γ- and δ-lactones with *Saccharomyces cerevisiae.* They also investigated the formation by *Sporobolomyces odourus* of 4-decanolide and *cis* 6-dodecen-4-olide in a series of experiments using ^{14}C-labelled compounds. From these results, it appears that saturated lactones may be formed by γ- and δ-hydroxylation of the corresponding saturated acids [20] while unsaturated lactones are obviously derived from unsaturated fatty acids [78].

2-Butenoic acid γ-lactone **75** (R^1 = R^2 = R^3 = H) was found in the RP's *Amadori* and *Heyns* intermediates of glucose-glycine, and fructose-glycine model systems [90]. It was also formed in the thermal degradation of *N*-formyl-*L*-lysine-*D*-lactose and in lactose-casein browning reactions [31]. The 4-methyl derivative **75** (R^1 = R^2 = H, R^3 = CH_3) was identified in the same systems as well as in the thermal degradation of glucose [141]. It was formed by degradation of the acid **74** (*see Scheme 18*).

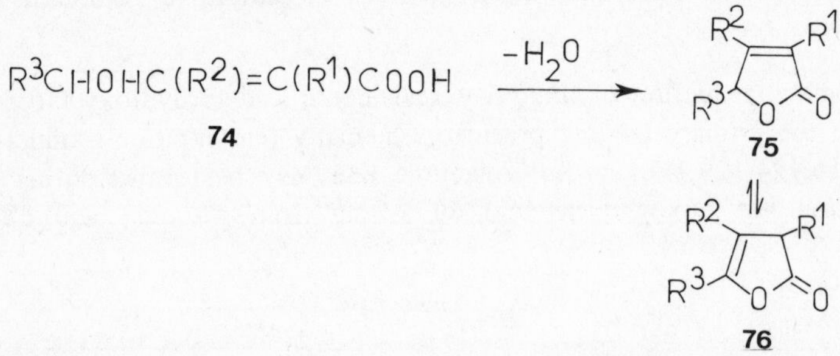

$R^3CHOHC(R^2){=}C(R^1)COOH$ ($\xrightarrow{-H_2O}$)

74 **75** ⇌ **76**

Scheme 18 – Formation of 2-butenoic acid γ-lactones from α,β-unsaturated γ-hydroxy acids.

Two other α,γ-unsaturated γ-lactones **75** (R^1 = R^2 = CH_3, R^3 = H and R^1 = H, R^2 = R^3 = CH_3) were also observed as well as 3-butenoic acid γ-lactone and its 4-methyl derivative **76** (R^1 = R^2 = H, R^3 = H, CH_3) or angelica lactone, in the lactose-casein system [31b]. According to Manley & Fagerson [71] α-angelica lactone is formed through carbohydrate degradation *via* 5-hydroxy-methyl-2-furan carboxaldehyde.

3-Hydroxy-4,5-dimethyl-2-(5*H*)-furanone **77** is obtained with *Saccharomyces cerevisiae* and is formed from oxobutyric acid **39a** arising from *L*-threonine [131] (*see Scheme 19*).

Benzofuran has been identified among the degradation products of glucose [46c] and its 2-methyl derivative in a reaction model between lactose and casein [31].

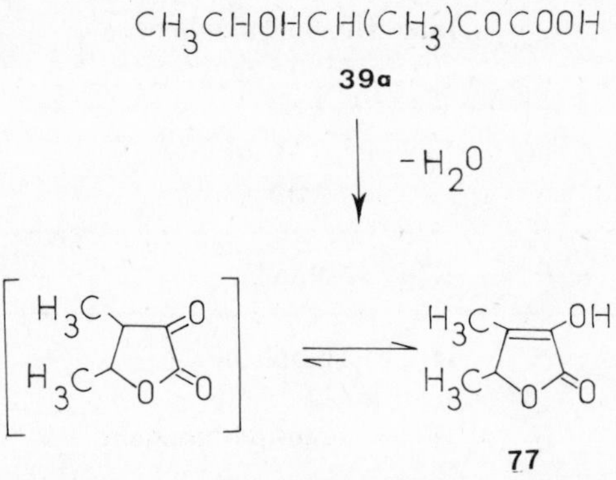

$CH_3CHOHCH(CH_3)COCOOH$

39a

$-H_2O$

77

Scheme 19 – Formation of 3-hydroxy-4,5-dimethyl-2-(5*H*)-furanone from 2-oxobutyric acid.

3.2 Thiophenes

Thiophene derivatives arise mainly from the thermal degradation of sulphur-containing amino acids alone [55d,73] or in the presence of reducing sugars: glucose [55c,55d], xylose [62a,84,139], ribose [79b] and pyruvaldehyde [55c]. Mussinan & Katz [84a] reported the formation of some thiophenes in a model system consisting of hydrolyzed vegetable protein (HVP)-*L*-cysteine (HCl)-*D*-xylose-water. They also result from the interaction of *D*-glucose or *L*-rhamnose with hydrogen sulphide with or without ammonia [114a,149].

Sugar degradation products such as 3-(2*H*)-furanones [89] and furfural [114c] can also react with hydrogen sulphide to afford thiophenes. It is well known that at high temperatures, thiophenes and tetrahydrothiophenes can be obtained by the action of hydrogen sulphide (or some other sulphur derivatives) upon the corresponding furans. Finally, thiophenes may arise in meat from dipeptides [73] and thiamin [24b]. In onion flavour alkylpropenyl disulphides [9–11] act as precursors of dimethylthiophenes, mainly the 3,4-isomer.

About thirty thiophenes **78** have been identified in these model systems. They are listed in *Table III–5*.

2-Alkylthiophenes, 2- and 3-thiophene carboxaldehydes, 2-mercaptothiophene and its *S*-methyl derivative, 2-thiophene carboxylic acid and its methyl ester, 2,3- and 2,5-disubstituted derivatives with alkyl and/or formyl and acetyl groups were among compounds found most often. Only four trisubstituted compounds have been reported [55d,104a].

2-Thiophene carboxaldehydes **79** and the corresponding alcohols **80** were formed in *Maillard's* reaction from α-dicarbonyl compounds **10** and hydrogen sulphide (*see Scheme 20, pathway a*).

They may also be obtained by the action of hydrogen sulphide on furfural [114c] as shown in *pathway b*. Other routes (*pathway c*) have been proposed by

Table III–5 Thiophenes

R3 R2 R4 S R1

110

110; R^1	R^2	R^3	R^4	Name	Model systems[a)]
H	H	H	H	Thiophene	*3,18,19,32,54, 65a,75,76*
CH_3	H	H	H	2-Methylthiophene	*3,18,19,54,60, 75,76*
C_2H_5	H	H	H	2-Ethylthiophene	*3,18,54,65a, 75,76*
n-C_3H_7	H	H	H	2-Propylthiophene	–
n-C_4H_9	H	H	H	2-Butylthiophene	–
CH_2COCH_3	H	H	H	1-(2′-Thienyl)-2-propanone	*18,75*
CH_2OH	H	H	H	2-Hydroxymethylthiophene	*75*
CH_2SH	H	H	H	2-Thiophenemethane thiol	*76*(tentative)
CHO	H	H	H	Thiophene-2-carboxaldehyde	*3,19,32,76,80*
$COCH_3$	H	H	H	2-Acetylthiophene	*3,19,75,78*
COC_2H_5	H	H	H	2-Propionylthiophene	*3,18*
$COCOCH_3$	H	H	H	1-(2′-Thienyl)-1,2-propanedione	*75*
COOH	H	H	H	2-Thiophenic acid	*65b*
$COOCH_3$	H	H	H	Methyl 2-thienyl carboxylate	*75*
2- (or 3-)OH	H	H	H	2- (or 3-)-Hydroxythiophene	*3*
SH	H	H	H	2-Mercaptothiophene	*3,76*
SCH_3	CH_3	H	H	2-Methylthiothiophene	*3*
H	CH_3	H	H	3-Methylthiophene	*3,18,19,65a,75, 76*
H	$COCH_3$	H	H	3-Acetylthiophene	*32*
CH_3	CH_3	H	H	2,3-Dimethylthiophene	*3,54,76*(tenta-tive)
C_2H_5	CH_3	CH_3	H	2-Ethyl-3-methylthiophene	–
n-C_3H_7	CH_3	H	H	2-Propyl-3-methylthiophene	*54*
CHO	CH_3	H	H	2-Formyl-3-methylthiophene	*32*
$COCH_3$	CH_3	H	H	2-Acetyl-3-methylthiophene	*32,75*
CH_3	H	CH_3	H	2,4-Dimethylthiophene	*76*
CH_3	H	H	CH_3	2,5-Dimethylthiophene	*3,18,19,65a.75, 76*
CHO	H	H	CH_3	2-Formyl-5-methylthiophene	*19,32,75*
$COCH_3$	H	H	CH_3	2-Acetyl-5-methylthiophene	*32,75,78*
CH_3	CH_3	H	CH_3	2,3,5-Trimethylthiophene	*18,54*

Table III–5 (*continued*)

110;R[1]	R[2]	R[3]	R[4]	Name	Model systems[a)]
CH_3	3- (or 4-)	C_2H_5	CH_3	2,5-Dimethyl-3 (or 4-) ethylthiophene	*54*
CH_3	H	CH_3	C_2H_5	2,4-Dimethyl-5-ethylthiophene	*54*
CH_3	CHO	H	CH_3	2,5-Dimethyl-3-formylthiophene	*75*

a) See the caption for Table III–3; *32* ribose-cysteine/cystine; *54* cysteine; *60* thiamin

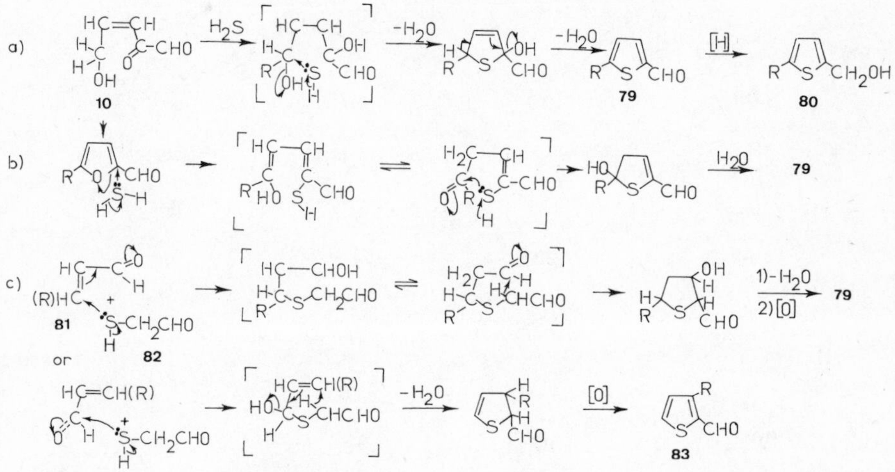

Scheme 20 – Various pathways for the formation of 2-formylthiophenes.

Mulders [79b] starting from α,β-unsaturated aldehydes **81** and α-mercaptoaldehydes **82**. If the reaction is initiated by *Michael* addition of the mercapto group to the α,β-unsaturated system, 2-formyl-5-substituted thiophenes **79** are obtained.

3-Substituted derivatives **83** were obtained by nucleophilic attack of the free doublet of the sulphur atom on the carbonyl group. Thiophene and its 2-methyl derivative were identified in the taurine and cysteine pyrolysis, respectively. 3-Alkyl derivatives found from cysteine pyrolysis may be formed by a mechanism similar to that proposed by Mulders. However, other pathways are also possible since 3-methyltetrahydrothiophene **84** (R = 3-CH_3) and 4,5-dihydro-2 (or -3)-ethylthiophene **85** (R = 2- (or 3)-C_2H_5) are formed in this reaction [55d]. The 2-ethyl derivative **85** (R = 2-C_2H_5) was also reported in *D*-glucose-hydrogen sulphide-ammonia system [114a].

4,5-Dihydro-2-(3*H*)-thiophenone **86** results from the interaction of cysteine with xylose. Tetrahydrothiophene **84** (R = H), the 4,5-dihydro-3-(2*H*)-thiophenone **87** (R = H) and their 2-methyl derivatives have been identified in the xylose-cysteine system [84a]. The compound **87b** (R = CH_3) a fermentation product of methionine, was also obtained by the reaction of pyruvaldehyde with cysteine [55c]. The corresponding unsaturated derivative **88** (R = CH_3) has been reported in a reaction between *D*-glucose and *L*-cysteine hydrochloride [105].

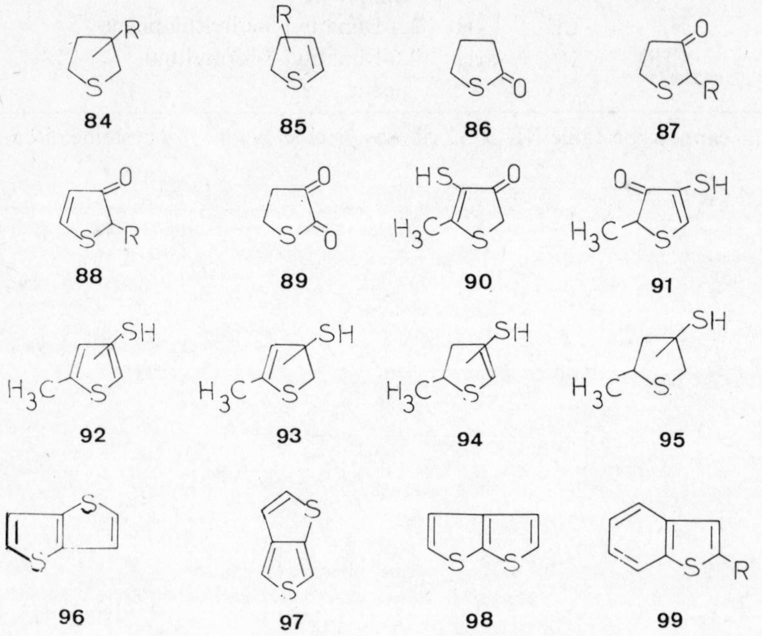

Fig. 1 – Thiophene derivatives in model systems.

Four thiophenes, i.e., 2-methyl-, 2-methyl-4,5-dihydro, 2-acetyltetrahydro-, and 2-methyl-4,5-dihydro-3-(2*H*)-thiophenone, have been identified among the pyrolysis products of thiamin [24b]. Although the reaction sequence of the formation of these products is not known [23b], it seems probable that an intermediate resulting from the homolytic cleavage of the $1_{1,2}$ and $1_{3,4}$ bonds can produce the furanone **87b** after rearrangement and cyclization (*see Scheme 21*).

Compounds **88** (R = H and CH_3) and the 2,3-dioxotetrahydrothiophene **89** were reported in the glucose-hydrogen sulphide model system [104a] among about twenty thiophene derivatives. Hydrogen sulphide reacts with 4-hydroxy-5-methyl-3-(2*H*)-furanone **60b** (R = H) to produce a mixture of 3- and 4-mercaptothiophenes **90** to **95** (see Fig. 1) which possess a characteristic meat aroma [89]. 2-Methyl-3-mercaptothiophene **92** (3–SH) was identified among these products by Evers [28] in a reaction mixture containing thiamin, cysteine, and hydrolyzed vegetable protein in water. The corresponding disulphide was also found in a relatively large amount.

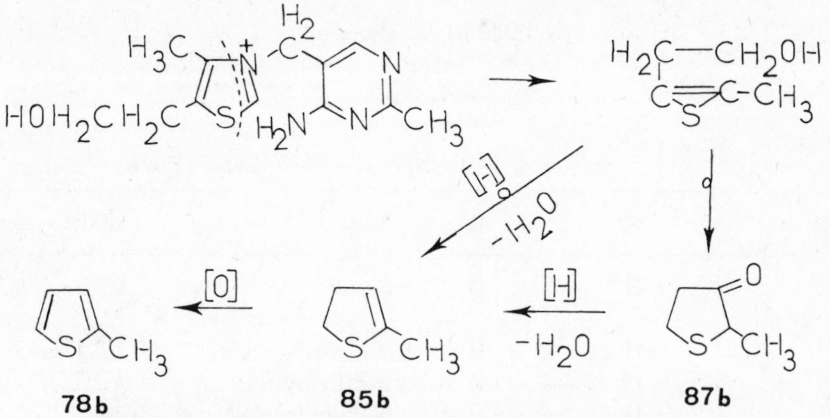

78b **85b** **87b**

Scheme 21 – 2-Methylthiophene formation from thiamin.

[3,2-b]- and [3,4-b]-Thienothiophenes **96** and **97** have been found to occur in various model reactions between cysteine (or its hydrochloride) and reducing sugars such as ribose [79b] and *D*-glucose [105]. With xylose only the isomer [2,3-b] **98** is formed [84a]. [3,2-b]-Thienothiophene may be formed from 3-thienylthioacetic acid [42]. Benzo[*b*]thiophene **99** (R = H) was found in the *D*-glucose-*L*-cysteine hydrochloride system [105].

3.3 Pyrroles

Many pyrrole derivatives found in food aromas and in tobacco arise *via Maillard* or *Strecker* reactions between sugars and amino acids during processing [19,63, 66]. To date, more than fifty pyrroles **100** including 1-, and 2-substituted 2,5-disubstituted derivatives, some pyrrolines, pyrrolidines and 2-pyrrolidinones (or γ-lactams) and various fused systems, have been identified in nearly twenty model systems (*see Table III–6*).

Amadori intermediates obtained by reacting proline and hydroxyproline with glucose are potent precursors in the formation of pyrroles [131]. Among them, 1-acetonylpyrrole **102** previously reported in glucose-hydroxyproline system [58b] is probably formed from hydroxyproline **101** and pyruvaldehyde (*see Scheme 22*).

Several *N*-substituted pyrroles **103** to **105** can be effectively obtained by reaction of α-dicarbonyl compounds, cyclic enolones, α,β-unsaturated aldehydes and 2-formylfurans with hydroxyproline [131]. *N*-Furfurylpyrrole **105** (R = CH_3) was also formed either from proline and 5-hydroxy-5,6-dihydromaltol [77] or by heating of glucose with *DL*-alanine [116b]. A large number of pyrroles, mainly 2-acyl derivatives, were identified in the following model systems: *D*-glucose-ammonia [115], *D*-glucose-acetic acid-methylamine [53], *L*-rhamnose-ammonia [114b], *L*-rhamnose-ammonia-hydrogen sulphide [149] and lactose-casein [31]. Japanese workers [116] suggested that compounds **107a** are formed by interaction of 3-deoxyosuloses and ammonia (or amines) arising from sugars and amino acids (*see Scheme 23*).

Table III–6 Pyrroles

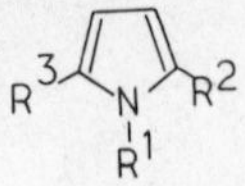

(structure **100**: pyrrole ring with R^1 on N, R^2 and R^3 at the 2- and 5-positions)

100; R^1	R^2	R^3	Name	Model systems[a)]
H	H	H	Pyrrole	*1,2,49,57b,58b, 71,77*
CH_3	H	H	*N*-Methylpyrrole	*1,2,57b,63*
C_2H_5	H	H	*N*-Ethylpyrrole	*1,2,35*
n-C_3H_7	H	H	*N*-Propylpyrrole	*2*
n-C_4H_9	H	H	*N*-Butylpyrrole	–
CH_2CH_2–N(pyrrolyl) 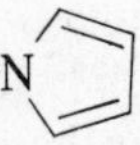	H	H	1,2-Di-(*N*-pyrrolyl)-ethane	*2*
$CH_2CH(CH_3)$–N(pyrrolyl) 	H	H	1-Methyl-1,2-di(*N*-pyrrolyl)-ethane	*2*
CH_2COCH_3	H	H	*N*-Acetonylpyrrole	*2*
$CH_2COC_2H_5$	H	H	1-(*N*-Pyrrolyl)-2-butanone	*2*
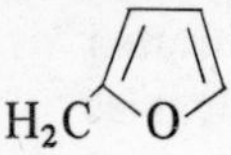 H_2C–(2-furyl)	H	H	*N*-Furfurylpyrrole	*2,71*
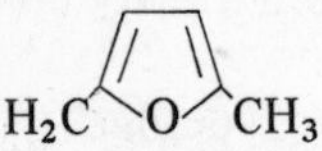 H_2C–(5-methyl-2-furyl), CH_3	H	H	*N*-(5′-Methylfurfuryl)-pyrrole	*2,71*
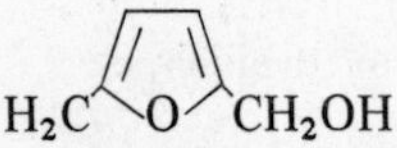 H_2C–(5-hydroxymethyl-2-furyl), CH_2OH	H	H	*N*-(5′-Hydroxymethyl)-pyrrole	*2*
i-C_3H_7	H	H	*N*-Isopropylpyrrole	*2*
$CH(CH_3)COCH_3$	H	H	*N*-(1′-Methylacetonyl)-pyrrole	*2*
$CH(CH_3)COC_2H_5$	H	H	1-(*N*-Pyrrolyl)-1-methyl-2-butanone	*2*
$CH(C_2H_5)COCH_3$	H	H	1-(*N*-Pyrrolyl)-1-ethyl-2-propanone	*2*
CHO	H	H	*N*-Formylpyrrole	*71,77*
H	CH_3	H	2-Methylpyrrole	*1,49,50,59b,71, 77*
H	C_2H_5	H	2-Ethylpyrrole	*1,77*

Table III–6 (*continued*)

100; R^1	R^2	R^3	Name	Model systems[a)]
H	n-C_3H_7	H	2-Propylpyrrole	*77*
H	n-C_4H_9	H	2-Butylpyrrole	–
H	CH_2COCH_3	H	2-Acetonylpyrrole	*77*
H	CHO	H	2-Formylpyrrole	*36,77,80*
H	$COCH_3$	H	2-Acetylpyrrole	*28,34–37,71,77, 78*
CH_3	CHO	H	*N*-Methyl-2-formyl-pyrrole	*28,37,63,80*
CH_3	$COCH_3$	H	*N*-Methyl-2-acetyl-pyrrole	*34,37,74*
CH_3	$COCH_2OH$	H	2-(2-Hydroxyacetyl)-1-methylpyrrole	*74*
C_2H_5	CH_3	H	*N*-Ethyl-2-methylpyrrole	*35*
C_2H_5	CHO	H	*N*-Ethyl-2-formylpyrrole	*36*
i-C_4H_9	CHO	H	*N*-Isobutyl-2-formyl-pyrrole	*67*
H	CH_3	CH_3	2,5-Dimethylpyrrole	*1,49,63*
H	CHO	CH_3	2-Formyl-5-methyl-pyrrole	*34,35,37,63,71*
H	CHO	C_2H_5	2-Formyl-5-ethylpyrrole	–
H	CHO	C_3H_7	2-Formyl-5-propylpyrrole	*71,77*
CH_7	CHO	CH_3	2-Formyl-1,5-dimethyl-pyrrole	–
C_2H_5	CHO	CH_3	*N*-Ethyl-2-formyl-5-methylpyrrole	*35,36*
C_2H_5	$COCH_3$	CH_3	*N*-Ethyl-2-acetyl-5-methylpyrrole	*36*

a) Model systems: 3-Methyl-4-ethylpyrrole was found from serine and threonine, respectively (55b); *1* glucose-*L*-proline; *2* glucose-hydroxyproline; *28* lactose casein; *34* RP's glucose-glycine; *35* RP's glucose-theanine; *36* dry RP's glucose-theanine; *37* wet RP's glucose-theanine; *37* RP's fructose-glycine; *49* serine; *50* threonine; *57b* proline or hydroxyproline; *58b* glycyl-serine; *59* alanine-serine; *63* trigonelline; *67* α-amino acids-furfural; *71* glucose-NH_4OH (5M); *74* glucose-AcOH-CH_3NH_2; *77* *L*-rhamnose-ammonia; *78* *L*-rhamnose-H_2S-NH_3; *80* furfural-H_2S-NH_3.
RP: reaction products.

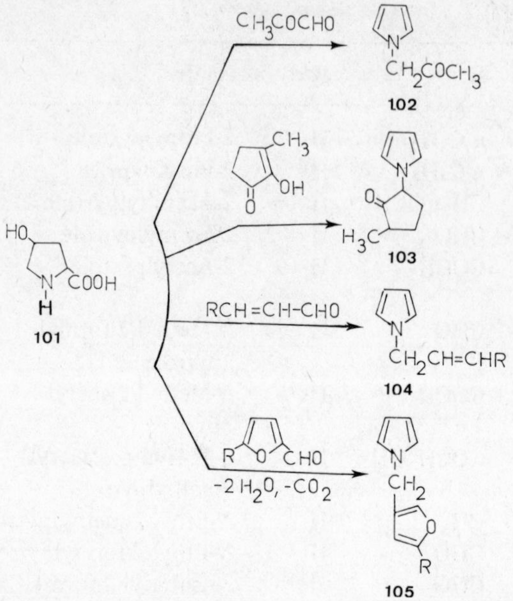

Scheme 22 – Formation of heterocyclic aroma compounds from hydroxyproline [58b,131].

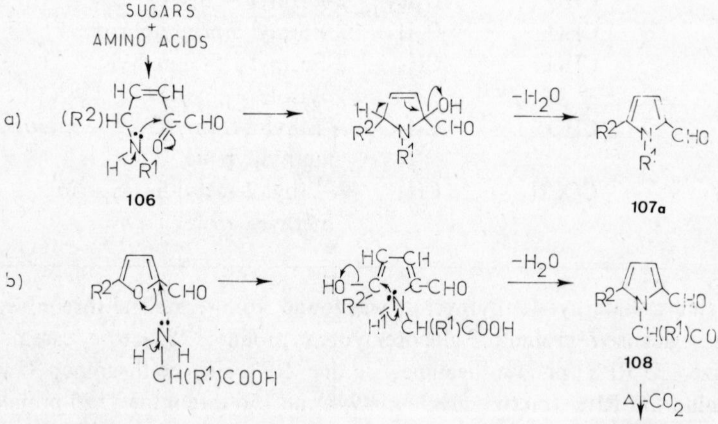

Scheme 23 – Formation of 1-alkyl-2-formylpyrroles from sugars or furfural and α-amino acids.

Rizzi [99b] reported that 1-alkyl-2-formylpyrroles **107b** can also be formed by reaction of furfural and its homologue with α-amino acids (*pathway b*). Thus, with valine *N*-isobutyl-2-formylpyrrole **107b** (R^1 = *i*-C_3H_7, R^2 = H) is obtained. Acids **108** have been identified in the reaction products of xylose with glycine and leucine, respectively [56]. The interaction of furfural with hydrogen sulphide and ammonia [114c] leads to 5-methyl-2-formylpyrrole among other products (*see Chapter V*).

1-β-Hydroxyethylpyrrole is formed in a browning model reaction of glycol aldehyde with amino ethanol [103]. Pyrrole and its 2-methyl derivative has been found among the pyrolysis products of proline and glycylalanine [55c,55d, 73]. *N*-Methylpyrrole arise from hydroxyproline pyrolysis [73].

Four pyrroles (*N*-methyl-, 2,5-dimethyl-, *N*-methyl-2-formyl-, and 5-methyl -2-formyl) have been observed among the degradation products of trigonelline [133]. In this case, they arise from the contraction of the pyridine ring. Partially or totally saturated pyrrole rings were also observed in model systems. Thus, by heating 1-deoxy-1-*L*-prolino-*D*-fructose [109] in vacuo at 240°C, Mills & Hodge [77] obtained variously substituted pyrrolines **110, 114** (R = CH_3, CH_2OH), and pyrrolidines **111** (R = CH_2OH, CHO), **112** (R = CHO, $COCH_3$, COC_2H_5), **113** (R = CH_3, CH_2OH), and **115**.

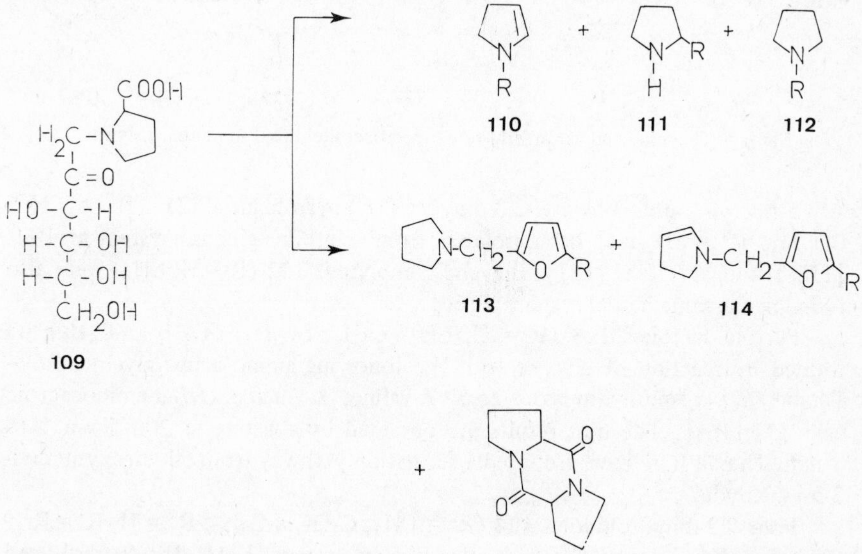

Scheme 24 – Pyrrolines and pyrrolidines by thermal decomposition of 1-deoxy-1-*L*-prolino-*D*-fructose [77].

1-Methylpyrroline **110** (R = CH_3) has been reported in the *L*-rhamnose-ammonia system [114b] and pyrrolidine **111** (R = H) by heating *D*-glucose and *L*-alanine (104°C, pH 9.6) [36b]. This latter compound was also found among the degradation products of the *Amadori* and *Heyns* rearrangement products of the fructose-glycine system [90]. The following *N*-substituted pyrrolidines **112** (R = CHO, CH_3CO, C_2H_5CO and *n*-C_3H_7CO) were characterized in the proline-glucose system [50,131].

Theanine **116** (*see Fig. 2*) is the precursor, *via Strecker* degradation of *N*-ethylsuccinimide **117** found in the black tea aroma [13]. This latter compound and γ-lactams **118** (R = H, C_2H_5) and **119** were characterized as degradation products of the glucose-theanine system [90]. *N*-Methyl-2-pyrrolidinone **118** (R = CH_3) was found in lactose-casein system [31]. In wine the γ-lactam **120** is formed from the corresponding diacid [81b].

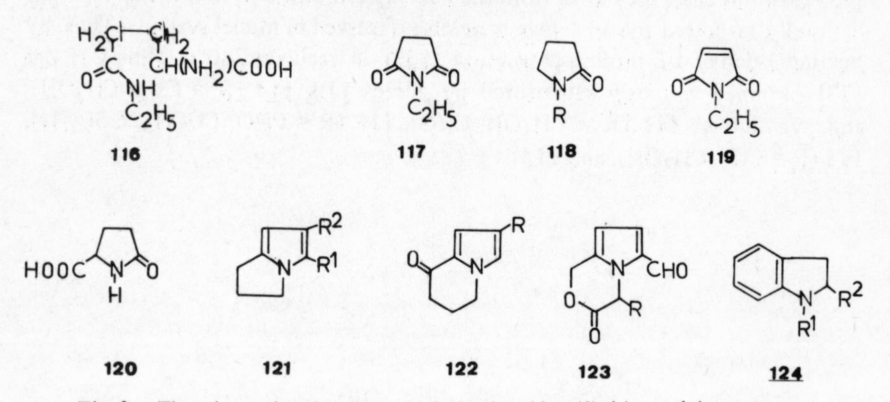

Fig. 2 – Theanine and various pyrrole derivatives identified in model systems.

5-Acetyl-, and 5-formyl-2,3-dihydro-(1*H*)-pyrrolizines **121** (R^1 = CHO, $COCH_3$, R^2 = H) have been isolated from a proline-glucose system at 100° [116c] and at 250°C [131]. Bicyclic compounds **122** (R = H, CH_3) were also found in the same model system [116c].

Pyrrole lactones **123** (R = H, CH_3, C_2H_5, *i*-C_3H_7, *i*-C_4H_9, *n*-C_4H_9) are formed by reaction of glucose with the following amino acids: glycine, *DL*-α-alanine, *DL*-α-amino-*n*-butyric acid, *L*-valine, *L*-leucine, *DL*-α-aminocaproic acid [37,116a]. The best results are obtained by roasting at 200°C for 1 to 3 min. The authors have proposed a formation pathway from 5-hydroxymethyl-2-furaldehyde.

Four 2,3-dihydroindoles **124** (R^1 = CH_3, C_2H_5, *n*-C_3H_7, R^2 = H; R^1 = R^2 = CH_3) have been identified in a proline-glucose system [131]. The *N*-ethyl-, and *N*-propyl derivatives were also characterized in hydroxyproline-glucose system [131]. The mechanism of their formation have not been discussed, but the presence in traces of indoles by heating furan and ammonia is known.

3.4 Oxazoles and Oxazolines

So far very few oxazoles and oxazolines have been identified in model systems [113b,115]. Possible pathways for the formation of alkyl-, and acyloxazoles in coffee aroma were discussed by Vitzthum & Werkhoff [137a] (*see Scheme 25*). The decarboxylation of hydroxylated amino acids **125** such as serine or threonine leads to β-amino alcohols **126**. Their condensation with aldehydes gives the corresponding Schiff bases **127**. Cyclization of these latter products affords the oxazolidines **128** and then the oxazoles **129** by oxidation. Thus, starting from pyruvaldehyde and serine, 2-acetyloxazole **129** (R^1 = $COCH_3$, R^2 = H) is obtained. 2,4-Dimethyloxazole can be formed by a similar mechanism from the α-amino alcohol $CH_3CH(NH_2)CH_2OH$ and acetaldehyde. However, it can also be formed by another pathway since it has been observed among the products resulting from the reaction of hydrogen sulphide and ammonia with furfural [114c]. In this case 2-hydroxymethylpyrrole formed by reduction of 2-formylpyrrole, a component of this reaction mixture, may act as precursor (*see Chapter V*).

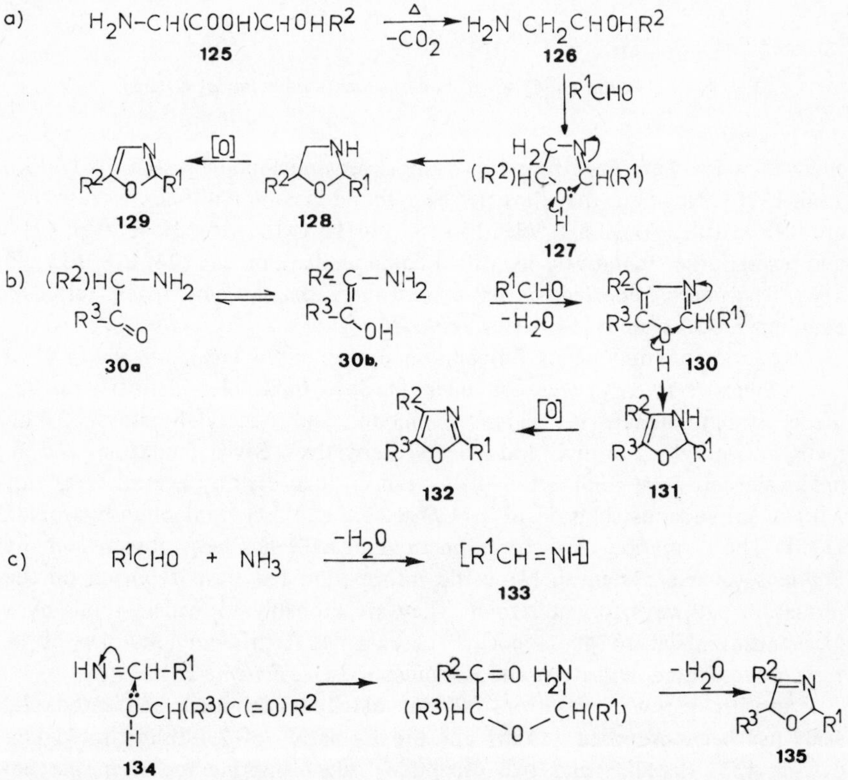

a) $H_2N{-}CH(COOH)CHOHR^2$ (**125**) $\xrightarrow[-CO_2]{\Delta}$ $H_2N\,CH_2CHOHR^2$ (**126**) $\xrightarrow{R^1CHO}$ **127** → **128** $\xrightarrow{[O]}$ **129**

b) **30a** ⇌ **30b** $\xrightarrow[-H_2O]{R^1CHO}$ **130** → **131** $\xrightarrow{[O]}$ **132**

c) $R^1CHO + NH_3 \xrightarrow{-H_2O} [R^1CH{=}NH]$ (**133**)

134 → → $\xrightarrow{-H_2O}$ **135**

Scheme 25 – Formation of oxazoles and oxazolines from aldehydes and hydroxylated amino acids, α-aminoketones and acyloins-ammonia model systems.

Trisubstituted oxazoles **132** can be obtained from aldehydes and α-amino-ketones **30**, *via* oxazolines **131** (*pathway b*). The oxazoles **132** ($R^1 = CH_3$, CHO, $COCH_3$, $R^2 = R^3 = H$ or CH_3) and 2-ethyl-4,5-dimethyloxazole found in *L*-rhamnose-ammonia [113b] and *D*-glucose-ammonia [115] model systems, respectively, may be formed by a similar mechanism from sugar degradation products and ammonia. Mussinan *et al.* [84b] have suggested that 2,4,5-trimethyl-3-oxazoline **135** ($R^1 = R^2 = R^3 = CH_3$) found in heated meat systems can also be formed from aldimine **133** ($R^1 = CH_3$) and acetoin **134** ($R^2 = R^3 = CH_3$) all of which have been reported to be present in cooked meat (*see Scheme 25, pathway c*). This compound has been observed by condensing alanine and diacetyl. 2-Ethyl-4 (or 5)-acetyl-2-oxazoline **136** has been identified among the wet degradation products of RP's of glucose-theanine system [90]. 5-Vinyl-2-oxazolidinone **137**, also called goitrin, was found among the products resulting from non-enzymatic breakdown of progoitrin, a component of rapeseed meal [43].

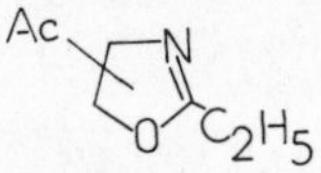

Fig. 3 – 2-Oxazoline and 2-oxazolidinone identified in model systems.

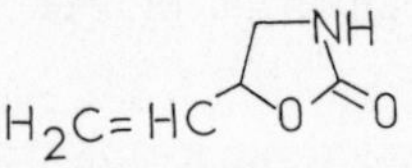

3.5 Imidazoles

Imidazoles **134** have been found in many sugar-amine model systems [61,113b, 114b,132b]. However, they have not been found in common foods because they are less volatile. Four imidazoles **134** ($R^1 = CH_3$, C_2H_5, $R^2 = H$, $R^3 = H$, CH_3) and several other tentatively identified compounds (mol. wt. 124, 138, 152 and 166) have been reported to be present in *L*-rhamnose-hydrogen sulphide-ammonia model system [149] (*see Table III–7*).

By action of ammonia on *L*-rhamnose, pyrazines (74.41%), pyrroles (5.43%) and imidazoles (4.89%) were obtained [113b]. Imidazoles identified in this model system include the parent compound and 1-acetyl-4-methyl-, 2,4-dimethyl-, 2-ethyl-, 2-formyl- and 2-acetyl derivatives. Seven imidazoles and an imidazole derivative (mol. wt. 138) have been identified by heating (100° for 2 h.) of an aqueous solution of 1 M *D*-glucose and 5 **N** ammonium hydroxide [115]. The formation of 4- (or 5)-methylimidazole has been observed in the *D*-glucose-glycine system [61]. Little information has been reported on the formation pathways for imidazoles. They are probably formed in foods by a mechanism similar to that reported for oxazoles from α-aminoketones **30** or their *N*-substituted derivatives and aldimines **133** (*see Scheme 26*).

Another pathway (b) from α-amino aldehydes-ammonia and carboxylic acids has been proposed [113b] for the formation of 2,4-disubstituted imidazoles **135**. Histidine and two dipeptides called anserine and carnosine can also play an important role in the formation of imidazoles in food aromas, particularly in red meats (*see Chapter I*).

Table III–7 Imidazoles

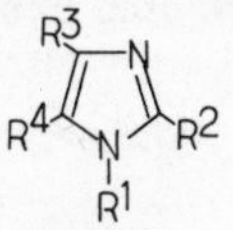

202; R^1	R^2	R^3	R^4	Name	Occurrence[a)]
H	H	H	H	Imidazole	*77*
CHO	H	H	H	Imidazole-1-carboxaldehyde	*71*(tentative)
H	CH_3	H	H	2-Methylimidazole	*71,77,78*
H	C_2H_5	H	H	2-Ethylimidazole	*71,77,78*
H	CHO	H	H	Imidazole-2-carboxaldehyde	*71*
H	$COCH_3$	H	H	2-Acetylimidazole	*77*(tentative)
H	H	4- (or 5-) CH_3		4- (or 5-)Methylimidazole	*5*
CH_3	CH_3	H	H	1,2-Dimethylimidazole	*71*
C_2H_5	$COCH_3$	H	H	1-Ethyl-2-acetylimidazole	*77*
CH_3	H	CH_3	H	1,4-Dimethylimidazole	*71*
$COCH_3$	H	CH_3	H	1-Acetyl-4-methylimidazole	*71,77*
H	CH_3	CH_3	H	2,4-Dimethylimidazole	*77*
H	CH_3	H	CH_3	2,5-Dimethylimidazole	*78*
H	C_2H_5	H	CH_3	2-Ethyl-5-methylimidazole	*78*

a) Model systems: *5 D*-Glucose-glycine; *71 D*-glucose-ammonia; *77 L*-rhamnose-ammonia; *78 L*-rhamnose-hydrogen sulphide-ammonia.

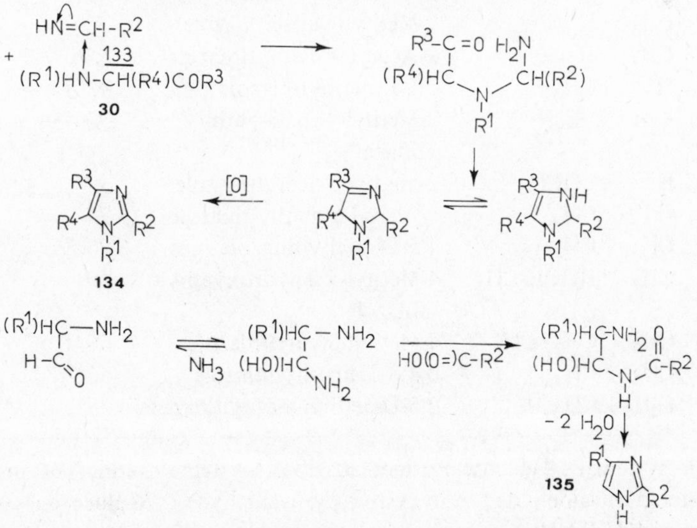

Scheme 26 – Formation of imidazoles from α-aminocarbonyl compounds.

3.6 Thiazoles

Thiazoles are relatively widely distributed in model systems (*see Table III–8*).

Table III–8 Thiazoles

(structure 141: thiazole ring with substituents R^1 at C-2, R^2 at C-4, R^3 at C-5)

141; R^1	R^2	R^3	Name	Model systems[a)]
H	H	H	Thiazole	*65a,76*
CH_3	H	H	2-Methylthiazole	*4,32,65b*
C_2H_5	H	H	2-Ethylthiazole	*65a,76,78*
n-C_3H_7	H	H	2-Propylthiazole	–
n-C_4H_9	H	H	2-Butylthiazole	–
$COCH_3$	H	H	2-Acetylthiazole	*32,78*
COC_2H_5	H	H	1-(2′-Thiazolyl)-1-propanone	*32*
COC_3H_7	H	H	1-(2′-Thiazolyl)-1-butanone	*32*
2-furyl (structure)	H	H	2-(2′-Furyl)thiazole	*32*
H	CH_3	H	4-Methylthiazole	*76*
H	4- (or 5-)C_2H_5		4- (or 5-)-Ethylthiazole	*65a*
H	H	CH_3	5-Methylthiazole	*76,78*
H	H	C_2H_5	5-Ethylthiazole	*32,65a*
CH_3	CH_3	H	2,4-Dimethylthiazole	*4,78*
CH_3	C_2H_5	H	2-Methyl-4-ethylthiazole	*78*
$COCH_3$	CH_3	H	2-Acetyl-4-methylthiazole	*65b*
CH_3	H	CH_3	2,5-Dimethylthiazole	*4,78*
CH_3	4- (or 5-)C_2H_5		2-Methyl-4-(or 5-)-ethylthiazole	*32,65b*
CH_3	H	$COCH_3$	2-Methyl-5-acetylthiazole	*32*
$COCH_3$	H	CH_3	2-Acetyl-5-methylthiazole	*78*
H	CH_3	CH_3	4,5-Dimethylthiazole	*76,78*
H	CH_3	$HOCH_2CH_2$	4-Methyl-5-β-hydroxyethylthiazole	*60*
H	CH_3	CH_2=CH	4-Methyl-5-vinylthiazole	*60*
CH_3	CH_3	CH_3	2,4,5-Trimethylthiazole	*4,32,78*
CH_3	C_2H_5	CH_3	2,5-Dimethyl-4-ethylthiazole	*78*

a) Model systems: *4* glucose-cystine; *32* ribose-cysteine/cystine; *60* thiamin; *65a* cysteine-pyruvaldehyde; *65b* cystine-pyruvaldehyde. *76* glucose-H_2S-NH_3; *78* *L*-rhamnose-H_2S-NH_3.

The thermal degradation of cysteine and cystine either alone [55d] or in the presence of reducing sugars such as glucose [55c], ribose [79b], and xylose [62a] is the best source of thiazoles. Thiazole itself was reported in the cysteine-pyruvaldehyde system [55c] and its 2-methyl- and 2-ethyl derivatives were obtained by pyrolysis of cysteine and cystine [55d].

Thiazolylketones **141** (R^1 = $COCH_3$, COC_2H_5, COC_3H_7-*n*, R^2 = R^3 = H) have been identified in the cysteine/cystine-ribose system [79b] along with the following thiazoles: 2-methyl-, 2-furyl-, 5-ethyl-, 2-methyl-4 (or 5)-ethyl-, 2,4,5-trimethyl-, and 2,4-dimethyl-5-ethyl derivatives. The reaction of hydrogen sulphide and ammonia with rhamnose gives twelve compounds of this category [149]. Thirteen thiazolines and ten thiazoles were also tentatively identified in this reaction. Five thiazoles estimated to be approximately 2% of the total yield of nitrogen heterocyclic compounds, were identified in the *D*-glucose-hydrogen sulphide-ammonia model system [114a].

Takken *et al.* [122] obtained heterocyclic compounds such as oxazoles, thiazoles, and thiazolines which had a meaty flavour in a model system consisting of α-dicarbonyl compound (or aldehydes), hydrogen sulphide, and ammonia. 4-Methyl-5-(β-hydroxyethyl)-thiazole is the major ether soluble in acidic and neutral thiamin solutions [24a]. Two other thiazoles **141** (R^1 = H, R^2 = R^3 = CH_3; R^1 = H, R^2 = CH_3, R^3 = CH_2=CH) were also found in this reaction mixture [23,24]. All arise from the cleavage of the thiazole moiety of thiamin (*see Scheme 21*).

According to Schutte [106a], 2-isobutylthiazole **138** (R^1 = *i*-C_4H_9) is formed from cysteamine **136** or cysteine and isovaleraldehyde. This aldehyde which can be derived from leucine occurs in relatively large amounts in fresh tomatoes (*see Scheme 27, pathway a*).

A similar mechanism has been used by Mulders [79b] to explain the formation of 2-acetylthiazole from cysteine and pyruvaldehyde. Thiazole derivatives found in reducing sugars-hydrogen sulphide-ammonia browning systems may be formed from other pathways such as those shown in *Scheme 27* (*pathways b and c*).

2-Methyl-2-thiazoline **144** (R^1 = CH_3, R^2 = R^3 = H) was obtained by pyrolysis of cysteine and cystine alone [55d] or in the presence of glucose [55c]. The 2-ethyl derivative was characterized only in the cysteine-pyruvaldehyde system [55c].

2,4,5-Trimethyl-3-thiazoline **143** (R^1 = R^2 = R^3 = CH_3) found in the 3-mercapto-2-butanone-acetaldehyde-ammonia system (*see Scheme 27, pathway c*) is formed by a mechanism similar to that proposed for the corresponding oxazoline [84b]. 2-Methylthiazolidine **137** (R^1 = CH_3) identified among the pyrolysis products of sulphur-containing amino acids [36,55d] may be formed from acetaldehyde and cysteamine (*see Scheme 27, pathway a*). Its *N*-nitroso derivative **139** (R^1 = CH_3) was isolated from a cysteamine-acetaldehyde-sodium nitrite model system [104b].

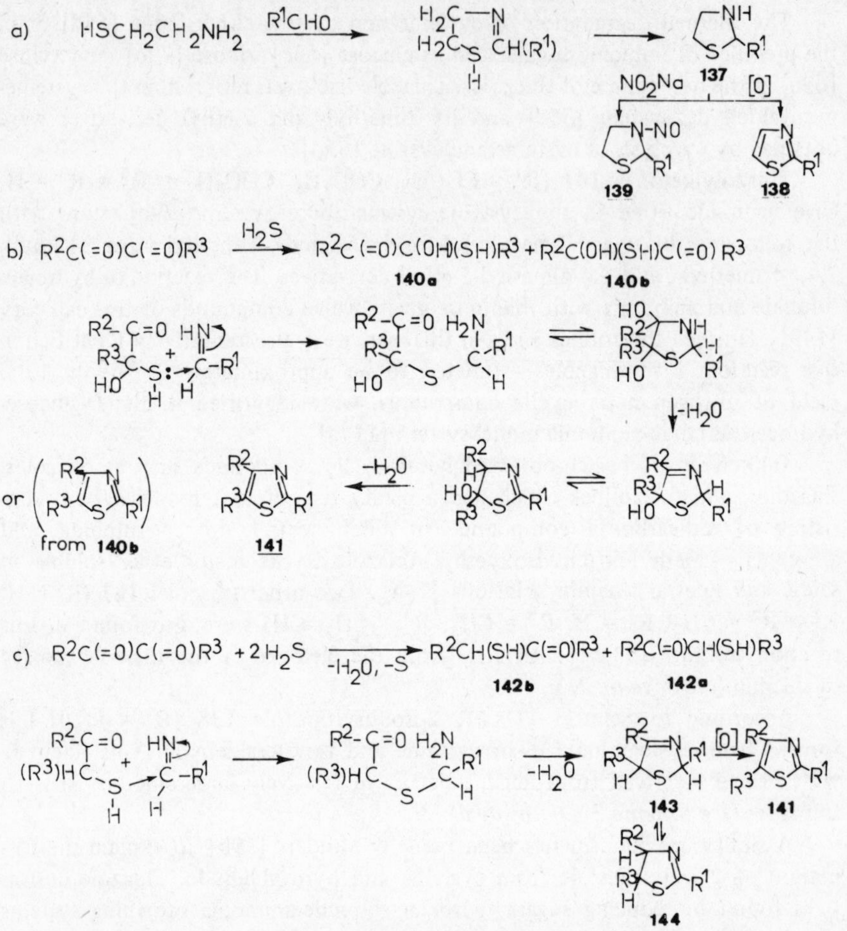

a) $HSCH_2CH_2NH_2$ $\xrightarrow{R^1CHO}$ H_2C—N, H_2C–S(H) CH(R^1) ⟶ thiazolidine (NH, S, R^1) **137**

137 —NO_2Na⟶ N-NO thiazolidine (R^1) **139**; **137** —[O]⟶ thiazole (R^1) **138**

b) $R^2C(=O)C(=O)R^3$ $\xrightarrow{H_2S}$ $R^2C(=O)C(OH)(SH)R^3$ + $R^2C(OH)(SH)C(=O)R^3$

140a **140b**

R^2–C=O, R^3–C(HO)–S:(H) + HN=C(H)–R^1 ⟶ R^2–C=O H_2N, R^3–C(HO)–S–C(H)–R^1 ⇌ HO, R^2, R^3, NH, S, H, HO, R^1

↓ $-H_2O$

or (R^2, R^3, N, S, R^1) from **140b** — R^2, R^3, N, S, R^1 **141** ⟵ $-H_2O$ — R^2, H, HO, N, S, R^3, R^1 ⇌ R^2, R^3, N, H, HO, S, R^1

c) $R^2C(=O)C(=O)R^3$ + 2 H_2S $\xrightarrow[-H_2O,\ -S]{}$ $R^2CH(SH)C(=O)R^3$ + $R^2C(=O)CH(SH)R^3$

142b **142a**

R^2–C=O, $(R^3)HC$–S(H) + HN=C(H)–R^1 ⟶ R^2–C=O H_2N, $(R^3)HC$–S–C(H)–R^1 $\xrightarrow{-H_2O}$ R^2, R^3, H, N, S, H, R^1 **143** $\xrightarrow{[O]}$ R^2, R^3, N, S, R^1 **141**

143 ⇅ R^2, H, N, R^3, H, S, R^1 **144**

Scheme 27 – Formation of thiazoles, thiazolines and thiazolidines.

No specific formation mechanism was proposed in the literature for benzothiazole **145** (R = H).

It seems that the most likely explanation would be the thermal interaction of an *Amadori* intermediate [13a] with hydrogen sulphide, ammonia and aldehydes as shown in *Scheme 28*.

3.7 Pyrans and pyrones

Maltol **148** and its 5-hydroxy derivative **149** have been identified in several model systems such as sugars-amine [76], *D*-glucose-acetic acid-methylamine [53] and RP's of glucose-glycine and fructose-glycine systems [90]. They were also formed by decomposition of 1-deoxy-1-*L*-prolino-*D*-fructose at 140°C [77], and by heating of 5-hydroxy-5,6-dihydromaltol **147** with *L*-proline. 5-Hydroxymaltol **149** was also found in the wet degradation of the reaction products of the glucose-theanine system [90].

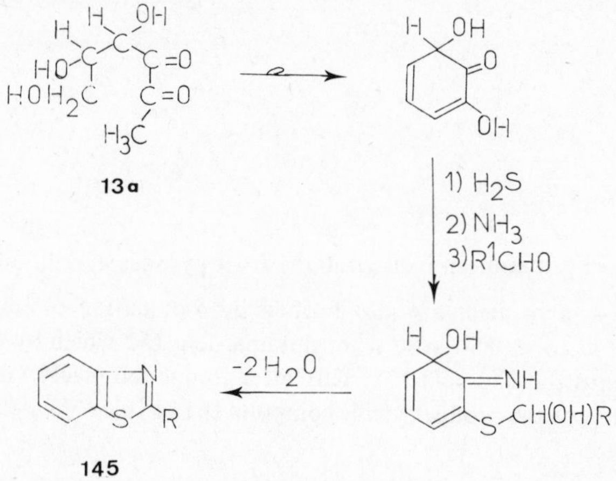

Scheme 28 – Formation of benzothiazoles from *Amadori* intermediates, hydrogen sulphide, ammonia, and aldehydes.

γ-Pyrones are assumed to arise from 4-*O*-substituted glucose derivatives (e.g., maltose, lactose, cellobiose, malto-oligosaccharides) by browning reactions with amino compounds from *Amadori* intermediates **146** as shown in *Scheme 29*.

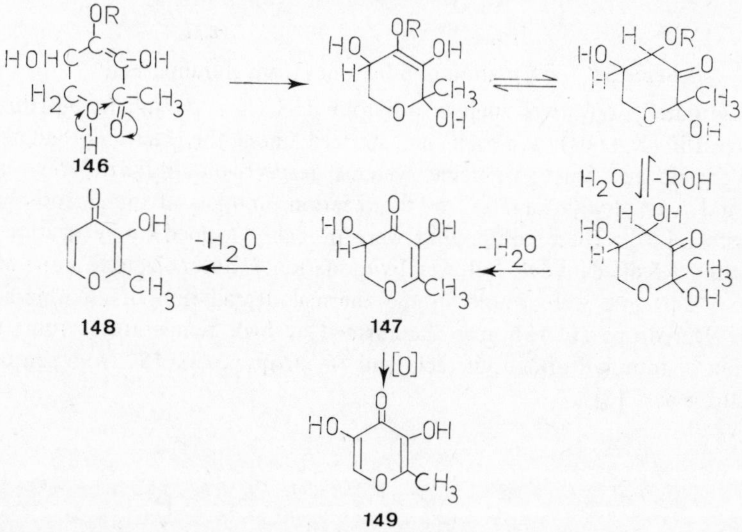

Scheme 29 – Formation of maltol and its 5-hydroxy derivative from *Amadori* intermediates.

The precursors of δ-lactones **151** in milk fat resulted from δ-hydroxy fatty acids esterified to a triglyceride **150** [20b] (*see Scheme 30*).

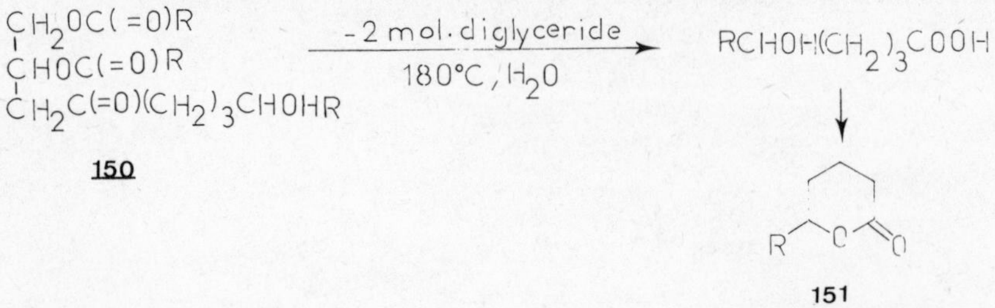

$CH_2OC(=O)R$
$CHOC(=O)R$
$CH_2C(=O)(CH_2)_3CHOHR$

150

$\xrightarrow[180°C,\ H_2O]{-2\ mol.\ diglyceride}$ $RCHOH(CH_2)_3COOH$ → **151**

Scheme 30 – Formation of 6-alkyltetrahydro-α-pyrones in milk fat [20b].

δ-Hydroxy fatty acids are also formed by δ-oxidation of saturated fatty acids. In wine lactones **154** arise from glutamic acid **152** which by fermentation is converted into the oxo acid **153**. This latter compound reacts with aldehydes, decarboxylation of the product yields compounds **154** (*see Scheme 31*).

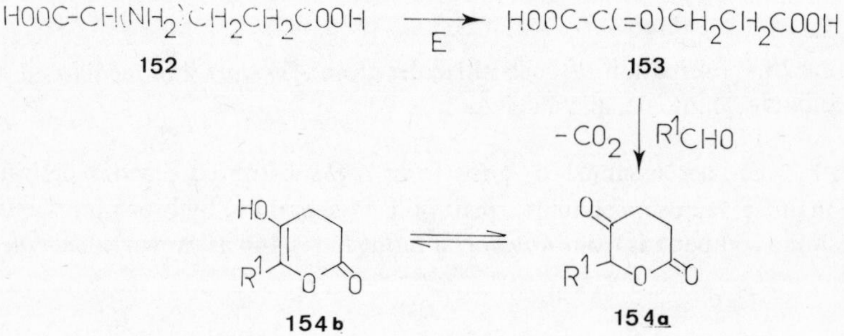

$HOOC\text{-}CH(NH_2)CH_2CH_2COOH$ (**152**) $\xrightarrow{E}$ $HOOC\text{-}C(=O)CH_2CH_2COOH$ (**153**)

$-CO_2$, R^1CHO

154b ⇌ **154a**

Scheme 31 – Formation of δ-lactones from glutamic acid.

Unsaturated derivatives such as α-pyrone **155** (X = H) and its 3-hydroxy derivative **155** (X = OH) have been characterized among the reaction products of glucose-glycine and fructose-glycine systems, respectively [90]. α-Pyrone was also found from dry degradation of the reaction products of the glucose-theanine system [90]. These compounds were probably formed by cyclization of unsaturated 1,5-dialdehydes, followed by oxidation. δ-Butenolactone along with 3-hydroxy-2-pyrone were found in the thermal degradation of ascorbic acid [111b]. Dihydropyran **156** may be formed at high temperature from ring expansion of tetrahydrofurfurylic acid and 4-benzopyranone **157** from pentoses and ascorbic acid [2].

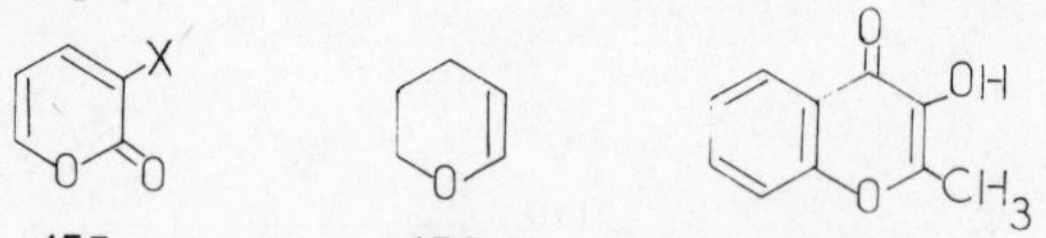

155 **156** **157**

Fig. 4 – Pyran derivatives identified from model systems.

3.8 Pyridines

More than twenty pyridine derivatives **158** have been characterized in various model systems (*see Table III–9*). They arise mainly from the thermal degradation of sulphur-containing amino acids alone [55d] or in the presence of glucose [55c]. They were also formed in the glucose-proline system [131], among the degradation products of the *Amadori* intermediates of the glucose-glycine system [90], and from pyrolysis of α and β-alanine [64].

The condensation reaction of amino acids with simple aldehydes leading to quaternary pyridinium betaines and the thermal decomposition of these salts to volatile pyridines has been recently studied [119].

Very little is known about the mechanism of the formation of pyridines in model systems. It seems evident that aldehydes and ammonia (or amino acids) play an important role as shown in *Scheme 32*.

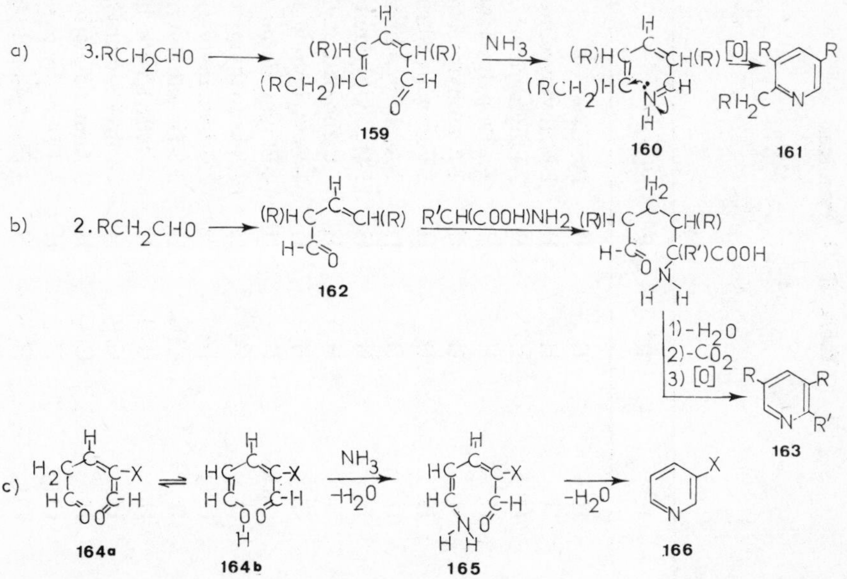

a) $3.RCH_2CHO$ → **159** $\xrightarrow{NH_3}$ **160** $\xrightarrow{[O]}$ **161**

b) $2.RCH_2CHO$ → **162** $\xrightarrow{R'CH(COOH)NH_2}$ → 1) $-H_2O$ 2) $-CO_2$ 3) [O] → **163**

c) **164a** ⇌ **164b** $\xrightarrow[-H_2O]{NH_3}$ **165** $\xrightarrow{-H_2O}$ **166**

Scheme 32 – Formation of pyridines from aldehydes (or 1,5-dicarbonyl compounds) and ammonia (or α-amino acids).

The aldolization product **159** reacts with ammonia to give the imino compound **160** which by cyclization and oxidation affords pyridines **161**. Condensation reaction of the aldolization product **162** with α-amino acids is another possibility (*see pathway b*). The conversion of 1,5-dialdehydes **164** (or 1,5-diketones) to 1,5-aminoaldehyde intermediate **165** by reaction with a nitrogen source, is a convenient synthetic method for pyridines **166** (*pathway c*).

Pyridines in foods can be formed also by thermal treatment of other heterocyclic compounds alone or in the presence of ammonia. Thus, furfurylamine,

Table III–9 Pyridines

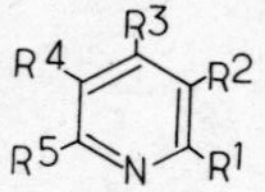

158

158; R^1	R^2	R^3	R^4	R^5	Name	Model systems[a)]
H	H	H	H	H	Pyridine	*1,3,4,63*
CH_3	H	H	H	H	2-Methylpyridine	*1,3,54,55,63*
C_2H_5	H	H	H	H	2-Ethylpyridine	*1*
n-C_3H_7	H	H	H	H	2-*n*-Propylpyridine	—
CH_2COCH_3	H	H	H	H	2-Acetonylpyridine	*77*
CHO	H	H	H	H	2-Formylpyridine	*1*
$COCH_3$	H	H	H	H	2-Acetylpyridine	*9,34*
H	CH_3	H	H	H	3-Methylpyridine	*1,3,4,9,57,63*
H	CO_2CH_3	H	H	H	3-Carbomethoxypyridine	*63*
H	H	CH_3	H	H	4-Methylpyridine	*63*
H	H	C_2H_5	H	H	4-Ethylpyridine	*63*
CH_3	CH_3	H	H	H	2,3-Dimethylpyridine	*1*
CH_3	H	CH_3	H	H	2,4-Dimethylpyridine	*57*
CH_3	H	H	CH_3	H	2,5-Dimethylpyridine	*69*
CH_3	H	H	C_2H_5	H	2-Methyl-5-ethylpyridine	*4,56*
C_2H_5	H	H	CH_3	H	2-Ethyl-5-methylpyridine	*54,55*
NH_2	H	H	CH_3	H	2-Amino-5-methylpyridine	*77*
H	CH_3	CH_3	H	H	3,4-Dimethylpyridine	*63.69*
H	C_2H_5	CH_3	H	H	3-Ethyl-4-methylpyridine	*63*

H	CH_3	H	CH_3	H	3,5-Dimethylpyridine	*57,68*
CH_3	CH_3	H	CH_3	H	2,3,5-Trimethylpyridine	*57*
C_2H_5	CH_3	H	CH_3	H	3,5-Dimethyl-2-ethylpyridine	*68*
H	CH_3	C_2H_5	CH_3	H	3,5-Dimethyl-4-ethylpyridine	*68*
H	$CONHCH_3$	$CONHCH_3$	H	H	Di-(*N*-methylcarboxamido)-3,4-pyridine	*63*

a) Model systems: *1* glucose-*L*-proline; *3* glucose-cysteine. *4* glucose-cystine; *9* glucose-alanine; *34* RP of glucose-glycine; *54* cysteine pyrolysis; *55* cystine pyrolysis; *56* *L*-α-alanine pyrolysis; *57* β-alanine pyrolysis; *63* trigonelline pyrolysis; *68* glycine-propanal; *69* glycine-propanal-crotonal; *77* *L*-rhamnose-ammonia.
RP: reaction product.

furfural, 2-acetylfurans, tetrahydrofurfuryl alcohol, *N*-substituted pyrroles, pyrylium salts and the *Diels-Alder* reaction of oxazoles and thiazoles with dienes can lead to the formation of pyridine derivatives.

A radical mechanism, similar to that of the *Ladenburg* rearrangement of *N*-alkylpyridinium salts [136] explains the formation of pyridines identified from the pyrolysis of trigonelline [133].

1-Ethyl-2,6-dioxotetrahydropyridine **167** and 1-ethyl-3-hydroxy-2-pyridinone **168** have been reported as dry and wet degradation products of the glucose-theanine system [90]. δ-Valerolactam **169** found in *N*-formyl-*L*-lysine-*D*-lactose system [31] is formed through the *Strecker* degradation of lysine, followed by oxidation and lactamization of the resulting δ-aminovaleric acid.

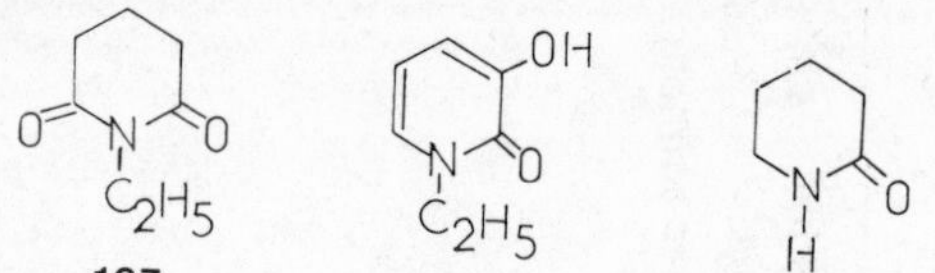

167 **168** **169**

Fig. 5 – δ-Lactams identified from model systems.

3.9 Pyrazines

Numerous pyrazines **170** were found among the products of *Maillard* reactions and related model systems (*see Table III–10*). While five alkylpyrazines were the products of the heating of glucose with cysteine, only methylpyrazine was present in the volatile products resulting from a mixture of glucose and cysteine [55c].

2,5-Dimethyl-, and 2,5-dimethyl-3-ethylpyrazine were detected in the cysteine (or cystine)-pyruvaldehyde system [55c]. Trimethylpyrazine was only found in the presence of cysteine. Ten pyrazines were characterized in the *D*-glucose-*L*-alanine system [37] and two others were tentatively identified (mol. wt. 164 and 178). The following pyrazines: the parent compound, 2-methyl-, 2,3-dimethyl-, 2,5-dimethyl-, 2-methyl-5 (or 6)-ethyl-, and trimethylpyrazine were found as the degradation products of the *Amadori* and *Heyns* rearrangement compounds or in the glucose-theanine system [90]. 2-Acetonylpyrazine was isolated from a *Maillard* model reaction [123]. Pyrolysis of α-hydroxyamino acids produced pyrazine and various alkyl derivatives [55b,73, 143]. The thermal degradation of dipeptides such as alanyl-serine [142] and glycyl-serine [73] also produces pyrazines. Twelve pyrazines have been identified by Shibamoto *et al.* [114b] from the reaction of ammonia with the following sugars: *D*-glucose, *D*-sorbitol, *D*-mannose, *D*-galactose, *D*-fructose, *D*-xylose, *L*-arabinose, *L*-rhamnose monohydrate, 2-deoxyglucose, *DL*-glyceraldehyde, glycerol, and 1,3-dihydroxyacetone. In their reaction of *D*-glucose and ammonia, these authors [115] identified 17 pyrazines and seven imidazoles in contrast to the *D*-glucose-hydrogen sulphide model system which produced

Table III–10 Pyrazines

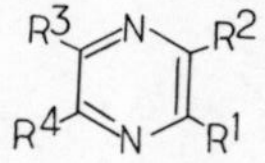

170

170; R^1	R^2	R^3	R^4	Name	Model systems[a) b)]
H	H	H	H	Pyrazine	*9,36,49–52,58,59,71,77*
CH_3	H	H	H	2-Methylpyrazine	*3,4,9,35,36,49–52,58,59,71,77,78*
C_2H_5	H	H	H	2-Ethylpyrazine	*49,51,59,71,77,78*
CH_2COCH_3	H	H	H	2-Acetonylpyrazine	—
CH_3	CH_3	H	H	2,3-Dimethylpyrazine	*9,37,49,52,71,77,78*
CH_3	C_2H_5	H	H	2-Methyl-3-ethylpyrazine	*3,71,77,78*
CH_3	H	CH_3	H	2,5-Dimethylpyrazine	*3,35,36,50,65a,71,77,78*
CH_3	H	C_2H_5	H	2-Methyl-5-ethylpyrazine	*9,35,36,37,52,71,77,78*
C_2H_5	H	C_2H_5	H	2,5-Diethylpyrazine	*71,77*
C_2H_5	H	5- (or 6-)	$COCH_3$	2-Ethyl-5-(or6-)acetylpyrazine	*78*
CH_3	H	H	CH_3	2,6-Dimethylpyrazine	*9,71,77,78*
CH_3	H	H	C_2H_5	2-Methyl-6-ethylpyrazine	*3,49,59,71,77,78*
C_2H_5	H	H	C_2H_5	2,6-Diethylpyrazine	*49,59,71,77,78*
CH_3	H	H	n-C_3H_7	2-Methyl-5-(or 6-)propylpyrazine	*50,71,77,78*
CH_3	CH_3	CH_3	H	Trimethylpyrazine	*9,37,50–53,65a,71,77,78*
C_2H_5	CH_3	CH_3	H	2-Ethyl-3,5-dimethylpyrazine	*31,77,78*
C_2H_5	C_2H_5	CH_3	H	2,3-Diethyl-5-(or 6-)methylpyrazine	*9,71,77*
C_2H_5	CH_3	C_2H_5	H	2,5-Diethyl-3-methylpyrazine	*71,77,78*
CH_2OH	CH_3	CH_3	H	2-Hydroxymethyl-3,5-dimethylpyrazine	*71,77*

Table III–10 (*continued*)

170;R^1	R^2	R^3	R^4	Name	Model systems[a) b)]
C_2H_5	CH_3	H	CH_3	2-Ethyl-3,6-dimethylpyrazine	*3,9,50,51,53,65a,65b,78*
C_2H_5	CH_3	H	C_2H_5	2,6-Diethyl-3-methylpyrazine	*9,49,78*
C_2H_5	C_2H_5	H	C_2H_5	Triethylpyrazine	*78*
CH_3	CH_3	CH_3	CH_3	Tetramethylpyrazine	*71,77,78*
C_2H_5	CH_3	CH_3	CH_3	Trimethylethylpyrazine	*71,77*
C_2H_5	CH_3	CH_3	C_2H_5	2,6-Diethyl-3,5-dimethylpyrazine	*71,77,78*
C_2H_5	CH_3	C_2H_5	CH_3	2,5-Diethyl-3,6-dimethylpyrazine	*71,77,78*

a) Several pyrazines of molecular weight: 136, 150, 164, 178 were also tentatively identified (37,55b,143).
b) Model systems: *3 D*-glucose-cysteine; *4 D*-glucose-cystine; *9 D*-glucose-*L*-alanine; *35* dry degradation of RP's glucose-theanine; *36* wet degradation of RP's glucose-theanine; *37* RP's fructose-glycine; *49* serine; *50* threonine; *51* ethanolamine; *52* glucosamine; *58* glycyl-serine; *59* alanyl-serine; *71* glucose-ammonia (5M); *77 L*-rhamnose-ammonia; *78 L*-rhamnose-hydrogen sulphide-ammonia.
RP: reaction products.

cyclic methylene polysulphides. They also studied the effect of time, temperature, and reactant ratio on the formation of pyrazines in this reaction [113a]. A very simple model system such as *L*-rhamnose-ammonia [113b] produces 39 pyrazine derivatives including cyclopentapyrazines **171** and tetrahydroquinoxalines **172** (*see Table III–11*). Some (5*H*)-cyclopentapyrazines **173**

Table III–11 6,7-Dihydro-(5*H*)-cyclopentapyrazines and 5,6,7,8-tetrahydroquinoxalines in model systems [113b,115,149]

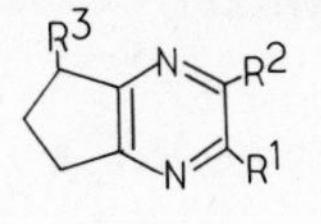

171

171; R^1	R^2	R^3	Model systems[a)]
H	H	H	*77*
H	H	CH_3	*71,77,78*
H	H	C_2H_5	*71,77,78*
CH_3	H	H	*77*
C_2H_5	H	H	*77,78*
CH_3	CH_3	H	*77*
CH_3	H	CH_3	*71,77,78*
C_2H_5	H	CH_3	*77*
$COCH_3$	H	CH_3	*77*
H	CH_3	CH_3	*71,77,78*
CH_3	CH_3	CH_3	*77,78*

a) Model systems: see Table III–10

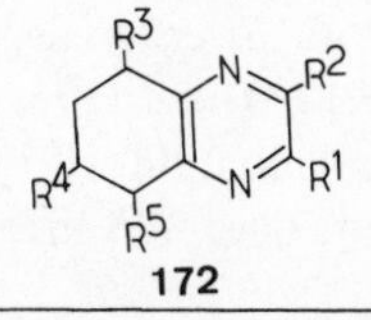

172

172; R^1	R^2	R^3	R^4	R^5	Model systems[a)]
H	H	H	H	H	*77*
CH_3	H	H	H	H	*77*
C_2H_5	H	H	H	H	*77,78*
H	H	CH_3	H	H	*77*
H	H	C_2H_5	H	H	*78*
CH_3	CH_3	H	H	H	*77*
C_2H_5	CH_3	H	H	H	*77*
H	H	CH_3	H	CH_3	*77,78*
H	H	CH_3	CH_3	H	*77*

a) Model systems: see Table III–10

(R^1 = H, C_2H_5, R^2 = CH_3 or H), 5,8-dihydroquinoxalines **174** (R^1 = C_2H_5, R^2 = H and CH_3) and quinoxalines **175** were also reported in this reaction and in many foods [33,140]. The same pyrazine derivatives were found in the presence of hydrogen sulphide [149]. 2,5-Diketo-3,6-dimethylpiperazine **176** was identified in the pyrolysis products of serine and threonine [55b,143] and a series of new pyrazinones 2-(3′-alkyl-2′-oxopyrazin-1′-yl) alkanoic acids **177** (R^1 = H, CH_3, *n*-C_4H_9, *i*-C_4H_9, *n*-C_5H_{11} and R^2 = H, CH_3, *i*-C_4H_9) were prepared from various dipeptides and glyoxal [16].

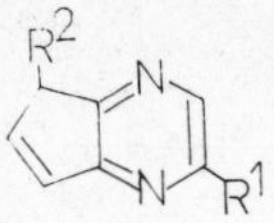

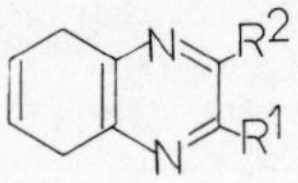

Fig. 6 – Pyrazine derivatives identified from model systems.

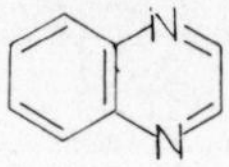

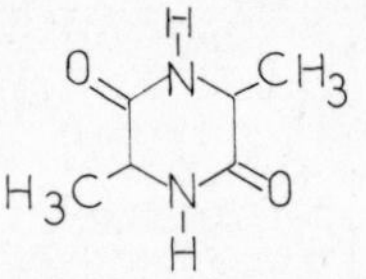

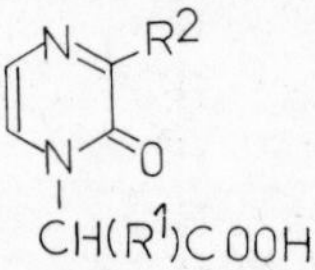

These products were assumed to increase the browning rate between dipeptides and α-dicarbonyl compounds.

Mechanistic hypotheses for the formation of pyrazines in food flavours and model systems have been described by several authors [18,50,59,68,71,99a, 115,140,142]. The condensation of α-diketones arising from sugar fragmentation with an amino acid *via Strecker* degradation form α-aminoketones (*see Scheme 5*, Section III–2.5.1). These latter compounds in subsequent steps of self-condensation (or by condensation with other aminoketones) and oxidation of the resulting dihydropyrazines **178** afford alkyl substituted pyrazines **179** (*see Scheme 33*).

With two different alkyl groups in α-diketones **30**, a mixture of two isomers is expected. Thus Rizzi [99a] obtained a mixture of 2,5-diethyl-3,6-dimethyl-, and 2,6-diethyl-3,5-dimethylpyrazine from 2,3-pentanedione and *DL*-alanine. The formation of acetylpyrazines may be the result of the condensation of α-amino methyl reductone with α-aminoketones.

The formation of 2-(2′-furyl)-pyrazines **181** may come from the condensation of a known *Maillard* browning product [31c], 1-(2′-furyl)-1,2-propanedione **180** with glyoxal and an amino acid as noted in *Scheme 33* [71]. Cyclic α-diketones such as 3-methyl-1,2-cyclopentanedione or cyclotene **182** (R^3 = CH_3) are responsible for the formation of a series of cyclopentapyrazines **171**. The self-condensation of the α-amino derivative of the carbonyl compound **182** (R^3 = CH_3) affords the *bis*-(methylcyclopenta) pyrazine **183** (R^3 = CH_3).

Starting from 2-hydroxy-2-cyclohex-2-one **184**, the dicyclohexapyrazine **185** was obtained [99a]. Pyrazines containing higher alkyl groups (ethyl-, propyl-, and isobutyl-) may be formed by another pathway [33]. Shibamoto *et al.* [115] have shown that under basic conditions, acetaldehyde arising from oxidation of ethanol can react with dihydropyrazine carbanions **187** to give ethylpyrazines **189** after dehydration of the alcohol **188**. Similarly, the formation of the alkenyl substituted pyrazines requires another pathway, probably through dehydration of the corresponding hydroxyalkylpyrazines [33].

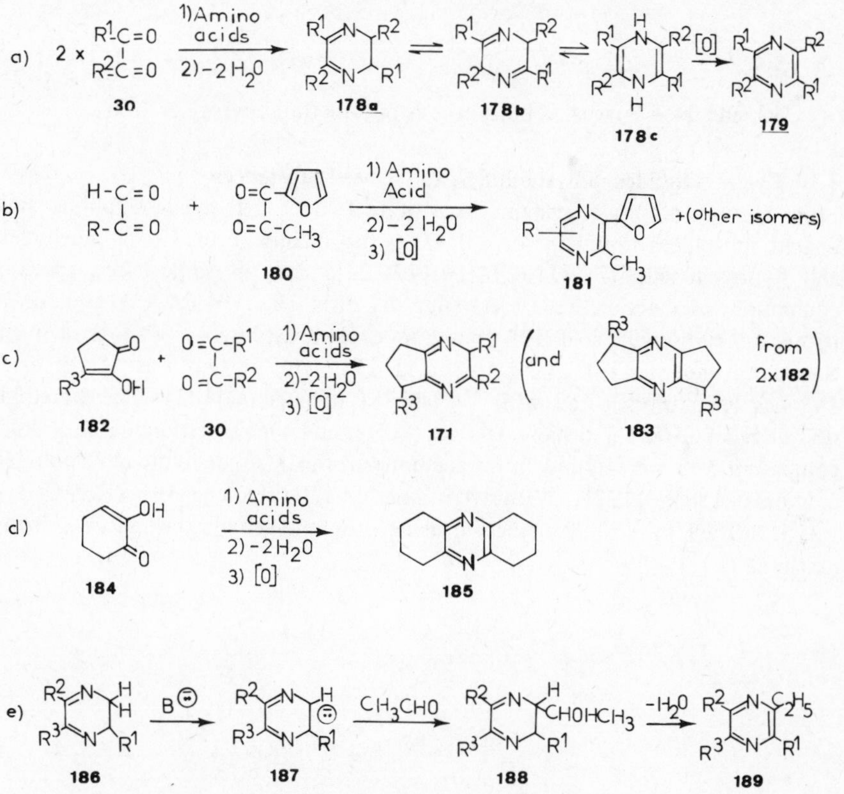

Scheme 33 – Formation of pyrazines.

Several comments on the possible biosynthesis of 3-alkyl-2-methoxypyrazines **192** in vegetables, have been suggested [82,83,86]. Murray *et al.* [82] reported that these compounds might be derived from 1,2-dicarbonyl compounds and α-aminoacids such as valine, isoleucine, and leucine, respectively. An hypothetical biosynthetic pathway is shown in *Scheme 34*.

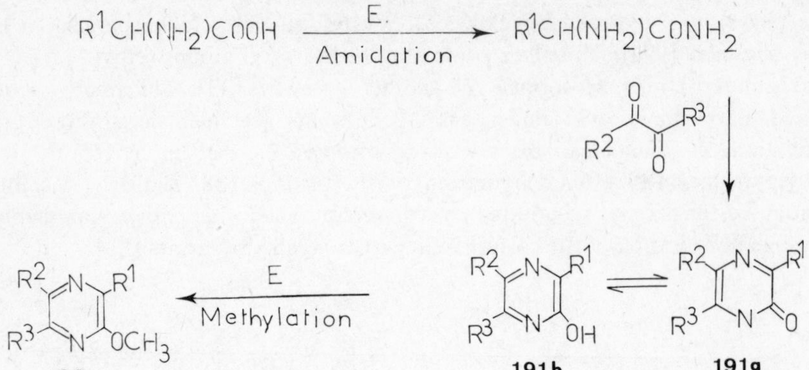

$R^1CH(NH_2)COOH \xrightarrow[\text{Amidation}]{E.} R^1CH(NH_2)CONH_2$

Scheme 34 – Suggested biosynthesis of 2-methoxypyrazines [82].

3.10 Cyclic sulphides, polysulphides, and related substances

From a study of model systems, it is quite evident that cyclic sulphides and related substances found in food flavours, are mainly formed from aldehydes and hydrogen sulphide [11,62b,106,122,128]. But, in some cases, specific compounds may act as precursors. Thus, 1,2-dithiolene **194** (R = H) and 1,2,3-trithiane-5-carboxylic acid **195** found in cooked asparagus [128] arise from thermal degradation of asparagusic acid **193**. The biosynthesis of the acid **195** has been investigated with ^{14}C-labelled precursors [128]. The 3-methyl derivative of (3*H*)-1,2-dithiole **194** (R = CH_3) and the corresponding saturated compound **196** were found in butanedione-crotonaldehyde-hydrogen sulphide-ammonia system [122]. 2-Ethyl-4,5-dimethyl-1,3-dithiolene **197** (R = C_2H_5) was identified in a similar reaction using propionaldehyde instead of crotonaldehyde [122].

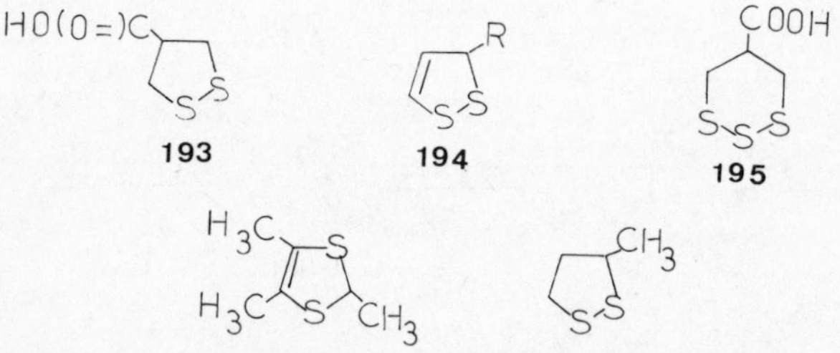

Fig. 7 – Cyclic polysulphides identified from model systems.

The diastereoisomeric 3,5-dimethyl-1,2,4-trithiolanes **199** (R = CH_3) found in boiled beef [15] and detected in the cysteine-xylose system [62a], can be formed from acetaldehyde and hydrogen sulphide *via* the sulphide intermediate **198** (*see Scheme 35*).

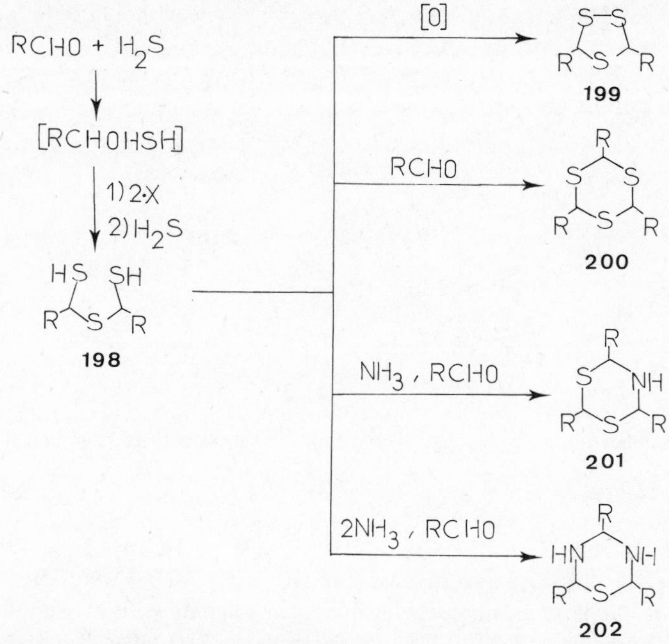

RCHO + H_2S → [RCHOHSH] → 1) 2·X 2) H_2S → **198**; **198** → [O] → **199**; RCHO → **200**; NH_3, RCHO → **201**; $2NH_3$, RCHO → **202**

Scheme 35 – Cyclic sulphides and polysulphides from aldehydes, hydrogen sulphide and ammonia [11,62b,122].

s-Trithiane **200** (R = H) was found in *D*-glucose-hydrogen sulphide system [104a]. Its trimethyl derivative **200** (R = CH_3), detected in cooked chicken flavour [93] can be formed by action of acetaldehyde upon the sulphide **198** (R = CH_3). In the presence of ammonia, the same reaction leads to the formation of 2,4,6-trimethyl-5,6-dihydro-1,3,5-dithiazine or thialdine **201** (R = CH_3) identified in beef broth [12], boiled meat [146] and to perhydro-1,3,5-thiadiazine **202** (R = CH_3). These two nitrogen-containing products were observed by Boelens *et al.* [11] from the reaction of ethanol, hydrogen sulphide, and ammonia.

The two dihydrovinyldithiins **204, 205** (R = H) found in cooked asparagus [128] are formed from acrolein and hydrogen sulphide arising from the *Strecker* degradation of *S*-methylmethionine and from cysteine, respectively (*see Scheme 36*).

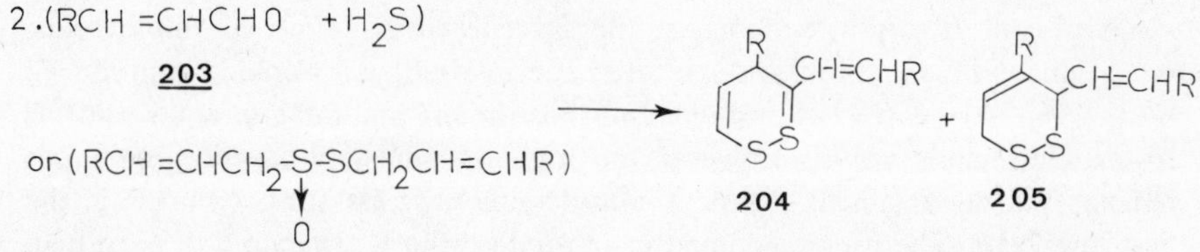

2.(RCH=CHCHO + H_2S) **203** or ($RCH=CHCH_2$-S(→O)-$SCH_2CH=CHR$) → **204** + **205**

Scheme 36 – Formation of dihydrodithiins from acrolein and hydrogen sulphide.

1,2-Dithiane **206** was formed by pyrolysis of cysteine with xylose [62a].

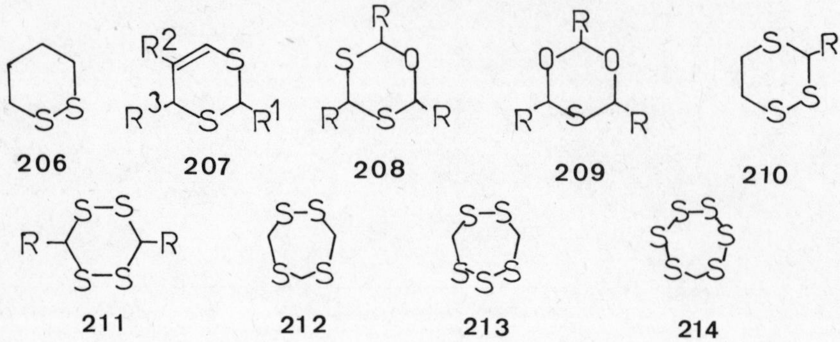

Fig. 8 – Sulphur-containing six and more membered rings identified from model systems.

1,3-Dithiins **207** ($R^1 = R^3 = CH_3$, C_2H_5; $R^2 = H$ or CH_3), 1,3,5-oxadithiane **208** ($R = C_2H_5$) and 1,3,5-dioxathianes **209** ($R = CH_3$, C_2H_5) were formed by interaction of aliphatic aldehydes with hydrogen sulphide with or without the presence of ammonia [11,62b,122]. 1,2,4-Trithianes **210** ($R = H$, CH_3) arise from thermal degradation of sulphur-containing amino acids [62c]. 3,6-Diethyl-1,2,4,5-tetrathiane **211** ($R = C_2H_5$) was observed in a reaction mixture of propionaldehyde, hydrogen sulphide and ammonia [62b]. Compounds **199**, **200** (R = H) and 1,2,4,6-tetrathiepane **212** have been identified among the reaction products of furfural with hydrogen sulphide and ammonia (*see Chapter V*). This latter compound and other polythiepans such as **213** and **214** will be formed in mushrooms (*Lentinus edodes*) by enzymatic degradation of lentinic acid *via* a thiosulphinate intermediate [152].

III–4 CONCLUSIONS

The material covered in the previous text of this chapter indicates that heterocyclic aroma compounds found in model systems arise mainly from interactions between mono- and dicarbonyl compounds, hydrogen sulphide and ammonia. The sources of carbonyl compounds include sugar fragmentation, rearranged *Amadori* and *Heyns* intermediates, the *Strecker* degradation of amino acids, oxidation of lipids, aldolization reactions, etc. Hydrogen sulphide is produced by thermal degradation of cysteine and thiamin and ammonia or amines arising from amino acids pyrolysis. However, the origin of some heterocyclic compounds remains yet unexplained, although considerable progress has been made in the past few years. The use of a computer-assisted synthesis program is of great help for the analysis of flavours, as shown in *Chapter V*.

III–5 REFERENCES

[1] Adams, C. R. & J. Falbe, *Brennstoff Chem.*, **47**, 184 (1966).

[2] Adams, C. R. & J. Falbe, *Neth. Pat.*, 6,512,874 (1964).

[3] Adrian, J., *Ind. Aliment. Agr.*, **89**, 1281 (1972); *idem, World Rev. Nutr. Diet.*, **19**, 71 (1974); *idem, Labo-Pharma-Probl. Tech.*, **23**, (244), 614 (1975).

[4] Amadori, M., *Atti Reale. Accad. Naz. Lincei.*, **2**, 337 (1925); *ibid*, **9**, *68*, (1929); *ibid*, **13**, 72 (1931).

[5] Anan, T., *J. Sci. Food. Agric.*, **30**, 906 (1979).

[6] Anet, E. F. L. J. & T. M. Reynolds, *Nature*, **177**, 1082 (1956); *Idem, Australian J. Chem.*, **10**, 182 (1957); Anet, E. F. L. J., *ibid*, **13**, 396 (1960).

[7] Arroyo, P. T. & D. A. Lillard, *J. Food Sci.*, **35**, 769 (1970).

[8] Baltes, W. & K. Franke, *Z. Lebensm.-Unters. Forsch.*, **167**, 403 (1978a); Baltes, W., *Lebensmittel. Ger. Chem.*, **34**, 39 (1980b).

[9] Boelens, M., P. J. De Valois, H. J. Wobben & A. Van Der Gen, *J. Agric. Food Chem.*, **19**, 984 (1971).

[10] Boelens, H. & L. Brandsma, *Rec. Trav. Chem.*, **91**, 141 (1972).

[11] Boelens, M., L. M. Van der Linde, P. J. De Valois, H. M. Van Dort & H. J. Takken, *J. Agric. Food Chem.*, **22**, 1071 (1974).

[12] Brinkman, H. W., H. J. Copier, J. J. M. De Leuw & S. B. Tjan, *J. Agric. Food Chem.*, **20**, 177 (1972).

[13] Cazenave, P., I. Horman, F. Mueggler-Chavan & R. Viani, *Helv. Chim. Acta*, **57**, 206 (1974).

[14] Chang, S. S., T. H. Smouse, R. G. Krishnamurthy, B. D. Mookherjee & B. R. Reddy, *Chem. Ind. (London)*, **1966**, 1926.

[15] Chang, S. S., C. Hirai, B. R. Reddy, K. O. Herz, A. Kato & G. Sipma, *Chem. Ind. (London)*, **1968**, 1639.

[16] Chuyen, N. V., T. Kurata & M. Fujimaki, *Agric. Biol. Chem.*, **37**, 327 (1973).

[17] Ciner-Doruk, M., & K. Eichner, *Z. Lebensm.-Unters. Forsch.*, **168**, 9 (1979).

[18] Dawes, I. W. & R. A. Edwards, *Chem. Ind. (London)*, **1966**, 2203.

[19] Dickerson, J. P., D. L. Roberts, C. W. Miller, R. A. Lloyd & C. E. Rix, *Tobacco*, **178**, 71 (1976).

[20] Dimick, P. S., N. J. Walker & S. Patton, *Biochem. J.*, **111**, 395 (1969a); *Idem, J. Agric. Food Chem.*, **17**, 649 (1969b).

[21] Drawert, F., *Proc. Int. Symp. Aroma Research*, **1975**, p. 13; Wageningen.

[22] Dubourg, J. & P. Devillers, *Bull. Soc. Chim. Fr.*, **1957**, 333; *ibid*, **1962**, 603.

[23] Dwivedi, B. K. & R. G. Arnold, *J. Agric. Food Chem.*, **19**, 923 (1971a); *Idem, J. Food Sci.*, **37**, 886 (1972b).

[24] Dwivedi, B. K., R. G. Arnold & L. M. Libbey, *J. Food Sci.,* **37**, 689 (1972a); *ibid,* **38**, 450 (1973b).

[25] Ellis, G. P. & J. Honeyman, *Advan. Carbohyd. Chem.,* **10**, 95 (1955a); Ellis, G. P., *ibid,* **14**, 63 (1959b).

[26] Erickson, C. E., *Progress in Flavour Research* (D. G. Land & H. E. Nursten, Eds.), Applied Science Publishers Ltd., London (1978), 159–174.

[27] Eskin, M. N. A., S. Grossman & A. Pinsky, *CRC Crit. Rev. Food Sci. Nutr.,* **9**, 1 (1977).

[28] Evers, W. J., *U.S. Pat.,* 3,666,495 (5.2.1969).

[29] Fagerson, I. S., *J. Agric. Food Chem.,* **17**, 747 (1969).

[30] Feather, M. S. & J. F. Harris, *Advan. Carbohyd. Chem. Biochem.,* **28**, 161 (1973).

[31] Ferretti, A., V. P. Flanagan & J. M. Ruth, *J. Agric. Food Chem.,* **18**, 13 (1970a); Ferretti, A. & V. P. Flanagan, *ibid,* **19**, 245 (1971b); *ibid,* **21**, 35 (1973c).

[32] Flament, I. *et al., 3rd Int. Coll. on Coffee Chemistry.,* Trieste; ASIC France (1967).

[33] Flament, I., M. Kohler & R. Aschiero, *Helv. Chim. Acta,* **59**, 2308 (1976).

[34] Flament, I., P. Sonnay, P. Willhalm & G. Ohloff, *Helv. Chim. Acta,* **59**, 2314 (1976).

[35] Forss, D. A., *The Chemistry and Physiology of Flavors* (H. W. Schultz, E. A. Day & L. M. Libbey, Eds.), The AVI Publ. Co. Inc., Westport, Connecticut, U.S.A. (1967), p. 492.

[36] Fujimaki, M., S. Kato & T. Kurata, *Agric. Biol. Chem.,* **33**, 1144 (1969a); Fujimaki, M., M. Tajima & H. Kato, *ibid,* **36**, 663 (1972b).

[37] Galliard, T., D. R. Phillips & J. Reynolds, *Biochim. Biophys. Acta,* **441**, 181 (1976a); Galliard *et al., J Sci. Food Agric.,* **28**, 863 (1977b).

[38] Garnero, J., *Riv. Ital. EPPOS,* **1980**, 253.

[39] Germs, A. C., *J. Sci. Food Agric.,* **24**, 7 (1973).

[40] Gianturco, M. A., A. S. Giammarino, P. Friedel & V. Flanagan, *Tetrahedron,* **20**, 2951 (1964).

[41] Gottschalk, A., *Biochem. J.,* **52**, 455 (1952a); *idem, Glycoproteins: Their Composition, Structure and Function*, Elsevier Publ. Co., Amsterdam (1972b), p. 141.

[42] Gronowitz, S. & P. Moses, *Acta Chem. Scand.,* **16**, 155 (1962).

[43] Gronowitz, S., L. Svensson & R. Ohlson, *J. Agric. Food Chem.,* **26**, 887, (1978).

[44] Grosch, W., G. Laskawy & K. H. Fischer, *Lebensm. Wiss. Technol.* **7**, 335 (1974).

[45] Hashiba, H., *Chomi Kagaku,* **18**, 66 (1971).

[46] Heyns, K. & H. Paulsen, *Ann.* **622**, 160 (1959a); *idem, Wiss. Veroeffentl. Deut. Ges. Ernaehrung.,* **5**, 15 (1960b); Heyns, K. *et al., Carbohydrate Res.,* **2**, 132 (1966c); *ibid,* **6**, 436 (1968d).

[47] Hodge, J. E., *J. Agric. Food Chem.*, **1**, 928 (1953a); *idem, Advan. Carbohyd. Chem.*, **10**, 169 (1955b).
[48] Hodge, J. E. & B. E. Fisher, *Methods in Carbohydrate Chemistry* (R. L. Whistler & M. L. Wolfram, Eds.), Vol. II, Academic Press, New York (1963).
[49] Hodge, J. E., *The Chemistry and Physiology of Flavors* (H. W. Schultz, E. A. Day & L. M. Libbey, Eds.), The AVI Publ. Co. Inc., Westport, Connecticut, USA (1967), 465–489.
[50] Hodge, J. E., F. D. Mills & B. E. Fisher, *Cereal Sci. Today*, **17**, 34 (1972).
[51] Ishizu, A., B. Lindberg & O. Theander, *Acta Chem. Scand.*, **21**, 424 (1967).
[52] Johnson, R. R., E. D. Alford & G. W. Kinzer, *J. Agric. Food Chem.*, **17**, 22 (1969).
[53] Jurch, G. R. & J. H. Tatum, *Carbohydrate Res.*, **15**, 233 (1970).
[54] Kallio, H., *Dissertation*, University of Turku, Finland (1975).
[55] Kato, S., T. Kurata, R. Ishitsuka & M. Fujimaki, *Agric. Biol. Chem.*, **34**, 1826 (1970a); Kato, S., T. Kurata & M. Fujimaki, *ibid*, **37**, 539 (1973b); Kato, S., T. Kurata, S. Ishiguro & M. Fujimaki, *ibid*, **37**, 1759 (1973c); Kato, S., M. Fujimaki & T. Kurata, *ibid*, **33**, 1144 (1969d).
[56] Kato, S. & M. Fujimaki, *J. Food Sci.*, **41**, 555 (1976).
[57] Katz, I., R. A. Wilson & C. J. Mussinan, *U.S. Pat.*, 3,713,848 (1973).
[58] Kobayashi, N. & M. Fujimaki, *Agric. Biol. Chem.*, **29**, 698 (1965a); *ibid*, **29**, 1059 (1965b).
[59] Koehler, P. F., M. E. Mason & J. A. Newell, *J. Agric. Food Chem.*, **17**, 393 (1969).
[60] Kort, M. J., *Advan. Carbohyd. Chem.*, **25**, 311 (1970).
[61] Langner, E. H. & J. Tobias, *J. Food Sci.*, **32**, 495 (1967).
[62] Ledl, F. & T. Severin, *Chem. Mikrobiol. Technol. Lebensm.*, **2**, 155 (1973a); Ledl, F., *Z. Lebensm.-Unters. Forsch.*, **157**, 28 (1975b); Ledl, F., *ibid*, **161**, 125 (1976c); Ledl, F. & T. Severin, *ibid*, **167**, 410 (1978d).
[63] Leffingwell, J. C., *Proc. Tobacco Chem. Res. Conf. 30th* (1978).
[64] Lien, Y. C. & W. W. Nawar, *J. Food Sci.*, **39**, 914 (1974).
[65] Lipparini, L. & M. R. Cavana, *Rass. Chem.*, **30**, 19 (1978).
[66] Lloyd, R. A., C. W. Miller, D. L. Roberts, J. A. Giles, J. P. Dickerson, N. H. Nelson, C. E. Rix & P. H. Ayers, *Tobacco*, **20**, 125 (1976); *ibid*, **178**, 43 (1976).
[67] Ludwig, E., *Nahrung*, **23**, 425 (1979).
[68] Mabrouk, A. F., *Phenolic, Sulfur and Nitrogen Compounds in Food Flavors* (G. Charalambous & I. Katz, Eds.), ACS Symposium Series No. **26**, 146–183 (1976).
[69] Maga, J. A. & C. E. Sizer, *J. Agric. Food Chem.*, **21**, 22 (1973a); Maga, J. A., *CRC Crit. Rev. Food Sci. Nutr.*, **8**, 1 (1976b).
[70] Maillard, L. C., *Compt. Rend. Acad. Sci. Paris*, **154**, 66 (1912).

[71] Manley, C. H., P. P. Vallon & R. E. Erickson, *J. Food Sci.*, **39**, 73 (1974).
[72] Mason, M. E., B. Johnson & M. C. Hamming, *J. Agric. Food Chem.*, **14**, 454 (1966).
[73] Merritt, C. & D. H. Robertson, *J. Gas Chromatog.*, **4**, 96 (1976).
[74] Micheel, F. & A. Klemer, *Chem. Ber.*, **84**, 212 (1951a); *ibid*, **85**, 1083 (1952b).
[75] Mills, F. D., B. G. Baker & J. E. Hodge, *J. Agric. Food Chem.*, **17**, 723 (1969a); *Carbohydrate Res.*, **15**, 205 (1970).
[76] Mills, F. D., D. Weisleder & J. E. Hodge, *Tetrahedron Letters*, 1243 (1970).
[77] Mills, F. D. & J. E. Hodge, *Carbohydrate Res.*, **51**, 9 (1976).
[78] Mizugaki, J., M. Uchiyama & S. Okui, *Biochem. J.*, **58**, 273 (1965).
[79] Mulders, E. J., *Z. Lebensm.-Unters. Forsch.*, **151**, 310 (1973a); *ibid*, **152**, 193 (1973b).
[80] Mulley, E. A., C. R. Stumbo & W. M. Hunting, *J. Food Sci.*, **40**, 985, 989 (1975).
[81] Muller, C. J., R. E. Kepner & A. D. Webb, *J. Agric. Food Chem.*, **20**, 193 (1972); *idem, Amer. J. Enol. Viticult.*, **24**, 5 (1973).
[82] Murray, K. E., J. Shipton & F. B. Whitfield, *Chem. Ind. (London)*, **1970**, 897.
[83] Murray, K. E. & F. B. Whitfield, *J. Sci. Food Agric.*, **26**, 973 (1975).
[84] Mussinan, C. J. & I. Katz, *J. Agric. Food Chem.*, **21**, 43 (1973a); Mussinan, C. J. *et al.*, *Phenolic, Sulfur and Nitrogen Compounds in Food Flavors* (G. Charalambous & I. Katz, Eds.), ACS Symposium Series No. **26**, 133 (1976b).
[85] Nishi, H. & I. Morishita, *Nippon Nogei Kagaku Kaishi*, **45**, 507 (1971).
[86] Nursten, H. E. & M. R. Sheen, *J. Sci. Food Agric.*, **25**, 643 (1974).
[87] Ohloff, G., in *Fette Funktionelle Bestandteile von Lebensmitteln* (J. Solms, Ed.), Forster Verlag, Zürich, 119 (1973).
[88] Ohloff, G., V. Rautenstrauch & K. H. Schulte-Elte, *Helv. Chim. Acta*, **56**, 1503 (1973).
[89] Ouweland Van Den, G. A. M. & H. G. Peer, *U.S. Pat.*, 3,697,295 (1.7. 1968a); *idem, J. Agric. Food Chem.*, **23**, 501 (1975b).
[90] Ouweland Van Den, G. A. M., H. G. Peer & S. B. Tjan, *Flavor of Foods and Beverages* (G. Charalambous & G. E. Inglett, Eds.), Academic Press, New York (1978), 131-144.
[91] Pepper, F. H. & A. M. Pearson, *J. Food Sci.*, **34**, 10 (1969).
[92] Piloty, M. & W. Baltes, *Z. Lebensm.-Unters. Forsch.*, **168**, 368, 374 (1979).
[93] Pippen, E. L. & E. P. Mecchi, *J. Food Sci.*, **34**, 443 (1969).
[94] Pokorný, J., H. Svobodová & G. Janiček, *Sb. Vys. Šk. Chem. Technol. Praze, Potraviny.*, E **49**, 521 (1977a); Pokorný, J. *et al.*, *Nahrung*, **23**, 921 (1979b).

[95] Praag Van, M., H. S. Stein & M. S. Tibbetts, *J. Agric. Food Chem.*, **16**, 1005 (1968).

[96] Reymond, D., *Agricultural and Food Chemistry, Past, Present, Future* (R. Teranishi, Ed.), AVI Publ. Co. Inc., Westport, Connecticut, USA, 315 (1978).

[97] Reynolds, T. M., *Advances in Food Res.*, **12**, 1 (1963a); *ibid*, **14**, 167 (1965b).

[98] Reynold, R. J., *U.S. Pat.*, 3,996,941 (14.12.1976).

[99] Rizzi, G. P., *J. Agric. Food Chem.*, **20**, 1081 (1972a); *ibid*, **22**, 279 (1974b).

[100] Rizzi, G. P., *Phenolic, Sulfur and Nitrogen Compounds in Food Flavors* (G. Charalambous & I. Katz, Eds.), ACS Symposium Series No. **26**, 122–132 (1976).

[101] Rohan, T. A., *J. Sci. Food Agric.*, **14**, 799 (1963a); *idem, J. Food Sci.*, **29**, 456 (1964b).

[102] Rooney, L. W., A. Salem & J. A. Johnson, *Cereal Chem.*, **44**, 539 (1967).

[103] Ruiter, A., *Lebensm. Wiss. Technol.* **6**, 142 (1973).

[104] Sakaguchi, M. & T. Shibamoto, *J. Agric. Food Chem.*, **26**, 1260 (1978a); *Agric. Biol. Chem.*, **43**, 667 (1979b).

[105] Scanlan, R. A., S. G. Kayser, L. M. Libbey & M. E. Morgan, *J. Agric. Food Chem.*, **21**, 673 (1973).

[106] Schutte, L., *Fenaroli's Handbook of Flavor Ingredients* (T. E. Furia & N. Bellanca, Eds.), 2nd edn, CRC Press Inc., Cleveland, Ohio, 132–183 (1975); *idem, Phenolic, Sulfur and Nitrogen Compounds in Food Flavors* (G. Charalambous & I. Katz, Eds.), ACS Symposium Series No. **26**, 96–113 (1976).

[107] Sekiya, J., T. Kajiwara & A. Hatanaka, *Phytochemistry*, **16**, 1043 (1977).

[108] Sessa, D. J., *J. Agric. Food Chem.*, **27**, 235 (1979).

[109] Severin, T. & W. Seilmeier, *Z. Lebensm.-Unters. Forsch.*, **137**, 4 (1968).

[110] Shankaranarayana, M., B. Raghavan, K. O. Abraham & C. P. Natarajan, *CRC. Crit. Rev. Food Technol.*, **4**, 395 (1974).

[111] Shaw, P. E., J. H. Tatum & R. E. Berry, *J. Agric. Food Chem.*, **16**, 979 (1968a); Tatum, J. H. *et al.*, *ibid*, **17**, 38 (1969b).

[112] Shaw, P. E. & R. E. Berry, *J. Agric. Food Chem.*, **25**, 614 (1977).

[113] Shibamoto, T. & R. A. Bernhard, *J. Agric. Food Chem.*, **24**, 847 (1976a); *ibid*, **26**, 183 (1978b).

[114] Shibamoto, T. & G. F. Russell, *J. Agric. Food Chem.*, **24**, 843 (1976a); *ibid*, **25**, 109 (1977b); Shibamoto, T., *ibid*, **25**, 206 (1977c).

[115] Shibamoto, T., T. Akiyama, M. Sakaguchi, Y. Enomoto & H. Masuda, *J. Agric. Food Chem.*, **27**, 1027 (1979).

[116] Shigematsu, H., T. Kurata, H. Kato & M. Fujimaki, *Agric. Biol. Chem.*, **35**, 2097 (1971a); *ibid*, **36**, 1631 (1972b); Shigetmatsu, H., S. Shibata, T. Kurata, H. Kato & M. Fujimaki, *J. Agric. Food Chem.*, **23**, 233 (1975c).

[117] Smouse, T. H. & S. S. Chang, *J. Amer. Oil Chemist. Soc.*, **44**, 509 (1967).

[118] Stetter, H., *Angew. Chem.*, **88**, 695 (1976); *idem, Angew. Chem. Int. Ed. Engl.*, **15**, 639 (1976).

[119] Suyama, K. & S. Adachi, *J. Agric. Food Chem.*, **28**, 546 (1980).

[120] Takahashi, K., M. Tadenuma & S. Sato, *Agric. Biol. Chem.*, **40**, 325 (1976).

[121] Takeoka, G. R., J. Coughlin & G. F. Russell, *Liquid Chromatography Analysis of Food Beverages* (G. Charalambous, Ed.), Academic Press, New York, San-Francisco, 179–214 (1979).

[122] Takken, H. J., L. M. Van der Linde, P. J. De Valois, J. M. Van Dort & M. Boelens, *Phenolic, Sulfur and Nitrogen Compounds in Food Flavors*, ACS Symposium Series No. **26**, 114 (1976); *idem*, 170th National Meeting of the American Chemical Society, Chicago, III, Aug. 1975, Abstract No. AGDF-15.

[123] Tas, A. C. & R. J. C. Kleipool, *Riechst. Aromen. Körperpflegem.*, **24**, 8, 10, 12 (1974).

[124] Timmer, R., R. Ter Heide, P. J. De Valois & H. J. Wobben, *J. Agric. Food Chem.*, **23**, 53 (1975).

[125] Tomomatsu, S., S. Kumamoto, A. & A. Komatsu, *Japan Tokkyo Koho*, 79,13500 (31.5.1979).

[126] Tonsbeek, C. H. T., E. B. Koenders, A. S. M. Van Der Zijden & J. A. Losekoot, *J. Agric. Food Chem.*, **17**, 397 (1969).

[127] Tressl von, R., T. Kossa & M. Holzer, *Geruch. Geschmackstoffe Int. Symp.*, 1974 (Publ. 1975) (F. Drawert, Ed.), Verlag Hans Carl, Nuernberg.

[128] Tressl von, R., D. Bahri, M. Holzer & T. Kossa, *J. Agric. Food Chem.*, **25**, 455 (1977).

[129] Tressl von, R., M. Apetz, R. Arrieta & K. G. Grünewald, *Flavor of Foods and Beverages* (G. Charalambous & G. E. Inglett, Eds.), Academic Press, New York, 145–168 (1978).

[130] Tressl von, R., *Monats, für Brauerei.*, 240 (1979).

[131] Tressl von, R., K. G. Grünewald, R. Silvar & D. Bahri, *Progress in Flavour Research* (D. G. Land & H. E. Nursten, Eds.), Applied Science Publ. Ltd., London, 193–213 (1978).

[132] Tsuchida, H. & M. Komoto, *Agric. Biol. Chem.*, **31**, 185 (1967a); Tsuchida, H., M. Komoto, H. Kato & M. Fujimaki, *ibid*, **39**, 1143 (1975b); Tsuchida, H., M. Komoto, H. Kato, T. Kurata & M. Fujimaki, *ibid*, **40**, 2051 (1976c).

[133] Viani, R. & I. Horman, *J. Food Sci.*, **39**, 1216 (1974).

[134] Veldink, G. A., J. F. G. Vliegenthart & J. Boldingh, *J. Progr. Chem. Fats*, **15**, 131 (1977).

[135] Velišek, J., J. Davídek, V. Kubelka, Z. Zelinková & J. Pokorný, *Z. Lebensm.-Unters. Forsch.*, **162**, 285 (1976).

[136] Vernin, G., *Parf. Cosm. Aromes,* **32**, 77 (1980); Vernin, G. &. J. Metzger, *Bull. Soc. Chim. Belge,* **90**, 553 (1981).

[137] Vitzthum, O. G. & P. Werkhoff, *Z. Lebensm.-Unters. Forsch.,* **156**, 300 (1974a); Vitzthum, O. G., *Kaffee und Coffein* (O. Eichler, Ed.), Springer-Verlag, Berlin (1975b).

[138] Vitzthum, O. G., P. Werkhoff & P. Hubert, *J. Food Sci.,* **40**, 911 (1975a); *idem, J. Agric. Food Chem.,* **23**, 999 (1975b).

[139] Vondenhof, T., K. W. Glombitza & M. Steiner, *Z. Lebensm.-Unters. Forsch.,* **152**, 345 (1973).

[140] Walradt, J. R., R. C. Lindsay & L. M. Libbey, *J. Agric. Food Chem.,* **18**, 926 (1970).

[141] Walter, R. H. & I. S. Fagerson, *J. Food Sci.,* **33**, 294 (1968).

[142] Wang, P. S., H. Kato & M. Fujimaki, *Agric. Biol. Chem. Japan,* **33**, 1775 (1969).

[143] Wang, P. S. & G. V. Odell, *J. Agric. Food Chem.,* **21**, 868 (1973).

[144] Watanabe, K. & Y. Sato, *Agric. Biol. Chem.,* **33**, 242 (1969a); *ibid.,* **34**, 464 (1970b).

[145] Whitlock, C. B. & W. W. Nawar, *J. Amer. Oil Chemist. Soc.,* **53**, 586 (1976a); *ibid,* **53**, 592 (1976b).

[146] Wilson, R. A., C. J. Mussinan, I. Katz & A. Sanderson, *J. Agric. Food Chem.,* **21**, 873 (1973).

[147] Wilson, R. A., M. H. Vock, I. Katz & E. J. Shuster, *Brit. Pat.*, 1,364,747 (1974).

[148] Wong, N. P., L. M. Libbey & R. C. Lindsay, *J. Dairy Sci.,* **52**, 888 (1969).

[149] Yamaguchi, K., S. Mihara, A. Aitoku & T. Shibamoto, *Liquid Chromatographic Analysis of Foods and Beverages* (G. Charalambous, Ed.), Academic Press, New York, 303-330 (1979).

[150] Yamamoto, K. & M. Noguchi, *Agric. Biol. Chem.,* **37**, 2185 (1973).

[151] Yamanishi, T., *Nippon Nogei Kagaku Kaishi,* **49**, R1 (1975).

[152] Yasumoto, K., K. Iwami & H. Mitsuda, *Agric. Biol. Chem.,* **35**, 2070 (1971).

CHAPTER IV

General Synthetic Methods for Heterocyclic Compounds used for Flavouring

Gaston VERNIN, University of Aix-Marseilles, France, and

A. K. EL-SHAFFEI, Assiut University, Sohag, Egypt

IV–1 INTRODUCTION

The very wide scope of the literature of heterocyclic chemistry precludes giving references to original papers. Consequently, we will limit this chapter to the most important secondary sources dealing with the general synthetic methods of heterocyclic compounds. Among them, the series 'Advances in Heterocyclic Chemistry' is by far of vital importance. All the reviews on heterocyclic compounds synthesis were collected up to 1966 by Katritzky & Weeds [20]. But further information was also made available in the following works: 'Heterocyclic Compounds' [9], 'The Chemistry of Heterocyclic Compounds' [48], 'The Chemistry of Carbon Compounds' [39], 'Annual Reports' [3], 'Organic Sulphur Compounds' [24], 'Specialist Periodical Reports' [43], etc. . .

In addition to the above series, numerous topics have been treated in monographs and review articles [4,21]. Lewis & Charnak [27] listed these secondary sources up to 1977. Finally, besides original papers given in the bibliography in the other chapters, the reader may refer to Maga's reviews (*see Chapter II*).

IV–2 FIVE-MEMBERED RINGS

2.1 Five-membered rings with one heteroatom: furans, thiophenes, pyrroles, and related reduced systems

The various types of ring closure for one heteroatom containing five-membered rings are reported in *Scheme 1*.

From all these possibilities, by far the two more important ones are those which involve C-X bond formation only and those which involve formation of the C_3-C_4 bond. But these compounds can also be obtained from other pathways.

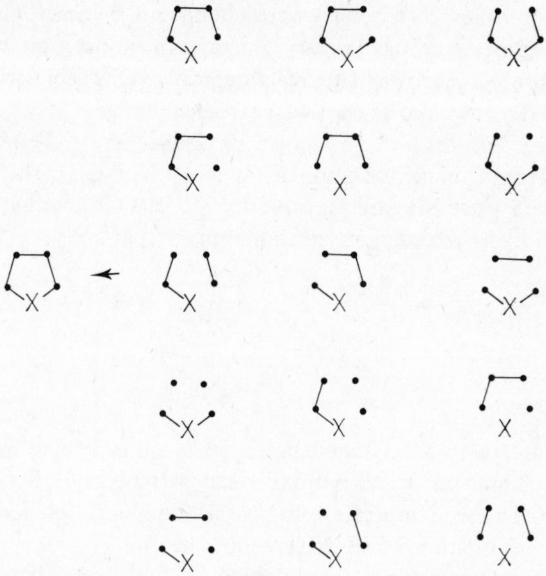

Scheme 1 – Various types of ring closure for one heteroatom containing five-membered rings (furans, thiophenes, pyrroles and reduced systems).

2.1.1 Formation of C–X bonds

2.1.1.1 From 1,4-diketones (Paal-Knorr *synthesis)* – This method is the most important one for furans [11], thiophenes [17] and pyrroles [19].

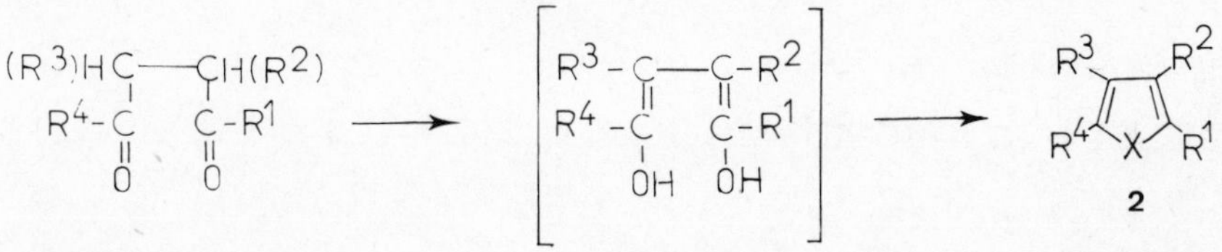

1a 1b 2 3

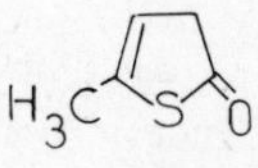

By treatment with H_2SO_4, P_2O_5, or $ZnCl_2$ enol intermediates **1b** lead to furans **2** (X = O). Substituted thiophenes **2** (X = S) are prepared by reacting **1a** with phosphorus pentasulphide. Thus 2,5-hexanedione **1a** ($R^1 = R^4 = CH_3$, $R^2 = R^3 = H$) and 2,3-dimethylsuccinic acid **1a** ($R^1 = R^4 = OH$, $R^2 = R^3 = CH_3$) produce 2,5-, and 3,4-dimethylthiophene, respectively.

If the cyclization of **1a** is carried out in the presence of ammonia, the corresponding pyrroles **2** (X = NH) were obtained. 2,5-Dimethylpyrrole is prepared from acetonyl acetone. Primary amines can be used in this reaction to obtain *N*-substituted pyrroles. This reaction was extensively used. 1,4-Diesters **1a** ($R^1 = R^4 = OR$) serve also as starting materials.

2.1.1.2 From 1,4-disubstituted butanes or 2-butenes – Tetrahydrofuran **5** (X = O) is obtained from dehydration of the diol **4** (X = OH). Thiolane **5** (X = S) and pyrrolidine **5** (X = NH) are prepared by reacting tetramethylene dibromide **4** (X = Br) with hydrogen sulphide and ammonia, respectively.

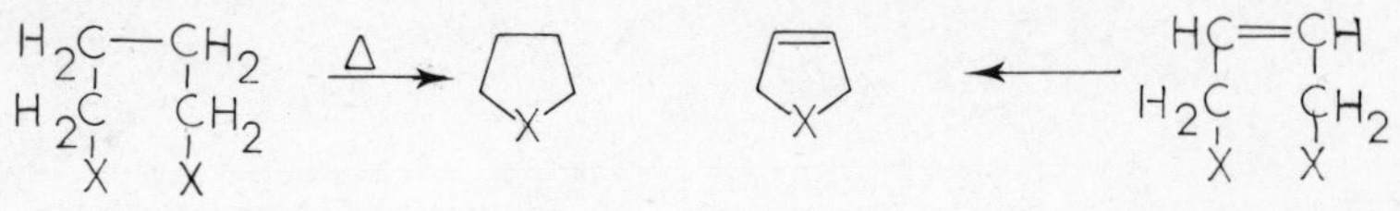

4 **5** **7** **6**

Compounds **7** with a 2,3 double bond can be similarly synthesized from the corresponding 2-butenes **6**. γ-Hydroxy-, and γ-thiol acids **8** (X = OH, SH) usually cyclize at room temperature to give lactones and thiolactones **9** (X = O, S), while lactam formation **9** (X = NH) requires heating.

α,β-Unsaturated γ-lactones are obtained by the same procedure from α,β-unsaturated acids or esters. Thus 2,3-dimethyl-2-butenoic acid γ-lactone **10** was synthesized by bromination of ethyl 2,3-dimethyl-2-butenoate followed by dehydrobromination and lactonization.

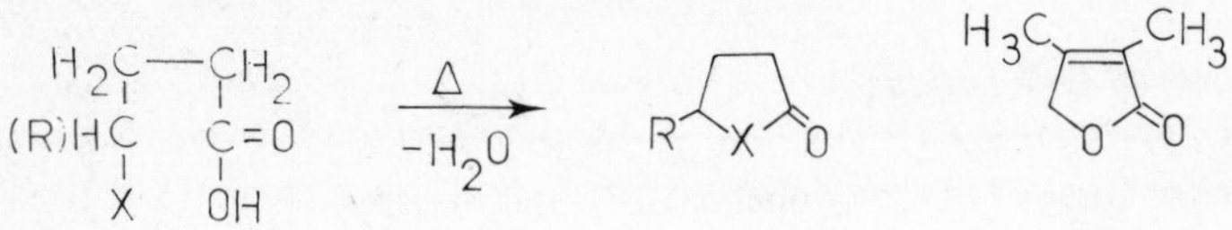

8 **9** **10**

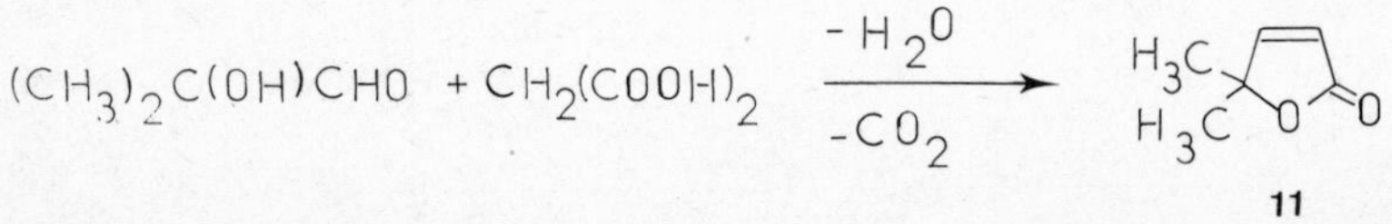

$$(CH_3)_2C(OH)CHO + CH_2(COOH)_2 \xrightarrow[-CO_2]{-H_2O} \mathbf{11}$$

4,4-Dimethylpent-2-en-5-olide **11** was prepared by condensing 2-hydroxy 2-methyl propanal with malonic acid. Intramolecular cyclizations of γ-keto acids **12** γ-keto alcohols **14** produce α,β-unsaturated γ-lactones **13** and 4,5-dihydrofurans **15**, respectively.

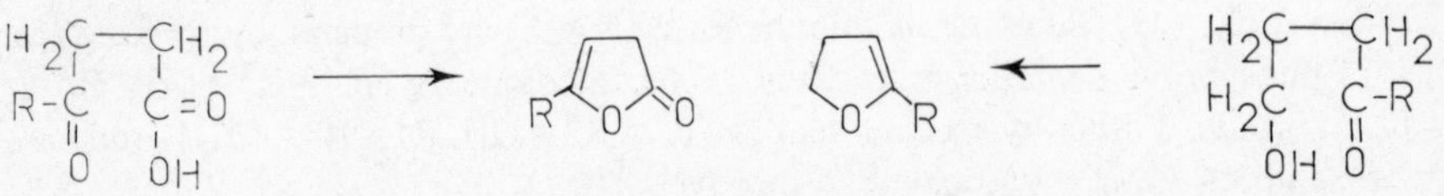

12 **13** **15** **14**

2,5-Dimethyl-4-hydroxy-3-(2*H*)-furanone or furaneol **17** has been prepared by a two-step synthesis, starting from readily available 3-hexyne-2,5-diol **16** [37].

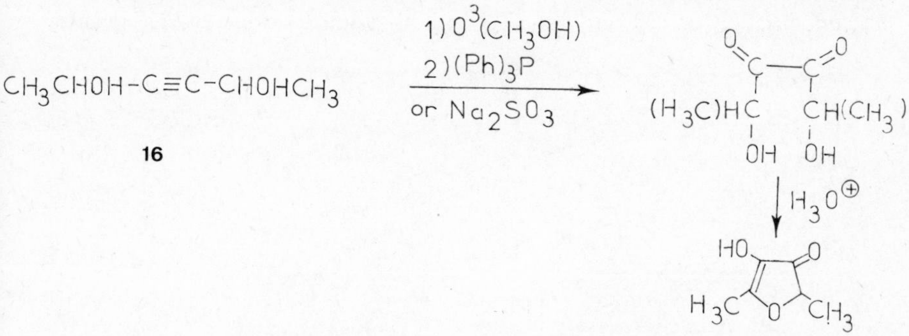

$$CH_3CHOH{-}C{\equiv}C{-}CHOHCH_3 \xrightarrow[\text{or } Na_2SO_3]{1)\,O_3(CH_3OH)\quad 2)\,(Ph)_3P} (H_3C)HC(OH){-}C(=O){-}C(=O){-}CH(OH)(CH_3) \xrightarrow{H_3O^{\oplus}} \mathbf{17}$$

16 **17**

The compound **17** was also prepared by heating methylglyoxal in acetic solution in the presence of zinc dust, followed by heating the resulting intermediate in acidic or basic medium.

The norcepanone **19** (R = n-C_6H_{13}) was prepared by cyclization of the corresponding diacetylene alcohols **18** using mercuric salts.

$$HC{\equiv}C{-}C{\equiv}C{-}CHOHR \xrightarrow{Hg^{++}} \mathbf{19}$$

18 **19**

Furfural, its 5-methyl-, and 5-hydroxymethyl derivatives **21** (R = H, CH_3, CH_2OH) were obtained from pentoses, methylpentoses and hexoses, respectively.

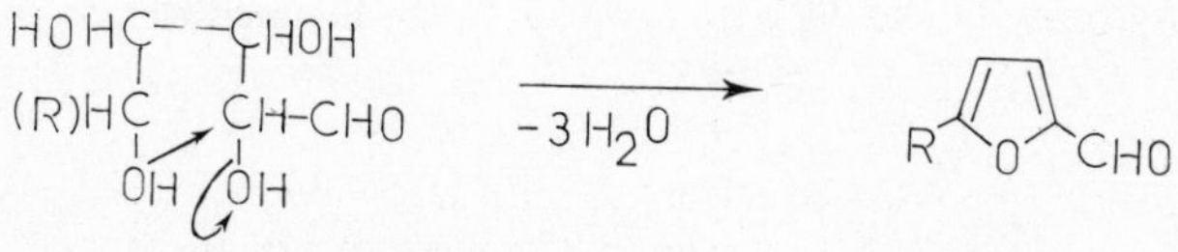

$$\mathbf{20} \xrightarrow{-3H_2O} \mathbf{21}$$

20 **21**

2.1.1.3 From pyrolysis of alkanes – Another example of ring closure is the thiophene **22** formation from pyrolysis of butane in the presence of sulphur.

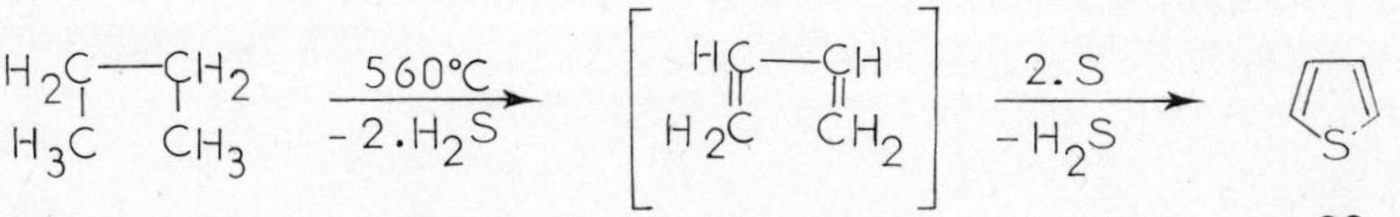

$$H_3C{-}CH_2{-}CH_2{-}CH_3 \xrightarrow[-2.H_2S]{560°C} [H_2C{=}CH{-}CH{=}CH_2] \xrightarrow[-H_2S]{2.S} \mathbf{22}$$

22

2.1.2 Formation of the C_3–C_4 bond

2.1.2.1 From β-dicarbonyl compounds and α-aminoketones (Knorr *pyrrole synthesis*) – This method which is the most important route to pyrroles, involves the condensation of a β-keto ester **23** with an α-aminoketone **24**. These latter

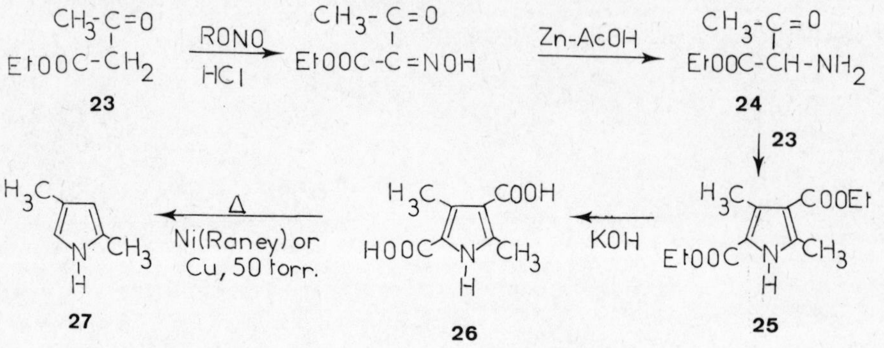

compounds are usually prepared *in situ* by isonitrosation and reduction of one mole of the β-keto ester **23**. The acetylacetone **28** reacts with the aminoketone **24** to give 3-acetyl-5-carboethoxy-2,4-dimethylpyrrole **29**. In a similar way, 3-alkyl-2,4-pentanediones lead to 2,3,4-trisubstituted pyrroles.

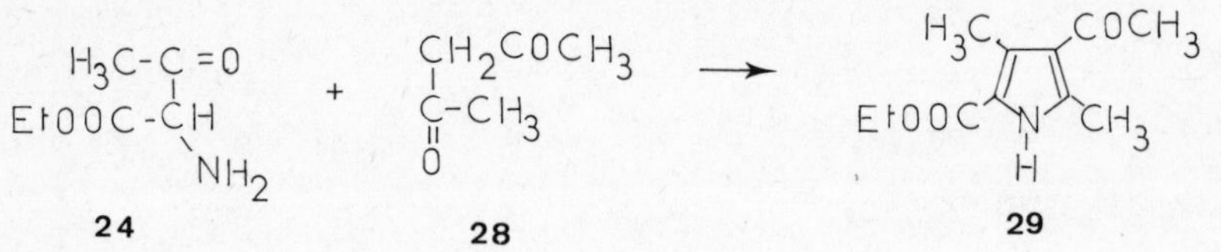

2.1.2.2 From α-halogenoketones and β-ketoesters (Hantzsch *synthesis*) – α-Halogeno ketones **30** react with β-keto esters **23** to give furans. In the presence of ammonia, pyrroles **33** and **34** were formed *via* vinylamine intermediates **32**.

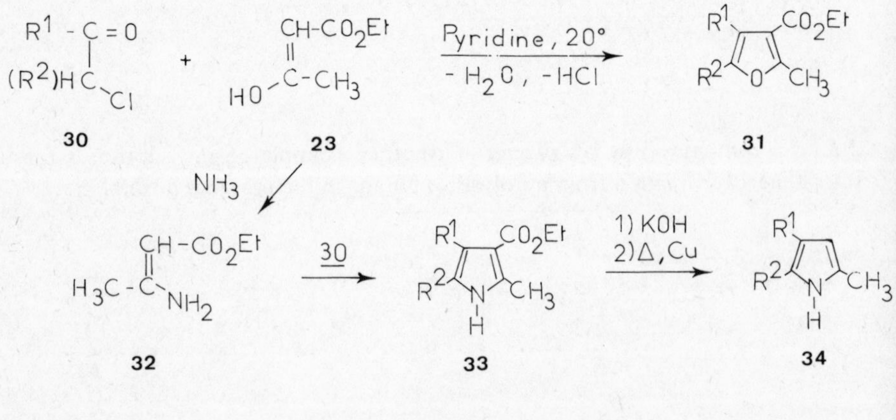

2.1.2.3 From α-mercaptoketones (MacIntosh *synthesis*) – A general synthetic method of trisubstituted alkylthiophenes **38** has been recently described [28]. They are synthesized by condensing vinylphosphonium salts **35** with α-mercapto ketones **36**. 2,5-Dihydrothiophene intermediates **37** are dehydrogenated by chloranil in *t*-butyl alcohol or pyridine as solvent.

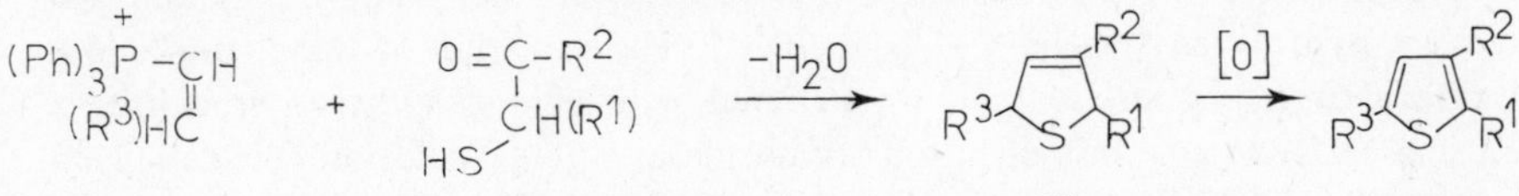

$(Ph)_3\overset{+}{P}-CH=CH(R^3)$ + $O=C(R^2)-CH(R^1)-SH$ $\xrightarrow{-H_2O}$ **37** $\xrightarrow{[O]}$ **38**

35 **36** **37** **38**

2.1.3 Formation of other bonds

Thiophenes **38** can also be obtained from β-halogeno-α,β-unsaturated aldehydes **39**, sodium sulphide and alkyl halides under basic catalysis conditions.

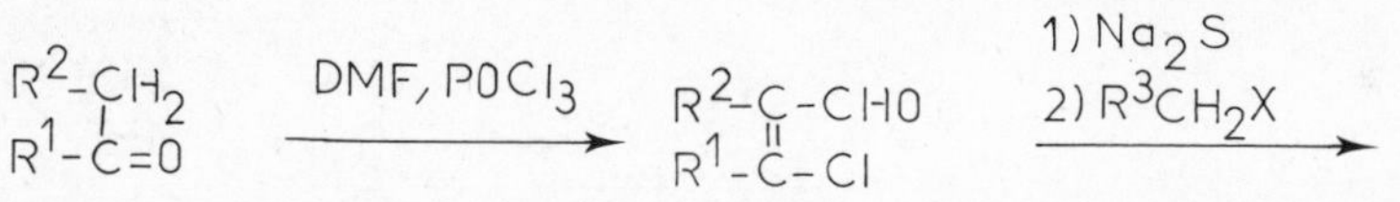

$R^2-CH_2-C(R^1)=O$ $\xrightarrow{DMF,\ POCl_3}$ $R^2-C(CHO)=C(R^1)-Cl$ $\xrightarrow{1)\ Na_2S\quad 2)\ R^3CH_2X}$ **38**

39

1,2-Disubstituted pyrrolidines **41** are prepared from *N*-chloro derivatives **40** *via* a radical mechanism.

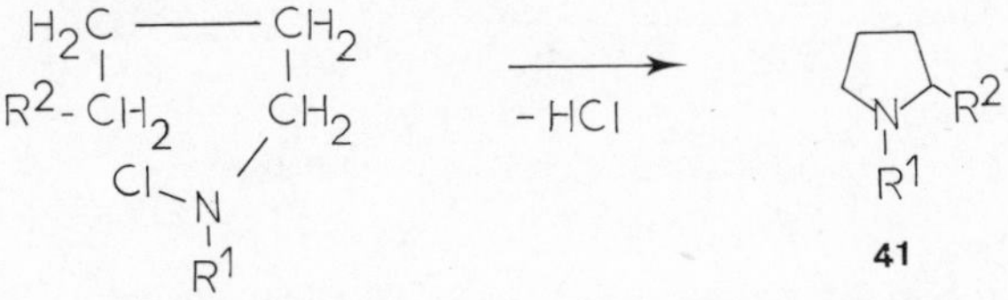

H_2C-CH_2, R^2-CH_2, CH_2, $Cl-N-R^1$ $\xrightarrow{-HCl}$ pyrrolidine ($N-R^1$, R^2)

40 **41**

2.1.4 Formation from other heterocyclic compounds

Furans **42** treated by hydrogen sulphide at 350–400°C afford the corresponding thiophenes **43**. In a similar way, the action of ammonia and steam in the presence of alumina as catalyst, leads to pyrrole formation **44**. Thus the 5-hydroxymethyl-1-methylpyrrole-2-carboxaldehyde was obtained in low yield by heating 24 h. at 140°C the corresponding furan in the presence of methylamine (40%, aqueous).

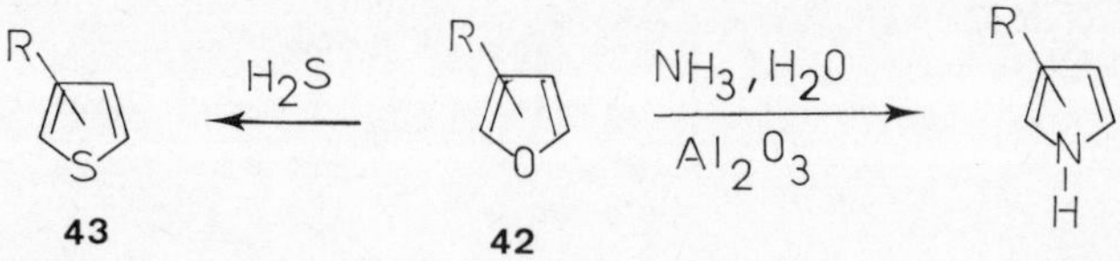

43 $\xleftarrow{H_2S}$ **42** $\xrightarrow{NH_3,\ H_2O;\ Al_2O_3}$ **44**

43 **42** **44**

2.1.5 *Variously substituted derivatives from parent compounds*

As shown in *Scheme 2*, numerous furan aroma compounds can be synthesized from parent compound. Furan and pyrrole react with acetic anhydride even without a catalyst to give the 2-acylated product. *N*-Substituted 2-acetylpyrroles have been prepared by acetylation of *N*-substituted pyrroles with acetyl chloride in the presence of anhydrous magnesium bromide in ether. These compounds are easily hydrolyzed (NaOH - H_2O at 20°C). The acylation of thiophenes under *Friedel-Crafts* conditions can be achieved with acid anhydrides or chlorides using catalytic amounts of boron trifluoride in ethereal solution. Yields ranged from 60 to 90%. Other mild catalysts (e.g., $SnCl_4$, $ZnCl_2$) may be used.

Heterocyclic 2-carboxaldehydes can be prepared by the *Vilsmeir-Haack* procedure by reaction of furans, thiophenes and pyrroles with dimethylformamide using $POCl_3$ as catalyst [42].

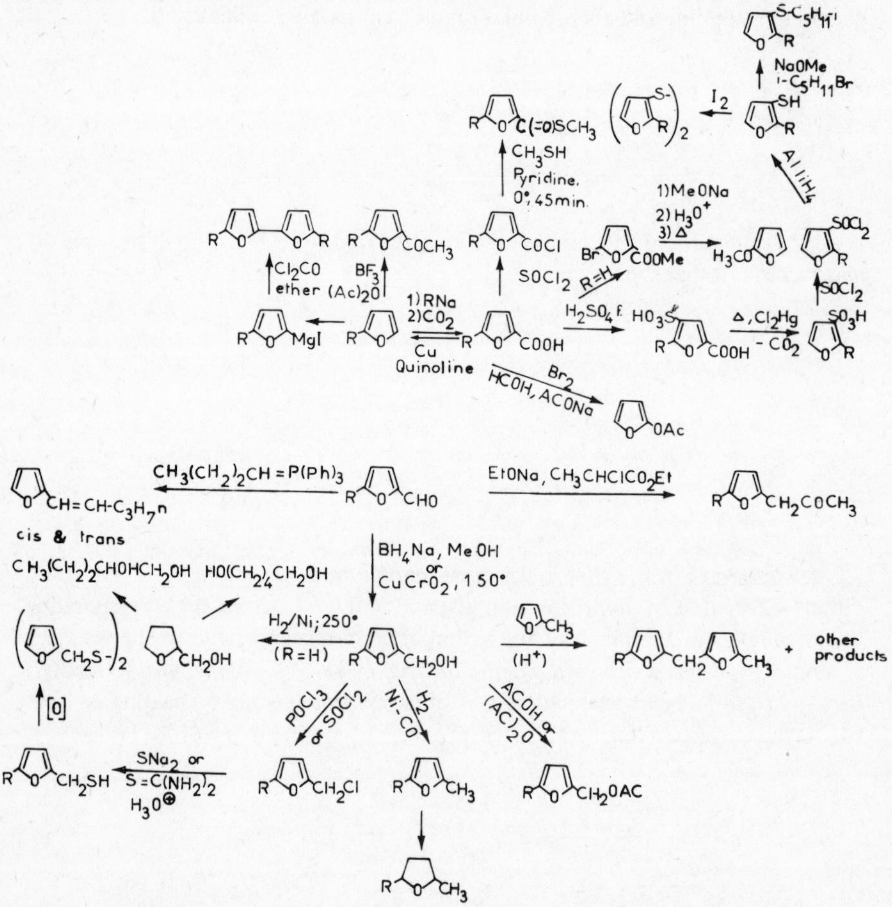

Scheme 2 – Main furan aroma compounds (R = H or CH_3).

2-Formylpyrroles can also be obtained by condensation of pyrroles with chloroform in the presence of potassium hydroxide according to the *Reimer-Tiemann's* reaction. A general method for preparing *N*-substituted pyrroles consists in treating pyrrylthallium salts with an alkyl halide.

2-Furfurylmethyl ketone was prepared by sodium ethylate catalysed condensation of 2-chloropropionate with furfural followed by alkaline hydrolysis and concomitant decarboxylation of the resulting glycidic ester.

2- and 3-Halogeno thiophenes react satisfactorily under photostimulation at 350 nm, in liquid ammonia, in the presence of potassium amide to afford the corresponding acetonyl derivatives. A radical nucleophilic unimolecular substitution reaction is involved [6,50].

Furyl-, and furfuryl alkylsulphides may be easily obtained by reacting the corresponding thiol compounds with alkyl halides under phase transfer catalysis conditions (NaOH, 50%, benzene and tetrabutylammonium bromide or triethylbenzylammonium chloride as catalysts) [23,47]. Finally, 2-alkyl-, 2-acetyl-, 2-aryl- and 2-heteroaryl derivatives can be obtained in low yields by free radical substitution [5,30].

2.2 Five-membered rings with two heteroatoms in the 1,3 positions: oxazoles, thiazoles, and imidazoles

The various methods of synthesizing these heterocyclic rings are collected in *Scheme 3*. Only six of them are of practical importance.

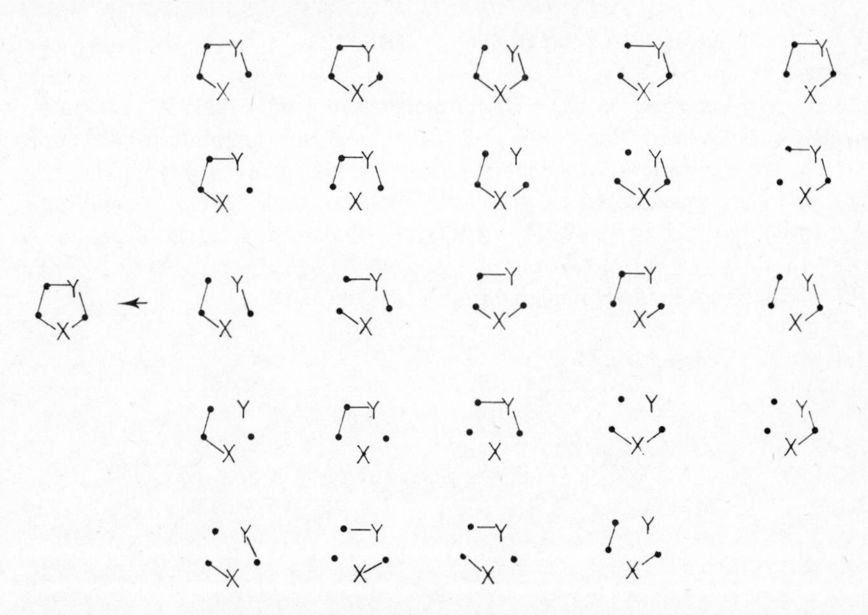

Scheme 3 – Various types of ring closure for the two non-adjacent heteroatoms containing five-membered ring systems.

2.2.1 From α-halogenoketones

α-Halogenoketones **30** react with amides, thioamides and amidines **45** to give oxazoles [49], thiazoles [34], and imidazoles **46** (X = O, S, NH) [41]. Thus the cyclization of **30** with formamide **45** (R^1 = H, X = O) leads to 4,5-disubstituted oxazoles in 50 to 70% yields **46** (X = O, R^1 = H). This method, known as *Hantzsch's* synthesis, is by far the most important one in the thiazole chemistry [46a,c]. The cyclization is carried out with a great variety of reactants including thioamides **45** (X = S, R^1 = alkyl), thioureas **45** (X = S, R^1 = NH_2), their mono-, and disubstituted derivatives, salts and esters of monothio- and dithiocarbamic acids **45** (X = S, R^1 = SNH_4, SR) leading to variously substituted thiazoles **46** (X = S). Thioamides are usually prepared *in situ* by condensing phosphorus pentasulphide with amides in dioxane solution at 40°C and in the presence of magnesium carbonate. The temperature is kept below 70°C during the addition of the halogenoketone **30** and then the reaction mixture is maintained at 100° for several hours. Alkylthiazoles were isolated by a double steam distillation.

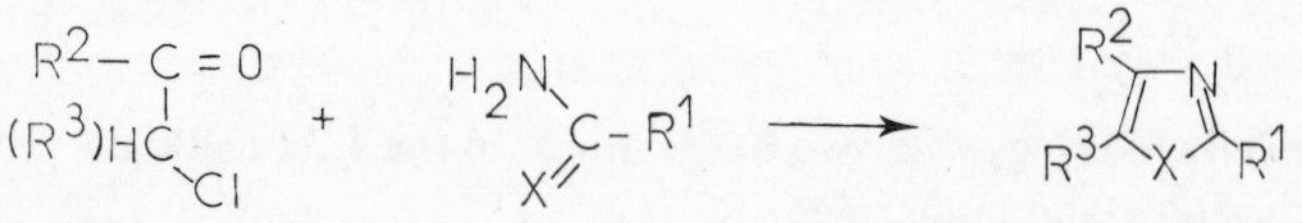

R^2– C = O ; (R^3)HC–Cl   + H_2N–C(=X)–R^1 ⟶   46

30   **45**   **46**

2.2.2 From α-acylaminoketones and α-acylamino acid esters (Robinson *and* Gabriel *synthesis*)

2,4,5-Trialkyloxazoles **46** (X = O) are prepared in good yields (65 to 95%) by cyclization of α-acylamino ketones **47** in the presence of concentrated sulphuric acid. If the cyclization is done in the presence of phosphorus pentasulphide, the corresponding thiazoles **46** (X = S) were obtained. Under the same conditions *N*-α-acylamino acid esters **47** (R^3 = OR) give 5-alkoxy oxazoles and thiazoles **46** (X = O, S, R^3 = OR). 2-Alkoxymethyl-, oxazoles and thiazoles **46** (X = O, S, R^1 = CH_2OR) are similarly prepared from **47** (R^1 = CH_2OR).

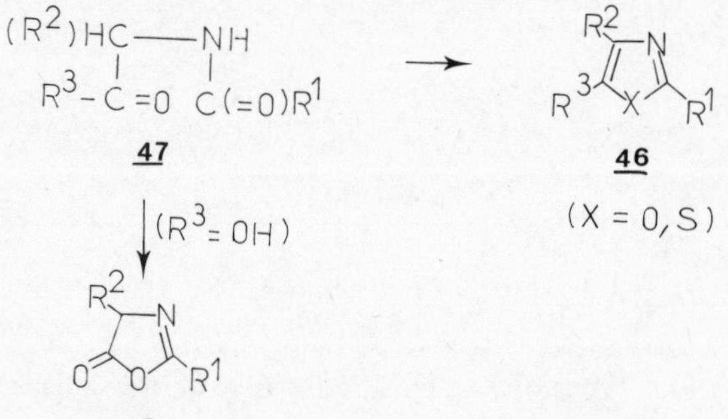

(R^2)HC——NH ; R^3–C=O   C(=O)R^1 ⟶   46

47   **46** (X = O, S)

↓ (R^3 = OH)

48

The cyclization of *N*-acryloyl-, and *N*-metacryloyl-α-alanine in the presence of phosphoric acid at 140°C during 4 h. gives 2-vinyl- and 2-isopropenyl-4,5-dimethyloxazoles. α-Acylamino carboxylic acids **47** (R^3 = OH) are converted into 5-(4*H*)-oxazolinones **48** by acid anhydrides.

In an analogous reaction, dithiolium salts **50** are prepared from α-dithioacyl ketones **49**.

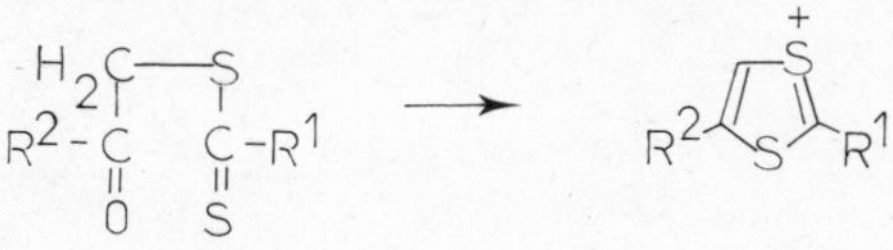

49 **50**

2.2.3 From α-aminoketones or α-dicarbonyl compounds

α-Aminoketones **51** react with imino esters **52** to give imidazoles **53** [40]. Compounds **53** can also be obtained by condensing α-diketones or α,β-keto aldehydes with aldehydes in the presence of ammonia.

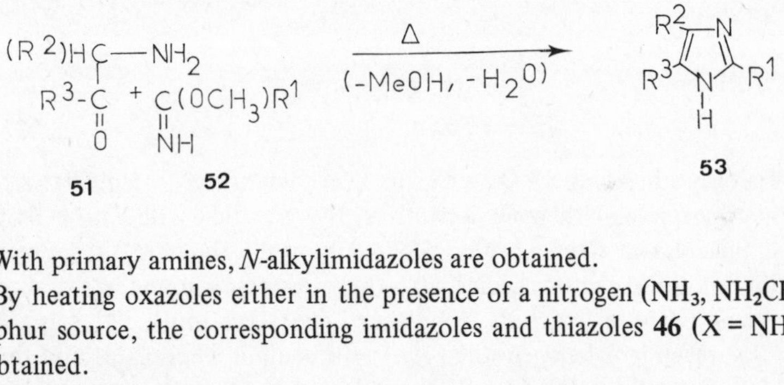

$(R^2)HC—NH_2$, R^3-C=O + $C(OCH_3)R^1$=NH —Δ, (-MeOH, -H_2O)→

51 **52** **53**

With primary amines, *N*-alkylimidazoles are obtained.

By heating oxazoles either in the presence of a nitrogen (NH_3, NH_2CHO) or a sulphur source, the corresponding imidazoles and thiazoles **46** (X = NH, or S) are obtained.

2.2.4 From acyl derivatives or α-hydroxy-, and α-mercaptoketones

Acyl derivatives of α-hydroxy, and α-mercapto ketones **54** (X = O, S) are cyclized in the presence of ammonium acetate and acetic acid to afford oxazoles and thiazoles **46** (X = O, S) *via* intermediates such as **55**. α-Mercaptoketones condense also with nitriles to give thiazoles.

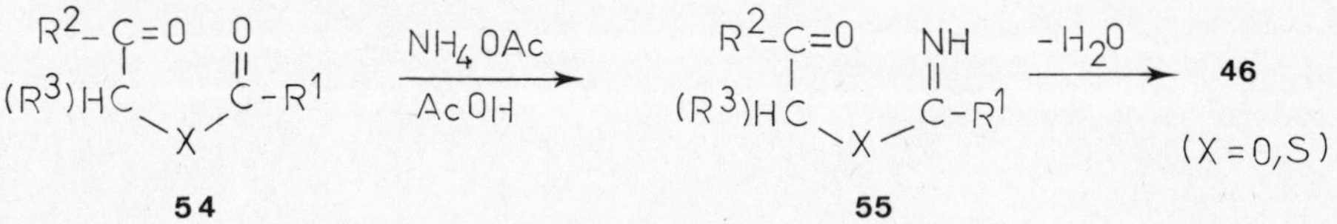

NH_4OAc / AcOH → ; $-H_2O$ → **46** (X = O,S)

54 **55**

2.2.5 From α-thiocyanoketones (Tcherniac *thiazole synthesis*)

The cyclization of α-thiocyanoketones **56** in the presence of halogeno acids in ether at 0° is another interesting thiazole synthesis.

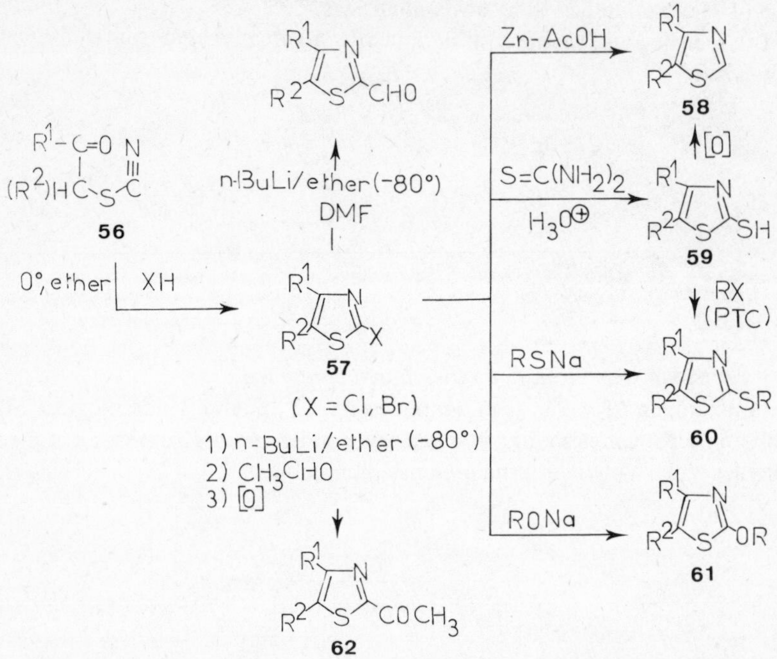

2-Halogenothiazoles **57** (X = Cl, Br, I) are important intermediates for the preparation of several thiazole derivatives. By reduction with Zn-AcOH unsubstituted thiazoles in the 2 position **58** are obtained. They are also prepared by oxidation of 2-mercaptothiazoles **59**. These latter compounds by alkylation under phase transfer catalysis conditions give alkyl-2-thiazolyl sulphides **60** [46b]. By reacting 2-halogenothiazoles with sodium alcoholates and thioalcoholates, 2-alkoxythiazoles **61** and 2-alkylthiothiazoles **60** were obtained, respectively. 2-Acetylthiazoles **62** were more difficult to obtain.

2.2.6 From β-hydroxy-, β-amino-, and β-mercapto acylamines

β-Substituted acylamines **63** cyclize to give 2-oxazolines, 2-imidazolines and 2-thiazolines **64** (X = O, S, NH). A procedure for 2-oxazoline rearomatization into oxazoles using 2,3-dichloro-5,6-dicyanobenzoquinone has been reported.

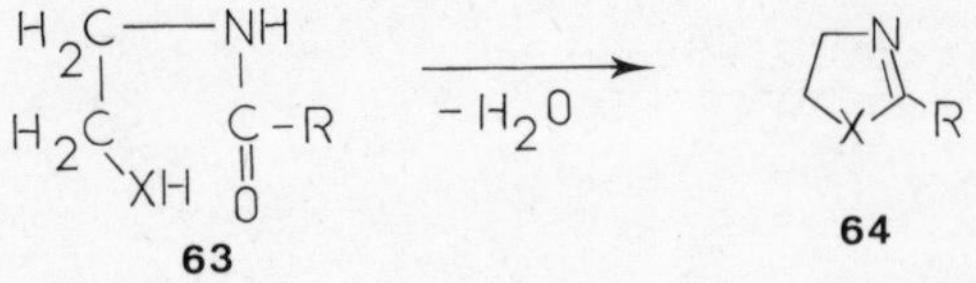

2.2.7 From 1,2-difunctional ethanes

1,2-Disubstituted ethanes **65** (X, Y = O, S, NH) react with aldehydes (or ketones) to form 2-alkyloxazolidines **66** and analogous derivatives.

R^1CHO ← $(R^2)HC{-}YH$ / $(R^3)HC{-}XH$ **65** → R^1COOH **68**; **66**; ↓ $COCl_2$ or $CO(OEt)_2$ **67**

1,3-Dioxolanes **66** (X = Y = O) are obtained by condensing α-glycols **65** (X = Y = O) with aldehydes or ketones. If the reaction is carried out in the presence of carbonyl chloride and carbonate esters, 2-oxazolidinones, 2-thiazolidinones and 2-imidazolidinones **67** are obtained. 1,2-Ethylenediamine **65** (X = Y = NH) condensed with mono- or dicarboxylic acids or their esters leads to 2-imidazolines **68** (X = Y = NH). 2-Thiazolines have been synthesized by reacting thioamides with 2-bromoethylamine.

2.2.8 Miscellaneous

An easy synthesis of 2,4,5-trimethyl-3-oxazoline from acetaldehyde, acetoin, and ammonia has been reported. Oxazoles can be also obtained from α-azido ketones, triphenylphosphine, and acid halides. Rearomatization of 2-oxazolines, using 2,3-dichloro-5,6-dicyanobenzoquinone as oxidizing agent, give oxazoles.

An interesting process for producing 2-methyl, or 2,4-dimethylthiazole by reacting diethylamine, ethylidene ethylamine or diisopropylamine with sulphur at 500° has been patented. The interaction of mercaptoacetaldehyde dimer, ammonia, and various aldehydes constitutes another efficient synthesis of 2-substituted thiazoles.

3-Thiazolines are prepared by bubbling a current of ammonia through a benzene or alcoholic solution of α-mercaptoketones or aldehydes according to *Asinger's* method. They are easily dehydrogenated into thiazoles in the presence of sulphur or quinones (chloranil). The catalytic reduction of monoacetyl derivative of α-aminonitriles leads to 2,5-disubstituted imidazolidines.

2.3 Five membered rings with two or more heteroatoms in the 1,2-, and 1,2,4 positions, respectively

The various types of ring closure for isoxazoles, isothiazoles, pyrazoles, 1,2-dithiolanes and 1,2,4-trithiolanes are reported in *Scheme 4*.

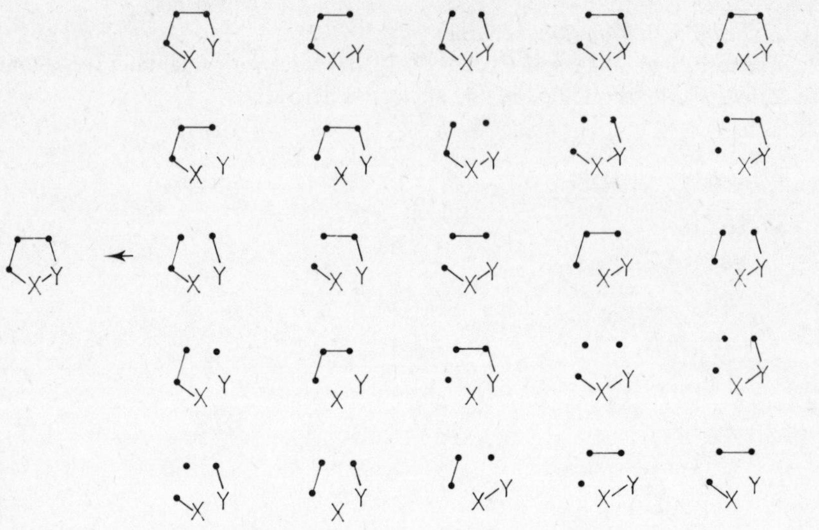

Scheme 4 – Various types of ring closure for five-membered ring systems with two heteroatoms in the 1,2 positions.

2.3.1 Isoxazoles and pyrazoles

The general synthetic methods for isoxazoles [36] and pyrazoles [26] **73** (X = O, NH, $R^1 = R^2$ = H, alkyl, CO_2Et) involve the reaction of β-dicarbonyl compounds **69** with hydroxylamine and hydrazine **70** (X = O, NH_2), respectively. These reactions take place under mild conditions. 1,3-Dicarbonyl compounds can be synthesized by reacting vinylketones. (RCOCH=CHCl), α-halogeno-1,3-diketones ($R^1COCHClCOR^2$) and 1,1,3,3-tetraalkoxypropanes. α,β-Unsaturated ketones **74** condensed with **70** to give 2-isoxazolines and 2-pyrazolines **75** (X = O, NH). 3-Isoxazolones and 3-pyrazolones **76** (X = O, NH) are similarly obtained from β-keto esters **69** (R^1 or R^2 = OEt).

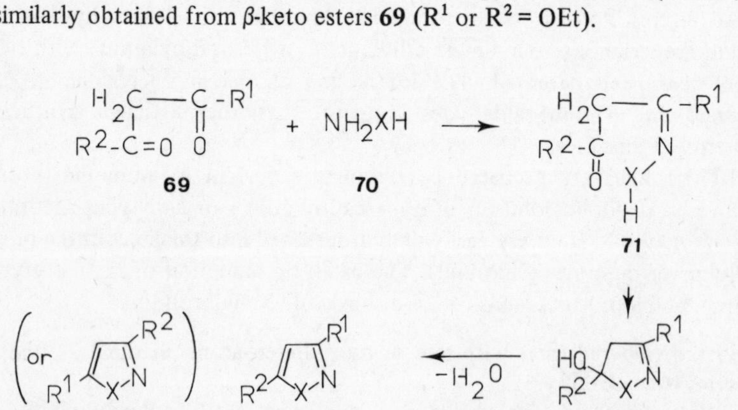

$H_2C—C-R^1$, $R^2-C=O$, O (**69**) + NH_2XH (**70**) → **71** → **72** $\xrightarrow{-H_2O}$ **73** (or regioisomer)

Saturated compounds **78** can be synthesized from 1,3-dibromides **77**. 3-Acetylisoxazole **80** is prepared from nitroketones **79** and the 3-formyl derivative **81** by reacting nitrogen dioxide with acetylene.

$$CH_3COCH_3 + HNO_3 \longrightarrow CH_3COC(NO_2){=}NOH \; (\mathbf{79}) \longrightarrow CH_3COC{\equiv}N\rightarrow O \xrightarrow{HC\equiv CH} \mathbf{80}$$

$$HC{\equiv}CH + N_2O_4 \rightarrow [O\leftarrow N{\equiv}C\text{-}CHO] \xrightarrow{HC\equiv CH} \mathbf{81}$$

Acetylene compounds **82** added to diazoalkanes give pyrazoles **83**. With acetylenic aldehyde **82** (R = CHO) 3-formylpyrazole **83** is obtained.

$$RC{\equiv}CH \; (\mathbf{82}) + {}^{\ominus}N{=}N{\overset{\oplus}{=}}CH_2 \longrightarrow \mathbf{83}$$

2.3.2 Isothiazoles

Few general and easy syntheses of isothiazoles are known [51]. However, they are prepared from various routes.

2.3.2.1 From 1,2-dithiolanes – Ammonia or amines added to 1,2-dithiolanes **84** give 5-isothiazolones **85**.

R^1 OEt R^2 S **84** — R^3NH_2 → R^1, OEt, S, R^2, NHR^3 — Br_2 → R^1, OEt, R^2, S, N, R^3 ↓ R^1, R^2, O, S, N-R^3 **85**

2.3.2.2 From β-thioxo imines – β-Thioxo imines **86** are cyclized into isothiazoles **87**.

H_2C—C-R^1, R 2-C=S, NH **86** → R^1, R^2, S, N **87**

2.3.2.3 From α-(O-tosyloximino) nitriles – By condensing α-mercaptoketones **36a** with nitriles **88**, 4-amino-5-acylisothiazoles **89** were obtained.

R(C=)C– CH_2–S–H **36a** + N≡C–C(=N–OTs)–R^1 **88** → N≡C–C(=N)–R^1, R(O=)C-CH_2–S → H_2N, R^1, R(O=)C, S, N **89**

2.3.2.4 From isoxazoles – Isothiazoles **91** can be obtained from readily available isoxazoles **90** by treating them with Raney nickel followed by the action of phosphorus pentasulphide and oxidation with chloranils.

R^2, R^1, R^3, O, N **90** — Ni(Raney) → R^2-C=C-R^1, R^3-C=O, NH_2 — 1) P_2S_5 2) [O] → R^2, R^1, R^3, S, N **91**

2.3.2.5 From β-crotonic esters and thiophosgene – Starting from 4-imino-2-pentanone **92**, the 4-acetyl-3-methylisothiazole **93** is obtained. This compound is used as intermediate in the preparation of 5-alkylated derivatives **94**.

Ac-CH_2—C(=NH)-CH_3 + Cl_2C=S **92** — -2HCl → Ac, CH_3, S, N **93** — 1) BuLi 2) RX → Ac, CH_3, R, S, N **94**

2.3.3 1,2-Dithiolanes

5-Methyl-(3*H*)-1,2-dithiole **96** and the corresponding saturated compound **97** are synthesized from crotonaldehyde **95** and hydrogen sulphide.

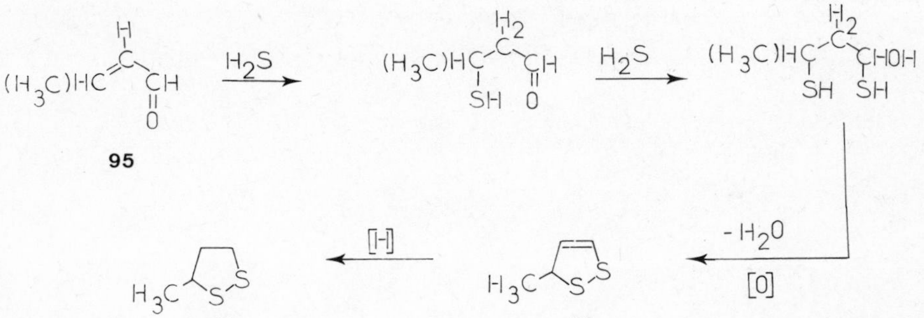

95 **97** **96**

2.3.4 1,2,4-Trithiolanes

3,5-Dimethyl-1,2,4-trithiolane **99** (R = CH_3) is easily prepared by reacting acetaldehyde with sulphur and hydrogen sulphide in the presence of a primary and secondary amine. 1-Mercaptoethyl sulphide **98** is a key product in this reaction. By treating an equimolar mixture of acetaldehyde and propionaldehyde (or isobutyraldehyde) with hydrogen sulphide in the presence of sulphur and diisobutylamine, 3-methyl-5-alkyl-1,2,4-dithiolanes **99** were synthesized.

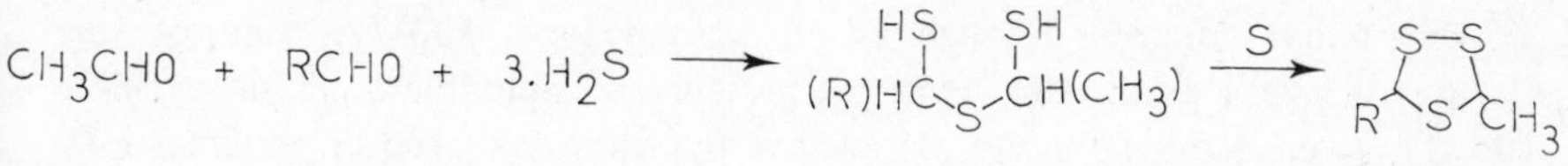

$$CH_3CHO + RCHO + 3.H_2S \longrightarrow (R)HC(SH)-S-CH(SH)(CH_3) \; (\mathbf{98}) \xrightarrow{S} \mathbf{99}$$

$$RCH_2\text{-}S\text{-}S\text{-}CH_2R \; (\mathbf{100}) \xrightarrow{[Cl]} 2.RCHClSCl \longrightarrow (RCH(Cl)S\text{-})_2 \xrightarrow[H_2S]{-2HCl}$$

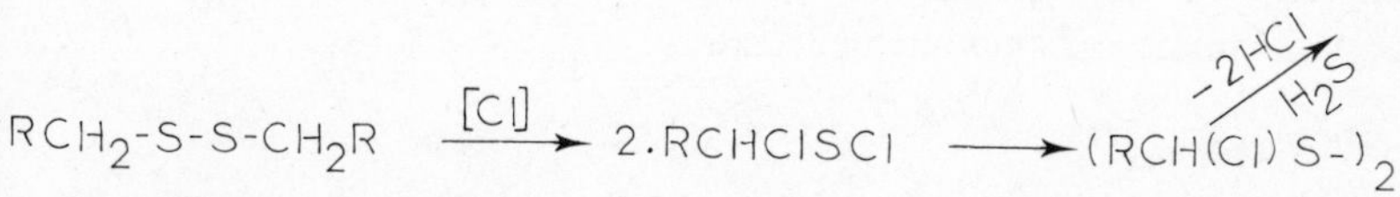

IV–3 SIX-MEMBERED RINGS

3.1 Six-membered rings with one heteroatom: pyrans [14], thiapyrans, pyridines [1], and related reduced systems

From the various types of ring closure reported in *Scheme 5* those with a C–X bond formed in the last stage are the most important ones.

The starting material of intermediate reaction contains a five-carbon atom chain with functional groups in the 1,5 positions. In the syntheses involving

C–C bond formation, those in which the C_3–C_4 bond is formed are more important than those involving C_2–C_3 bond formation. These compounds can be also formed from each other or from other heterocyclic compounds.

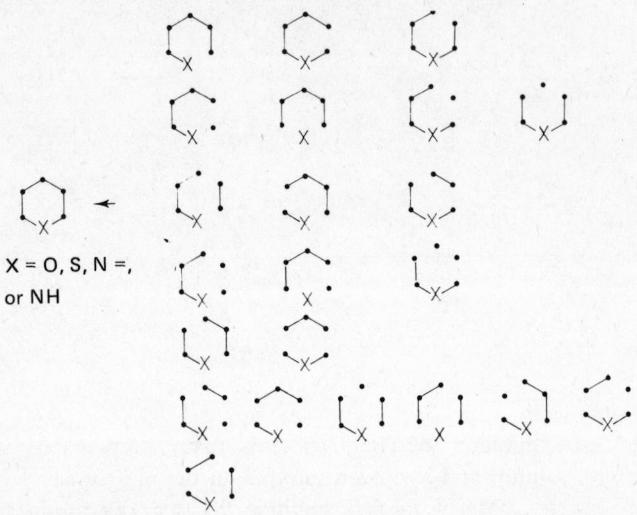

X = O, S, N =, or NH

Scheme 5 – Various types of ring closure for one heteroatom containing six-membered ring systems.

3.1.1 From 1,5-dicarbonyl compounds and derivatives

1,5-Dicarbonyl compounds such as the pentanedione **103** can undergo ring closure to give a pyran **104**, or in the presence of ammonia a dihydropyridine **106**. These products are easily oxidized in pyrylium salts **105** or pyridines **107**. Pyrylium salts are transformed in pyridines by heating in acetic acid in the presence of ammonia and ammonium acetate.

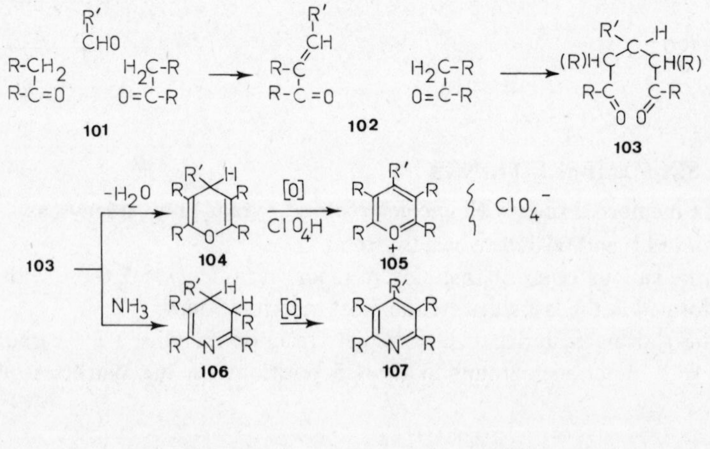

101 102 103

103 $-H_2O$ 104 [O] ClO_4H 105 ClO_4^-

NH_3 106 [O] 107

The pentane-1,5-dione **103** is usually formed *in situ* by aldol or *Michael*-type reactions (**101** → **102** → **103**).

1,5-Dialdehydes **110** and **113** can be prepared by hydrolysis of dihydropyrans **109** and furanopyran **112** respectively. These compounds are synthesized according to the following sequences: **108** → **109** and **111** → **112**.

108 + CH_2=CHOR′ → **109** —(H_2O, −R′OH)→ **110** → 4-R-pyridine

111 + H_2C=CH–CH=O → **112** —(H_2O)→ **113** → HOH_2CH_2C-pyridine

Glutaconic aldehyde **114** ($R^1 = R^2 = H$) cyclizes in the presence of amines or in acidic medium to give pyridine, pyridine *N*-oxide, and the pyridinium and pyrilium cations **115–118** ($R^1 = R^2 = H$).

114 —(NH_3)→ **115**; **114** —(NH_2OH)→ **116**; **114** —(RNH_2)→ **117**; **114** —(H^+)→ **118** —(NH_2OH)→ **115**

Ring closure of 1,5-keto acids **120** or **114** (R^1 = OH) gives α-pyrones **121** by dehydration according to the following sequence: **119** → **120** → **121**. Aldol-type condensation of two molecules of β-keto acids **122** followed by decarboxylation of **123** gives compounds **121**.

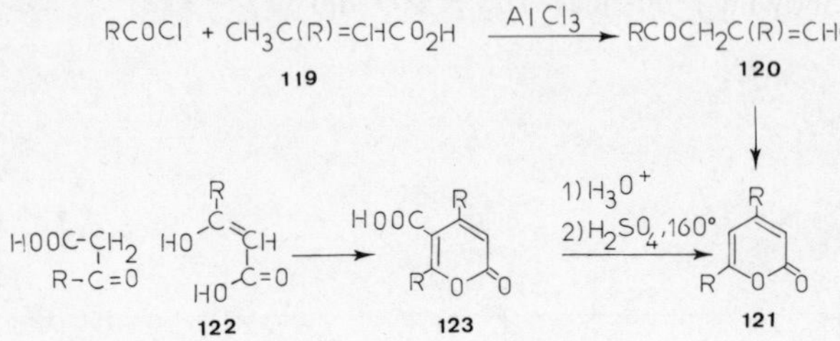

$$RCOCl + CH_3C(R){=}CHCO_2H \xrightarrow{AlCl_3} RCOCH_2C(R){=}CHCO_2H$$

119 → **120** → **121**

1) H_3O^+ 2) H_2SO_4, 160°

122 → **123** → **121**

α-Pyranones **125** can also be synthesized from 1,5-acetylenic acids **124**.

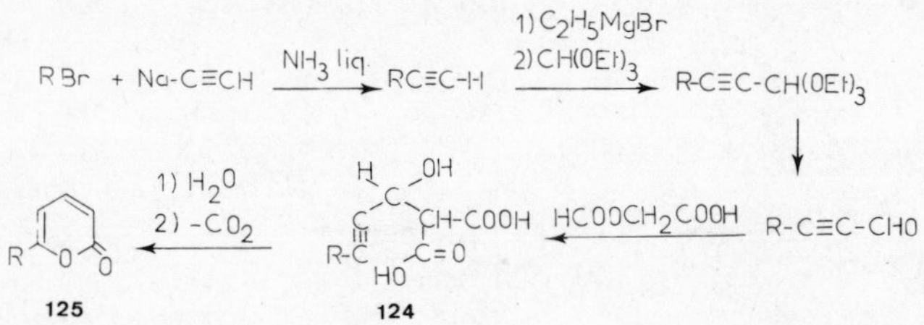

$$RBr + Na{-}C{\equiv}CH \xrightarrow{NH_3\ liq.} RC{\equiv}C{-}H \xrightarrow{1)\ C_2H_5MgBr,\ 2)\ CH(OEt)_3} R{-}C{\equiv}C{-}CH(OEt)_3$$

$$\downarrow$$

$$R{-}C{\equiv}C{-}CHO \xrightarrow{HCOOCH_2COOH} \mathbf{124} \xrightarrow{1)\ H_2O,\ 2)\ -CO_2} \mathbf{125}$$

Starting from 1,5-keto nitriles **126**, pyridines **127** ($R^1 = R^2 = CH_3, C_2H_5$) were obtained.

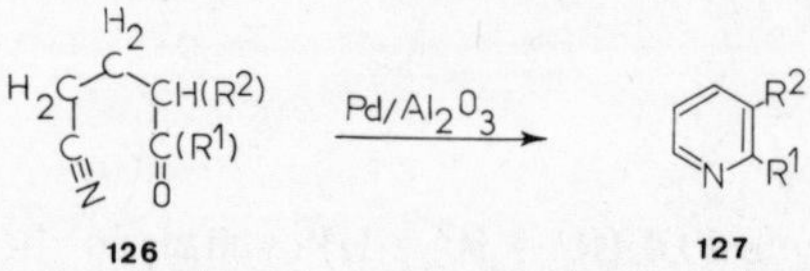

$$\mathbf{126} \xrightarrow{Pd/Al_2O_3} \mathbf{127}$$

4,5-Dihydro-α-pyrones **129** are obtained from unsaturated 1,5-hydroxy acids **128** according to the following scheme adapted from Nobuhara [31b].

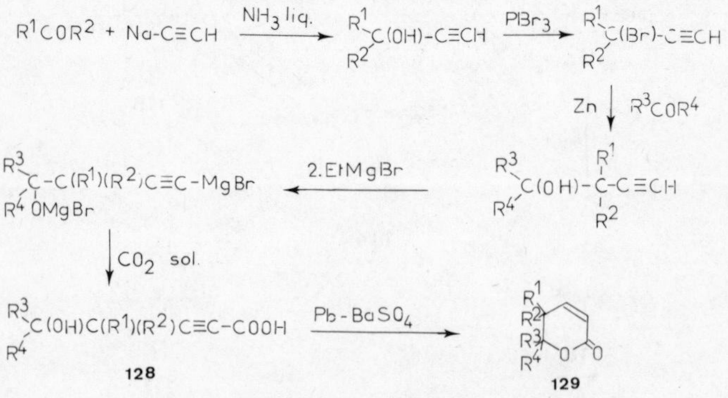

$$R^1COR^2 + Na{-}C{\equiv}CH \xrightarrow{NH_3\ liq.} R^1R^2C(OH){-}C{\equiv}CH \xrightarrow{PBr_3} R^1R^2C(Br){-}C{\equiv}CH$$

$$\xrightarrow{Zn,\ R^3COR^4} R^3R^4C(OH){-}C(R^1)(R^2){-}C{\equiv}CH$$

$$\xrightarrow{2.\ EtMgBr} R^3R^4C(OMgBr){-}C(R^1)(R^2)C{\equiv}C{-}MgBr$$

$$\xrightarrow{CO_2\ sol.} R^3R^4C(OH)C(R^1)(R^2)C{\equiv}C{-}COOH\ (\mathbf{128}) \xrightarrow{Pb-BaSO_4} \mathbf{129}$$

In a similar way saturated 1,5-hydroxy acids **131** give δ-lactones **132** themselves synthesized from the cyclopentanone **130** according to the procedure of Boldingh & Taylor [5b].

$$\mathbf{130} \xrightarrow[B^{\ominus}]{RCHO} \text{RHOHC-cyclopentanone} \xrightarrow[-H_2O]{(HCO_2)_2} \text{R-HC=cyclopentanone} \xrightarrow{Ni,\ H_2} \text{RH}_2\text{C-cyclopentanone} \xrightarrow[{[O]}]{(CrO_3)} RCH_2(O{=})C(CH_2)_3CO_2H \xrightarrow{[H]} RCH_2CHOH(CH_2)_3CO_2H\ (\mathbf{131}) \xrightarrow{\Delta} \mathbf{132}$$

2,5-Dioxo compounds **133** are obtained from saturated 1,5 diacids.

133

Methods starting from 1,5-dibromopentane **134** afford tetrahydropyran, pentamethylene sulphide, or piperidine **135** (X = O, S, NH) according to experimental conditions. δ-Hydroxy-, δ-thiohydroxy- and δ-amino ketones **136** (X = O, S, NH) lead to similar products **137** to **139** to those described in the five-membered ring series.

$$BrCH_2CH_2CH_2CH_2CH_2Br\ (\mathbf{134}) \xrightarrow[-BrH]{[XH_2]} HXCH_2CH_2CH_2CH_2CH_2Br \xrightarrow{-BrH} \mathbf{135}$$

136 (HX–$CH_2CH_2CH_2CH_2$–C(=O)–R) → **137** (X = O, S) or **138** (from **136**, X = NH); R = OH → **139**

1,5-Amino ketones **136** (X = NH) are obtained by hydrolysis of the phthalimide **140** in acidic medium.

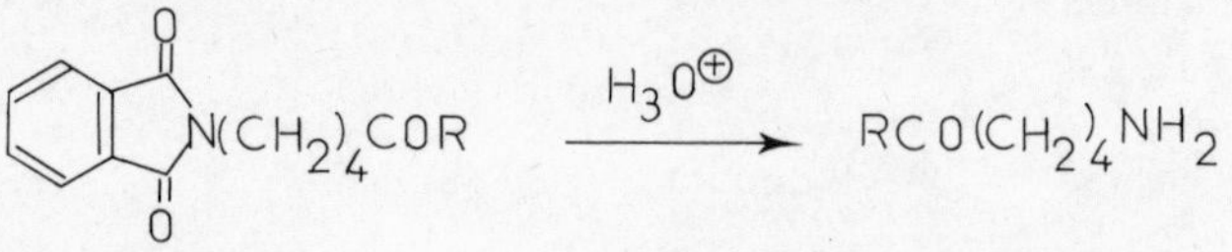

$$\text{140: } C_6H_4(CO)_2N(CH_2)_4COR \xrightarrow{H_3O^{\oplus}} RCO(CH_2)_4NH_2$$

140

3.1.2 From β-keto esters, aldehydes and ammonia (Hantzsch *pyridine synthesis*)
This reaction, the subject of numerous patents, involves the condensation of two molecules of a β-keto ester with an aldehyde and ammonia **141** → → → **142**. In

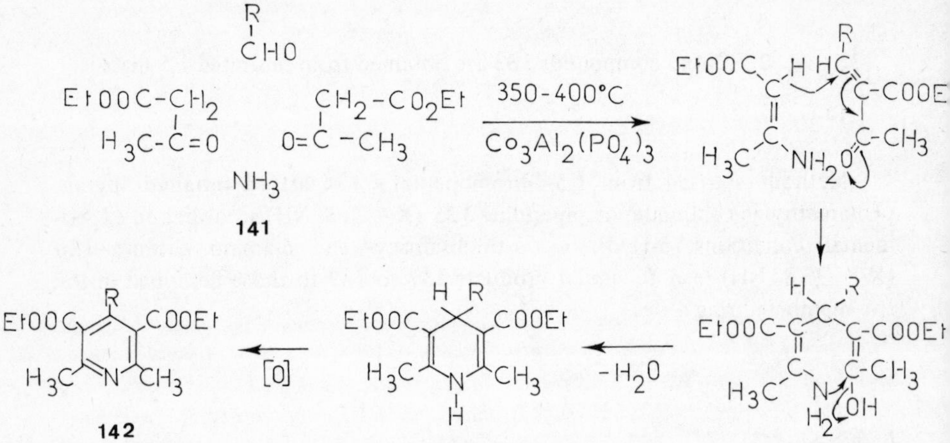

$$\text{EtOOC-CH}_2\text{-C(=O)-CH}_3 + \text{RCHO} + \text{CH}_3\text{-C(=O)-CH}_2\text{-CO}_2\text{Et} + NH_3 \xrightarrow[Co_3Al_2(PO_4)_3]{350\text{-}400°C} \ldots \xrightarrow{-H_2O} \ldots \xrightarrow{[O]} \ldots$$

141 **142**

the above reaction, one can use one molecule of a β-keto ester, two molecules of aldehydes and ammonia. Under drastic conditions, paraldehyde, acetaldehyde, acrolein and crotonaldehyde condense with ammonia to give alkylpyridines. By cyclizing acrolein with hydrogen sulphide **143** in methylene chloride containing Cu turnings and triethylamine at −10°C, 3-formyldihydropyran **144** was obtained in a good yield.

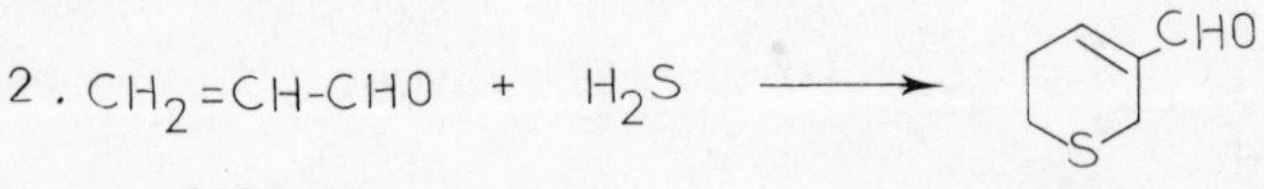

$$2 \cdot CH_2{=}CH\text{-}CHO + H_2S \longrightarrow \text{144}$$

143 **144**

3.1.3 From 1,3,5-tricarbonyl compounds
2,6-Dimethyl-γ-pyrone **146** (X = O, R = CH_3) can be obtained from dehydration of the 1,3,5-pentanetrione **145**. The parent compound **146** (X = O, R = H) is

prepared from the diester **145** (R = OEt). Some γ-pyrones **148** can be synthesized from dehydration of β-keto esters **147**.

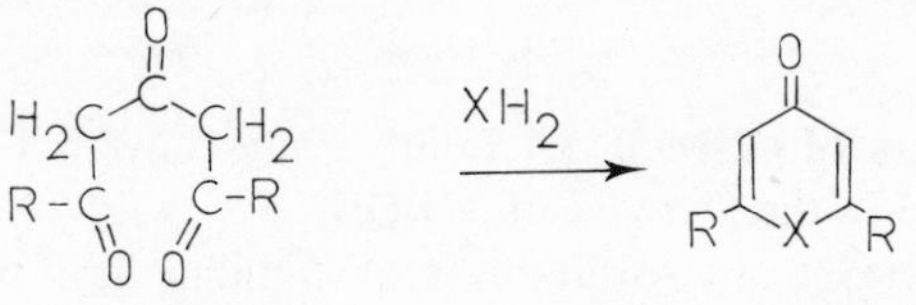

145 **146**

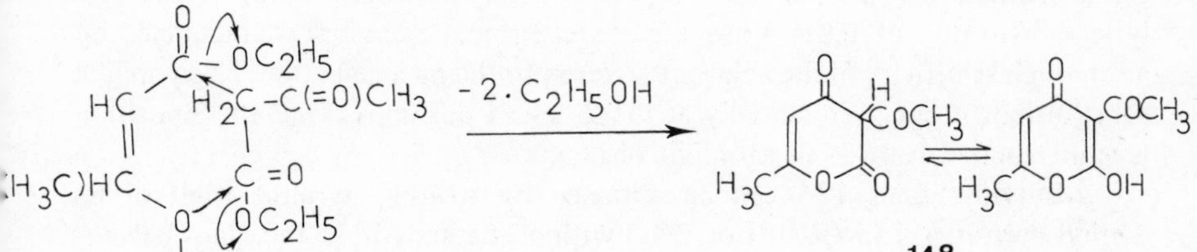

147 **148**

Owing to their mesomeric structures γ-pyrones possess a pronounced aromatic character. γ-Pyridones **146** (X = NH) are obtained by the action ammonia on 1,3,5 triones **145**.

3.1.4 Pyridines from other sources

Pyridines **150** and **152** can be obtained from *Diels-Alder* condensation of dienes with either nitriles **149**, or oxazoles and thiazoles **151** (X = O, S).

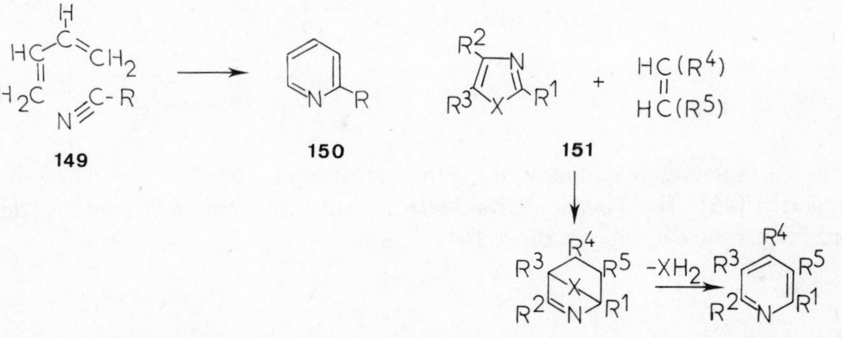

149 **150** **151**

152

1,5-Dialdehydes **110** and **113** (*see* § 3.1.1) on the action of ammonia or hydroxylamine give the corresponding 4-alkyl-, and 3-(β-hydroxyethyl)-pyridine. In a similar way, tetrahydropyrans **153**, yield 3-alkoxymethylpyridines **154**.

$$\text{153} \xrightarrow[\text{Cat. 300-500°C}]{NH_3\,(\text{or } NH_2OH)} \text{154}$$

153 154

The pyrilium salts **118** (*see* § 3.1.1) are easily converted into the corresponding pyridines **115** in the presence of NH_4OH.

Numerous pyridines can be synthesized from pyridine itself or its substituted derivatives *via* nucleophilic, electrophilic, and radical substitution.

The homolytic alkylation of pyridine and its 4 substituted derivatives in acidic medium is an original route to prepare 2-alkylpyridines [5,30]. The homolytic substitution of these protonated heteroaromatic bases is characterized by a very high selectivity of the adjacent position to the nitrogen atom. Nucleophilic substitutions occur preferentially at the 2- and 4 positions, while the 3 position is more reactive towards electrophilic reactants.

4-Ethylpyridines **156** can be obtained by treating pyridine itself or its 3-ethyl derivative **155** (R = H or CH_3) with acetic anhydride and zinc (*Wibaut-Arens'* reaction). In the same conditions, 3,5-dimethylpyridine **157** gives the 4-acetyl derivative **158**.

$$\text{155} \xrightarrow{AC_2O,\,Zn} \text{156} \qquad \text{158} \longleftarrow \text{157}$$

155 156 158 157

By dry fusion, in the presence of calcium acetate and calcium oxide, 2-acetylpyridine **160** can be obtained from 2-pyridine carboxylic acid **159**.

$$\text{159} \xrightarrow[\Delta]{(AcO)_2Ca,\ OCa} \text{160}$$

159 160

An interesting synthesis of methyl nicotinates **163** has been recently reported [45]. The 3-cyano derivative **162** is obtained from ethyl acetyl acetic acid esters and α-cyano acetamide **161**.

3.2 Six-membered rings with two heteroatoms in the 1,4 positions

3.2.1 Pyrazines

Pyrazines are the more important aroma compounds in this category, and a great number of them have been patented. They can be obtained either by direct ring closure or from cyclized compounds.

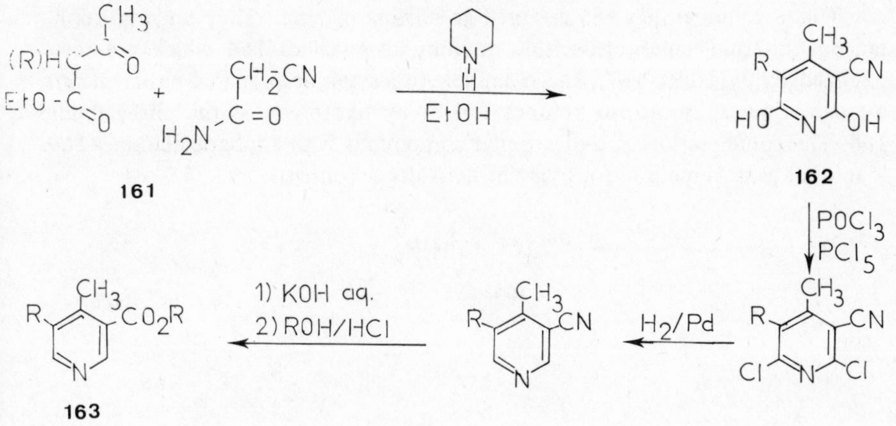

CH_3 (R)HC=O EtO-C=O + CH_2CN H_2N-C=O EtOH CH_3 R CN HO N OH

161

162

$POCl_3$ PCl_5

CH_3 R CO_2R N 1) KOH aq. 2) ROH/HCl CH_3 R CN N H_2/Pd CH_3 R CN Cl N Cl

163

3.2.1.1 By direct ring closure – Among the various types of ring closure (*see Scheme 6*), the main route to obtain pyrazines involves reaction of type (2 · X – C – C or X – C – C – X + C – C).

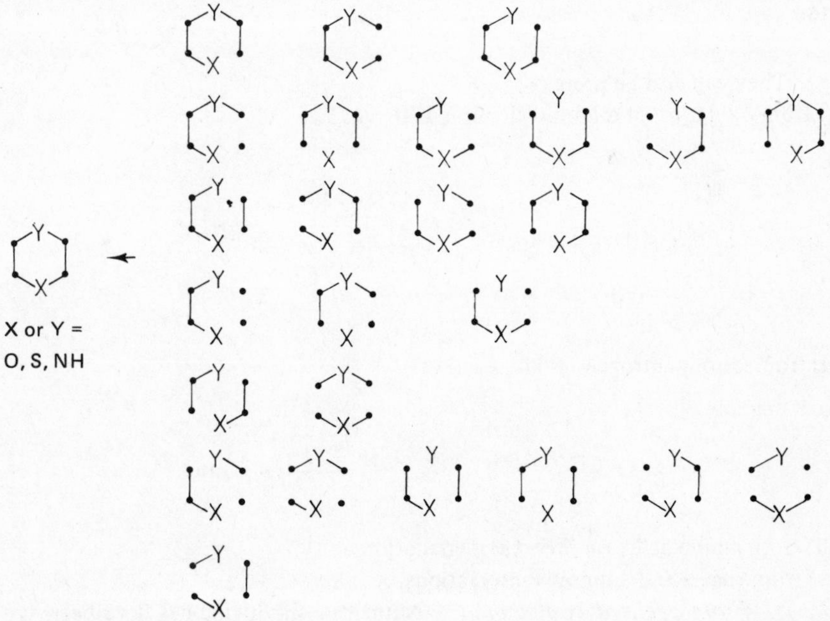

X or Y =
O, S, NH

Scheme 6 – Various types of ring closure for 1,4 two heteroatoms containing six-membered heterocyclic systems.

The α-aminoketones **165** are used as starting material. They undergo spontaneous intermolecular cyclization to dihydropyrazines **166** which are then oxidized to pyrazines **167**. The α-aminoketones can be prepared either *in situ* by reduction of isonitroso ketones **164** or by hydrolysis of the phthalamide **168**. The condensation of α-dicarbonyl compounds with aliphatic diamines **169** is another general method for pyrazine derivatives synthesis.

$$R^1COCH_2R^2 \xrightarrow{HOON} R^1COC(=NOH)R^2 \; (\mathbf{164}) \xrightarrow[\text{Zn, AcOH}]{[H]\ \text{SnCl}_2\ \text{or}} (R^2)HC(NH_2)-C(=O)R^1 \; (\mathbf{165})$$

$$\text{Potassium phthalimide (NK)} \xrightarrow[-XK]{R^1COCHXR^2} \text{phthalimido-}N\text{-CH}(R^2)COR^1 \xrightarrow{H_3O^{\oplus}} \mathbf{165}$$

$$2\ \mathbf{165} \longrightarrow \mathbf{166} \xrightarrow{[O]} \mathbf{167}$$

$$R^2-C{=}O,\ R^1-C{=}O\ (\mathbf{168}) + H_2N-CH(R^1),\ H_2N-CH(R^2)\ (\mathbf{169}) \xrightarrow[-2\,H_2O]{[O]} \mathbf{167}$$

They can also be prepared:

i) from oxidation of α-amino alcohols **170** → **171**,

$$2\ R^1CH(NH_2)CH_2OH\ (\mathbf{170}) \xrightarrow{1)\ Cu,\ 300°C\quad 2)\ [O]} \mathbf{171}$$

ii) from ketone-nitrogen iodide reaction:

$$R^1CH_2COR^2 \xrightarrow{I_2,\ NH_3} \mathbf{167}$$

iii) from amino acids *via Strecker* degradation,

iv) from sugars and ammonia interactions.

3.2.1.2 From cyclized compounds – Numerous alkylpyrazines have been synthesized by alkylation of methyl-, and dimethylpyrazines with a suitable alkylhalide in sodamide/liquid ammonia. Yields do not normally exceed 50% [31]. Similarly, a dimethyl- or the trimethylpyrazine **174** ($R^1 = R^2 = R^3 = CH_3$) when treated by an organolithium compound affords the corresponding alkyl derivative **175**. The yields are generally below 40% [38].

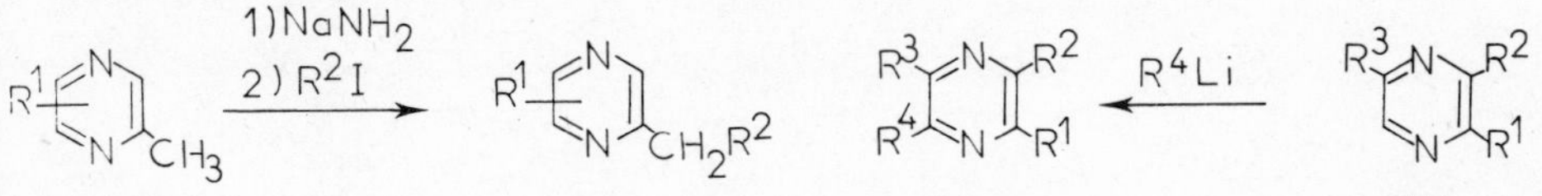

1) $NaNH_2$
2) R^2I

172 **173** **175** **174**

R^4Li

Vinyl- and isopropenylpyrazines can be prepared from methyl- and ethylpyrazine, respectively *via Mannich* bases. A great variety of pyrazine derivatives have been synthesized using an halopyrazine alkali metal type reaction. Pyrazine isomers were formed by this technique. Pyrazine and its 2-methyl, and 2,5-dimethyl derivatives **176** yielded mainly mono oxides **177** when treated with one equivalent of hydrogen peroxide in hot acetic acid. By reacting them with phosphorus oxychloride a mixture of non-halogenated pyrazines **178** is obtained. These latter compounds can react with sodium methoxide or thiomethoxide to give the corresponding methoxy or methylthiopyrazines **179** (XR′ = OCH_3, SCH_3).

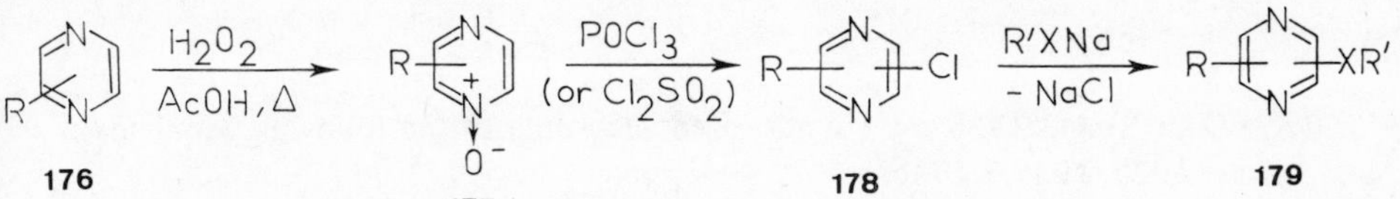

H_2O_2 / AcOH, Δ

$POCl_3$ (or Cl_2SO_2)

R′XNa / - NaCl

176 **177** **178** **179**

The 2-hydroxymethylpyrazines **182** can be obtained as shown **180** → **181** → **182**.

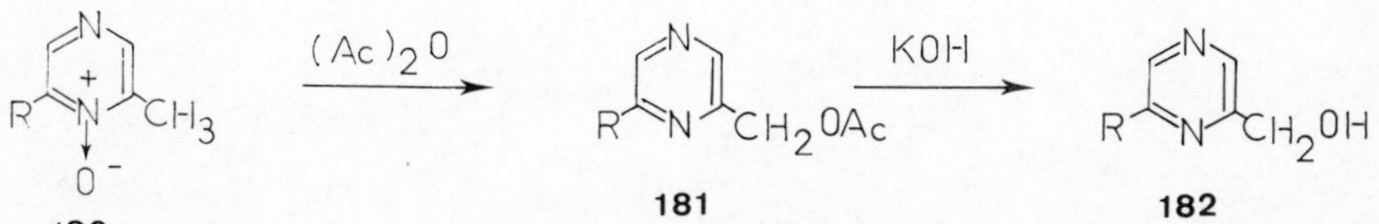

$(Ac)_2O$ KOH

180 **181** **182**

2-Methoxy-3-alkylpyrazines **185** with strong bell pepper-like odour were prepared by methylation of the 2-hydroxy derivatives **184** with diazomethane [41].

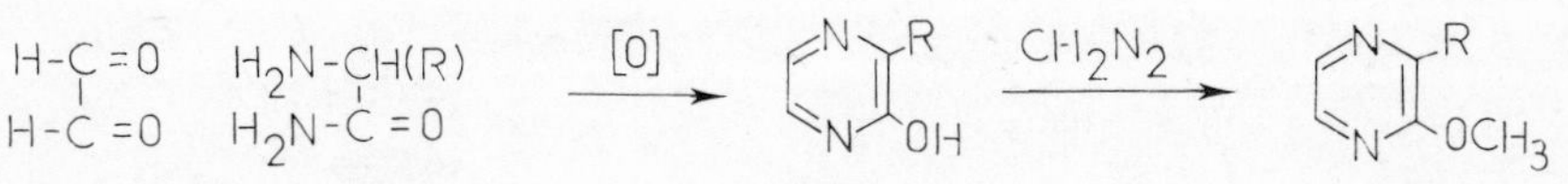

[O] CH_2N_2

183 **184** **185**

2-Hydroxypyrazines **184** were synthesized from amino acid amides and glyoxal. 2-Acetylpyrazine **187** (R = H) is obtained from amide **186** (R = H) with a 50% overall yield.

186 $\xrightarrow{POCl_5}$ (2-cyanopyrazine) → **187**

The oxidation of piperazines with copper chromate affords also pyrazines.

3.2.2 Dioxanes, dithianes, piperazines and 1,4-oxathianes

Compounds **189** (X = O, S, NH) can be prepared as shown **188** → **189**. 1,4-Oxathianes **191** (R = H or CH_3) were obtained by cyclization of halogeno ethers **190** in the presence of an alkaline sulphide. The unsaturated derivatives **192b**

$H_2C{-}XH$ / $H_2C{-}Br$ + $BrCH_2$ / $HXCH_2$ or $(2.BrCH_2CH_2Br + 2.XH_2)$ (**188**) → **189**

191 $\xleftarrow[alc., \Delta]{K_2S}$ $(XCH_2CHR{-})_2O$ (**190**)

(X = O or S) and **193b** were synthesized according to the following sequences: **192a** → **192b**, **193a** → **193b**.

192a + $Cl{-}CH_2{-}CH(OR)_2$ $\xrightarrow{-HCl}$ $H_2C(SCH_2CH(OR)_2){-}H_2C{-}XH$ $\xrightarrow{-ROH}$ (2-alkoxy ring) $\xrightarrow[-ROH]{P_2O_5}$ **192b**

$BrCH{=}CHBr$ (**193a**) $\xrightarrow{2.SNa_2}$ **193b**

The S-(β-hydroxyethyl)-thioglycolic acid **194** affords the 2-oxo-1,4-oxathiane **195**.

194 → **195**

3.3 Six-membered rings with two heteroatoms in the 1,2- and 1,3 positions

3.3.1 1,2-Oxazines, 1,2-pyridazines and 1,2-dithiins

Although oxazines and pyridazines have not yet been reported in food flavours, their presence appears very probable. The most important synthetic method for pyridazine derivatives **197** (X = NH) involves condensation of hydrazine with a 1,4-dioxygenated carbon chain **196** or **198** [2].

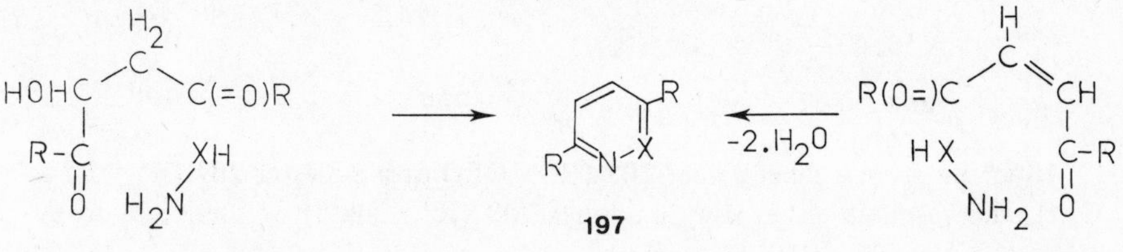

196 **197** **198**

Replacing hydrazine by hydroxylamine, oxazines **197** (X = O) were obtained. Pyridazines **200** can be prepared from furans **199** as shown below:

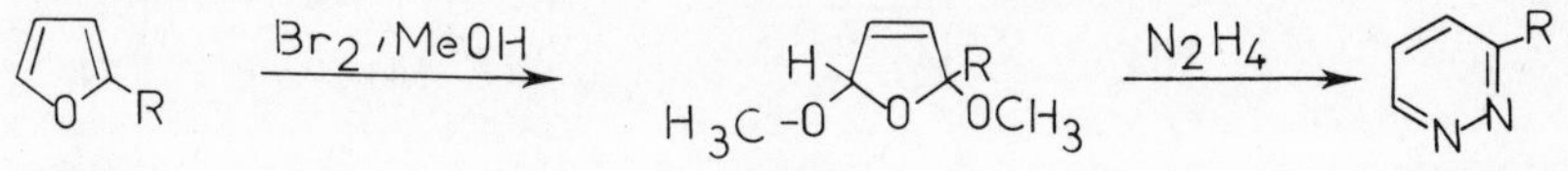

199 **200**

Pyridazones and 1,2-oxazones **202** (X = NH, O) can be synthesized from the γ-keto acid **201**, hydrazine and hydroxylamine, respectively.

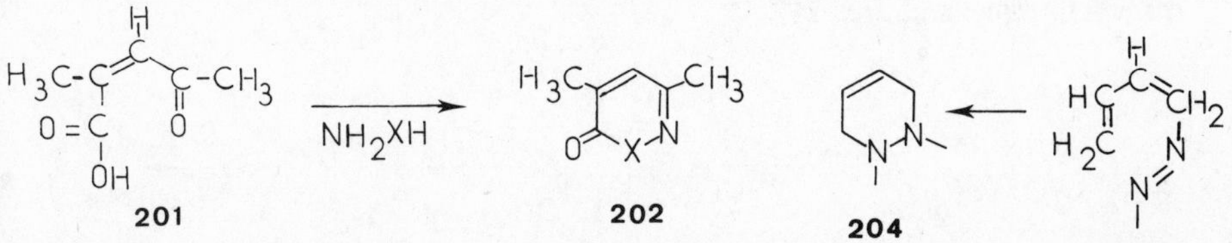

201 **202** **204** **203**

Diels-Alder type reactions from butadiene and azo compounds lead to reduced pyridazines **204**. Dithiin was recently obtained (in 30% yield) by reacting myrcene with elemental sulphur in pyridine/DMF, in the presence of ammonia at 100° [32] **205** → **206** (R = $(CH_3)_2C = CH(CH_2)_2$).

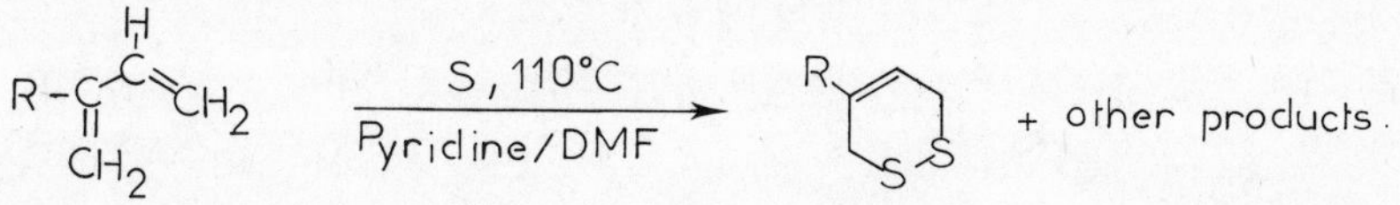

205 **206**

3.3.2 Pyrimidines

Compounds **208** are mainly synthesized by reaction of a 1,3-dicarbonyl compound (β-keto esters or β-diketones) with amidines **207** in the presence of a basic catalyst (NaOEt–EtOH).

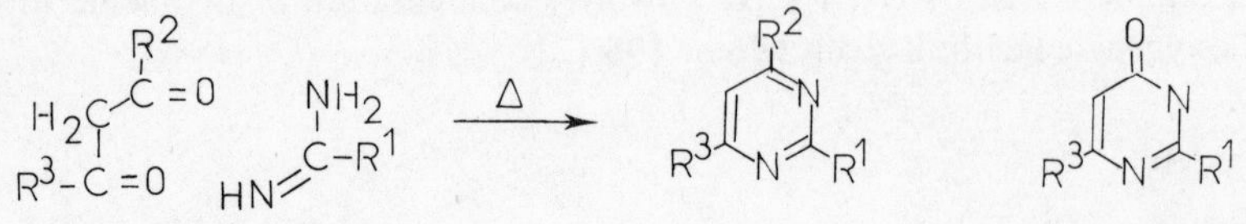

207 **208** **209**

Starting from a β-keto ester **207** (R^2 = OEt) or a β-diester **207** ($R^2 = R^3$ = OEt), the result is either 4-pyrimidones **209** (R^3 = alkyl) or their 6-hydroxy derivatives **209** (R^3 = OH), respectively.

By heating guanidine with ethyl acetyl acetic acid ester **210** 2-aminopyrimidine derivatives **211** were obtained.

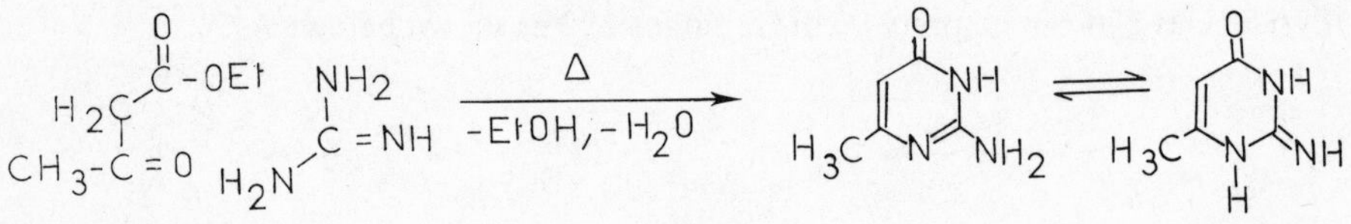

210 **211**

4,6-Dimethylpyrimidine **215** (R = CH_3) has been synthesized by catalytic dehalogenation of 2-chloro-4,6-dimethylpyrimidine **214** (R = CH_3) prepared from β-diketone and urea **212**.

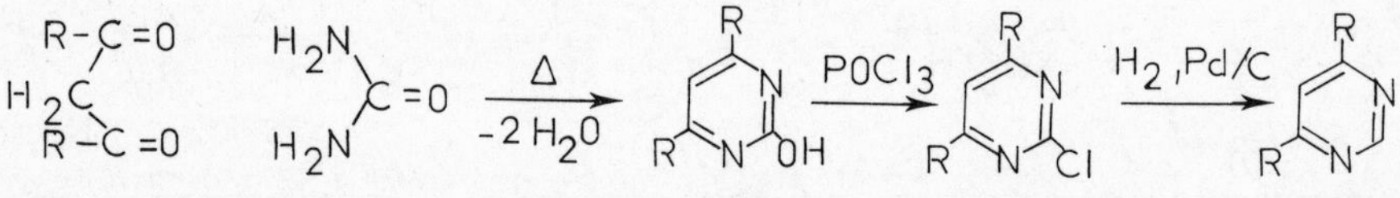

212 **213** **214** **215**

4-Acetyl-2-methylpyrimidine **217** synthesis was carried out according to the following sequence: **216** → **217** [12].

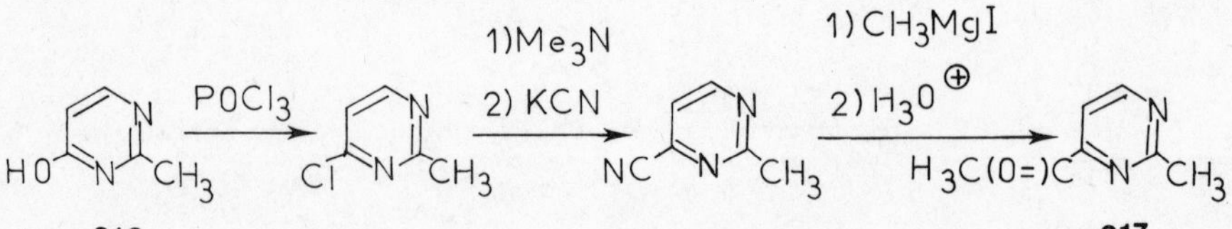

216 **217**

3.3.3 1,3-Dioxanes, 1,3-thioxanes, 1,3-dithianes, reduced pyrimidines, 1,3-oxazines and 1,3-thiazines

The cyclization of a γ-difunctional alkane in the presence of a carbonyl compound **218** gives the title compounds **219** (X, Y = O, S, NH).

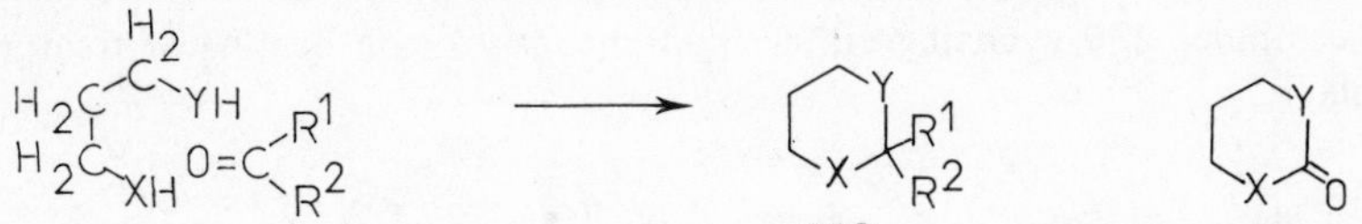

218 219 220

If the above reaction is carried out in the presence of ethyl carbonate **218** ($R^1 = R^2 = OEt$) the corresponding 2-oxo derivatives **220** are obtained.

3.4 Six-membered rings with more than two heteroatoms

3-Methyl-1,2,4-trithiane **222** was synthesized by reacting acetaldehyde, hydrogen sulphide and 1,2-dimercapto ethane **221** [12].

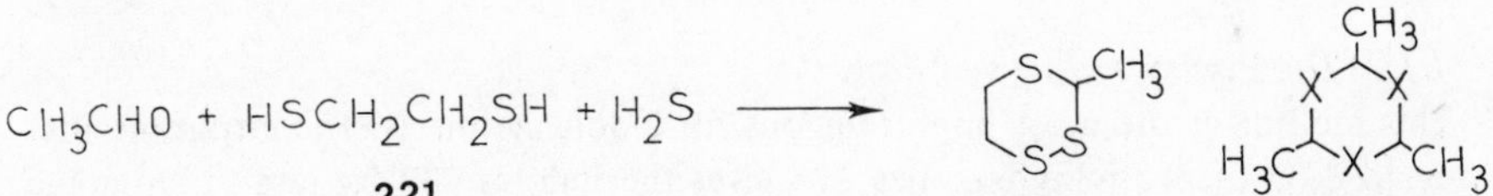

$$CH_3CHO + HSCH_2CH_2SH + H_2S \longrightarrow$$

221 222 223

The symmetrical saturated heterocyclic systems **223** (X = O, S) are obtained by trimerization of acetaldehyde alone, or in the presence of hydrogen sulphide. 1,2,4-, and 1,3,5 Triazines have not yet been identified in food flavours. They can be prepared from amino-guanidines and α-dicarbonyl compounds **224** → **225**, and from trimerization of the monomer R–C≡N, → **226**, respectively.

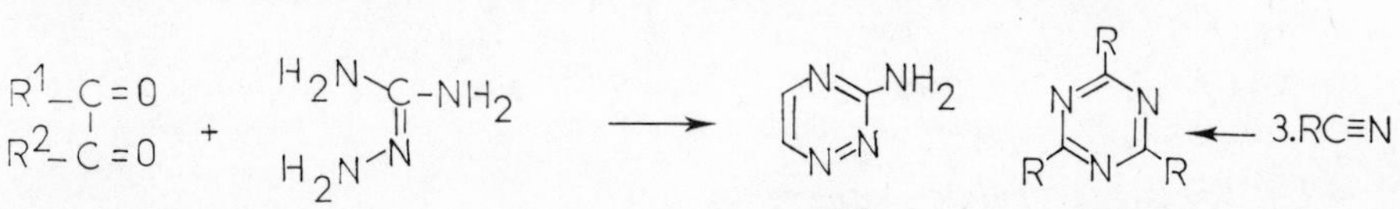

224 225 226 ← 3.RC≡N

IV–4 FUSED RING SYSTEMS

4.1 Benzofurans [16], benzothiophenes [15], and indoles [44]

4.1.1 General methods

Compounds **228** can be prepared from cyclization of ketones of type **227**.

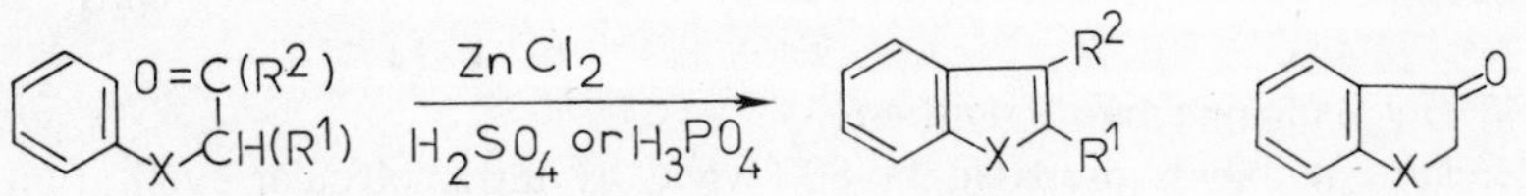

$$\xrightarrow[H_2SO_4 \text{ or } H_3PO_4]{ZnCl_2}$$

227 228 229

Dimethylacetals [$C_6H_5XCH_2CH(OMe)_2$] can replace ketones **227** in this reaction. Intramolecular cyclization of α-substituted acetic acids **227** (R^1 = H, R^2 = OH) in the presence of $NaNH_2$ (for X = NH), P_2O_5 (for X = O) and H_2SO_4 (for X = S) gives the corresponding 3-oxo derivatives **229**.

2,3-Dihydro derivatives **231** were obtained from *o*-substituted β-phenylethyl bromides **230** cyclizing either spontaneously or on heating or treatment with alkali.

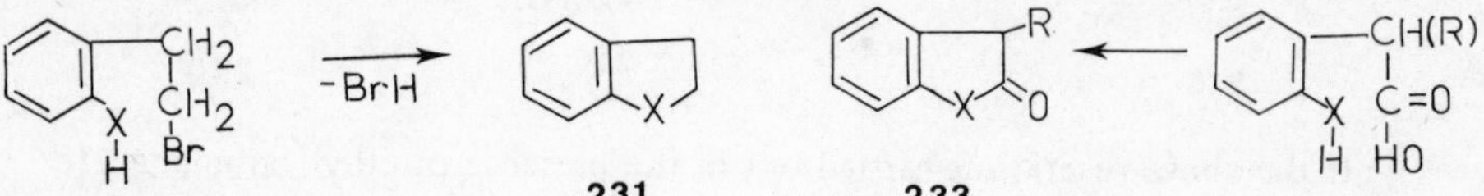

-BrH

230 231 233 232

2-Oxo compounds **233** were likewise obtained from acids **232**. Oxindoles **233** (X = NH) can also be prepared from phenylhydrazides ($C_6H_5NHNHCOCH_2R$) at 200°C in the presence of CaO as catalyst.

4.1.2 The Fischer *indole synthesis*

This method is the most important one for indole synthesis. The intramolecular cyclization of phenylhydrazones **235** gives the indoles **237** by loss of ammonia from rearranged intermediates **236**. The reaction is carried out in the presence of an acid catalyst (e.g., $ZnCl_2$, CuCl, BF_3, H_2SO_4, H_3PO_4, HCl–H_2O) at 100–200°C.

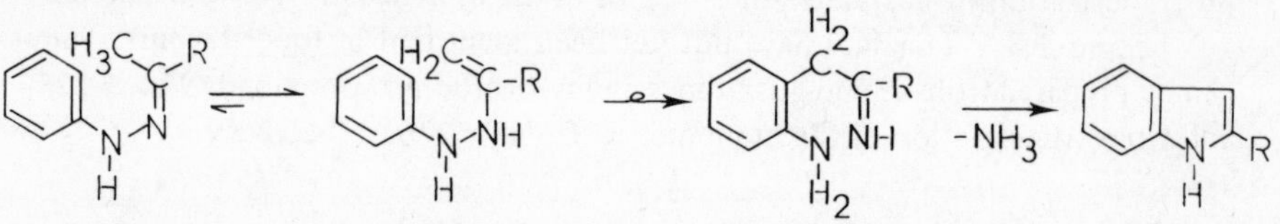

$-NH_3$

234 235 236 237

4.1.3 The Reissert *indole synthesis*

The reduction of pyruvic ester **238** gives the corresponding amino derivative which spontaneously cyclizes to 2-ethoxycarbonyl indole.

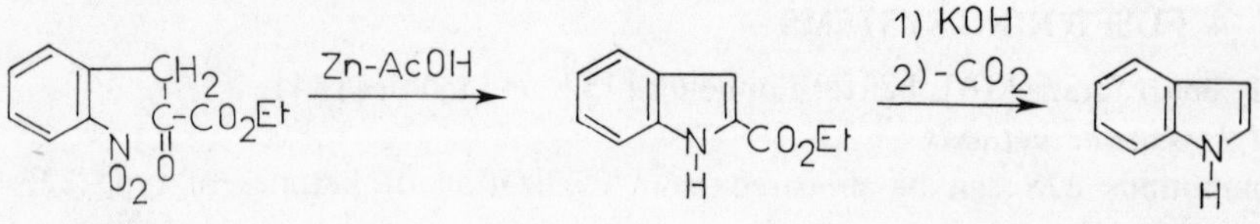

Zn-AcOH

1) KOH 2) $-CO_2$

238 239 240

4.1.4 The Madelung *indole synthesis*

2-Mercapto indole is prepared in 80% yield by intramolecular cyclization of *o*-acylaminotoluene in basic medium ($B^{\ominus}$ = $FtO^{\ominus}$, $^{\ominus}NH_2$, *t*-$BuO^{\ominus}$).

241 → 242 ($B^{\ominus}$, $-H_2O$)

4.1.5 *The* Bischler *indole synthesis*

By reacting an aromatic amine with an halogeno ketone **243**, a mixture of comparable amounts of 2- **244** and 3- **245** substituted indoles is obtained.

$PhNH_2$ + $RCOCH_2Br$ (**243**) → **244** + **245**

4.1.6 *Benzothiophene synthesis*

2-Alkylbenzothiophenes **247** can be synthesized by intramolecular cyclization of *o*-alkylthiophenols **246** in the presence of a catalyst (e.g., CuO – Cr_2O_3). The benzothiophene itself **247** (R = H) is obtained by action of hydrogen sulphide on styrene **248** (R = H).

246 → **247** (450°C, CuO – Cr_2O_3); **248** → **247** (H_2S, 600°C, Al_2O_3–SFe)

4.1.7 *Acyl derivatives*

Benzofuran reacts similarly as furan in acylation reactions to give the corresponding 2-formyl-, and 2-acetyl derivatives **249** (R = H, CH_3). In the same conditions, indole is converted into its 1,3-diacetyl derivative **250**.

249 **250** **251**

2-Methylindole undergoes the *Gatterman* aldehyde synthesis and is converted into 2-methylindole-3-carboxaldehyde **251** upon treatment with HCl and HCN.

4.2 Benzoxazoles, benzothiazoles, benzimidazoles, indazoles and benzisothiazoles

Ortho-substituted anilides **253** (X = O, S, NH) cyclize under mild conditions to benzoxazoles [8], benzothiazoles [29], and benzimidazoles [52] **254** (X = O, S, NH), respectively.

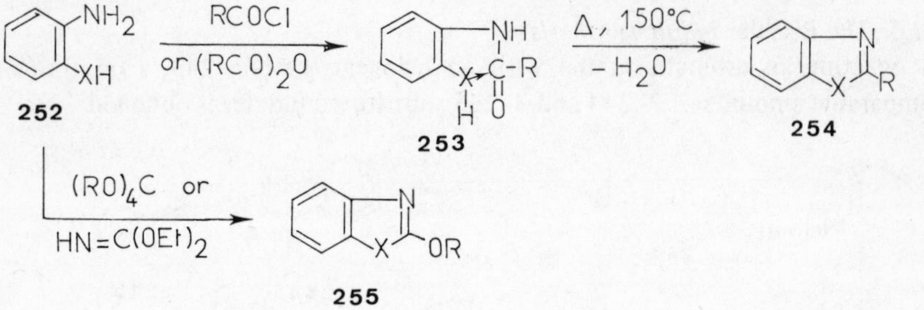

252 **253** **254** **255**

The anilides **253** are prepared *in situ* by heating *o*-substituted anilines with a carboxylic acid, anhydride, acid chloride, ester, nitrile, amidine, etc. . . Thus 2-methylbenzoxazole **254** (X = O, R = CH_3) is obtained in 74% yield by condensing aniline with acetic anhydride.

2-Alkoxy derivatives **255** were obtained from anilines **252** and tetraalkoxy methanes. 2-Oxo compounds **256** can be either prepared from *o*-substituted anilines and carbonic acid derivatives or from a *Curtius* rearrangement of azido compounds **257**. 2-Benzoxazolines, 2-benzothiazolines and 2-benzimidazolines

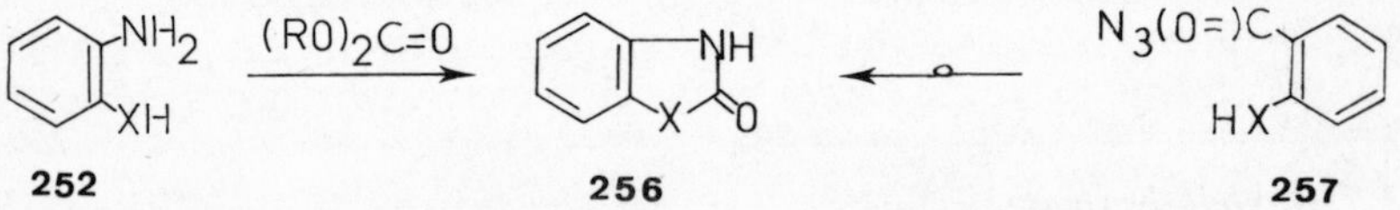

252 **256** **257**

can be prepared by reacting *o*-substituted anilines with aldehydes **258** → **260**.

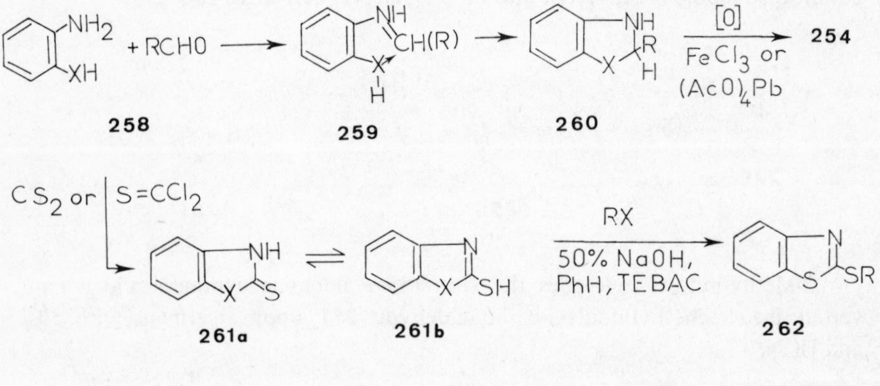

258 **259** **260** **261a** **261b** **262**

2-Mercaptobenzoxazole and 2-mercaptobenzothiazole **261** (X = O, S) were obtained from anilines **258** and carbon disulphide. These compounds are easily transformed into 2-alkylthio derivatives by alkylation under phase transfer catalysis conditions.

The Jacobson *benzothiazole synthesis*

The intramolecular cyclization of *N*-arylthioamides **264** affords 2-variously substituted benzothiazoles **265** (R = H, alkyl, alkoxy, carbomethoxy, carbamoyl).

NHC(=O)R $\xrightarrow{P_2S_5}$ NHC(=S)R $\xrightarrow[NaOH]{K_3Fe(CN)_6}$

263 **264** **265**

Indazoles **267** were formed by spontaneous cyclization of *o*-acylphenylhydrazines **266** and benzisothiazoles **268b** were prepared from sulphenyl chlorides **268a** and ammonia.

$\xrightarrow{-H_2O}$ $\xleftarrow{NH_3}$

266 **267** **268b** **268a**

4.3 Quinolines and isoquinolines

Quinolines can be synthesized from the following routes [10]:

i) from *o*-aminobenzaldehyde and ketones **269** → **270** (*Friedlaender* synthesis).
ii) from aniline and acetaldehyde **271** → **270** (*Doebner-Miller* synthesis).
iii) from an arylamine and either a 1,3-diketone **272** → **274** (*Combes* synthesis) or an α,β-unsaturated carbonyl compound **275** → **277**.
iv) from aniline and β-keto ester **278**. Two alternative orientations are possible leading either to 2-quinolone **280** or 4-quinolone **282**. 2-Quinolone was also prepared by reduction of *o*-nitrocinnamic acid with $(NH_4)_2S$.
v) from arylamines and acrolein (or glycerol) **283** → **287** (*Skraup* synthesis). Arsenic pentoxide is used as oxidizing agent in the last step.
vi) Finally, quinolines are also obtained from indoles **288**. By the action of $CHCl_3$ or $CHBr_3$ under phase transfer catalysis conditions, they give 3-halogenoquinolines **289** which subsequently are reduced to quinolines **290** (*Scheme 7*).

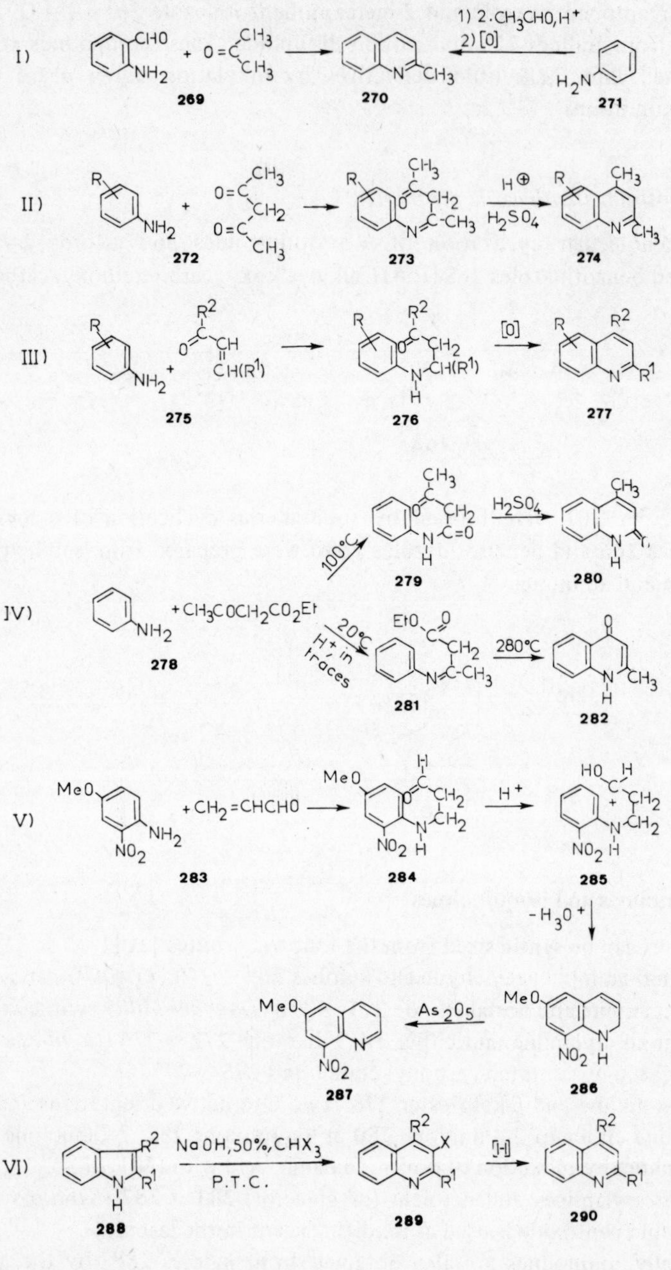

Scheme 7 – Main routes for the preparation of quinolines

Isoquinolines are mainly synthesized:

i) by ring closure of an *o*-substituted benzaldehyde **291** → **292**.

ii) by intramolecular cyclization of either *N*-acylated 2-phenethylamines **293** → **294** (*Bischler-Napieralski* synthesis), or *N*-acylated 2-hydroxyphenethylamines **295** → **296** (*Pictet-Grams* synthesis).

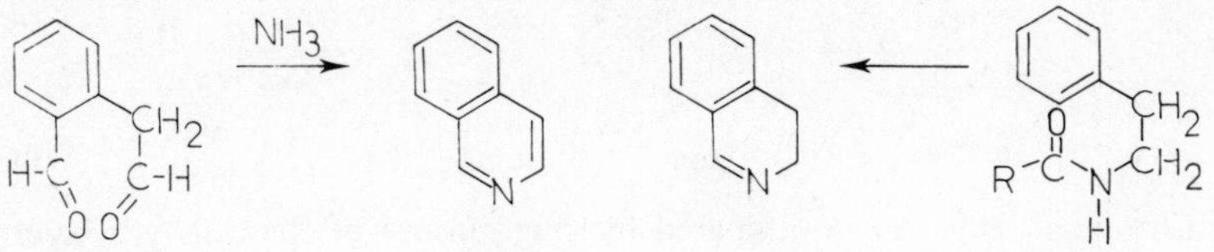

291 **292** **294** **293**

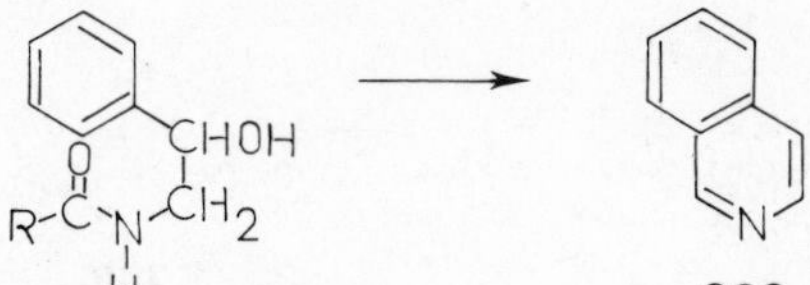

296

295

iii) from a *β*-phenethylamine *via* a *Mannich*-type reaction **297** → **298** (*Pictet-Spengler* synthesis).

iv) from a benzylamine **299** → **300** (*Pomeranz-Fristsch* synthesis).

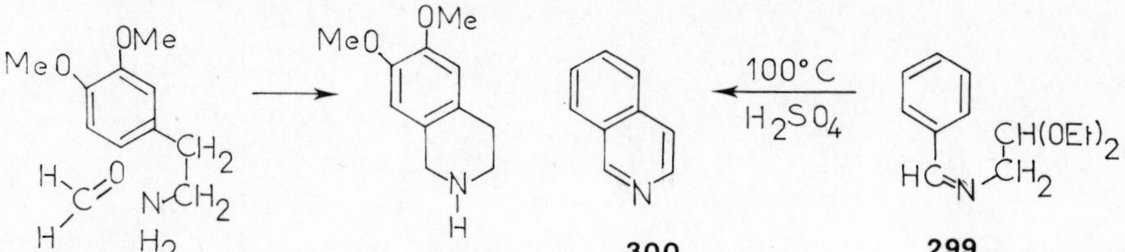

297 **298** **300** **299**

4.4 Benzopyrans and derivatives

By condensing phenols and *β*-keto esters **301** in the presence of P_2O_5 or H_2SO_4, chromones **302** (X = O) and coumarins **303** (X = O) are obtained, respectively.

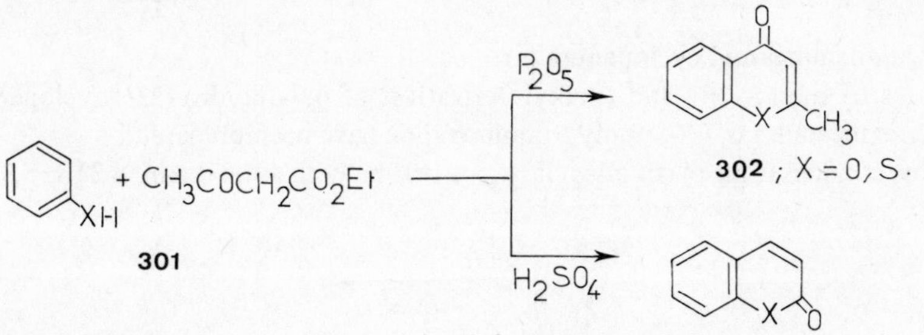

301

302 ; X = O, S.

303 ; X = O, S.

Thiophenols give thiochromones **302** (X = S), and thiocoumarins **303** (X = S).

The condensation of an α-acylphenol with an anhydride in the presence of the corresponding sodium salt can lead to coumarins **305** or chromones **307**.

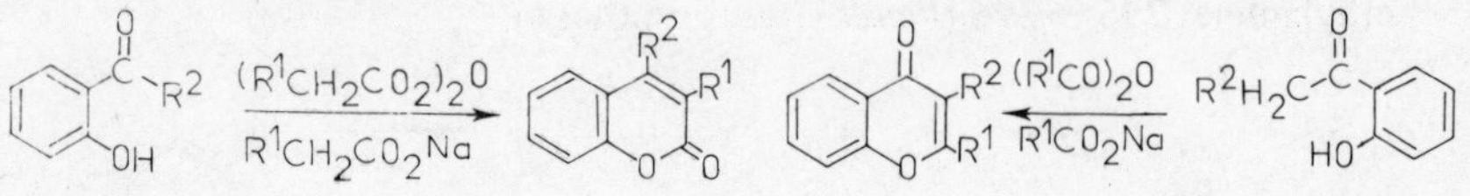

$(R^1CH_2CO_2)_2O$, $R^1CH_2CO_2Na$; $(R^1CO)_2O$, R^1CO_2Na

304 **305** **307** **306**

Chromans **309** can be prepared by ring closure of the *o*-β-hydroxyethylphenol **308**.

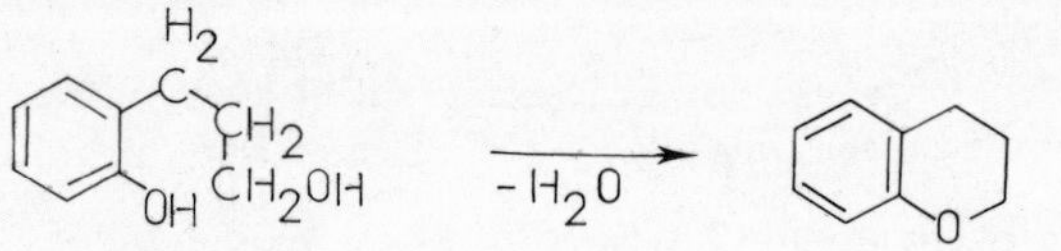

$-H_2O$

308 **309**

Bicyclic pyrans and lactones **311** and **312** have been obtained by condensing malonic acid with 2,2-dimethyl-3-hydroxypropanal **310** in the presence of acetic anhydride [18].

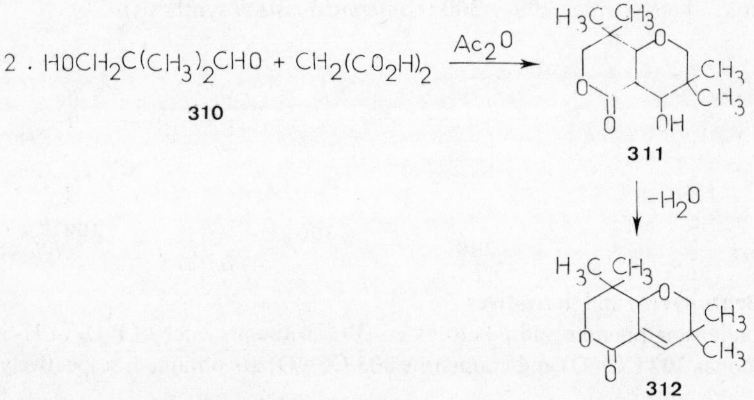

$$2 \cdot HOCH_2C(CH_3)_2CHO + CH_2(CO_2H)_2 \xrightarrow{Ac_2O}$$

310 **311**

$-H_2O$

312

4.5 Quinoxalines and cyclopentapyrazines

A series of mono-, di-, and trialkyl derivatives of 6,7-dihydro-(5*H*)-cyclopenta-(b)-pyrazine and 5,6,7,8-tetrahydroquinoxaline have been prepared:

i) by condensation of an alicyclic α,β-diketone with a α,β-diamine **313** → **314**,

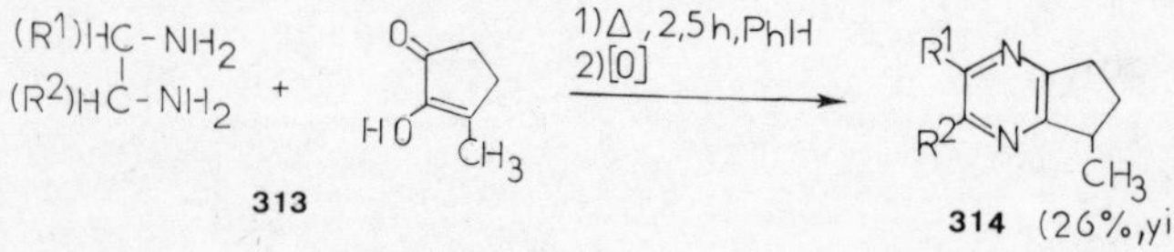

$(R^1)HC-NH_2$ / $(R^2)HC-NH_2$ + ; 1) Δ, 2,5 h, PhH 2) [O]

313 **314** (26%,yield)

ii) by condensation of an alicyclic α,β-diamine with a α,β-dicarbonyl compound **315** → **316**,

iii) by alkylation of 5,6,7,8-tetrahydroquinoxaline in liquid ammonia **316** → **317** [13,33].

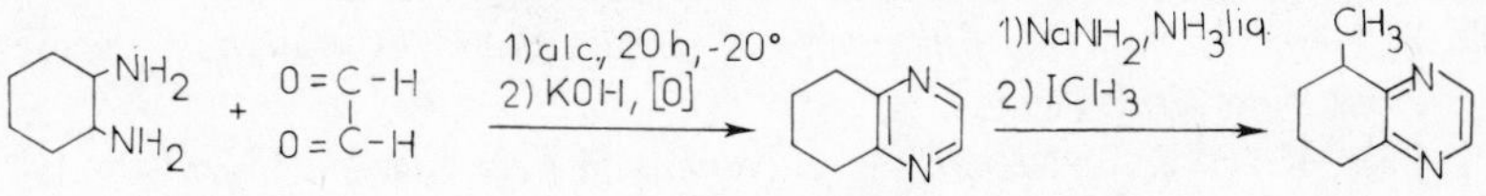

1) alc, 20 h, -20° 2) KOH, [O]; 1) $NaNH_2$, NH_3 liq. 2) ICH_3

315 316 (25%, yield) 317 (50%, yield)

Quinoxalines [7] **319** are obtained by reacting *o*-phenylenediamines with α-diketones **318** (*Körner-Hinsberg* synthesis).

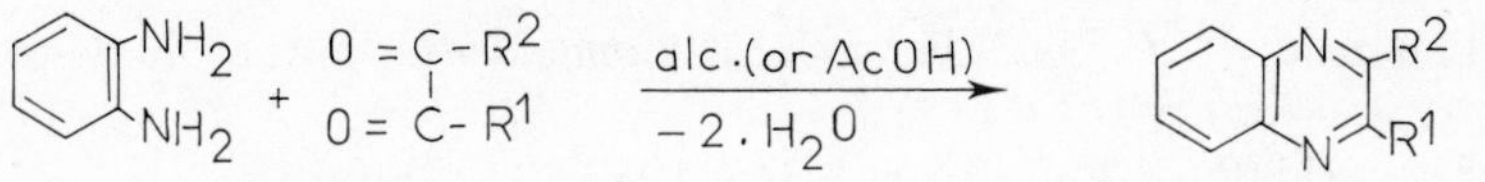

alc. (or AcOH) $-2 \cdot H_2O$

318 319

4.6 Miscellaneous

Benzopyrimidines **321** are prepared similarly as the corresponding five-membered ring systems from *o*-aminomethyl aniline and aldehydes **320**. By replacing aldehydes by ethylcarbonate the parent compound **321** (R = H) is obtained. Synthesis of some bicyclic pyrimidines has been reported by Katz *et al.* [22].

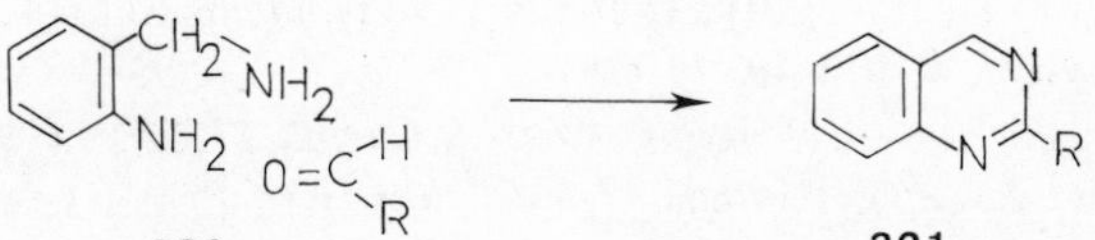

320 321

o-Ethylenic **322** *o*-acetylenic **324**, and *o*-diazonium ions, cyclize spontaneously to give cinnolines **323** or cinnolone **325**.

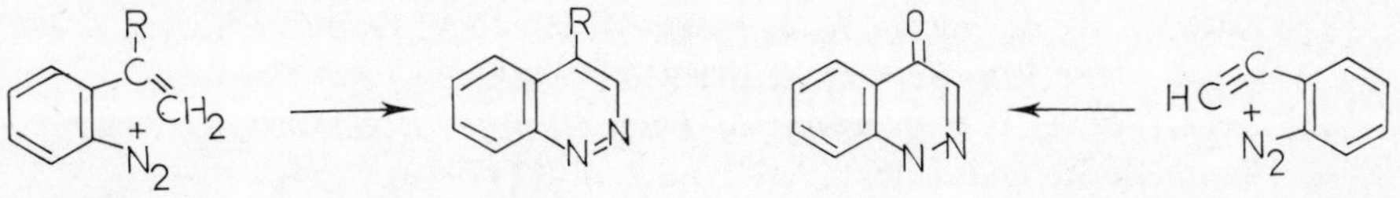

322 323 325 324

IV–5 REFERENCES

[1] Abramovitch, R. A., Ed., *Heterocyclic Compounds: Pyridine and its Derivatives*, Wiley-Interscience, New York, Vol. **14** (1974).

[2] Aldous, D. L. & R. N. Castle, Ref. [48], **28**, 219-352 (1973).

[3] Annual Reports (The Chemical Society, Ed.), London.

[4] Badger, G. M., *The Chemistry of Heterocyclic Compounds*, Academic Press, New York (1961).

[5] Bass, K. C. & P. Nababsing, *Advances in Free Radical Chemistry* (G. H. Williams, Ed.), **5**, 1-47 (1972).

[5b] Boldingh, J. & R. J. Taylor, *Nature,* **194**, 909 (1962).

[6] Bunnett, J. F. & B. F. Gloor, *Heterocycles,* **5**, 377 (1976); Bunnett, J. F. & J. E. Sundberg, *J. Org. Chem.*, **41**, 1702 (1976); *idem, Chem. Pharm. Bull.,* **23**, 2620 (1975).

[7] Cheeseman, G. W. H. & E. S. G. Werstiuk, Ref. [20], **22**, 368-425 (1978).

[8] Cornforth, J. W., Ref. [9], **5**, 418 (1957).

[9] Elderfield, R. C., Ed., *Heterocyclic Compounds*, J. Wiley & Sons Inc. (1950-1961).

[10] Elderfield, R. C., Ref. [9], **4**, 1-343 (1952).

[11] Eugster, C. H. & D. P. Bosshard, *Advances in Heterocyclic Chemistry,* Ref. [20], **7**, 377-490 (1966).

[12] Flament, I., B. Willhalm & G. Ohloff, *Flavor of Foods and Beverages* (G. Charalambous & G. E. Inglett, Eds.), Academic Press, New York, 15-32 (1978).

[13] Flament, I., (Firmenich SA), *Swiss Pat.*, 540,016 (1973); *Swiss Pat.*, 559,518 (31.1.1975). Flament *et al., Helv. Chim. Acta,* **60**, 1872 (1977).

[14] Fried, J., Ref. [9], **1**, 343-396 (1950).

[15] Fukushima, D. K., Ref. [9], **2**, 145-163 (1951).

[16] Gagniant, P. & D. Gagniant, Ref. [20], **18**, 338-473 (1975).

[17] Gronowitz, S., Ref. [20], **1**, 1-124 (1963); *idem*, Ref. [43], **2**, 352-496 (1973); *idem,* **3**, 400-493 (1975).

[18] Johnson, R. N. & N. V. Riggs, *Austr. J. Chem.,* **27**, 2519 (1974).

[19] Jones, R. A. & G. P. Dean, *The Chemistry of Pyrroles*, Academic Press, New York (1977).

[20] Katritzky, A. R. & S. M. Weeds, *Advances in Heterocyclic Chemistry* (A. R. Katritzky & A. J. Boulton, Eds.), Academic Press, New York (1966).

[21] Katritzky, A. R. & J. M. Lagowski, *The Principles of Heterocyclic Chemistry*, Methuen & Co. Ltd., London (1967).

[22] Katz, I., R. A. Wilson, W. J. Evers, M. H. Vock, G. W. Verhoeven & J. Sieczkowski, *West Ger. Pat.*, 2,141,916 (1972).

[23] Keller, W. E., *Compendium of Phase Transfer Reactions and Related Synthetic Methods* (Fluka, A. G., Ed.), 78-92 (1979).

[24] Kharasch, W., Ed., *Organic Sulphur Compounds*, Pergamon Press, Oxford.

[25] Komin, A. P. & J. F. Wolfe, *J. Org. Chem.*, **42**, 2481 (1977).
[26] Kost, A. N. & I. I. Grandberg, Ref. [20], **6**, 347–429 (1966).
[27] Lewis, D. A. & P. Charnak, *Index of Reviews in Organic Chemistry*, 2nd Cumulative edn. 1976; The Chemical Society, London (1977).
[28] MacIntosh, J. M., H. B. Goodbrand & G. M. Masse, *J. Org. Chem.*, **39**, 202 (1974); MacIntosh *et al.*, *Can. J. Chem.*, **52**, 1934 (1974); *idem, J. Org. Chem.*, **40**, 1294 (1975); *idem, Can. J. Chem.*, **53**, 209 (1975).
[29] Metzger, J. & H. Plank, *Chim. & Ind.*, **75**, 929–939 (1956); *ibid*, **75**, 1290–1303 (1956).
[30] Minisci, F. & O. Porta, *Advances in Homolytic Substitution of Heteroaromatic Compounds* in *Advances in Heterocyclic Chemistry*, **16**, 123 (1974).
[31] Nakel, G. M. & B. M. Dirks, *U.S. Pat.*, 3,579,353 (18.5.1971).
[31b] Nobuhara, A., *Agric. Biol. Chem.*, **32**, 1016 (1968); *ibid*, **33**, 225, 1264 (1969).
[32] Peppard, T. L. & J. A. Elvidge, *Chem. Ind. (London)*, **16**, 552 (1979).
[33] Pittet, A. O., R. Muralidhara, J. P. Walradt & T. Kinlin, *J. Agric. Food Chem.*, **22**, 273 (1974).
[34] Pittet, A. O. & D. E. Hruza, *U.S. Pat.*, 3,769,040 (1973).
[35] Pratt, Y. T., Ref. [9], **6**, 377–454 (1956).
[36] Quilico, A. & G. Speroni, Ref. [48], *The Chemistry of Heterocyclic Compounds*, **17**, 1–232 (1962).
[37] Re, L., B. Maurer & G. Ohloff, *Helv. Chim. Acta*, **56**, 1882 (1973).
[38] Rizzi, G. P., *J. Org. Chem.*, **33**, 1333 (1968).
[39] Rodd, E. H., Ed., *The Chemistry of Carbon Compounds*, Elsevier, Amsterdam (1957–1960).
[40] Schipper, E. S. & R. A. Day, Ref. [9], **5**, 194–297 (1957).
[41] Seifert, R. M., R. G. Buttery, D. G. Guadagni, D. R. Black & J. G. Harris, *J. Agric. Food Chem.*, **18**, 246 (1970).
[42] Silverstein, R. M., E. E. Ryskiewicz, C. Willard & R. C. Koehler, *J. Org. Chem.*, **20**, 668 (1955); *idem, Org. Synth. Coll. IV*, Wiley, New York, 831 (1963).
[43] Specialist Periodical Report, Vol. **1–9** (1970–1979), *Organic Compounds of Sulphur, Selenium and Tellurium* (The Chemical Society, Ed.), Burlington House, London.
[44] Sumpter, W. C. & F. M. Miller, Ref. [48], **8**, 1–288 (1954).
[45] Toyoda, T., S. Muraki & T. Yoshida, *Agric. Biol. Chem.*, **42**, 1901 (1978).
[46] Vernin, G., *Heterocyclic Compounds. Thiazole and its Derivatives* (J. Metzger, Ed.), J. Wiley & Sons, New York, **1**, 165–335 (1979a); *idem*, Unpublished results (b); *idem, Riv. Ital. EPPOS*, **53**, 190 (1981c).
[47] Weber, W. P. & G. W. Gokel, *Phase Transfer Catalysis in Organic Synthesis*, Springer Verlag, Berlin, Heidelberg, New York (1977).

[48] Weissberger, A., Ed., *The Chemistry of Heterocyclic Compounds*, J. Wiley & Sons, Interscience, New York.

[49] Wiley, R. H., *The Chemistry of the Oxazoles, Chem. Rev.*, **37**, 401-442 (1945).

[50] Wolfe, J. F. & D. R. Carter, *Organic Preparations and Procedures INT.*, **10**, 225-253 (1978).

[51] Wooldrige, K. R. H., Ref. [20], **14**, 1-41 (1972).

CHAPTER V

Computer Application of Non-Interactive Program of Simulation of Organic Synthesis in Maillard's Reaction: A Proposition for New Heterocyclic Compounds for Flavours

R. BARONE and M. CHANON, University of Aix-Marseilles, France

---

V–1 INTRODUCTION

The main characteristic of the *Maillard* reaction is its great compexity (*see Chapter III* and reference [13]). This point is further substantiated when the products possess high flavour values resulting from the low threshold concentration (e.g., ppm to ppb level). It is sometimes difficult to guess the structures of the various components present in these mixtures. Therefore the aim of this chapter is to show how the computer may be of use in this task.

Some years ago, following the pioneering work by Corey & Wipke [5], we developed in our group a program to find routes in the synthesis of heterocyclic compounds [2]. The wide variety of known reactions of organic compounds was classified as far as possible into what is presumed to be the basic minimum reactions or key steps. Representations of these reactions were stored in the computer. Rules dealing with the application and manipulation of these steps or building blocks were also written into the computer program. This program was then used to suggest possible routes for the synthesis of any heterocyclic compound. Clearly, the computer may not suggest a new reaction step, but lacking prejudice and by relentless systematic searching it may suggest mechanisms and synthetic routes which have escaped the attention of even the most vigilant chemist.

Such kinds of programs usually work in the *retrosynthetic* direction. For a given target, the program proposes a given set of precursors. Each of these precursors is then examined as if it were a target and so on recurrently until satisfactory starting materials are obtained [2,15] (*see Scheme 1a*). The *synthetic* strategy is also amenable. In this case, the written program contains the logic of reaction mechanisms. When a reaction mixture is entered into the computer program, then the computer will predict the products which could be expected at the end of the reaction (*see Scheme 1b*). In this approach, the advantage of the computer is that it examines very rapidly all the possibilities. This point is very important in the field of flavouring compounds because very small amounts of products may play a determinant role in the flavouring properties of food.

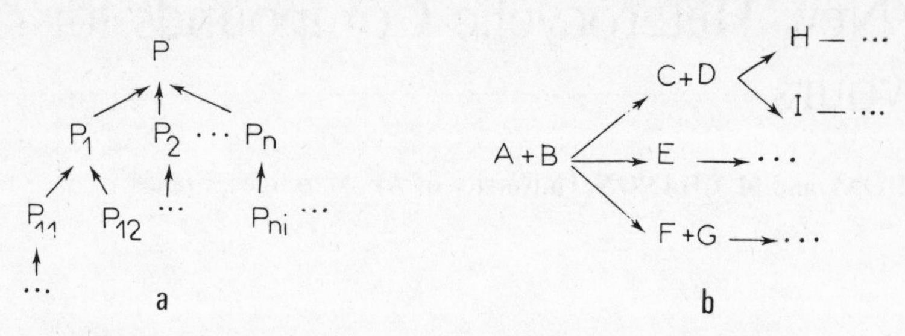

Scheme 1 – Programs working in the retrosynthetic (a) and synthetic strategy (b).

V–2 PRINCIPLE OF THE METHOD

Two problems must be solved when a program of computer-assisted synthesis is to be written: i) representation of molecules, and ii) representation of reactions.

2.1 Representation of molecules

Connectivity tables such as the one illustrated in *Table V–1* provide a simple solution to the problem of the representation of molecular structure. The program recognizes in the structure chemically significant information such as functional groups, rings, nucleophilic centres, electrophilic centres . . . and will use them later in solving chemical problems. From this connectivity table and approximate coordinates of the atoms calculated from a subroutine, the program may draw the molecule on the plotter.

2.2 Representation of reactions

Mechanistic considerations underly the representation of the reactions. Only the reactive centres are coded. For example, a nucleophilic substitution is written:

$$\underset{1\ \ 2}{C\text{-}L} + \underset{3\ \ 4}{Nu\text{-}H} \longrightarrow \underset{1\ \ \ 3}{C\text{-}Nu} + \underset{2\ \ 4}{L\text{-}H}$$

with L = leaving group and Nu = nucleophilic center.

When the program meets the structural situation C—L/Nu—H, it replaces it by C—Nu/L—H by automatically modifying the connectivity table. A transformation is associated for every substructure which represents a mechanism.

Table V–1 Connectivity table containing the structural information associated with hydroxyacetone

$$HOCH_2-C(=O)-CH_3$$

2 3 4 5 6

N	Atoms	Bonded atoms	Nature of the bonds
1	H		
2	O	1 3	1 1
3	C	2 1 1 4	1 1 1 1
4	C	3 5 6	1 2 1
5	O	4	2
6	C	4 1 1 1	1 1 1 1

In the foregoing nucleophilic substitution, the program suppresses bonds 1–2 and 3–4 and connects atoms 1–3 and 2–4. A bibliographic study of the main heterocyclization reactions showed that the most frequent schemes involve nucleophilic substitution, addition and elimination (*see Chapter IV*) [1].

2.3 How does the program work?

The various steps of this program are reported in *Scheme 2*.

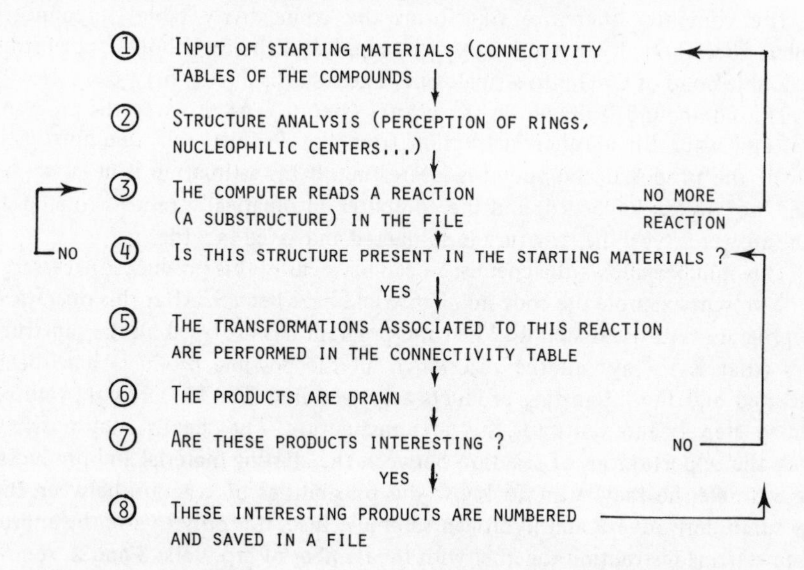

Scheme 2 – Different steps of the simulated organic synthesis program.

The chemist writes down the connectivity table for the compounds to be tested and then he punches the corresponding cards to communicate the starting information to the computer (step 1.). This analyses the information content of the connectivity table and saves this information in memory (step 2.). During the third and fourth steps, the program compares the first substructure taken from the file with the studied structure. If the structural block is present in the structure the corresponding chemical transforms are performed. For example, if compounds **1** and **2** are given to the computer:

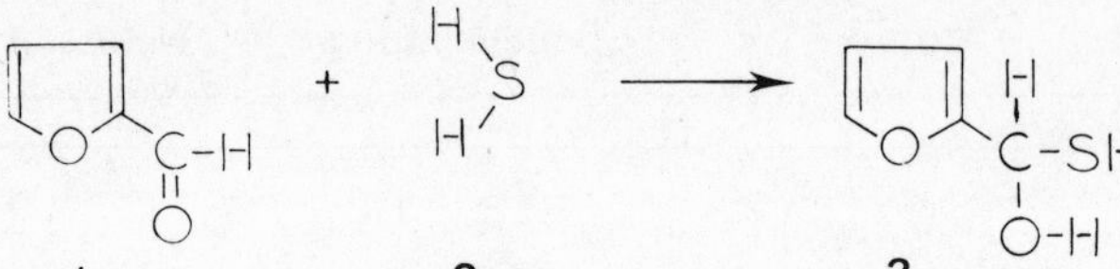

1 + **2** ⟶ **3**

when the substructure to be searched is that of nucleophilic addition:

Nu—H + C=Nu (step 3)

This substructure is found: SH + C=O (step 4)

and the mechanistic result is:

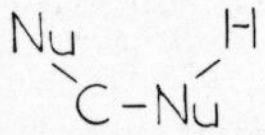

The computer therefore transforms the connectivity table: it connects sulphur to carbon, hydrogen to oxygen, suppresses the S—H bond, transforms the double bond of C=O into a single one. The result is **3** (step 5.).

This compound is drawn on the plotter (step 6.). At this step the program stops and waits for another instruction from the chemist; thus one must estimate if the proposed compound is interesting. If his estimation is negative, he types 'no' on the keyboard, and the computer automatically returns to step 4. If the answer is 'yes' the structure is numbered and saved in a file.

This number allows the chemist to call back easily this product if necessary. In the present example the code number would have been **3**. After this operation the program returns to step 4. When the program has analysed all the substructures (that is to say: all the reactions), all the possible products have been generated and the interesting products are saved in a file. The program returns then to step 1. and waits for the next instruction. The chemist may possibly check the opportunities of reaction between the starting material and products. For example, he may want to know the possibilities of reaction between the new furan derivative **3** and hydrogen sulphide. If so, it is only to give the appropriate starting instruction together with the number of products: **3** and **2**. A new cycle of operations then starts, according to previously described lines.

V–3 RESULTS

3.1 From 4-hydroxy-3 (2*H*)-furanones, hydroxyacetone, glucose, maltol and isomaltol

The number of possible combinations utilizing the starting material in the *Maillard* reaction is fairly high. We limited our analysis to the structures originating from 4-hydroxy-3-(2*H*)-furanones, hydroxyacetone, glucose, maltol and isomaltol [3]. Less probable structures such as 3- and 4-membered heterocyclic systems such as epoxides were neglected in this preliminary approach and are not presented here but were predicted by the computer. They could be reexamined without much work if careful analytical studies happen to evidence flavouring structures whose presence can only be explained through the intermediacy of such strained heterocyclic compounds.

Some 2-R-, and 5-R substituted 4-hydroxy-3-(2*H*)-furanones (R = H, alkyl) are important flavour contributors of many foods (meat, nut, coffee, soy sauce, berries, etc. . .) (*see Chapter II*). They could after their formation dimerize according to various reaction schemes. Some detailed mechanistic steps given by the program are shown in *Scheme 3*.

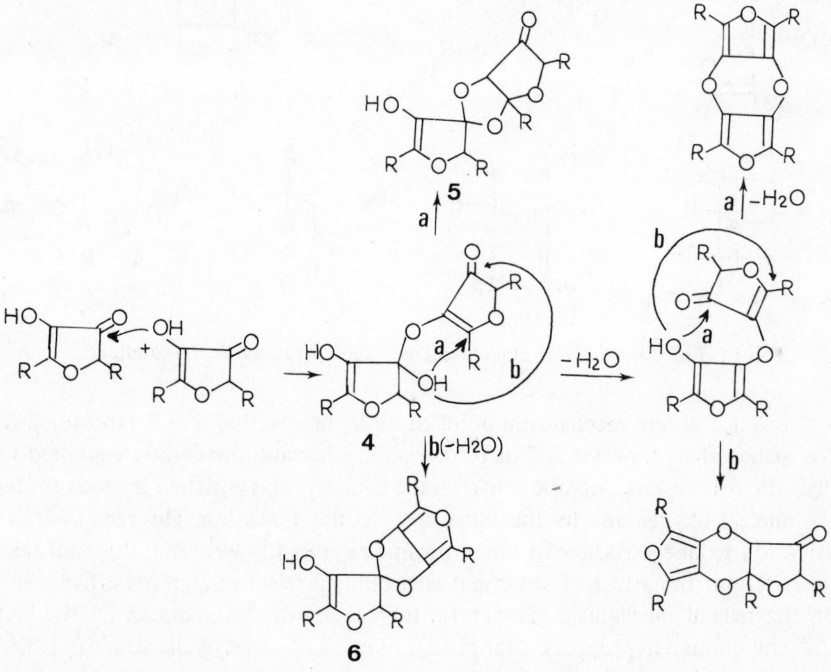

Scheme 3 – Some mechanistic steps dealing with the dimerization of 4-hydroxy-3-(2*H*)-furanones.

The other products predicted for this reaction are listed in *Fig. 1*. It should be noticed that compounds **5** and **6** bear some resemblance to 3′,4-, and 3′,5-dioxa-2,8-dithiaoctanespirocyclopentanes which have been patented for their meat-like flavour.

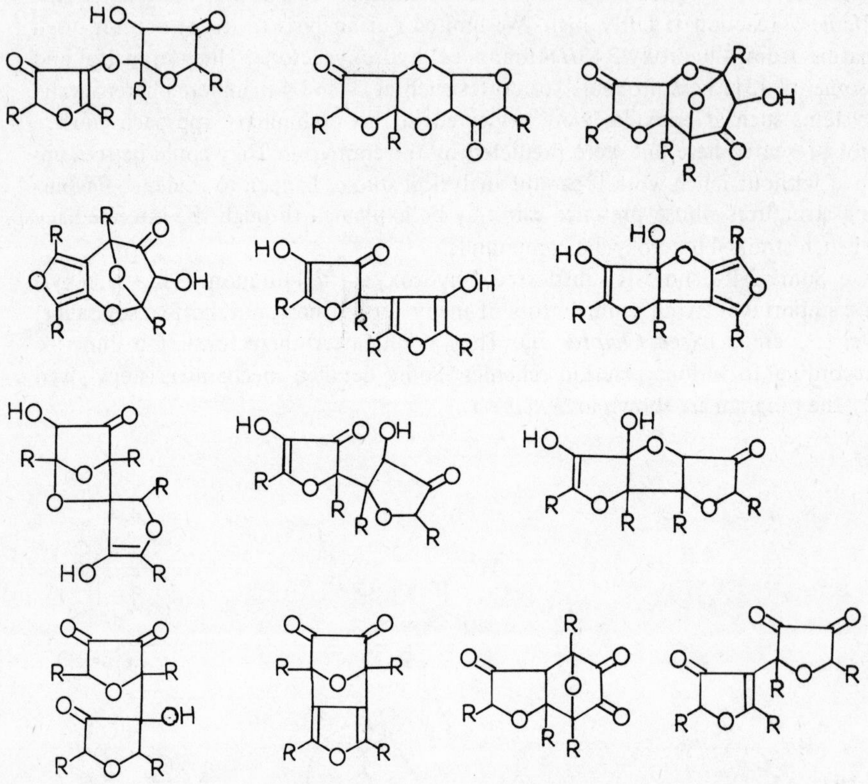

Fig. 1 – Fused heterocyclic rings predicted from 4-hydroxy-3-(2*H*)-furanones.

From a purely mechanistic point of view the reaction 4 → 5 (according to the arrow a) is, however, of little probability because the carbon supposed to play the role of electrophilic centre in the cyclization is enriched in electrons by the nuclear oxygen and by the substituent in the 4 position. This remark draws attention to one weakness of our program: as presently written it does not take into account the effect of structural environment (electronic, steric effects etc.) on the overall mechanism. Therefore, there are two main reasons for the fact that the computer proposes a far greater number of heterocyclic structures than the analytical methods. The first one is that classical methods are not efficient enough to separate the various components of a given flavour. The second reason lies into an overproduction of structures by the computer because of its lack of perception of structural variation consequences on the basic mechanistic scheme. The programmation of this perception is, however, feasible and has been realized in programs devised for computer-assisted synthesis [2,15].

The program applied to the self-condensation of hydroxyacetone proposes the formation of heterocyclic structures listed in *Fig. 2*.

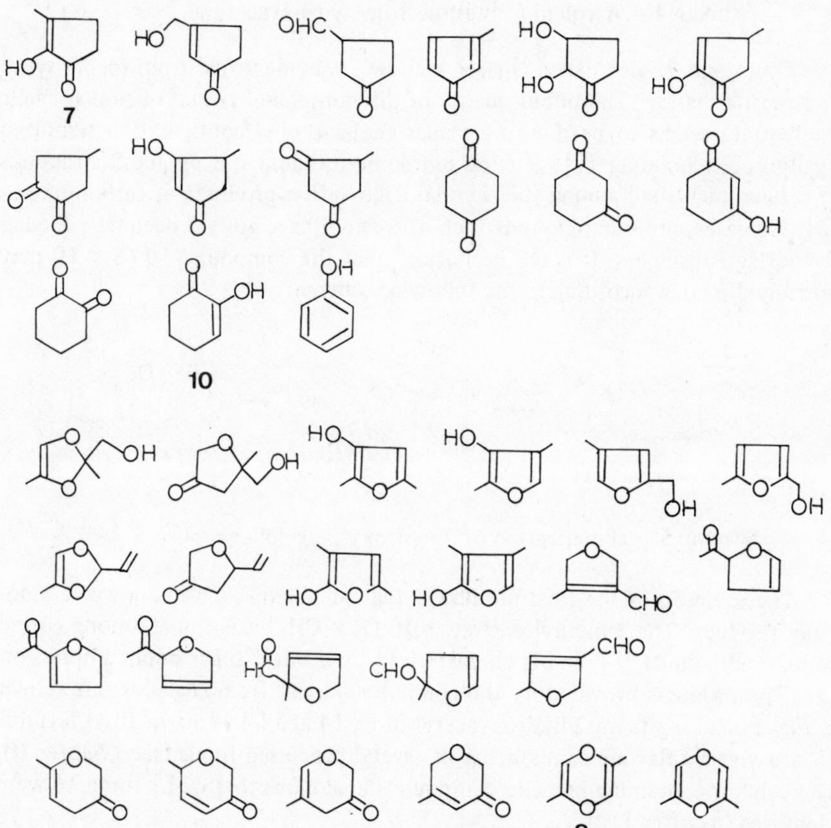

Fig. 2 – Heterocyclic aroma compounds from hydroxyacetone.

Hydroxyacetone is considered as a mixture of several starting reagents in this approach (*see Scheme 4*).

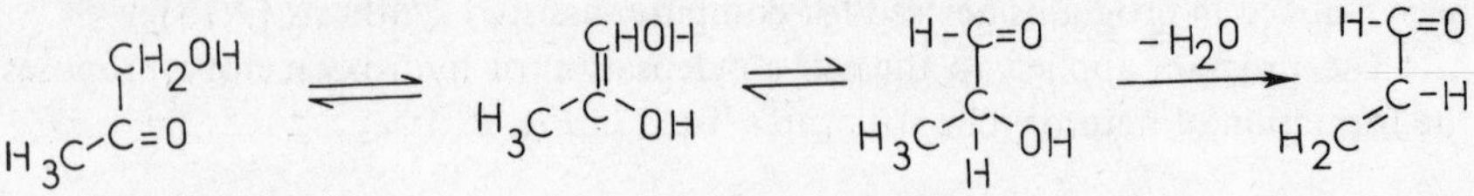

Scheme 4 – Acrolein formation from hydroxyacetone.

Cyclotene **7**, also called 'maple lactone', was identified from maple syrup and roasted barley. The odour quality of this compound as that of similar cyclic α-diketones seems to be due to a planar enol-carbonyl configuration stabilized by hydrogen bonding [9]. Phenol and some furanoid and pyranoid structures have been identified among the thermal degradation products of carbohydrates [6]. However, other compounds such as **8** and **9** have not yet been described as flavouring substances. It could be noticed that the compound **10** (R = H) may possibly dimerize according to the following scheme:

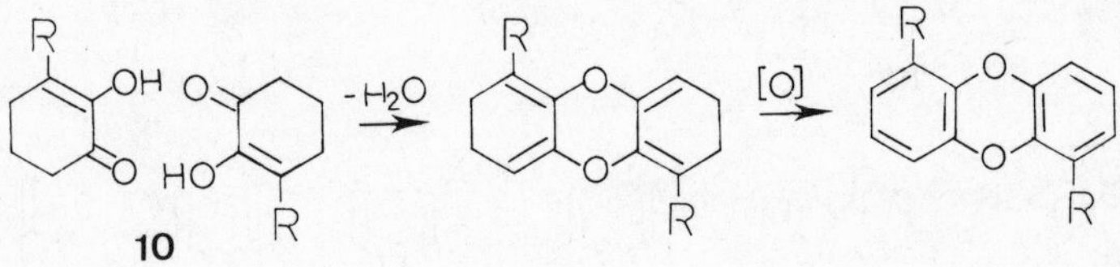

Scheme 5 – Dimerization of 2-hydroxy-2-cyclohexenones.

The skeleton of the final product is that of 'dioxin', a well-known carcinogenic product. The 3-methyl derivative **10** (R = CH_3) was found among coffee aroma compounds [8] and is characterized by a burnt sugar odour impression [9]. The products provided by the computer starting from *D*-glucose are shown in *Fig. 3*. Among them 2-hydroxyacetyl furan **14** and 5-hydroxymethyl furfural **15** are volatile flavour components of several processed foods (*see Chapter II*). The sulphur-containing bicyclic compounds analogous to [3,2-b] furan **16** were identified in coffee [12].

The rather small range of heterocyclic structures presented here results from the fact that degradation reactions were not taken into account. Only the cyclization reactions leading to five or six-membered heterocyclic compounds were analysed. Compounds **11** to **13** may further react according to a *Diels-Alder* reaction (*see Scheme 6*).

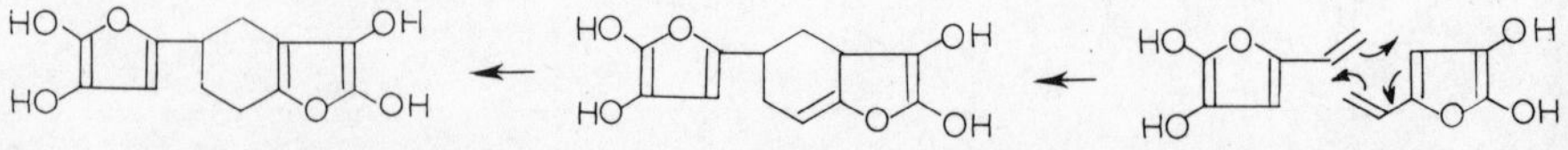

Scheme 6 – *Diels-Alder* self-condensation of 2,3-dihydroxy-5-vinylfuran.

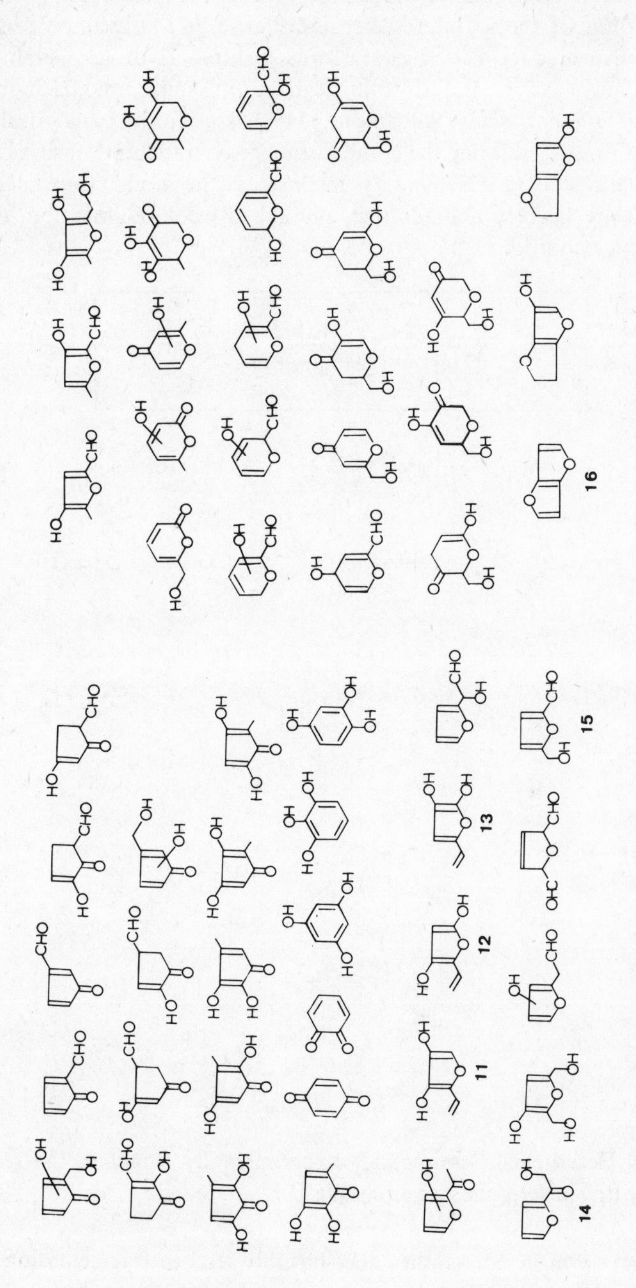

Fig. 3 – Carbon and oxygen containing rings predicted by computer for the thermal glucose degradation.

3.2 From furfural-ammonia-hydrogen sulphide model system[10]

This reaction is of special interest because it arises in nearly all processed foods. The thermal degradation of sugars affords furfural. Hydrogen sulphide results from cysteine, cystine, and thiamin degradation, and ammonia is formed by pyrolysis of α-amino acids. Shibamoto [11] has performed analytical work on this system, thus facilitating the comparison between theoretical suggestions and experimental results (*see Scheme 7*). In this case the mechanisms taught to the computer were nucleophilic addition and substitution, elimination, oxidation, and some cycloadditions.

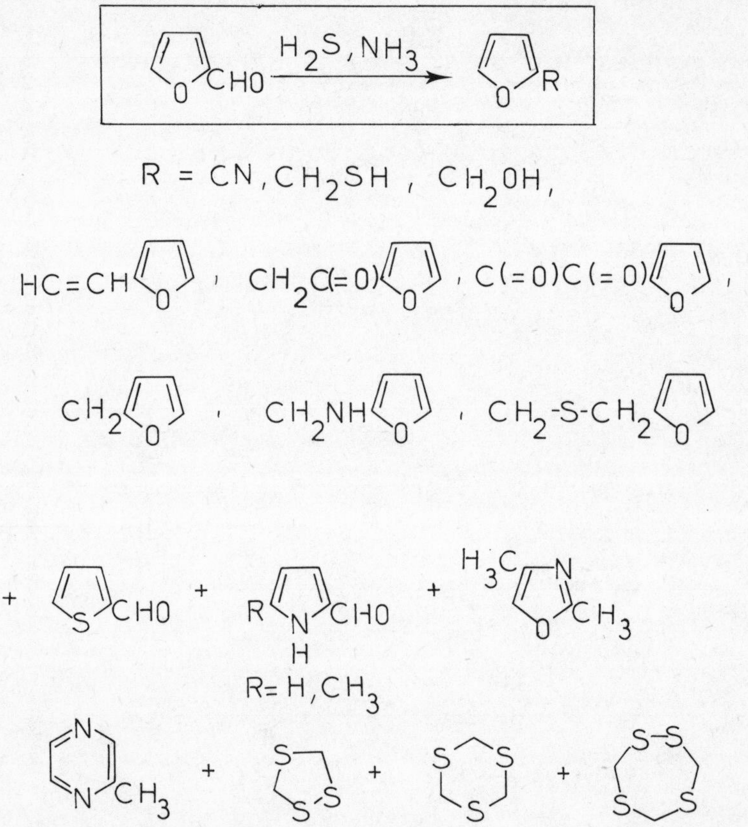

H_2S, NH_3

R = CN, CH_2SH, CH_2OH,

HC=CH–, $CH_2C(=O)$–, C(=O)C(=O)–,

CH_2–, CH_2NH–, CH_2-S-CH_2–

R= H, CH_3

Scheme 7 – Heterocyclic compounds experimentally found in the reaction of furfural, hydrogen sulphide and ammonia [11].

Furfural at high temperature may possibly lead to fragmentation products displayed in *Scheme 8*. These fragments are often evidenced during the decomposition of sugars.

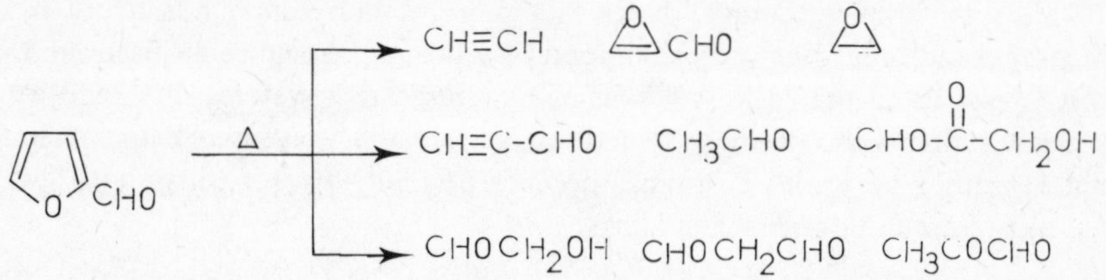

Scheme 8 – Predicted products resulting from the fragmentation of furfural [10].

When the computer was faced with the problem of the prediction of possible reactions between furfural, these fragments, hydrogen sulphide, and ammonia, it proposed more than a thousand structures [4]. Not only were all the compounds identified by Shibamoto found by the computer, but also most of the products identified in the *Maillard* reaction including furans, pyrroles, thiophenes, oxazoles, thiazoles, imidazoles, pyrans, pyridines, pyrazines, etc. . . (*see Fig. 4*).

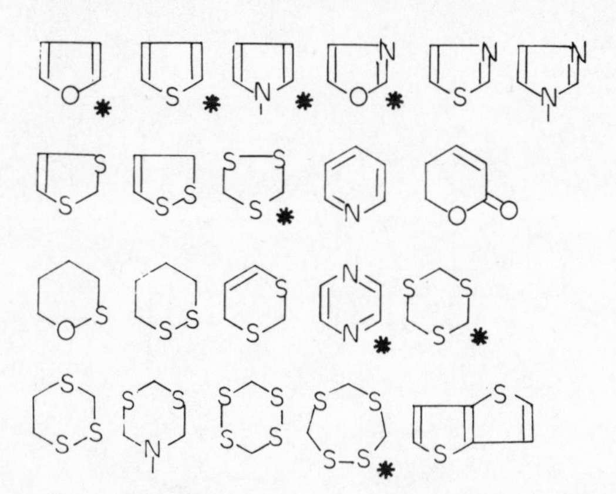

Fig. 4 – Basic skeletons of the heterocyclic compounds predicted by computer for the reaction: furfural–NH_3–H_2S. Substituents are not given. (* Indicates a structure identified by Shibamoto.)

Reinvestigating this model reaction, Vernin [14] has identified in a preliminary experiment, some pyridine derivatives not previously reported by Shibamoto. Among them was the 2-pyridinemethanethiol **17** patented [7] as a key ingredient in a synthetic meat flavour like roasted pork and dimethylpyridine isomers.

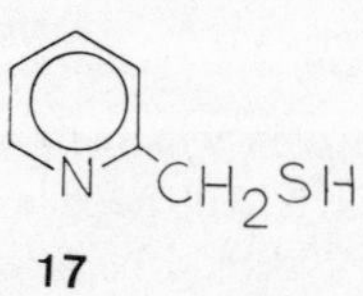

17

This very simple example shows that most of the compounds found by computer could indeed be present in food flavours and remain to be discovered. In *Fig. 5* are listed the basic skeletons of new fused ring systems. It should be appreciated that some of these structures, particularly polyfused heterocyclic systems, cannot be found in aromas because of their low volatility, but they could be present in tars or melanoidins.

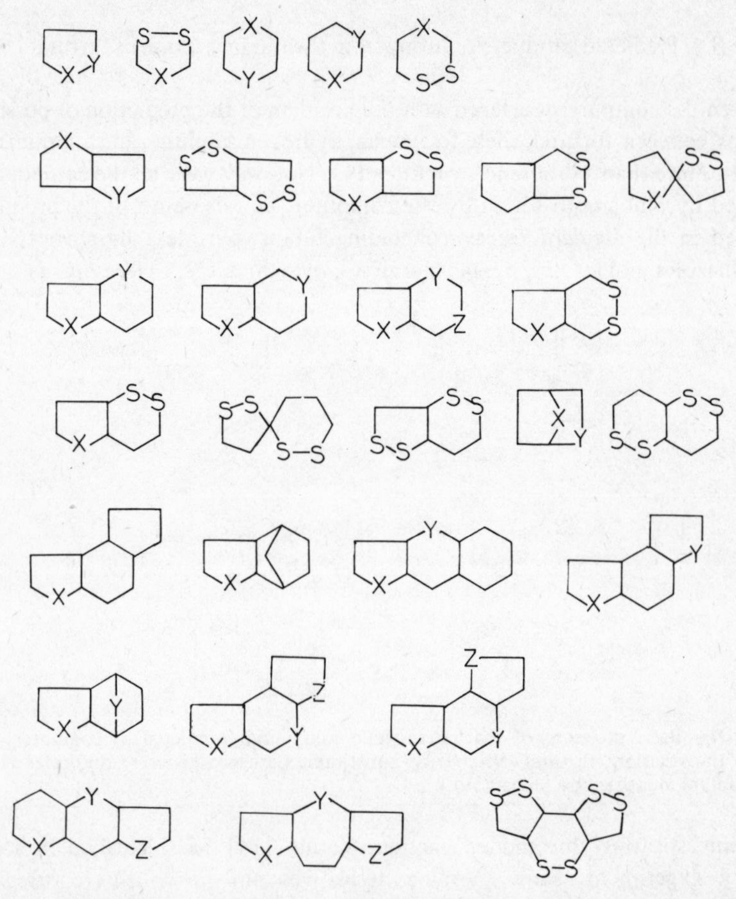

Fig. 5 – Some basic skeletons of heterocyclic aroma compounds which may be present in food flavours (X, Y, Z = O, S, NH) predicted for the reaction: furfural $-NH_3-H_2S$.

V–4 CONCLUSION

These results show the directions in which flavour chemistry may expect help from the computer. Much work is still needed to add a satisfactory discrimination to the present program. It should be stressed, however, that in the field of flavours the problem is somewhat special. In normal chemistry, from which most of the known reactivity rules are extracted, products formed in amounts of less than 1% are often neglected. In the field of food chemistry, the trace amounts of an heterocyclic compound may be the basis of a flavour or a toxic material. Flavour chemistry should go far beyond simple gastronomic considerations.

V–5 REFERENCES

[1] Barone, R., *The Use of Computer in Organic Synthesis*, Ph.D. (French), Marseille (1976).

[2] Barone, R., M. Chanon & J. Metzger, *Chimia,* **32**, 216 (1978); Barone, R. & M. Chanon, *Nouveau J. Chimie.,* **2**, 659 (1978).

[3] Barone, R., M. Chanon, G. Vernin & J. Metzger, *Riv. Ital. EPPOS,* **62**, 136 (1980); *idem,* VIIIth International Congress on Essential Oils, Cannes, October 1980.

[4] Barone, R., M. Chanon, G. Vernin & J. Metzger, *Parf. Cosm. Arômes,* **38**, 71 (1981).

[5] Corey, E. J. & W. T. Wipke, *Science,* **166**, 178 (1979).

[6] Fagerson, I. S., *J. Agric. Food Chem.,* **17**, 747 (1969).

[7] Feldman, J. R. & J. H. Berg, *U.S. Pat.*, 3,803,330 (09.4.1974).

[8] Gianturco, M. A., A. S. Giammarino & P. Pitcher, *Tetrahedron,* **19**, 205 (1963).

[9] Ohloff, G. & I. Flament, in *Fortsch. Chem. Org. Naturst.,* **36**, 231-283; W. Herz, H. Grisebach & G. W. Kirby, Eds.; Wien, Springer Verlag (1979).

[10] Petitjean, M., G. Vernin, J. Metzger, R. Barone & M. Chanon, in *The Quality of Foods and Beverages*, Vol. **2**, pp. 253-268 (G. Charalambous & G. Inglett, Eds.), Academic Press, New York (1981); *idem*, 2nd International Flavor Conference, July 1981, Athens, Greece.

[11] Shibamoto, T., *J. Agric. Food Chem.,* **25**, 206 (1977).

[12] Stoll, M., M. Winter, F. Gautschi, I. Flament & B. Willhalm, *Helv. Chim. Acta,* **50**, 628 (1967).

[13] Vernin, G., *Parf. Cosm. Arômes,* **29**, 77 (1979); *idem, Parf. Cosm. Arômes,* **32**, 77 (1980); *idem, Ind. Aliment. Agric.,* **5**, 433 (1980); *idem, Riv. Ital. EPPOS,* **58**, 2 (1981).

[14] Vernin, G., Personal communication.

[15] Wipke, W. T. & W. J. Howe, *Computer Assisted Organic Chemistry*, ACS Symposium Series **61** (1977).

CHAPTER VI

Recent Techniques in the Analysis of Heterocyclic Aroma Compounds in Foods

L. PEYRON, Lautier-Aromatiques Company, Grasse, France

VI–1 INTRODUCTION

The present chapter is devoted to the discussion of isolation and identification techniques of heterocyclic compounds possessing flavouring (odour- and/or taste-producing) properties and present in natural sources which are often very complex. Thus, a glossary of at least some relevant definitions seems appropriate.

(1) Flavour foodstuffs are a part of normal human food consumption.
(2) Spices, fragrances, and condiments isolated from plants or animals can be used as such or processed, and contain no additives or foreign substances.
(3) Natural flavour materials and flavour concentrates are fragrances and flavourings obtained exclusively by physical procedures (extraction, squeezing, steam distillation, distillation, etc.) from aromatic[†] materials, spices, flavourful foodstuffs, essential oils, resins, oleoresins, balsams, etc.
(4) Flavouring substances are aromatic compounds defined by their respective chemical formulas, occurring in natural flavour materials, spices, or flavour foodstuffs, and isolated by physical methods or other procedures.
(5) Flavouring compositions are mixtures containing one or several natural or synthetic flavouring substances, without or with preservatives, colours, solvents, supports, etc., intended for use as flavouring substances for foods, tobacco, etc.

Flavouring substances can be divided into three groups: (1) odour-producing substances (volatile materials); (2) taste-producing substances (with the exception of sugars and acids) whose volatility is normally low, and (3) substances which are both odour- and taste-producing. Flavouring substances are frequently

[†] In this chapter, the word 'aromatic' has its original meaning (flavourful) and does not indicate a benzene-like or delocalized π-electron system.

complex mixtures used in food products as such or, in some cases, they are isolated from their respective sources by extraction which is often quite laborious.

The skeletons of the molecular structures of these substances can be simple or complex, with carbon chains or with heteroatoms. Organic compounds containing heteroatoms are very common among aroma compounds. The significant role which they exhibit on the odour and flavour of these substances is due mainly to their wide occurrence rather than to their unique structures [73]. Among the heteroatomic substances, heterocyclic compounds constitute the most important group.

The most recent listing of TNO [20] contains 790 heterocyclic compounds (lactones excluded), among which 35% are oxygen heterocyclic compounds, 8% contain sulphur or sulphur and oxygen, and 57% contain nitrogen, nitrogen and oxygen, or nitrogen and sulphur. This information refers to 124 most common foods, such as alcoholic beverages, bread and bakery products, dairy products, eggs, fish, red meat, poultry, fruits, spices, condiments, vegetables, and various other products (cocoa, coffee, tea, etc.).

In the U.S., the FEMA/GRAS listings contain also artificially obtained heterocyclic systems not found in natural sources, either because they are only synthetic and not natural products, or because the present analytical methods are not sensitive enough to detect them. The complexity of naturally occurring materials, the high level of dilution of flavouring substances, and the lability and reactivity of certain heterocyclic systems make their study a delicate task. Problems connected with the isolation and identification of these compounds are due to the following factors:

(1) Complexity of the mixtures in which they are present.
(2) Large amounts of various constituents present in these mixtures and masking the minor components from the organoleptic point of view.
(3) Important differences in the chemical composition of aromas due to morphological or genetic variations.
(4) Mixtures boiling within a relatively wide range of temperatures and combinations of products sometimes simulating azeotropic mixtures.
(5) The polar character, reactivity or nonreactivity of certain constituents.

The goal of this chapter is not to compile an exhaustive list of references [28–47, 99–102] but to concentrate on those separation and analytical methods for heterocyclic compounds present in aromas which are important, characteristic, and essential. The objectives of work within this area fall into two categories: isolation and physical, chemical, and organoleptic characterization of new components in aromas is a task for researchers, whereas the identification of constituents already known and their analytical determination following the isolation are the task of industrial experts.

In this chapter, the following two major topics will be considered:

(1) Isolation of heterocyclic compounds from complex media.

(2) Identification, characterization, and determination of heterocyclic compounds.

The relationship between the structures and organoleptic properties of heterocyclic compounds identified in naturally occurring flavour materials and their formation are discussed in detail in *Chapters II and III*.

VI–2 ISOLATION OF HETEROCYCLIC COMPOUNDS FROM COMPLEX MEDIA

2.1 General

The considerable dilution of organoleptically active compounds coupled with the complexity of the media in which they are present (aqueous or lipidic) are the reason why a preliminary treatment of these systems is usually necessary. In some cases, a direct determination of a substance is possible in a concentrate; in other cases, the active constituent can be identified and characterized after its isolation by fractionation [71]. Three factors have to be considered: (1) the medium; (2) the constituent in question, and (3) the isolation technique.

(1) The media can be solid, liquid, or gaseous and include natural or processed foods, flavoured foods, primary flavouring substances in natural or refined form, and aromatic substances obtained as reaction products.
(2) Flavour constituents are usually quite numerous and are present together with other nonodorous volatile and nonvolatile substances. Thus, the enormous variety of structures of heterocyclic compounds present in flavour substances and their different physicochemical properties mean that their isolation is often a difficult and lengthy procedure. In certain cases, however, when the heterocyclic compound in question is one of the predominant components, the isolation is relatively easy.
(3) The choice of a suitable isolation method depends mainly on the physicochemical characteristics of the respective heterocyclic compounds such as concentration, the range of boiling points, polarity, chirality, and chemical reactivity.

These characteristics, which can be responsible for large differences in organoleptic properties, depend also on the complexity of the medium, its nature, and its physical state of aggregation. This is true of food products with or without alcohol, lipidic or nonlipidic, solid or liquid, with or without fatty materials. Precautions have to be taken to ensure that the components remain unchanged during the isolation or concentration process. It is essential to avoid the formation of artefacts which would lead to erroneous conclusions. Thus, the work has to be carried out at low temperatures in the case of thermolabile substances, whereas oxidatively unstable compounds have to be handled in the absence of oxygen. Also, it is necessary to avoid enzymatic reactions during the processing, chopping, grating, or pressing of food products and other materials which are complex mixtures containing a number of constituents which can react with each other and generate various volatile and other substances.

It is very important to avoid the introduction of contaminants or losses of important components due to the use of insufficiently selective or inefficient methods. Erroneous extrapolations have to be avoided as well. When the composition in the vapour phase is not the same as in the liquid mixture, the relative amounts of components in concentrates obtained from mixtures are not the same as in the vapour phase.

Thus, the different operational schemes can be summarized as follows: (*see Scheme VI–1*).

In summary, the isolation consists of two steps. The primary step, (A), represents the isolation of a mixture rich in the component under study from the starting raw material, i.e., an extract most representative or specific with respect to a particular flavour. The second step, (B), is the isolation of the component in a desirable state of purity.

The first step is often the most delicate if not the most laborious one. There are a number of general or specific methods available for the extraction of components from their mixtures which can be utilized in one of the necessary steps. There is no universal method; however, usually it is possible to use a judicious combination of physical and chemical methods, each of which has its advantages and disadvantages. Therefore, the choice of the most suitable approach is not always easy.

Physical methods are based on the differences in vapour pressure, solubility, polarity, or molecular size of the heterocyclic compounds which are to be isolated and of other components in the mixture. The following several methods will be discussed in some detail:

(1) Methods based on vapour pressure
Distillation, rectification, sublimation
Steam distillation
Gas extraction (cryogenic absorption of a substance by a stream of gas)

(2) Methods based on solubility
Extraction with volatile or nonvolatile solvents
Absorption of vapour by a solvent
Extraction combined with steam distillation (*Likens-Nickerson*)

(3) Methods based on polarity
Adsorption
Chromatography

(4) Methods based on molecular size
Dialysis, electrodialysis, reverse osmosis
Chromatography

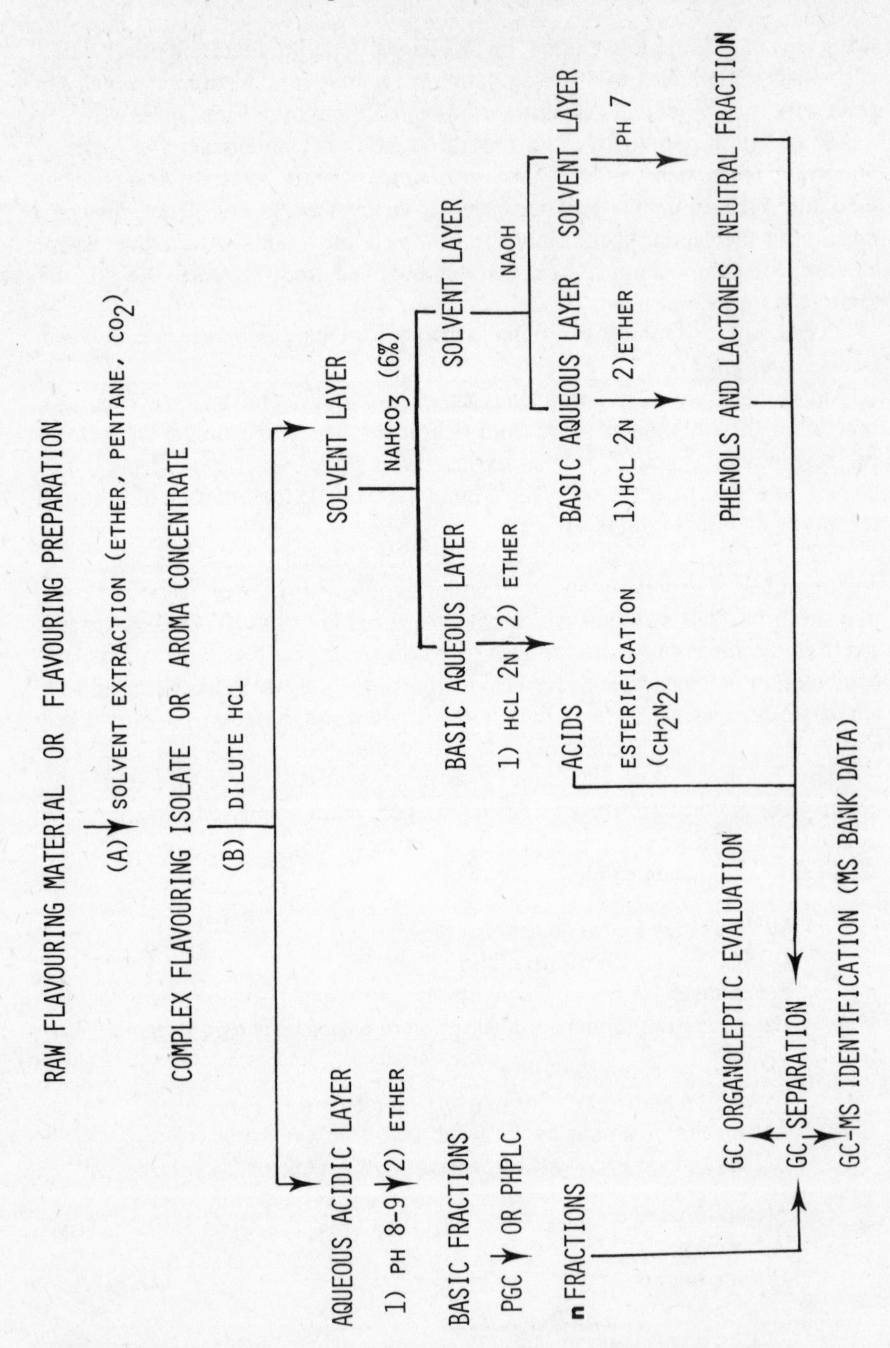

Scheme 1 – Summary of Separation and analytical Techniques for flavouring compounds in foods

Chemical methods are based on reactions of the heterocyclic rings and the functional groups (formation of salts, etc.).

Finally, at least some simple primary methods should be mentioned, such as squeezing, centrifugation, and filtration which are used for essential oils, in the preparation of fruit juices and dairy products, etc.

2.2 Isolation and purification

Various physical methods currently used in primary isolation processes leading to concentrated extracts as well as to the separation and purification of the respective heterocyclic compounds will be discussed here. A combination of these methods normally gives a better separation, and cases where a single method is used are quite rare. Several examples of separation procedures will be discussed in the subsequent text.

2.2.1 Physical methods [61]

Methods based on vapour pressure:

Simple distillation at atmospheric pressure. A three-necked flask equipped with a stirrer, a nitrogen inlet tube, and an adaptor with a thermometer is immersed in an oil bath and connected to a condenser with a receiving flask. The receiving flask is connected through a side tube of the adaptor to two traps cooled to −20 to −40°C. The condensates in both traps are extracted by commonly used solvents.

Simple distillation under reduced pressure. The apparatus used is the same as above except that a stirrer is replaced with a finely drawn out capillary tube to prevent uncontrolled bumping. A third trap cooled to −80°C completes the set-up. The distillation is normally carried out at 10–20 torr.

Cold finger distillation. Volatile aromatic substances present in dehydrated foodstuffs or in liquid or solid food products (degassed) are distilled under reduced pressure. The most volatile components are condensed in traps cooled with liquid nitrogen. Less volatile compounds are distilled under higher vacuum and condensed on a 'cold finger' cooled with liquid nitrogen. In the case of water-containing products, the *Dumont-Adda* apparatus is used. The sample is dehydrated, frozen, transferred into a flask placed in a bath, and collected in three traps cooled with liquid nitrogen. The distillation is carried out in vacuum obtained using a diffusion pump (10^{-4} torr).

Distillation through a spinning-band column (short pathway thin layer). Molecular distillation. The liquid to be distilled forms a thin layer on a spinning-band heated to a suitable temperature by internal circulation of a liquid. Reduced pressure of the order of 10^{-2} to 10^{-5} torr is used. This ensures a practically instantaneous evaporation of the volatile components. The distillate and the distillation residue are treated separately.

Fractional distillation at atmospheric pressure. This technique makes it possible to remove volatile solvents present in extracts obtained by solvent extraction, and it can also be used for fractionation of certain essential oils, flavours, and fragrances, etc. A Vigreux-type distillation column (30–40 cm long) or a column packed with Raschig rings (25–35 cm) or wire gauze (Multiknit type) is used with a reflux condenser. Traps maintained at −20°C and −40°C are normally used.

Fractional distillation under reduced pressure. The well-known apparatus with a capillary tube and an inert gas inlet is employed, in combination with traps cooled to −20°C and with liquid nitrogen.

Steam distillation at atmospheric pressure, under reduced pressure, or at 1–3 bar pressure. This classical technique used for the isolation of essential oils makes it possible to recover very volatile components, although not quite efficiently. Under certain conditions, large amounts of artefacts can be formed.

A two-step distillation system with steam at atmospheric pressure utilizing three cooled traps has been introduced by *Forst Holloway*. It can be adapted for distillation under reduced pressure as well.

Cryogenic absorption of vapours using an inert carrier gas. This method is used with unrefined materials for isolation of highly volatile flavouring substances. A stream of nitrogen is passed through the mixture, and the vapours are condensed in a cooled trap or dissolved in suitable solvents.

Methods based on solubility:
Extraction methods can be utilized for direct isolation of volatile components or for less volatile substances present in flavouring materials in their raw or refined form. They can also be used to recover these substances from aqueous distillates. The extraction methods are based on favourable values of the partition (distribution) coefficients of the components in question between the solvent and the mixture to be extracted. These mixtures can be solid or liquid. A number of common solvents were tested and utilized for this purpose, such as gases in their supercritical states (carbon dioxide), liquefied gases (carbon dioxide, butane, the freons), liquids with a low boiling point (30–80°C), liquids with boiling points above 80°C, without or in combination with volatile liquids.

The first two groups mentioned above require the use of a sophisticated apparatus which can withstand high pressures. The existing laboratory apparatus of this type are the continuous extractor by Koller [50] and that by Blakesley (freon 12). The third group is the group most commonly used, and a large number of solvents possessing more or less specific characteristics are at our disposal for this purpose. One can use a continuous or simple extraction [61].

Simple extraction of liquids which are water-soluble or insoluble. The classical method is represented by shaking the mixture and the solvent in a separatory funnel.

Continuous extraction with a flow of water-soluble or insoluble liquids utilizing light solvents (pentane, isooctane). Also extractors for heavy solvents are currently in use.

Extraction of water-soluble or insoluble liquids with light solvents. Volatile flavouring substances are separated by distillation and extracted in continuous fashion with an organic solvent. The sample is placed into a flask heated to boiling and with an extractor containing water and pentane above the flask. An efficient coolant is used. Volatile flavouring substances are condensed from the vapour phase and extracted with pentane.

Extraction of flavouring substances insoluble in water with light solvents. A Soxhlet extractor is used.

Extraction of solid substances rich in lipids. The material is thoroughly mixed with Celite 545 and introduced into a chromatographic tube (30 × 500 cm) with a frit, stopcock, and a receiving flask. Acetonitrile is used for elution and 10, 20, or 50 ml fractions are collected. These fractions are then extracted with pentane.

The system 'Extrelut' makes it possible to extract lipophilic substances from aqueous solutions by passage through a cartridge packed with kieselguhr with large pores, or granulated consistency (increased specific surface).

Rotary extractor for liquid-liquid extraction. The solvent is continuously evaporated in a heated flask and condensed in a reflux condenser. Owing to the centrifugal force, the solvent passes in the form of tiny droplets through the liquid to be extracted and thus ensures an excellent contact between the two liquid phases. Separation takes place in the separatory part of the extractor and the solvent returns to the distillation flask. This system can be used both for light and heavy solvents. A number of other liquid-liquid extractors for continuous extraction have been described and/or are commercially available, as well as the apparatus for counter-current extraction.

Passage of solvent vapours through an organic solvent with a low or high boiling point. In the first case, the traps are cooled, whereas in the latter case it is often possible to work at room temperature (Vaseline, phthalate, etc.) [67].

Crystallization. – Centrifugation. – Cryoconcentration. All these procedures find very specific uses (e.g., for the isolation of cineole).

Methods based on polarity:

Adsorption. Adsorption methods are currently used especially for the recuperation of very diluted volatile substances present in the air or in an inert gas. The apparatus is simple, the adsorbents can be easily transported and are stable, and the recuperation of volatile substances is easy. In the past, activated carbon and stationary phases used in gas chromatography were commonly employed. However, their capacity was limited. More recently, porous synthetic polymers have been found which are of interest from this point of view. The vapours pass through a tube containing several hundred milligram quantities of the polymer. The materials are desorbed by heat or with small amounts of a solvent. These tubes can be used directly as pre-chromatographic columns (e.g., Porapak, Tenax, Chromosorb, etc.). The use of conveniently chosen pre-chromatographic columns enables selective adsorption of certain groups of solutes which can be either removed or analysed in this fashion. It is also possible to utilize several pre-columns, each with a different selectivity. This technique makes it easy to obtain the individual samples.

Preparative chromatography. There are different chromatographic techniques available for the analysis of aromas. They not only complement but actually tend to replace the classical methods. The principles of chromatography are quite well known, and the importance of chromatographic techniques is so widely accepted that no attempt will be made to discuss them in detail in this chapter. Several isolation techniques as well as their application to heterocyclic compounds with or without functional groups are mentioned in section VI–3.

Adsorption thin-layer chromatography [41]. A number of texts describe the possibilities of this method. Different preparative arrangements are available which enable separation of 10–50 mg quantities of materials on stationary phases whose thickness is 0.5–1.0 mm.

Preparative liquid chromatography. This is a separation method based on the partition (distribution) of components of a mixture between two phases, a stationary liquid or solid phase and a mobile liquid phase. The order of elution of the individual components is a function of their respective affinities for the two phases. Considerable progress achieved in recent years has led to a spectacular improvement of separations with very short elution times, mostly thanks to the use of pressure which increases the velocity of the mobile phase flow. Efficient preparative systems make it possible to isolate satisfactory amounts of heterocyclic compounds from complex mixtures.

One should not forget to mention *column chromatography* which is widely employed as a fractionation and separation technique for relatively large amounts of components present in complex mixtures and especially for compounds with different functional groups.

Preparative gas chromatography. Gas-liquid chromatography is indispensable for the analysis of complex mixtures, and especially for isolation and identification of flavour-producing substances. On the preparative scale, this technique makes possible laboratory and industrial separations, similar to distillation of substances with close boiling points. It also allows one to remove impurities, to separate azeotropic mixtures and, in certain cases, to separate thermally unstable compounds. Various chromatographic equipments for laboratory and industrial separations on laboratory scale are commercially available.

Methods based on molecular size:

Dialysis, electrodialysis, reverse osmosis, and *microfiltration* are used for separations of constituents with large molecules which interfere with the isolation of small heterocyclic components.

Exclusion chromatography. In this procedure, molecules of the solute are slowed down in their permeation through the pores of a gel. The pores are filled with a solvent compatible with the nature of the solute. This solvent then represents the stationary phase. Depending on their size, the molecules penetrate more or less into the pores and are more or less slowed down in their passage through the column. The solutes are eluted in the order of decreasing molecular weight. This procedure allows the removal of certain large molecules interfering with the isolation of heterocyclic compounds.

Miscellaneous physical methods:

Steam distillation combined with extraction. The technique of continuous distillation combined with extraction is widely used. There are several different models of apparatus available [9], such as those by *Likens-Nickerson* (1964), *Williams, Schultz-Maarse, MacLeod, Groenen*, etc. In the *Likens-Nickerson* apparatus, an aqueous solution of the compounds is placed in a flask connected to the right-hand side arm of the apparatus, whereas a flask attached on the left contains the solvent. The two flasks are heated separately. Steam and vapours of the solvent condense together in the central cooled section of the apparatus. The solvent phase and the aqueous phase which do not mix are separated in a U-shaped part of the apparatus and are returned to their respective flasks. The result is a continuous distillation with simultaneous extraction. This means that a small volume of the solvent can be used to extract a significant quantity of aromas [61].

Steam distillation – partial freezing – extraction. The distillate is gradually cooled under stirring so that it may form a liquid cone surrounded by a cooling-jacket. The coolant is maintained at −25°C. About 10 to 20-fold enrichment is obtained in the liquid phase. The liquid phase is removed and extracted with an organic solvent [61].

2.2.2 Chemical methods

The procedure of obtaining a certain substance from a mixture depends on its volatility and on whether it is a liquid or a solid. This is true in the case of chemical methods as well.

Selective separation based on functional groups is an approach that has been known for a long time. It is laborious but it has its good points. However, it normally requires larger amounts of liquid or solid aromatic materials, and it is time-consuming. The usual scheme [61] consists of successive separation of the following groups of compounds by using specific reagents:

(1) Acids. The pentane extract is extracted with a 6% aqueous $NaHCO_3$ solution in a separatory funnel. After acidification with 2*N* HCl, the organic acids are extracted from the aqueous solution with diethyl ether. The ether is evaporated until a residue of about 5 ml is left. Subsequently the free acids in this solution can be esterified with diazomethane.

(2) Phenols, coumarins, and lactones. The sample free from the acids is submitted to an extraction with a solution of 6% NaOH in water. Phenols and lactones isolated as phenolates and sodium salts of hydroxy acids, respectively, are obtained by acidification of the alkaline solution with 2*N* HCl, followed by extraction with diethyl ether.

(3) Organic bases: pyridines, pyrazines, pyrroles, thiazoles. Organic bases are separated from the extract (*n*-pentane or ether) by extracting the sample with aqueous HCl in a separatory funnel followed by transformation of the hydrochlorides into free bases with 10% aqueous KOH and extraction with diethyl ether.

(4) Carbonyl compounds can be isolated from the neutral fraction using 2,4-dinitrophenylhydrazine, *Girard's* and *Sandulesco's* reagents, sodium sulphite and hydrosulphite.

(5) Alcohols (3,5-dinitrobenzoyl chloride).

(6) Hydrocarbons and esters: column adsorption chromatography on silica gel or alumina using apolar or highly polar solvent as eluents.

The components in each group are then identified by gas chromatography-mass spectrometry coupling (GC–MS).

In the case of volatile compounds, selective absorption in traps by specific reagents used for the various functional groups is especially useful. Absorption of pyrazines discharged by a cocoa plant using *Pavelka's* trap with 1 *N* sulphuric acid can serve as an example. Similar procedures are currently used to trap acids, bases, carbonyl compounds, and sulphur-containing compounds.

It is also possible to use a pre-column packed with reagents which form the desired derivatives, in combination with liquid or gas chromatography.This approach enhances the specificity, modifies the chromatographic properties (e.g., it can decrease the polarity of certain types of molecules), and thus improves the resolution.

Ion-exchange chromatography is based on the fixation of the solute on ionic sites on a support and is excellent for the chemical separation of reactive heterocyclic compounds (e.g., nitrogen heterocyclic compounds).

2.3 Examples of isolation of heterocyclic compounds from complex flavouring materials

2.3.1 Oxygen-containing heterocyclic compounds

5-Methylfurfuryl ketone and furfuryl alcohol in tomatoes [81]. The condensate obtained after the treatment of tomato juice is extracted with methylene chloride in a 'cyclone'-type apparatus.

Acetylfuran, 5-methylfurfural, and 5-methyl-2-furyl-(2-furyl)methane in popcorn [95]. The condensate after steam distillation *in vacuo* is extracted with diethyl ether.

2-Pentylfuran in potatoes [15]. Treatment with steam is followed by continuous extraction with hexane.

2-Acetylfuran, 2-methyltetrahydrofuran-3-one, and epoxycaryophyllene in beer [90].

2-Methyltetrahydrofuran-3-one in boiled beef [37]. Flash vaporization is followed by evaporation from a continuously heated thin film under reduced pressure. The volatiles are collected in eight traps connected in series. Separation into acidic and nonacidic components is obtained by extraction with 10% aqueous solution of sodium carbonate.

2.3.2 Sulphur-containing heterocyclic compounds

1,2-Dithiacyclopentene, 3-vinyl-3,6-dihydro-1,2-dithiin, and 2-formylthiophene in asparagus [89]. A filtrate of an asparagus homogenate is refluxed with a phosphate buffer and then extracted with a pentane-diethyl ether mixture (1:1).

2-Formyldimethylthiophene and 2-formylthiophene in dried mushrooms (*Boletus edulis*) [86]. The mushrooms are macerated with water and then heated on a bath for two hours. This is followed by a week-long extraction with pentane.

2,5-Dimethylthiophene, 2,4- and 3,4-dimethylthiophenes, and 3,4-dimethyl-2,5-dioxo-2,5-dihydrothiophene in onions (*Allium cepa*) and leeks (*Allium porrum*) [3]. A slurry of onions or leeks with water was extracted with freon 11.

3,6-Dimethyl-1,2,4,5-tetrathiane in boiled mutton [66]. This is isolated by simultaneous distillation and extraction in the *Likens-Nickerson* apparatus followed by evaporation of the solvent.

2,5-(2,4- and 3,4-)Dimethylthiophenes and 3,4-dimethyl-2,5-dihydrothiophen-2-one in onions (*Allium cepa*) [11]. Freshly chopped onions are extracted in the Soxhlet extractor with methylene chloride.

2,5-Dihydro-3,4-dimethylthiophen-2-one in leeks (*Allium porrum*) [80]. Isolation from a dilute solution of volatile components obtained by headspace condensation in an original apparatus, with methylene chloride as the solvent. The apparatus consists of a hemispherical vessel cover (diameter 40 cm, height about 30 cm) placed on a 1 cm thick glass plate and equipped with a cold finger containing methanol kept just above 0°C (cryostat) in order to avoid freezing of the concentrate. About 1 kg of freshly harvested and chopped leeks was spread over the glass plate. At the centre of the plate a small conical flask is placed into which any condensate from the cold finger was free to run. After 24 to 36 hours' condensation, 250 ml of diluted aqueous solution of volatiles were obtained which was continuously extracted for 18 hours with 50 ml of methylene chloride.

Alkylthiophenes, alkylthiophenecarbaldehydes, furans, and pyrazines from pressure-cooked pork liver [63]. They were obtained by filtration, steam distillation at atmospheric pressure, and continuous extraction with diethyl ether in a *Williams'* apparatus.

2.3.3 Nitrogen-containing heterocyclic compounds

Guaipyridine from patchouli [36]. Patchouli flowers are steam-distilled and the essential oil obtained is extracted with diethyl ether and the solution treated with 0.2*N* hydrochloric acid. The same procedure is used for epiguaipyridine and patchouli pyridine [14].

Alkylpyridines and alkylpyrazines in the steam distillate of rice bran [91]. The entire steam-volatile concentrate is separated into neutral and basic fractions.

Alkylpyridines and alkylpyrazines in whisky fusel oils [65]. Three-litre portions of whisky are fractionally distilled under atmospheric pressure to yield fusel oils. A distillation column (2.5 cm inner diameter, 2 m long) packed with stainless steel helices is used. Distillation of the oil (2.5 mm inner diameter Heli-Grid column, 1.5 m long) removes most of the amyl alcohol. The remaining fractions are divided into basic, acidic, and phenolic portions by successive extractions with 5*N* hydrochloric acid, aqueous sodium bicarbonate, and 2*N* sodium hydroxide.

Alkylpyridines and alkylpyrazines in potato chips [17]. The essential oil from potato chips is isolated by steam distillation and extraction in the *Likens-Nickerson* apparatus (vacuum and atmospheric pressure, extraction with diethyl ether). The basic fraction is obtained by extraction of the obtained volatile oil with 3*N* hydrochloric acid.

Pyridine and alkylpyrazines in potato chips [27]. An aqueous slurry of potato chips is distilled and the volatile substances collected in a series of traps consisting of a water-cooled *Friedrich* condenser, four traps cooled with dry ice, and two traps cooled with liquid nitrogen. The combined condensate collected in the cooled traps is saturated with sodium chloride and extracted with diethyl ether. The ether extract is separated into acidic and nonacidic fractions by extraction with a 10% aqueous solution of sodium carbonate.

Pyridine and alkylpyrazines from roasted barley [24] are obtained by steam distillation and isolation of the basic fractions. The steam distillate is made alkaline with sodium bicarbonate and continuously extracted with diethyl ether for 42 hours. The extract is washed three times with one third its volume of 1*N* sulphuric acid. The combined acid washings are extracted with diethyl ether to remove neutral compounds.

Pyridine and alkylpyrazines from roasted pecans [98] are obtained by steam distillation under reduced pressure. The distillate saturated with sodium chloride is extracted three times with diethyl ether. This is followed by the separation into basic, acidic, and noncarbonyl basic fractions by 5% hydrochloric acid.

Alkylpyridines and alkylpyrazines in roasted lamb fat [18]. Steam distillation and simultaneous extraction in the *Likens-Nickerson* apparatus with a head modified in such a way that the condensed water does not return directly to the 12-litre flask but is led off into a 1-litre flask from which it is pumped by a Zenith pump back into the 12-litre flask through one of its side arms. The basic fraction is isolated.

Alkyl- and arylpyridines, dihydrocyclopentapyrazines, tetrahydroquinoxalines, and alkylpyrazines in roasted cocoa [93]. The procedure employed here is the extraction of cocoa with carbon dioxide in its supercritical state.

Other papers report on the compounds obtained from roasted cocoa by well-known isolation techniques, such as organic solvent extraction, distillation under high vacuum, pumping of volatile compounds under high vacuum followed by isolation using cold finger traps, etc. Steam distillation at atmospheric pressure combined with solvent extraction of the obtained steam distillate or co-distillation with propylene glycol or ethanol and subsequent extraction of the distillate diluted with water using pentane have also been reported.

Alkylpyrrole carbaldehydes, methyl 2-pyrrolyl ketone, and alkylpyrazines in roasted peanuts [96]. Steam distillation at atmospheric pressure is followed by saturation of the distillate with sodium chloride and continuous extraction in a liquid-liquid extractor with diethyl ether. The ether extract is concentrated and then separated by preparative gas chromatography.

Alkylpyrazines and alkylquinoxalines in roasted sesame seeds [60]. Distillation and extraction are carried out in an extractor similar to the *Likens-Nickerson* apparatus (diethyl ether). The concentrate is fractionated by separation according to the functional groups.

Piperine and piperamine from black pepper [88]. The aqueous residue after the steam distillation of the pepper oleoresin is continuously extracted with diethyl ether.

Separation by column chromatography on silica gel:

2-Methoxy-3-*sec*-butylpyrazine in the galbanum essential oil (*Ferula galbaniflua*) [12]. Steam distillation of the resin is followed by fractional distillation of the essential oil and preparative liquid chromatography.

Various pyrazines in roasted peanuts [45]. The volatile compounds are isolated by vacuum distillation of the oil pressed from whole roasted peanuts. The aqueous vacuum distillate is separated into basic and neutral fractions by methylene chloride extraction at pH 0.5 and 8.5.

Alkylpyrazines in Gouda cheese [83]. The volatile components of Gouda cheese are obtained by low-temperature vacuum distillation. The distillate is separated into polar and nonpolar components by solvent extraction. The nonpolar components are then used to obtain a fraction containing higher levels of the minor components by reversed phase adsorption chromatography on Amberlite XAD-2 followed by analysis of the polar extract.

Alkylpyrazines in processed American cheese [56]. The cheese is kept in a freezer overnight and ground into a powder with a Waring blender. The powder is packed into 4 × 60 cm Pyrex glass chromatographic columns. The lower end of the column is connected to a Tenax-GC tube and a Poropak Q-tube, and the head of the column is connected to a nitrogen cylinder. Nitrogen is passed through the cheese column and through the adsorption tubes. The flavour-containing sample is then transferred from the tubes into the gas chromatograph.

Twenty-three alkylpyrazines in roasted cocoa [94]. The essential oil is obtained from a commercial cocoa powder by co-distillation with ethanol and subsequent extraction of the water-diluted distillate with pentane.

1,4,5,6-Tetrahydro-2-acetylpyridine in bread aroma [39]. A methylene chloride extract of freshly baked bread is gently shaken with 100 ml of 5*N* aqueous solution of sodium bisulphite.

Alkylpyrazines, dihydrocyclopentapyrazines, and alkylacetylpyridines in cooked beef [62]. Flavour concentrates are isolated from beef cooked at elevated pressure at 162°C by simultaneous steam distillation and continuous solvent extraction.

Alkylpyrazines in cocoa beans [77]. Cocoa beans and Celite are pulverized and the mixture eluted on a chromatographic column with diethyl ether. The pyrazines are extracted from the eluate with an aqueous solution of sodium chloride and hydrochloric acid. The pyrazines are liberated by addition of potash and extraction with methylene chloride.

Alkylpyrazines, cyclopentapyrazines, and quinoxalines in roasted filberts (*Corylus avellana*) [49]. Uncondensed volatiles obtained during the steam distillation at atmospheric pressure are passed into a stainless steel tube packed with Porapak Q.

2-Acetyl-2-thiazoline in beef broth [87]. Preparation of a flavour concentrate by continuous extraction of clear beef broth with diethyl ether.

3-Acetylthiazoline and 2,4,5-trimethyl-3-oxazoline in roasted lean ground round beef [75]. Simultaneous steam distillation at atmospheric pressure and continuous extraction with diethyl ether in *Williams'* apparatus.

2,4,6-Trimethylperhydro-1,3,5-dithiazine and 3,5-dimethyl-1,2,4-trithiolane from beef broth [13]. A special apparatus is used for the isolation of the head-space volatiles. The materials are condensed in three cooled traps.

Alkylthiazoles, benzothiazole, and alkylpyrazines in *Tamarindus indica* [55]. Steam distillation under reduced pressure in a *Büchi* rotary evaporator. The distillate is collected in traps cooled with liquid nitrogen.

Oxazoles and thiazoles in coffee aroma [92]. After steam distillation, the condensate and the uncondensed volatiles are passed through a glass tube packed with Porapak Q.

Thiazoles and pyrazines in volatile flavours of baked potatoes [23]. Direct removal and subsequent condensation of the volatile flavour materials in the headspace.

Chemical extraction:

Alkylpyrazines in grilled meat flavour [29]. The basic components of the pyrolysate are extracted with 5% hydrochloric acid and liberated by addition of dilute sodium hydroxide, extracted with diethyl ether, and distilled at 0.2 torr.

Alkyldihydro-6,7-(5*H*)-cyclopenta[*b*]pyrazines in grilled meat flavour [30]. The dried extract of water-soluble substances from the meat is heated to 160–170°C/10^{-3} torr and the volatile products collected in a series of traps cooled with ice, dry ice, and liquid nitrogen. This is followed by the separation of basic components.

VI–3 IDENTIFICATION. DETERMINATION. CHARACTERIZATION

3.1 General

There are three principal objectives to be discussed here:

(1) Identification of a known heterocyclic compound in a more or less complex mixture, i.e., a compound whose properties are well described.

(2) Isolation of an apparently pure substance whose properties do not match the properties of compounds already described.

(3) Quantitative determination of a heterocyclic compound isolated from a complex mixture or still present in a more or less complex mixture.

The practical approaches are obviously different in the three different situations described above, although often analogous methods are used. All of them possess at least some characteristics which make them quite often some of the most advanced methods used today. These methods, in principle, do not represent specific methods for heterocyclic molecules. Thus, one can find a source of inspiration among all the various groups of organic compounds and in various media.

The methods utilized for the three above purposes are as follows:

(1) Organoleptic methods which rely on the olfactory and/or taste-producing properties (e.g., the bitter taste) [70].

(2) Physical methods, with a whole arsenal of classical old methods or the modern spectroscopic methods, as well as the chromatographic techniques [26].

(3) Chemical methods for functional groups and structural determinations, generally well known, as well as the biochemical methods which are somewhat less developed in this particular field.

However, all these studies require an essential condition to be met, i.e., the availability of reference compounds and standards, and data banks as complete and up to date as possible. Unfortunately, the data banks are still quite a rarity in the case of heterocyclic compounds, and information concerning their physical properties, spectral data, and chromatographic characteristics is often either scattered in the literature, or incomplete and uncertain.

3.2 Identification and determination

The identification and determination of aromas often require a preliminary isolation of the compound in question, in more or less pure state, but in some cases their identification and determination are possible in rather complex mixtures.

Both the identification which may be of a somewhat more general nature and the characterization of heterocyclic aromas will be discussed here.

3.2.1 Identification

Usually the use of banks of bibliographical data (physical constants) or the actual reference samples provide sufficient information necessary for the identification of the heterocyclic compound under study whose characteristics are known and well established. The organoleptic methods of identification, as well as the physical, chemical, and biochemical methods are used.

3.2.1.1 Organoleptic methods. Most heterocyclic compounds present in naturally occurring aromas are odour-producing or taste-producing, although many of them are not extremely powerful. However, some of the more exceptional compounds possess bitter, burnt, or sweet taste. As far as these compounds are concerned, heterocyclic compounds possessing sweet taste (oses, osides) and bitter taste (lactones) and present in naturally occurring materials have been discussed in Chapter II.

The problem of correlations between the structure and organoleptic properties should be approached very carefully. Here, we shall discuss only some general principles and a few practical examples.

First of all, it is possible to say that each of the individual groups of heterocyclic compounds (oxygen, sulphur, and nitrogen) as well as the mixed heterocyclic compounds with oxygen, sulphur and/or nitrogen combined in the same molecule possesses certain odour-producing properties. It is almost certain that the presence of various functions and functional groups which are normally odoriferous, plays a predominant role. This significance is even more pronounced in the case of the intensity of the odour and odour threshold values.

For example, in the case of nitrogen heterocyclic compounds a replacement of the pyridine ring with a pyrazine ring noticeably enhances the odour character and its strength. Odoriferous intensity is also enhanced by alkyl and/or alkoxy substituents. Di- and trisubstituted pyrazines are very strongly odoriferous. Replacement of an alkyl group by an alkoxy group or a thioalkyl group causes a slight modification of the odour character (for more details on this topic, see the papers by Takken [84] and Calabretta [19]). All methoxypyrazines possess green odour. Isobutyl- and isopropyl substituted pyrazines are much more odoriferous than the corresponding di- and trialkyl derivatives.

As an example of heterocyclic compounds with several different heteroatoms, one can mention thiazoles whose odour character is described as green, nutty, and vegetable-like [53].

Another factor to be considered is the odour and/or taste threshold value [59] of an aromatic heterocyclic compound. This value can be affected by isomerism or its place in the homologous series. The threshold values are different in water and in oil. It seems worth mentioning that the threshold values reported in the literature should be considered carefully, because their estimation is a very delicate task often affected by various factors. Some of these threshold values for oxygen, sulphur, and nitrogen heterocyclic compounds are presented in *Tables VI–1 and VI–2*.

The concept of odour units (U_o) was introduced in 1966 by Guadagni *et al.* [18b]. It is of value for heterocyclic compounds which are often present in aromas in minute traces, but from the organoleptic point of view they represent the character impact constituents. These values are obtained by dividing the concentration of the substance in the medium expressed as F_c, by its threshold value T_c (in parts per 10^9 parts of water). For example, the values found for

2-pentylfuran and furfural identified among the volatile aroma compounds of cooked artichoke are shown below [18b]:

Compound	T_c (parts per 10^{-9} parts of water)	F_c (relative % of whole oil)	$U_o \times 10^{-6}$
Whole artichoke oil	0.6	100	1690
2-Pentylfuran	6	0.9	1.5
Furfural	3000	2	0.007

Table VI–1 Odour threshold values of oxygen- and sulphur-containing Heterocyclic compounds [7]

Compound	Threshold value[a)]	Compound	Threshold value[a)]
2-Methylfuran	3 500	2-[(Ethyldithio)-methyl]-furan	0.04
2-Ethylfuran	8 000	3-Methyl-2-butenylpyran (rose oxide)	0.5
2-Propylfuran	6 000	2-Methyl-3-hydroxy-pyranone (maltol)	35 000
2-Pentylfuran	1 500	4,4-Dimethyl-1,3-dioxane	2 500
1,3,3-Trimethyl-2-oxabi-cyclo-[2.2.2]octane (1,8-cineole)	12	3,4-Dimethylthiophene	1.3
2-Vinylfuran	1 000	5-Methyl-2-formyl-thiophene	1
2-Hydroxymethylfuran	3 000		
2-Formylfuran	6	3,5-Dimethyl-1,2,4-tri-thiolane	10
5-Methyl-2-formylfuran	10 000		
5-Hydroxymethyl-2-formyl-furan	200 000		
2-Acetylfuran	110 000	Lenthionine	400
2-Hydroxyacetylfuran	200 000		
2,5-Dimethyl-4-hydroxy-3(2*H*)-furanone	0.04		
2,5-Dimethyl-4-methoxy-3(2*H*)-furanone	0.03		

a) In μg/litre of water (ppb).

Table VI–2 Odour threshold values of some heteroaromatic bases [7]

Compound	Threshold value[a)]	Compound	Threshold value[a)]
5-Methyl-2-formylpyrrole	110 000	Trimethylpyrazine	400
2-Formyl-*N*-ethylpyrrole	2 000	2-Ethyl-3-methylpyrazine	130
2-Acetylpyrrole	200 000	2-Ethyl-5-methylpyrazine	100
		Tetramethylpyrazine	1 000
2-Isobutylthiazole	2	2-Ethyl-3,5-dimethylpyrazine	15 000
4-Butyl-5-propylthiazole	0.003		
5-Acetylthiazole	10	2-Ethyl-3,6-dimethylpyrazine	0.4 and 43 000
2-Acetyl-2-thiazoline	1.3	2,5-Diethylpyrazine	20
2-Methylthiobenzothiazole	5	2,6-Diethylpyrazine	6
Pyridine	30	Pentylpyrazine	1 000
2-Methyl-5-ethylpyridine	19 000	2-Isobutyl-3-methylpyrazine	35
2-Methyl-5-vinylpyridine	40	Methoxypyrazine	700
2-Pentylpyridine	0.6	2-Methoxymethylpyrazine	150
2-Acetylpyridine	19	2-Methyl-3-methoxypyrazine	4
		2-Methyl-5-methoxypyrazine	3
Pyrazine	500 000	2-Ethyl-3-methoxypyrazine	0.4
2-Methylpyrazine	100 000	2-Propyl-3-methoxypyrazine	0.006 and 10
2-Ethylpyrazine	60 000 and 22 000		
2,3-Dimethylpyrazine	400	2-Isopropyl-3-methoxy-pyrazine	0.002
2,5-Dimethylpyrazine	1 800		
2,6-Dimethylpyrazine	1 500 and 54 000	2-Isopropyl-5-methoxy-pyrazine	10
		2-Isobutyl-3-methoxy-pyrazine	0.002
2-Isobutyl-5-methoxy-pyrazine	10	2-Methyl-3-thiomethoxy-pyrazine	1
2-Hexyl-3-methoxypyrazine	0.001	2-Methyl-5-thiomethoxy-pyrazine	4
2-Isobutyl-3-methoxy-5-methylpyrazine	0.3	2-Thiomethoxymethyl-pyrazine	20
2-Isobutyl-3-methoxy-6-methylpyrazine	2.6	2-Methyl-3-thiofurfuryl-pyrazine	<1
2-Isobutyl-3-methoxy-5,6-dimethylpyrazine	315	2-Methyl-5-thiofurfuryl-pyrazine	<1
2-Ethyl-3-ethoxypyrazine	11	2-Thiofurfurylmethoxypyrazine	<1
		2-Acetylpyrazine	62

a) In μg/litre of water (ppb).

3.2.1.2 Physical methods. The methods employed in the analysis of food aromas are not specifically designed for heterocyclic compounds; they are physicochemical and spectroscopic methods used in organic chemistry in general. These techniques often require technically advanced instrumentation and a very sophisticated infrastructure [46].

Classical physicochemical methods. These commonly used techniques, although known for a long time, have retained their importance. They include the determination of specific weight (density), refractive index, and specific rotation. All these techniques are well known and there is no need to discuss them in detail.

Spectroscopic methods. The most widely used techniques at the present time are the ultraviolet spectroscopy, infrared spectroscopy, nuclear magnetic resonance (NMR), and mass spectrometry coupled with gas chromatography (GC–MS). X-ray diffraction permits the determination of absolute configuration of crystalline compounds. Katritzky's *Physical Methods in Heterocyclic Chemistry* is one of the recent most comprehensive treatises on this topic [48].

Ultraviolet spectroscopy [35]. The ultraviolet spectroscopy permits the identification of characteristic chromophores (with delocalized electrons) and condensed ring systems (e.g., indoles, quinolines, benzothiazoles, etc.). In some cases, the differences in the experimental absorption curves (extinction coefficient, ϵ, plotted against the wavelength, λ, in nm) make it possible to identify the class of a heterocyclic system to which a particular compound belongs. However, the principal advantage of this technique is its use for the detection and quantitative determination of heterocyclic compounds separated by HPLC (high-performance liquid chromatography). As an example, the absorption maxima and the respective intensities for several parent heterocyclic systems are presented in *Table VI–3.*

Table VI–3 Electronic absorption spectra of some parent heterocyclic systems [103]

Compound	λ_{max}, nm	ϵ	Solvent
Furan	207	9100	Cyclohexane
Thiophene	231	7100	Cyclohexane
Pyrrole	208	7700	Hexane
Oxazole	205	4100	Water
Thiazole	209	2750	Heptane
	232	3550	
Imidazole	206	4800	Water
Pyridine	198	6000	Hexane
	251	2000	
	270	450	Cyclohexane
Pyrazine	194	6100	Hexane
	260	6000	
	328	1040	Cyclohexane

Pyrrole and furan exhibit very similar ultraviolet absorption spectra, whereas a bathochromic shift is observed in the spectrum of thiophene. Alkyl and alkoxy groups do not exhibit any pronounced effect upon these spectra. Substituents conjugated with the heterocyclic ring (e.g., a formyl or acetyl group) cause a significant bathochromic shift and hyperchromic effect. The electronic spectra of six-membered ring systems (pyridine, pyrazine, etc.) resemble the spectrum of benzene, but they also exhibit a weak absorption band corresponding to an $n \rightarrow \pi^*$ transition. A number of researchers have used ultraviolet spectroscopy to determine the position of alkyl groups present as substituents on the pyrazine ring.

Infrared spectroscopy [22,32,94]. Infrared spectroscopy represents one of the methods which make it possible to identify the functional groups (OH, NH=, NH_2, C=O, COOH, C—O—C, C=C, C=N—, aromatic rings, etc.) present in a compound. Identification of these group frequencies in an infrared spectrum often narrows down the number of possible structures to a few systems whose spectra can be quickly found by using an empirical formula index of standard collections of infrared spectra. High-resolution instruments, fast response, and scale expansion are now commonly used. The instruments can be coupled directly to gas chromatographs. Carbon tetrachloride and carbon disulphide can be conveniently used as eluents for heterocyclic compounds trapped after gas chromatography. A variety of microcells are commercially available which are adequate for most analyses. Infrared spectra can be obtained with 0.1 μl of materials purified by gas chromatography. Potassium bromide crystals can also be used as the carrier. The substance after gas chromatography is then condensed directly on the crystals which are pressed to form a clear pellet which can be easily handled and placed in the infrared spectrometer. Infrared spectra of pyrazines found in food aromas have been reported [22,32,94]. They are characterized by four stretching vibrations in the 1600–1370 cm^{-1} region, with an absorption of variable intensity between 1600 and 1575 cm^{-1}. The 1550–1520 cm^{-1} band is medium weak, and the bands at 1500–1465 cm^{-1} and 1420–1370 cm^{-1} are moderately strong bands.

^1H and ^{13}C nuclear magnetic resonance (NMR). ^1H and ^{13}C-NMR have proved to be very valuable tools for the elucidation of heterocyclic structures. ^1H-NMR spectra show the number and the nature of protons as well as their relative positions in the molecule. Proton chemical shifts (in δ, ppm/TMS) in the α-, β-, and γ-positions of some five- and six-membered heterocyclic compounds are summarized in *Fig. 1*.

It can be seen from these values that the deshielding effect of protons in the α-position with respect to heteroatoms is decreasing in the order furan > thiophene > pyrrole. In the case of the β-protons, the order is furan > pyrrole > thiophene. Variations of the chemical shifts of aromatic protons in the furan, thiophene, thiazole and cycloalkylpyrazine series are presented in *Tables VI–4, VI–5 and VI–6*.

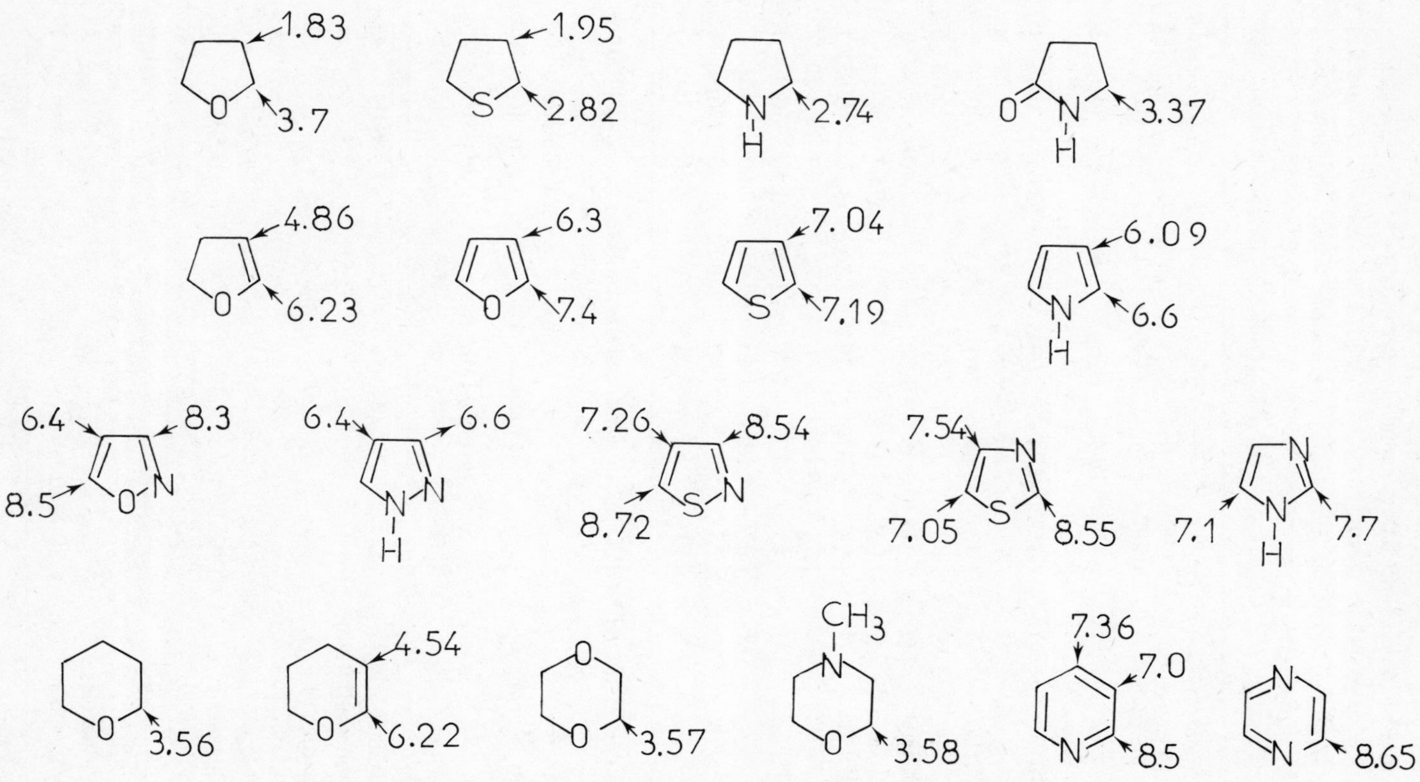

Fig. 1 – Proton chemical shifts in the α-, β-, and γ-positions of some five- and six-membered ring heterocyclic compounds [4,44].

Table VI–4 ¹H-NMR data for some 2-substituted furans (X = O) and thiophenes (X = S) [33]

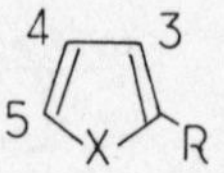

Substituent, R	X	Chemical shifts δ(ppm/TMS, H-3	CD₃COCD₃) H-4	H-5	H-R	Coupling constants J (Hz) $J_{3,4}$	$J_{3,5}$	$J_{4,5}$
CHO	O	7.45	6.74	7.94	7.72	3.45	0.83	1.73
	S	7.93	7.30	7.96	9.98	3.79	1.22	4.75
$COCH_3$	O	7.32	6.65	7.81	2.41	3.58	0.76	1.74
	S	7.80	7.17	7.80	2.52	3.74	1.09	5.07
COOH	O	7.24	6.59	7.76	–	3.53	0.88	1.72
	S	7.80	7.15	7.78	–	3.67	1.16	5.03
$COOCH_3$	O	7.23	6.61	7.79	3.86	3.50	0.83	1.74
	S	7.74	7.09	7.66	3.85	3.69	1.20	4.97
SCH_3	O	6.43	6.39	7.75	2.38	3.26	0.88	1.98
	S	7.07	6.96	7.40	2.45	3.60	1.26	5.36
CH_2OH	O	6.25	6.31	7.40	6.90	3.30	0.86	1.80
	S	6.91	6.88	7.21	4.67	3.40	1.11	5.14

Table VI–5 ¹H-NMR data for mono- and disubstituted thiazoles [100]

Compound	Chemical shifts δ(ppm/TMS) H-2	H-4	H-5	Coupling constant J (Hz)	Solvent
2-Alkylthiazoles	–	7.44–7.58	7.0–7.1	3.15±0.05	$CDCl_3$
2-Hydroxymethylthiazole	–	7.76	7.51	–	Acetone-d_6
2-Methoxythiazole	–	7.05	6.62	–	CCl_4
2-Methylthiothiazole	–	7.48	7.03	–	CCl_4
4-Methylthiazole	8.5	–	6.74	1.90	CCl_4
5-Methylthiazole	8.38	7.39	–	0.50	CCl_4
5-Methoxythiazole	8.35	7.23	–	–	Acetone-d_6
5-Methylthiothiazole	9.08	7.88	–	–	Acetone-d_6
2,4-Dialkylthiazoles	–	–	6.50–6.55	–	CCl_4
2,5-Dialkylthiazoles	–	7.03–7.16	–	–	CCl_4
4,5-Dialkylthiazoles	8.20–8.30	–	–	–	CCl_4

Table VI–6 ^{1}H-NMR data of some alkylpyrazines [49b] and cycloalkylpyrazines [93b]

Compounds	δ (ppm/TMS), CCl_4
2,3-Dimethyl-5-n-pentylpyrazine	8.06 (s, 1H); 2.62 (t, 2H); 2.4 (s, 6H); 1.85–1.45 (m, 2H); 1.45–1.15 (m, 4H); 0.85 (t, 3H, J=6 Hz)
2,3-Dimethyl-5-isopentylpyrazine	8.18 (s, 1H); 2.67 (t, 2H); 2.45 (s, 6H); 1.8–1.3 (m, 3H); 0.95 (d, 6H, J=5,7 Hz)
2,3-Dimethyl-5-(2-methylbutyl)-pyrazine	8.07 (s, 1H); 1.7–1.5 (m, 2H); 2.45 (s, 6H); 2.1–1.7 (m, 1H); 1.4–1.1 (m, 2H); 0.88 (t, 3H); 0.85 (d, 3H, J=7,5 Hz)
2,3-Dimethyl-5-neopentylpyrazine	8.08 (s, 1H); 2.57 (s, 2H); 2.46 (s, 6H); 0.93 (s, 9H)
2,3-Dimethyl-5-(1,2-dimethylpropyl)-pyrazine	8.08 (s, 1H); 2.7–2.3 (m, 1H); 2.45 (s, 6H); 2.2–1.7 (m, 1H); 1.23 (d, 3H, J=7.2 Hz); 0.93 (d, 3H, J=7.2 Hz); 0.75 (d, 3H, J=7.2 Hz)
2,3-Dimethyl-5-(1-methylbutyl)-pyrazine	8.08 (s, 1H); 3.0–2.6 (m, 1H); 2.43 (s, 6H); 1.8–1.4 (m, 2H); 1.23 (d, 6H,J=7.5 Hz); 1.3–1.1 (m, 3H); 0.85 (t, 2H)
2,3-Dimethyl-5-(1-ethylpropyl)-pyrazine	8.04 (s, 1H); 2.43 (s, 6H); 2.6–2.2 (m, 2H); 1.8–1.4 (m, 2H); 0.93 (t, 6H, J=7.2 Hz)
2-Isobutyl-3,5,6-trimethylpyrazine	2.56 (d, 2H, J=6.9 Hz); 2.41 (s, 9H); 2.4–1.9 (m, 1H); 0.92 (d, 6H, J=7.2 Hz)
2-(2-Methylbutyl)-3,5,6-trimethyl-pyrazine	2.8–2.4 (m, 2H); 2.43 (s, 9H); 2.1–1.6 (m, 1H); 1.5–1.1 (m, 2H); 0.87 (t, 3H); 0.83 (d, 3H, J=7.2 Hz)
2,3-Dimethyl-5-(1,5-dimethyl-4-hexenyl)-pyrazine	8.08 (s, 1H); 5.1–4.9 (m, 1H); 2.9–2.6 (m, 1H); 2.45 (s, 6H); 1.47–1.18 (m, 1H); 1.93–1.47 (m, 4H); 1.57 (d, 6H, J=12 Hz); 1.23 (d, 3H, J=7.2 Hz)
2,3-Dimethyl-5-(3,7-dimethyl-7-heptenyl)-pyrazine	8.11 (s, 1H); 5.08 (t, 1H); 2.67 (t, 2H); 2.45 (s, 6H); 2.1–1.9 (m, 1H); 1.8–1.2 (m, 6H); 1.63 (d, 6H, J=6 Hz); 0.95 (d, 3H, J=5.4 Hz)
6,7-Dihydro-(5*H*)-cyclopentapyrazine	8.10 (*s*, 2H); 2.97 (*t*, 4H); 2.12 (*quint*, 2H)

Table VI–6 (*continued*)

Compounds	δ (ppm/TMS), CCl_4
2-Methyl-(5*H*)-cyclopentapyrazine	7.94 (*s*, 1H); 2.90 (*t*, 4H); 2.40 (*s*, 3H); 2.10 (*quint*, 2H)
5-Methyl-(5*H*)-cyclopentapyrazine	3.40–2.80 (*m*, 3H); 2.70–2.12 (*m*, 1H); 2.0–1.55 (*m*, 1H); 1.35 (*d*, 3H)
5,7-Dimethyl-(5*H*)-cyclopentapyrazine	6.16 (*s*, 2H); 3.5–1.65 (*m*, 4H); 1.35 and 1.29 (2*d*, 6H, two isomers)
2,3-Dimethyl-(5*H*)-cyclopentapyrazine	2.90 (*t*, 4H); 2.38 (*s*, 6H); 2.05 (*quint*, 2H)
2-Methyl-3-ethyl-(5*H*)-cyclopentapyrazine	3.1–2.48 (*m*, 6H); 2.42 (*s*, 3H); 2.10 (*m*, 2H); 1.24 (*t*, 3H)
2,3,5-Trimethyl-(5*H*)-cyclopentapyrazine	2.5–2.0 (*m*, 1H); 2.0–1.50 (*m*, 1H); 1.30 (*d*, 3H)
2,5,7-Trimethyl-(5*H*)-cyclopentapyrazine	8.08 (*s*, 1H); 2.30 (*s*, 3H); 3.65–1.85 (*m*, 4H); 1.30 and 1.25 (2*d*, 6H, two isomers)
5,6,7,8-Tetrahydroquinoxaline	8.16 (*s*, 2H); 3.0–2.76 (*m*, 4H); 2.03–1.80 (*m*, 4H)
2-Methyltetrahydroquinoxaline	8.02 (*s*, 1H); 2.95 (*m*, 4H); 2.41 (*s*, 3H); 2.0–1.75 (*m*, 4H)
2-Ethyltetrahydroquinoxaline	8.05 (*s*, 1H); 2.98–2.75 (*m*, 4H); 2.72 (*q*, 2H); 2.0–1.75 (*m*, 4H); 1.30 (*t*, 3H)
2,3-Dimethyltetrahydroquinoxaline	2.92–2.70 (*m*, 4H); 2.42 (*s*, 6H); 2.05–1.78 (*m*, 4H)

^{1}H-NMR spectra also enable us to identify alkyl and alkenyl groups in various environments. Their chemical shifts are not affected by the structures of the respective heterocyclic compounds to any considerable extent.

Modern, high-resolution (360 MHz) spectrometers make it possible to obtain spectra of good quality using only a few micrograms of the substance (*Fourier* transform) [25].

More detailed information about the structure of a compound (the number and nature of the carbon atoms present in the molecule) can be obtained from ^{13}C-NMR spectra. Another important application of the ^{13}C-NMR spectroscopy for food analysis is the direct use of the sample of a food product as such. This approach allows one to obtain information about the nature of the components present in the sample which can be later separated by using a suitable procedure.

Mass spectrometry. Mass spectrometry of heterocyclic compounds will be discussed in *Chapter VII*.

Chromatographic methods. The use of chromatographic techniques (HPLC, HPTLC, GC) [46] for identification of heterocyclic compounds is so important that it is now impossible to carry out any study in this field without using one or several chromatographic methods. Unfortunately, these procedures can occasionally lead to changes in the original composition or to destruction of some materials.

High-performance chromatography. In high-performance thin-layer chromatography (HPTLC), heterocyclic compounds are separated according to the distance travelled by capillary action along a coated plate. In high-performance liquid chromatography (HPLC), the compounds are separated according to the time required for a diluted solution of a component to travel in a pumped solvent stream through a column packed with small particles with a large surface area. Pressures up to and sometimes exceeding 6000 psi are used. Column elution times or capacity factors (k') are the specific characteristics of the individual components of a mixture under a given set of conditions used in the separation, such as the solvent, the packing material, size of its particles, its type, other components present in the mixture, etc. Microparticulate analytical columns are available in lengths from 100 to 500 mm, with mean particle sizes 5 to 10 μm. The usual columns are 250 mm long with an inner diameter of 4.6 mm. The superior resolution capability of HPLC as compared with HPTLC can be shown by the fact that the individual TLC spots can often be resolved into multiple peaks by HPLC.

In adsorption chromatography on silica gel, gradient elution techniques have been successfully employed.

The main advantage of reversed-phase *vs*. adsorption chromatography is the simplicity of the mobile phase. In many cases, a mixture of water and methanol will lead to satisfactory separation. Water, methanol, acetone, chloroform, and *n*-heptane form an eluotropic series of solvents. Also, the aqueous component of the mobile phase can be replaced with an organic component (methanol/tetrahydrofuran or acetonitrile/tetrahydrofuran).

HPLC is especially well suited for rapid analysis of the various components in foods and beverages.† However, only a few reports on the separation of volatile heterocyclic compounds have been published in the literature. The technique has been recently used to separate the heterocyclic compounds (thiophenes, pyrazines, pyrroles, thiazoles, furans, and imidazoles) from a mixture obtained by heating an *L*-rhamnose-hydrogen sulphide-ammonia model system [103]. The separation was carried out on a Lichrosorb Si-100, 10 μm column using gradient elution from the mobile phase A (hexane-methylene chloride, 98:2) to the mobile phase B (ethanol-hexane-methylene chloride, 78:20:2). The order of elution, furan $\simeq$ thiophene, pyrrole, thiazole, oxazole, pyrazine, and imidazole, suggests that the separation of heteroaromatic bases depends on the electron density on the nitrogen atom in the heterocyclic ring, i.e., on its basicity [$k' = f(pK_a)$]. However, the elution times can also be influenced by steric effects of the alkyl groups in positions adjacent to the nitrogen atom. An increase in the number of alkyl groups decreases the polarity of the molecule and thus reduces the elution time.

Capillary gas chromatography (CGC). Since 1963, many heterocyclic components of cooked foods have been studied and characterized by CGC [91b]. This technique represents one of the very powerful methods useful in the analysis of aromas. These analyses are usually carried out on 50 m $\times$ 0.25 mm (or 0.5 mm, inner diameter) coated open tubular (WCOT) glass columns (0.2 to 0.4 μm film thickness). The efficiency of such columns is of the order of 3000–6000 plates/m. Stainless steel columns are not suitable for some analyses, such as, e.g., higher-boiling sulphur compounds, probably because of adsorption on stainless steel. On the other hand, some compounds, such as furaneol, do not pass through the column when only a small amount is present. Numerous liquid phases (OV-1, OV-101, SE-52, SF-96, squalane, DEGS, Carbowax 20M, PPE) are now available.

Because of the complexity of mixtures to be analysed (several hundred components in some cases), some analyses must be carried out on several different columns with different polarity. However, in most cases it is preferable to do a preliminary separation of the mixture either by preparative chromatography (LC or TLC) or by steam distillation, first in an acidic medium, and then in a basic medium, in order to separate the heteroaromatic bases. This also facilitates the interpretation of the results.

The use of coupling techniques in combination with other physical methods (TLC, ultraviolet or infrared spectroscopy, mass spectrometry) is always necessary. With flame ionization detectors (FID), a splitter is required. Selective gas chromatographic detectors represent a valuable aid for rapid identification and determination of nitrogen- and sulphur-containing heterocyclic compounds of interest

† For more details on this topic see *Liquid Chromatographic Analysis of Food and Beverages* (G. Charalambous, Ed.), Academic Press, New York (1979).

in complex mixtures. In the case of sulphur compounds, the flame photometric detector (FPD) is by far the easiest and the most reliable one. Its sensitivity is about 10^{-10} g (S) sec^{-1}. The detectors most commonly used for nitrogen heterocyclic compounds are the alkali flame ionization detector (AFID or thermionic) whose normal sensitivity is of the order of 10^{-13} g (N) sec^{-1}, and the electrolytic conductivity detector (ECD). These detectors are used in addition to the usual flame ionization detector (FID, sensitivity $\sim 10^{-12}$ g (C) sec^{-1}) and a sensory detector such as a particularly perceptive expert human nose.

For the assignment of chemical structures, retention indices are used in conjunction with mass spectral data.†† The *Kováts* indices I_K values (with respect to alkanes) or the I_E values (with respect to ethyl esters of *n*-alkanoic acids) in programmed temperature gas chromatography have been reported for furans, thiophenes, pyrroles, thiazoles, oxazoles, γ- and δ-lactones, pyridines, pyrazines, indoles, etc., on Carbowax 20M and SF-96 capillary columns [49,62,63, 96]. In most cases, these values were quite close to those obtained on standard packed columns [96]. Retention indices of some pyrazines, thiazoles and pyrroles listed in *Tables VI–7, VI–8* and *VI–9* are determined according to the method of Van der Dool and Kratz using the following formula [36]:

$$I_E = n + \frac{(t'_R)_x - (t'_R)_n}{(t'_R)_{n+1} - (t'_R)_n}$$

where n refers to the number of carbon atoms of the acid moiety of aliphatic acid esters and t'_R, $(t'_R)_n$ and $(t'_R)_{n+1}$ are the reduced retention times ($t'_R = t_R - t_0$) of the substance x, and of the corresponding ethyl alkanoic acid esters with n and n+1 carbon atoms, respectively.

The difference between the retention index on two columns of different polarity (ΔI) constitutes valuable information on the nature of the various classes of heterocyclic compounds. Other workers used the retention values relative to a standard such as furfural [95]. But these values are not always reproducible, and for a good degree of certainty of chemical structure one must rely on chemical and physical properties.

Headspace technique. In the case of volatile materials present in the gas phase, the headspace technique can be used under certain conditions [21,36, 47]. Only the more abundant and more volatile components are present in detectable levels in the samples to be directly injected. Also, it is often necessary to pre-concentrate a sample on activated carbon. Furthermore, the composition of the mixture can change depending on the method of trapping.

†† The retention indices of about 1200 compounds including heterocyclic compounds have been reported by W. Jennings & T. Shibamoto in *Qualitative Analysis of Flavor and Fragrance Volatiles by CGC*, Academic Press, New York (1980).

Table VI–7 Retention index of some pyrazines [49,96]

Compound	I_E Values	
	a)	b)
Pyrazine	5.87 - 5.77	3.75
Methylpyrazine	6.42 - 6.41	4.38
2,5-Dimethylpyrazine	6.86	5.17
2,6-Dimethylpyrazine	6.98	–
Ethylpyrazine	7.05 - 7.06	5.21
2,3-Dimethylpyrazine	7.18 - 7.10	5.26
Isopropylpyrazine	7.33 - 7.34	–
2-Ethyl-6-methylpyrazine	7.55 - 7.53	6.06
2-Ethyl-5-methylpyrazine	7.61 - 7.57	6.07
Trimethylpyrazine	7.73 - 7.68	6.07
2-Ethyl-3-methylpyrazine	7.72 - 7.67	6.17
Methylisopropylpyrazine	7.82	–
Propylpyrazine	7.87 - 7.75	6.19
Vinylpyrazine	7.80 - 8.08	–
2-Ethyl-3,6-dimethylpyrazine	8.14 - 8.10	6.83
2-Ethyl-3,5-dimethylpyrazine	8.27	–
Methylpropylpyrazine	8.39	–
Diethylpyrazine	8.40 - 8.24	6.88
Tetramethylpyrazine	8.42 - 8.46	6.91
Isopropenylpyrazine	8.55	–
2-Methyl-6-vinylpyrazine	8.58	–
2-Methyl-5-vinylpyrazine	8.63	–
2,3-Diethyl-5-methylpyrazine	8.62 - 8.65	–
Diethylmethylpyrazine	8.87	7.64
Acetylpyrazine	9.92	6.31

a) These values were obtained on 500 ft × 0.03 in. i.d. Carbowax 20 M capillary column at programed temperature from 70° to 190°C using standards of ethyl esters of *n*-aliphatic acids.

b) On a 500 ft × 0.03 in. i.d. SF-96 open tubular column. The temperature programme was of 60–198°C at 2°C per min.

Table VI–8 Retention index of some thiazoles

Thiazoles	a)	b)
2-Methylthiazole	6.15	3.81
2-Ethylthiazole	6.83	4.87
2-Propylthiazole	7.65	5.91
2-Isopropylthiazole	6.99	5.51
2-Isobutylthiazole	7.89	6.41
2-*sec*-Butylthiazole	7.79	6.37
4,5-Dimethylthiazole	7.55	5.31
5-Ethyl-4-methylthiazole	8.17	6.19
2,4,5-Trimethylthiazole	7.60	5.94
5-Ethyl-2,4-dimethylthiazole	8.25	6.80
2,5-Diethyl-4-methylthiazole	8.85	7.64
2-Acetylthiazole	10.15	6.27
4-Acetylthiazole	11.56	6.50
4-Acetyl-2,5-dimethylthiazole	11.00	7.96
5-Acetyl-4-methylthiazole	12.46	7.62
5-Acetyl-2,4-dimethylthiazole	12.47	8.40
2-Methoxythiazole	7.18	4.80
2-Ethoxythiazole	7.74	5.60
2-Butoxythiazole	9.64	7.65
5-Methoxythiazole	9.51	5.62
5-Ethoxythiazole	9.95	6.49
5-Methoxy-2-methylthiazole	9.42	6.27
4-Isobutyl-5-methoxythiazole	10.83	8.41
4-Isobutyl-5-methoxy-2-methylthiazole	10.43	8.77
4-Isobutyl-5-ethoxythiazole	11.00	9.02
4-Isobutyl-5-ethoxy-2-methylthiazole	10.65	9.40

a) These values were obtained on 500 ft × 0.03 in. i.d. Carbowax 20 M capillary column at programmed temperature from 70° to 190°C using standards of ethyl esters of *n*-aliphatic acids [49,96].

b) On SE-30 column, reported by Pittet & Hruza, *J. Agric. Food Chem.*, **22**, 264 (1974).

Table VI–9 Retention index of some alkyl- and furfurylpyrroles [90b]

Compounds	I_K Values (Carbowax 20 M)
N-Methylpyrrole	1148
N-Ethylpyrrole	1195
N-Propylpyrrole	1245
N-Isobutylpyrrole	1265
N-(2-Methylbutyl)-pyrrole	1375
N-Isoamylpyrrole	1391
N-Acetonylpyrrole	1677
N-(2-Butanoyl)-pyrrole	1770
N-Furfurylpyrrole	1801
N-Methyl-2-methylpyrrole	1216
N-Methyl-3-methylpyrrole	1273
N-Ethyl-2-methylpyrrole	1260
N-Ethyl-3-methylpyrrole	1297
N-Methyl-2-ethylpyrrole	1303
N-Propyl-2-methylpyrrole	1314
N-Isobutyl-2-methylpyrrole	1328
N-Isobutyl-3-methylpyrrole	1368
N-(2-Methylbutyl)-2-methylpyrrole	1440
N-(2-Methylbutyl)-3-methylpyrrole	1474
N-Isoamyl-2-methylpyrrole	1457
N-Isoamyl-3-methylpyrrole	1489
N-Isobutyl-2,5-dimethylpyrrole	1420
N-(2-Methylbutyl)-2,5-dimethylpyrrole	1517
N-Isoamyl-2,5-dimethylpyrrole	1532
N-Methyl-2-formylpyrrole	1584
N-Methyl-2-acetylpyrrole	1614
N-(5-Methylfurfuryl)-pyrrole	1863
N-Furfuryl-2-methylpyrrole	1876
N-Furfuryl-3-methylpyrrole	1878
N-(3-Methylfurfuryl)-pyrrole	1888
N-(5-Methylfurfuryl)-2-methylpyrrole	1930
N-Furfuryl-2,5-dimethylpyrrole	1934
N-Furfuryl-2,3-dimethylpyrrole	1948
N-Furfuryl-2-ethyl-5-methylpyrrole	2011
N-Furfuryl-2,3,5-trimethylpyrrole	2040
N-Furfuryl-2-formylpyrrole	2212
N-(5-Methylfurfuryl)-2-formylpyrrole	2263
N-(Furfuryl-2-acetyl)-pyrrole	2232
N-(Methylfurfuryl)-2-acetylpyrrole	2270

Combinations of chromatographic and spectrometric techniques. All chromatographic techniques, when used alone, fall short in their ability to provide sufficiently accurate qualitative information. Similarly, spectroscopic (spectrometric) techniques alone are frequently inadequate when the materials to be analysed are mixtures of structurally related or different compounds, or when the compounds of interest are contaminated. Combinations of the above techniques provide much more valuable information.

Thus, a combination of gas chromatography and infrared spectroscopy makes it possible to scan and analyse hundreds of compounds in a few hours. The effluent from a gas chromatograph is passed into an infrared cell and then into a mass spectrometer. Similarly, a sample from a TLC plate, in the form of a potassium bromide disc, can be directly transferred into an infrared spectrometer (*Merck* technique).

Combinations of mass spectrometry, infrared and ultraviolet spectroscopy, and fluorescence spectroscopy with liquid chromatography. There are certain limitations:

(1) In LC/MS, the requirements for a sample transfer and solvent removal impose severe restrictions as far as the suitable instrumentation is concerned. At the present time, there is only one HPLC/MS interface on the market, and it is more expensive than the HPLC itself.

(2) The window requirements in infrared spectroscopy have generally restricted the use of an LC/IR hook-up to those wavenumbers where the compounds themselves absorb but the solvents do not.

(3) In the combination LC/on-line fluorescence spectrophotometry, the relatively small number of compounds which fluoresce limits the applicability of this technique.

Much of the current practical liquid chromatography work involves a combination of solvent delivery systems, columns, and on-line spectroscopic detectors. The present availability of digital scanning accessories for inherently stable modern single-beam LC detectors makes possible a successful use of the spectral scanning techniques on components present in minute concentrations.

A combination of LC or other separation methods and ^{13}C-NMR allows one to investigate flavour-producing compounds. This technique seems especially suited for studies of taste-producing substances.

Capillary gas chromatography coupled with mass spectral analysis is clearly the most sensitive technique used in analysis of complex mixtures at present [50b]. Contemporary mass spectrometers possess such pumping capabilities as to make possible a direct and complete introduction of the effluent from a gas chromatograph into a mass spectrometer without any expected resolution loss ($\sim$ 15 000). When an effluent is observed as a peak on a GC/MS chromatogram, or an oscilloscope screen (a combination of both is also possible), its complete mass spectrum is recorded on a high-speed oscillograph, a magnetic tape, or a photographic plate. The spectra are numbered, and these numbers are written on the GC/MS chromatogram while the mass spectrum is being recorded (4000 spectra per analysis can now be registered).

The instruments are equipped with a chemical ionizing source (methane, ammonia, etc.). The ions formed transfer their energy to the molecules with which they collide, thus causing their fragmentation. This also makes it possible to determine the molecular weight of the respective compounds. The results are compared with the standard collections of mass spectral data which provide a convenient and quick means of identification. Compilations of mass spectral data on certain types of compounds, e.g., the aromas (data on about 5000 compounds) considerably augment the efficiency of the method.

Specific structural information can be obtained from computerized GC/MS analysis. Computer-generated output from a GC/MS analysis of a complex mixture provides plots for certain selected masses. This type of analysis is especially helpful for the differentiation between specific classes of components from a more general analysis, as shown on an example below.

Class of compounds	Specific ion plots (m/z)
Acylthiophenes	111 → [2-thienyl–CO]$^{+\bullet}$
Acylfurans	95 → [2-furyl–CO]$^{+\bullet}$
Alkylfurans	81 → [2-furyl–CH_2]$^{+\bullet}$
Alkylmercaptans and sulphides	47 → [SCH_3]$^{+\bullet}$

However, the identification of a particular heterocyclic compound cannot be considered definitive unless its mass spectrum and its retention time (index) are identical with those of a reference sample.

Identification by physical methods – examples

Oxygen-containing heterocyclic compounds. Furoates and acetylfuran in brandy [36]. Headspace technique was used by passing 1000 l of nitrogen through 2 l of brandy. The compounds were trapped on column A (Tenax GC), then back-flushed at 200°C into column B (Tenax GC), and repeatedly back-flushed at 200°C by helium with trapping at liquid nitrogen temperature. Capillary column gas chromatography coupled with mass spectrometry followed.

2-Ethoxymethylfuran and ethyl 2-furoate in the aroma of red wine. The headspace volume was 2 l and the aroma complex was collected on Tenax GC (60–80 mesh). The GC/MS analysis was carried out with a 50 ft × 0.02 in. i.d. Carbowax 20M column.

Sulphur-containing heterocyclic compounds. Mono- and polyacrylic thiophene derivatives in shade oil [79]. An enriched fraction of sulphur compounds was separated on a glass column (20 m × 0.3 mm i.d.) coated with SE–52, with the temperature programmed from 40°C to 60°C at 10°C/min and then from 60°C to 250°C at 2°C/min. Hydrogen served as the carrier gas (100 cm/sec) and an FPD detector was used.

4-Methylthio-1,2-dithiolane and 5-methylthio-1,2,3-trithiolane in algae (*Chara globularis*) [6]. The green algae were extracted using a *Likens-Nickerson* apparatus with pentane, and the extract was submitted to steam distillation. Analyses were carried out by coupled GC/MS followed by the measurement of ultraviolet, infrared, ^{1}H-NMR, and mass spectra recorded for identification purposes.

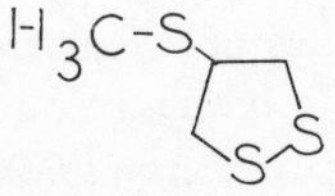

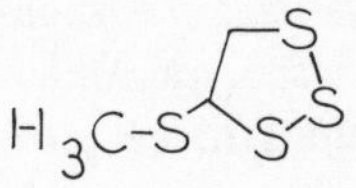

Nitrogen-containing heterocyclic compounds. Phenanthridine and 1,2,7,8-dibenzacridine in rue oil [54]. A capillary glass column (20 m × 0.3 mm i.d.) coated with SF–2340 was used. The temperature was programmed from 100°C to 250°C at 4°C/min, and an NPD detector was employed.

Pyrazines in roasted peanuts [45]. Retention indices (I_E values) were determined by the method of *Van der Dool* and *Kratz* on a glass capillary column (500 ft × 0.03 in. i.d.) packed with Carbowax 20M.

2-Acetyl-2-thiazoline [87]. The pentane extract of beef broth, after removal of free acids and fats, was separated by preparative chromatography on a Diatoport S column (60/80 mesh, 1 m × 0.3 mm i.d., 20% Carbowax 20M). Condensation was accomplished in a cooled trap packed with 60/80 mesh Diatoport S coated with 10% Apiezon L and 1% Carbowax 20M. The trap preceded a second GC column (3 m × 0.3 cm i.d.) packed with 10% Apiezon L and 1% Carbowax 20M on 60/80 mesh Diatoport S. The material corresponding to the peak from the first column was separated into two major components and several minor impurities. The aroma was condensed in a cooled trap packed with purified 60/80 mesh Diatoport S. The trapped material was used for the structure elucidation by infrared spectroscopy (trap connected to KBr and then to the infrared spectrometer) and by mass spectrometry (trap connected to the ionization source of an AEI MS–9 mass spectrometer).

3.2.1.3 Chemical and biochemical methods. The well-known chemical methods have been in use for a relatively long time and are based on the presence of various functional groups in heterocyclic compounds under study (acids, phenols, lactones, bases, carbonyl compounds, etc.). One can use various colour tests or fluorescence [8], precipitation of crystalline derivatives (e.g., indole picrate), etc. This is a very large area and, because the methods are well known and documented, they will not be discussed here. The biochemical methods have their limitations; however, the biochemical methods used for homocyclic series of compounds can be used for heterocyclic compounds as well.

3.2.1.4 Combination of physical and chemical methods. Use of derivatives [10,52,53,57,72]. Conversion of compounds into their derivatives based on the presence of various functional groups in the original compounds carried out before or after chromatographic separation can considerably augment the potential of chromatographic methods and facilitate the resolution of complex mixtures by increasing the specificity, by modification of chromatographic properties of the system (e.g., by lowering the polarity of certain molecules), or by improved resolution when chromatography is followed by mass spectrometry. Conversion into derivatives before or after chromatography has its advantages and disadvantages. Although the techniques using chromatography in combination with a spectroscopic or spectrometric method and using derivatives of the compounds to be separated and identified are relatively new, a considerable number of papers have been devoted to this topic. Although these techniques have not been explored in much detail, the number of papers simply reflects the fact that the area of potential applications is a vast one, and that any of the available chromatographic techniques can be used. Most of the bibliographic references are devoted to different types of organic compounds, whereas the number of papers devoted to heterocyclic compounds is relatively low. These methods can be used for any type of organic compounds.

Thin-layer chromatography (TLC) and high-performance thin-layer chromatography (HPTLC) [51]. One of the most remarkable advantages of TLC and HPTLC in comparison to all other chromatographic techniques is the possibility of simple derivatization of compounds directly in the thin layer.

Dinitrophenyl derivatives are often used to separate and analyse compounds containing a hydroxy, mercapto, amino, or carbonyl group. The derivatives can be obtained before or after TLC separation. Aldehydes, ketones, and sugars can be derivatized to form *p*-nitrobenzyloxime (PNBO) derivatives which absorb in the ultraviolet region.

Liquid chromatography (LC). The popularity of chemical derivatization approach in liquid chromatography has shown a dramatic increase in the past few years. In most papers published on this topic, the compounds are converted into their derivatives before chromatographic separation. To detect the substances after HPLC, chemical reactions can be used. In the past year, special attention has been paid to the detection of micro-traces of compounds. The methods used for the detection utilize the index of refraction, ultraviolet absorption, fluorescence, electrical conductivity, polarography, and radioactivity. The major advantage of this approach is the absence of the formation of artefacts which is normally a serious drawback in pre-column derivatization techniques.

Gas chromatography (GC). Analysis of any substance by gas chromatography can be carried out under the assumption that these compounds are thermally stable and inert with respect to the chromatographic system. This is not the case with many polar substances which cannot be chromatographed unless they are chemically modified (derivatization) prior to their chromatographic separation. Examples of suitable derivatives include acyl derivatives, esters, and silyl derivatives. Trimethylsilyl derivatives or trialkylsilyl ethers can be used for heterocyclic compounds containing a hydroxy, mercapto, carbonyl, or amino group. Cyclization is another derivatization reaction which can be conveniently used for heterocyclic compounds with several functional groups.

Pyrolysis before the chromatographic separation represents a more drastic approach which is of interest in certain situations.

3.2.2 Determination

Determination of heterocyclic compounds possessing organoleptic properties and isolated from complex mixtures is often a delicate task. There are numerous analytical methods which can be employed. They are described in various analytical publications, especially those devoted to gas chromatography [2,43] which contain both a general description and detailed discussions of special applications.

Several examples selected here represent cases of determination of oxygen, sulphur, and nitrogen heterocyclic compounds present either in large quantities or, on the other hand, in trace amounts.

Oxygen-containing heterocyclic compounds. Rapid HPLC determination of hydroxymethylfurfural in tomato paste [5]. The clear liquid obtained after the ultracentrifugation of tomato paste was analysed by HPLC using a reverse phase Lichrosorb C_8 column, an ultraviolet detector operating at 284 nm, and elution with water.

Eucalyptol in essential oils [74]. The temperature of crystallization of a mixture of the tested essential oil and *o*-cresol was determined. The result was compared with the data in a table relating the percentage content of eucalyptol in the essential oil and the temperature of crystallization.

Piperine in oleoresin from pepper [85]. Determination of absorbance at 345 nm by ultraviolet spectroscopy of a solution of the oleoresin in chloroform. The amount of piperine (in %) was calculated by comparison with a calibration curve obtained by measuring the absorbance for known concentrations of a reference sample of pure piperine.

3.3 Characterization

If the isolated substance cannot be related to an already known compound, it is necessary to determine its elemental composition and its structure. The classical elemental analysis has to be carried out, and the functional groups and the structure of the skeleton have to be determined by chemical or physical methods which normally make it possible to propose one or several more or less likely structures. The final proof is the total synthesis or an unambiguous determination of the structure by comparison with another compound whose structure is known.

This, of course, represents an ideal situation and in most published papers the assignment of the structure does not go beyond the first stage described above.

3.3.1 Physical and chemical analytical methods used to establish a likely structure

Numerous available methods have already been mentioned in the part devoted to the identification of heterocyclic compounds. The interpretation of ultraviolet, infrared, NMR, and mass spectra and the measurement of other physical properties of apparently new compounds allow one to propose a possible structure.

Chemical methods. A considerable progress has been achieved in elemental analysis (empirical formula) as well as in the functional group and structural analysis, especially from the point of view of the quantities required for these determinations. Thus, thanks to the ultramicroanalytical methods, it is possible to determine very small amounts of aromatics present in heterocyclic substances.

3.3.2 Confirmation by total synthesis [78]

The final proof of the structure of an apparently new heterocyclic compound cannot be accomplished without a total synthesis and a comparison of the physical, chemical, and organoleptic properties of the synthesized compound and of the compound to be identified. This is not an easy task because natural products usually possess very complex structures. As a rule, it is a tedious and lengthy process. Some of these synthetic methods have been reviewed in Chapter IV.

A correlation of a natural product with a known substance is a much more common procedure. It is possible to start with a compound of known structure and, using a sequence of a few unambiguous reactions, to convert it into the compound whose structure is being established. On the other hand, it is also possible to convert a compound of unknown structure into another compound with a known structure, e.g., a crystalline material whose absolute configuration can be obtained by X-ray diffraction.

3.3.3 Examples

The several examples which follow should be sufficiently characteristic.

Oxygen-containing heterocyclic compounds. Theaspirone [45] (*cis*- and *trans*-1-oxa-8-oxo-2,6,10,10-tetramethyl-*spiro*(4,5)-6-decene).

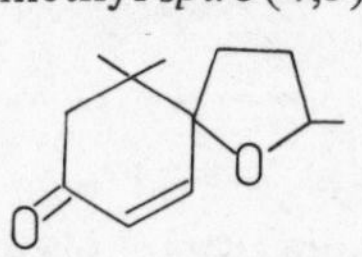

This ketone was isolated from tea by extraction with methanol followed by preparative gas chromatography. One of the fractions obtained reacted with the *Girard-Sandulesco* T reagent. After preparative thin-layer chromatography, a new ketone was isolated. The probable structure has been proposed on the basis of its ultraviolet, infrared, NMR, and mass spectra. No stereochemical investigation has been carried out. The confirmation of the structure [62] has been accomplished by the synthesis of the *cis*- and *trans*-isomers corresponding to the proposed structure. Reduction of β-ionone gave an alcohol which, upon epoxidation, was converted into cyclic ethers. Those were converted into the corresponding cyclohexene by thionyl chloride in pyridine. Oxidation yielded ketones (*cis*- and *trans*-) with properties identical with those of theaspirone. The odour of the *cis*-isomer is typical of tea whereas that of the *trans*-isomer is earth-like.

Nitrogen-containing heterocyclic compounds. α-Guaiapyridine [34] is an alkaloid isolated from the basic fraction of the essential oil of *Pogostemon patchouli Pellet*. The proposed structure is

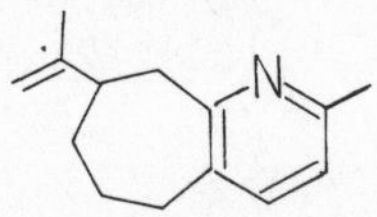

N

The configuration was determined by synthesis starting from isoguaiol which was converted into the corresponding ozonide, reduced to a hydroxydiketone which was treated with hydroxylamine, converted into the pyridine derivative, and dehydrated with thionyl chloride in pyridine. The product exhibited the same spectra as the compound isolated from patchouli.

The structure of epiguaiapyridine, also isolated from patchouli, was confirmed by synthesis starting from α-gurjunene.

3,5-Dimethyl-2-propylpyrazine from the aroma of grilled meat [29]. The basic extract from grilled beef meat yielded a pyrazine whose structure was proposed and confirmed by synthesis. 2,6-Dimethylpyrazine was alkylated with *n*-propyllithium. The properties of the product were the same as those of the pyrazine isolated from grilled beef meat.

VI–4 CONCLUSIONS

Isolation, identification, determination, and characterization of organoleptically important heterocyclic compounds that are present in various natural media used in the food industry is a procedure which is often lengthy and complicated and which requires scientific methods and work by competent groups of scientists.

The progress in analytical methods (especially physical methods) made in the last few years has considerably aided in these studies and has led to a substantial increase in the number of new heterocyclic compounds isolated from naturally occurring materials.

New heterocyclic systems are constantly being synthesized. The use of computers for studies of model systems will undoubtedly lead to the synthesis of a very large number of new heterocyclic compounds (*cf.* Chapter V). It is possible that some of them will be found in natural materials as well when the analytical methods become even more efficient than they are today.

VI–5 REFERENCES

[1] Adda, J. & P. J. Eriksson, *Cah. Nutr. Diet.*, 115 (1980).

[2] Afnor, ISO/TC 54-CHR-doc., 91 (1980).

[3] Albrand, M., P. Dubois, P. Etievant, R. Gelin & B. Tokarska, *J. Agric. Food Chem.*, **28**, 1037 (1980).

[4] Aldrich, *Library of NMR Spectra*, C. J. Pouchert & J. R. Campbell (1974).

[5] Allen, B. H. & H. B. Chin, *J. Assoc. Analyt. Chemists,* **63**, 1074 (1980).

[6] Anthoni, U., C. Christophersen, J. O. Madsen, S. Wium-Andersen & N. Jacobsen, *Phytochemistry,* **19**, 1228 (1980).

[7] ASTM DS 48 *Compilation of Odor and Taste Threshold Values Data*, W. H. Stahl, Philadelphia (1973).

[8] Bartos, J. & M. Pesez, *Techniques de l'Ingénieur,* **7**, 3255a, 3255-1, 3255-2, 3255-3; (Ed. Techniques de l'Ingénieur) Paris (1980).

[9] Bemelmans, J. M. H., *Review of Isolation and Concentration Techniques* in *Progress in Flavour Research* (D. G. Land & H. E. Nursten, Eds.), Applied Sci. Publ., London (1978).

[10] Blau, K. & G. King, *Handbook of Derivatives for Chromatography*, (Heyden), London (1977).

[11] Boelens, M., P. J. De Valois, H. J. Wobben & A. Van Der Gen, *J. Agric. Food Chem.,* **19**, 984 (1971).

[12] Bramwell, A. F., J. W. K. Burrell & G. Riezebos, *Tetrahedron Letters,* **37**, 3215 (1969).

[13] Brinkman, H. W., H. Copier, J. J. M. De Leuw & S. Boen Tjan, *J. Agric. Food Chem.,* **20**, 177 (1972).

[14] Buchi, G., I. M. Goldman & D. W. Mayo, *J. Amer. Chem. Soc.,* **88**, 3109 (1966).

[15] Buttery, R. G., R. M. Seifert & L. C. Ling, *J. Agric. Food Chem.*, **18**, 538 (1970).

[16] Buttery, R. G., R. M. Seifert, D. G. Guadagni & L. C. Ling, *J. Agric. Food Chem.,* **19**, 524 (1971).

[17] Buttery, R. G., R. M. Seifert, D. G. Guadagni & L. C. Ling, *J. Agric. Food Chem.,* **19**, 969 (1971).

[18] Buttery, R. G., L. C. Ling, R. Teranishi & T. R. Mon, *J. Agric. Food Chem.,* **25**, 1227 (1977a); R. G. Buttery, D. G. Guadagni & L. C. Ling, *ibid,* **26**, 791 (1978b).

[19] Calabretta, P. J., *Cosmetics & Perfumery,* **90**, 6, 74 (1975).

[20] Central Institute for Nutrition and Food Research TNO, Zeist, *Volatile Compounds in Food* (Van Straten, Ed.) (1977-1980).

[21] Charalambous, G., Ed., *Analysis of Foods and Beverages – Headspace Techniques*, Academic Press, New York (1978).

[22] Cheeseman, G. W. H. & E. S. G. Verstink, *Advances in Heterocyclic Chemistry,* **14**, 108, Academic Press, New York (1972).

[23] Coleman, E. C. & Chi Tang Ho, *J. Agric. Food Chem.,* **28**, 66 (1980).

[24] Collins, E., *J. Agric. Food Chem.,* **19**, 533 (1971).

[25] Cooper, J. W., *The Computer in Fourrier Transform 1H-NMR,* **2**, 392 (1976); Topics in Carbon 13C-NMR Spectroscopy.

[26] Debbrecht, F. J., *Modern Practice of Gas Chromatography*, 166, (J. Wiley & Sons, Eds.) (1977).

[27] Deck, R. E., J. Pokorny & S. S. Chang, *J. Food Sci.,* **38**, 345 (1973).

[28] Elderfield, R. C., *Heterocyclic Compounds*, Chapman & Hall, London (1950).

[29] Flament, I. & G. Ohloff, *Helv. Chim. Acta,* **54**, 1911 (1971).

[30] Flament, I., M. Kohler & R. Aschiero, *Helv. Chim. Acta,* **59**, 2308 (1976).

[31] Frei, R. W., *Res. Development,* **22**, 42 (1971).

[32] Friedel, P., V. Krampl, T. Radford, J. A. Renner, F. W. Shephard & M. A. Gianturco, *J. Agric. Food Chem.,* **19**, 530 (1971).

[33] Fringuelli *et al., Acta Chim. Scand.,* **28**, 175 (1974).

[34] Gen, Van Der, A., L. M. Van Der Linde & J. G. Witteveen, *Rec. Trav. Chim.,* **91**, 1433 (1972).

[35] Grammaticakis, P., Ed., *Spectres d'Absorption UV de Composés Organiques Azotés et Corrélations Spectrochimiques*, 1977, 1978, 1979, **1**, **2**, **3**; Techniques et Documentation, Paris.

[36] Heide, R. Ter, P. J. De Valois, J. Visser, P. P. Jaegers & R. Timmer, *Analysis of Foods and Beverages – Headspace Techniques* (G. Charalambous, Ed.), 249 (1978).

[37] Herz, K. O., *Diss. Abstr.,* **29**, 1398-B (1968).

[38] Hirai, C., K. O. Herz, J. Pokorny & S. S. Chang, *J. Food Sci.,* **38**, 393 (1973).

[39] Hunter, I. R., M. K. Walden, J. R. Scherer & R. E. Lundin, *Cereal Chem.,* **46**, 189 (1969).

[40] Ina, K. & Y. Sakato, *Tetrahedron Letters,* **23**, 2777 (1968).

[41] Iofi, *Analytical Procedure for a General Method for Thin-Layer Chromatography, Flavour*, Nov./Dec. (1975).

[42] Iofi, *Analytical Procedure for a General Headspace Method, Flavour*, March/April (1975).

[43] Iofi, *Analytical Procedure for a Method for Gas Chromatography, Flavour*, March/April (1976).

[44] Jackman, L. M., *Applications of Nuclear Magnetic Resonance Spectroscopy in Organic Chemistry*, Pergamon Press, Oxford (1959).
[45] Johnson, B. R., G. R. Waller & A. L. Burlingame, *J. Agric. Food Chem.*, **19**, 1020 (1971).
[46] Kaiser, M., Ed., *Qualitative and Quantitative Analysis by Gas Chromatography: Modern Practice of Gas Chromatography,* **4**, 151 (1977); J. Wiley & Sons.
[47] Katritzky, A. R., Ed., *Advances in Heterocyclic Chemistry*, Vol. **1–25** (1980), Academic Press.
[48] Katritzky, A. R., Ed., *Physical Methods in Heterocyclic Chemistry,* **4**, 265–434 (1971), Academic Press.
[49] Kinlin, T. E., A. O. Pittet, R. Muralidhara, A. Sanderson & J. P. Walradt, *J. Agric. Food Chem.*, **20**, 1021 (1972).
[49b] Kitamura, K. & T. Shibamoto, *J. Agric. Food Chem.*, **29**, 188 (1981).
[50] Koller, W. D., *Lebensm-Wiss. Technol.*, **13**, 49 (1980).
[50b] Kolor, M. G., in *Mass Spectrometry* (C. Merritt & C. N. MacEwen, Eds.), M. Dekker Inc., New York (1979), pp. 67–117.
[51] Kreuzig, F., *Chromatographia,* **13**, 238 (1980).
[52] Lawrence, J. F. & R. W. Frei, *Chemical Derivatization in Liquid Chromatography*, Elsevier, Amsterdam (1976).
[53] Leathard, D. A. & B. S. Shurlock, Eds., *Identification Techniques in Gas Chromatography*, Wiley-Interscience, London (1970).
[54] Lee, M. L. & B. W. Wright, *J. Chromatog. Sci.*, **18**, 345 (1980).
[55] Lee, P. L., G. Swords & G. L. K. Hinters, *J. Agric. Food Chem.*, **23**, 1195 (1975).
[56] Lin, S. S., *J. Agric. Food Chem.*, **24**, 1252 (1976).
[57] Ma, T. A., Athanastos & S. Ladas, *Organic Functional group Analysis by Gas Chromatography*, Academic Press, London (1976).
[58] Maga, J. A. & C. E. Sizer, *Fenaroli's Handbook of Flavor Ingredients*, 2nd edn., Vol. **1**, 47 (1970); CRC Press, Cleveland.
[59] Maga, J. A., *Fenaroli's Handbook of Flavor Ingredients*, 2nd edn., Vol. **1**, 228 (1970); CRC Press, Cleveland.
[60] Manley, C. H., P. P. Vallon & R. E. Erickson, *J. Food Sci.*, **39**, 73 (1974).
[61] *Manuel Suisse des Denrées Alimentaires, Substances Aromatisantes*, 5th edn., Vol. **2–3**, Ch. 43 (1978), Ed. Office Central Federal Imprimés et Matériel, Berne.
[62] Mussinan, C. J., R. A. Wilson & I. Katz, *J. Agric. Food Chem.*, **21**, 871 (1973a); R. A. Wilson, C. J. Mussinan, I. Katz & A. Sanderson, *ibid*, **21**, 873 (1973b).
[63] Mussinan, C. J. & J. P. Walradt, *J. Agric. Food Chem.*, **22**, 827 (1974).
[64] Nakatani, Y. & T. Yamanishi, *Tetrahedron Letters,* **24**, 1955 (1969).
[65] Nishimura, K. & M. Masuda, *J. Food Sci.*, **36**, 819 (1971).

[66] Nixon, L. N., E. Wong, C. B. Johnson & E. J. Birch, *J. Agric. Food Chem.*, **27**, 355 (1979).
[67] Nonaka, M., *Food Technol. Champaign.*, **25**, 45 (1971).
[68] Ohloff, G. & I. Flament, *Progress in the Chemistry of Organic Natural Products*, 36, Springer Verlag, Wien (1978).
[69] Ohloff, G. & I. Flament, *Heterocycles,* **11**, 663 (1978).
[70] Peyron, L., *Parf. Cosm. Arômes,* **1**, 53 (1975).
[71] Peyron, L., *Riv. Ital. EPPOS,* **LV**, 620 (1973).
[72] Peyron, L., *Spectra 2000,* **10**, 31 (1975).
[73] Peyron, L., *Perf. and Flav.*, 61 (1980). Special, VIIIth Essential Oil Congress, Cannes (1980).
[74] *Pharmacopée Francaise*, 8e Ed., 1551 (1965).
[75] Pokorny, J., *Prum. Potravin.*, **21**, 262 (1970).
[76] Poole, C. F. & A. Zlatkis, *J. Chromatog. Sci.*, **17**, 115 (1979).
[77] Reineccius, G. A., P. G. Keeney & W. Weissberger, *J. Agric. Food Chem.*, **20**, 202 (1972).
[78] Riddell, F. G., *The Conformational Analysis of Heterocyclic Compounds*, Academic Press (1980).
[79] Saluste, S., I. Klesment & S. Kivirahk, *Eesti. NSV. Tead. Akad. Toim. Keem.*, **28**, 7 (1979).
[80] Schreyen, L., P. Dirinck, F. Van Wassenhove & N. Schamp, *J. Agric. Food Chem.*, **24**, 1147 (1976).
[81] Seck, S. & J. Crouzet, *Phytochemistry,* **12**, 2925 (1973).
[82] Shankaranarayana, M. L., B. Raghavan, K. O. Abraham & G. P. Natarajan, *Fenaroli's Handbook of Flavor Ingredients*, 2nd edn., Vol. **1**, 184 (1970); CRC Press, Cleveland.
[83] Slott, D. & P. D. Harkes, *J. Agric. Food Chem.*, **23**, 356 (1975).
[84] Takken, H. J., L. M. Van Der Linde, M. Boelens & J. M. Dort, *J. Agric. Food Chem.*, **23**, 638 (1975).
[85] Tausig, F., J. I. Suzuki & E. Ray, *Food Technol.*, **10**, 157 (1956).
[86] Thomas, A. P., *J. Agric. Food Chem.*, **21**, 955 (1973).
[87] Tonsbeek, C. H. T., H. Copier & A. J. Plancken, *J. Agric. Food Chem.*, **19**, 1014 (1971).
[88] Traxler, J. T., *J. Agric. Food Chem.*, **19**, 1135 (1971).
[89] Tressl, R., D. Bahrs, M. Holzer & T. Kossa, *J. Agric. Food Chem.*, **25**, 459 (1977).
[90] Tressl, R., L. Friese, F. Fendesack & H. Köpler, *J. Agric. Food Chem.*, **26**, 1422 (1978a); Tressl, R., K. G. Grünewald & R. Silvar, *Chem. Mikrobiol. Technol. Lebensm.*, **7**, 28 (1981b).
[91] Tsugita, T., T. Kurata & M. Fujimaki, *Agric. Biol. Chem.* (Japan), **42**, 643 (1978).
[91b] Vernin, G., *Parf. Cosm. Arômes*, 77 (1981).
[92] Vitzthum, O. G. & P. Werkhoff, *J. Food Sci.*, **39**, 1210 (1974).

[93] Vitzthum, O. G., P. Werkhoff & P. Hubert, *J. Food Sci.*, **40**, 911 (1975a); O. G. Vitzthum & P. Werkhoff, *J. Agric. Food Chem.*, **23**, 510 (1975b).

[94] Wal, Van der, B., D. K. Kettenes, J. Stoffelsma, G. Sipma & A. T. J. Semper, *J. Agric. Food Chem.*, **19**, 276 (1971).

[95] Walradt, J. P., R. C. Lindsay & L. M. Libbey, *J. Agric. Food Chem.*, **18**, 926 (1970).

[96] Walradt, J. P., A. O. Pittet, T. E. Kinlin, R. Muralidhara & A. Sanderson, *J. Agric. Food Chem.*, **19**, 972 (1971).

[97] Wakeham, S. G., *Environ. Sci. Technol.*, **13**, 1119 (1979).

[98] Wang, P. S. & G. V. Odell, *J. Agric. Food Chem.*, **20**, 207 (1972).

[99] Weissberger, A. & E. C. Taylor, Eds., *The Chemistry of Heterocyclic Compounds*, 28 vol. (1950); Interscience, J. Wiley, New York.

[100] Weissberger, A. & E. C. Taylor, Eds., *The Chemistry of Heterocyclic Compounds*, Vol. **34**, tome 1, Interscience, J. Wiley, New York (1979).

[101] Wise, S. A. & W. E. May, *Res. Development,* **28**, 45 (1977).

[102] Withycombe, D. A., J. P. Walradt & A. Hruza, *Phenolic, Sulfur and Nitrogen Compounds in Food Flavors* (G. Charalambous & I. Katz, Eds.) (1976), 85; ACS Symposium Series No. **26**.

[103] Yamaguchi, K., S. Mihara, A. Aitoku & T. Shibamoto, *Liquid Chromatographic Analysis of Food and Beverages* (G. Charalambous, Ed.), Academic Press, New York, Vol. **2**, 303 (1979).

CHAPTER VII

Mass Spectrometry of Heterocyclic Compounds used for Flavouring

Gaston VERNIN and **Michael PETITJEAN,** University of Aix-Marseilles, France

VII–1 INTRODUCTION

Mass spectral fragmentation modes are extremely valuable in the interpretation of the mass spectra of heterocyclic compounds found in food flavours and model systems. In many cases, molecular weight data, high-resolution measurements, ^{13}C- and deuterium labelling experiments, isotopic abundance data, metastable ions,* chemical ionization and the nitrogen rule can be of great help to obtain a good description of the molecule if not the exact structure itself. Thus sulphur-containing compounds are easily recognized because of the natural abundance of the S^{34} isotope (4.2%) which gives rise to a peak two mass units higher than the molecular ion. Nitrogen-containing ring systems are evident from the nitrogen rule which states that the molecular weight of a molecule will be an odd number only if there is an uneven number of nitrogen atoms in the molecule. Furthermore, simple heterocyclic systems and certain groups such as alkoxyl, acyl, acid, ester, thiol, etc., give very simple and characteristic fragmentation modes which render them easily recognizable.

The common fragmentation patterns and rearrangement processes that are encountered in heterocyclic chemistry are described in several books [6,10,22, 25,33]. It is beyond the scope of this chapter to discuss them in detail. We will confine ourselves to giving here some general and useful information dealing with the main features associated with the fragmentation of each main category of heterocyclic compounds.†

Mass spectra listed in Tables proceed either from more representative heterocyclic compounds naturally occurring in food flavours, or from artificial aroma compounds synthesized in our laboratory or elsewhere.

* in the schemes indicates a process supported by the presence of an appropriate metastable ion.

† According to IUPAC rules, ion fragments are given at m/z but most authors used m/e for mass-charge ratios.

VII–2 FURANS, THIOPHENES AND PYRROLES

The mass spectra of furans [10,15a,18], thiophenes [3,17], and pyrroles [7,36] are generally quite simple and give information not only on the type of substituents, but also on their relative location. The fragmentation modes of parent compounds resemble each other. The molecular ion is the base peak (except for furan), thus reflecting the great stability of the aromatic ring. The most abundant is at m/z 39 corresponding to the cyclopropenyl ion ($C_3H_3^+$) which is the base peak in furan (*see Scheme 1*).

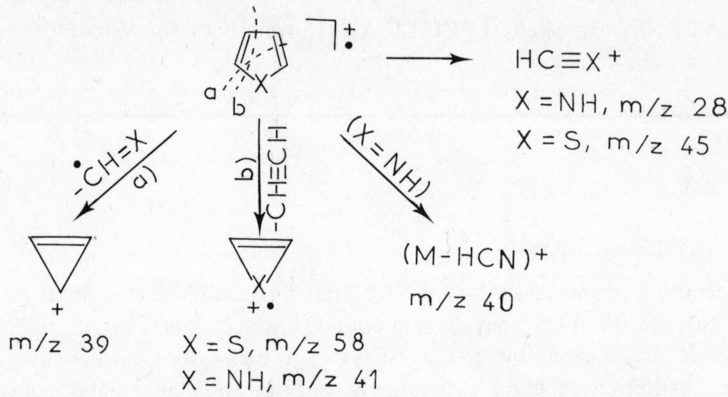

Scheme 1 – Fragmentation of furan, thiophene, and pyrrole.

The only major difference between these fragmentation modes lies in the occurrence of an abundant fragment ($C_2H_2X^+$) at m/z 58 (X = S) and 41 (X = NH) in the spectra of thiophene and pyrrole respectively. It arises from a loss of acetylene from the molecular ion. In the mass spectra of thiophene and pyrrole another abundant ion is due to the $HC{\equiv}X^+$ cation at m/z 45 (X = S) and 28 (X = NH). The peak at m/z 40 in pyrrole results from the loss of hydrogen cyanide, (*see Table VII–1a* and *VII–1b*).

Table VII–1a Mass spectra of some furans, thiophenes and pyrroles

Compound and (Molecular formula)	Ref.	Principal fragments (relative intensity %)[a]
Furan (C_4H_4O)	[10a]	**68**(62),42(7),40(12),39(100),38(17), 37(11),29(16),24(4),14(4),13(3)
2-*n*-Pentylfuran ($C_9H_{14}O$)	[36]	**138**(15),123(1),109(2),95(4),82(24), 81(100),71(4),67(2),53(11),41(6),27(8)
Furfuryl alcohol ($C_5H_6O_2$)	[36]	99(6),**98**(100),97(57),96(12),95(15), 81(49),70(25),69(26),53(39),42(46), 41(55),39(51),31(12),29(22)

Table VII–1a (*continued*)

Compound and (Molecular formula)	Ref.	Principal fragments (relative intensity %)[a)]
Furfuryl mercaptan (C_5H_6OS)	[36]	116(1.2),115(1.6),**114**(25),81(100), 53(49),51(8),45(15),39(8),27(18)
Furfuryl thioethers		*See Table VII–2*
Furfural ($C_5H_4O_2$)	[36]	97(6),**96**(100),95(91),67(4),51(1),42(3), 39(55),38(14),29(14)
2-Acetylfuran ($C_6H_6O_2$)	[36]	111(3),**110**(40),95(100),67(4),43(17), 39(21),29(2)
Methylfuroate ($C_6H_6O_3$)	[10b]	127(2.8),**126**(39),96(9),95(100),39(1), 38(13),37(5)
2,5-Dimethylfuran (C_6H_8O)	[10a]	97(6),**96**(88),95(74),81(37),53(57), 51(20),50(17),43(100),39(15),27(30), 15(26)
5-Hydroxymethylfurfural ($C_6H_6O_3$)	[35b]	127(2.7),**126**(37),119(13),97(73), 69(41),53(24),41(100),39(56)
2,5-Dimethyl-4-hydroxy-3-(2*H*)-furanone ($C_6H_8O_3$)	[25]	129(3),**128**(45),85(22),72(9),58(5), 57(76),56(15),55(20),54(5),53(4), 43(100),40(22),39(9)
2-Methyl-3-tetrahydrofuranone ($C_5H_7O_2$)	[43]	**100**(26),78(6),72(37),55(6),45(19), 44(31),43(100),42(16),41(5),39(5), 31(9),29(31),28(43),27(19),26(6), 15(13),14(7)
Thiophene (C_4H_4S)	[10a]	86(4.6),85(5.3),**84**(100),83(6),69(7), 58(65),57(13),50(6),45(55),39(28), 38(8),37(7)
2-Formylthiophene (C_5H_4OS)	[36]	114(4.5),113(10.8),**112**(90),111(100), 83(12),69(2),58(9),57(8),45(15), 39(25),29(2)
2-Acetylthiophene (C_6H_6OS)	[36]	**126**(34),111(100),83(9),69(2),57(5), 50(2),45(8),43(12),39(22),27(1)
2-Thienylmercaptan ($C_4H_4S_2$)	[10a]	118(9),117(6),**116**(100),72(14),71(93), 69(15),58(14),57(9),45(46),39(16)
2-Methyl-3-tetrahydrothiophenone (C_5H_8OS)	[36]	118(2),117(3),**116**(57),101(1),88(9), 74(2),60(100),59(19),45(20),27(11)
Pyrrole (C_4H_5N)	[10a]	68(5),**67**(100),66(8),41(63),40(54), 39(68),38(25),37(17),33(7),28(51), 26(6)
N-Methylpyrrole (C_5H_7N)	[10a]	82(6),**81**(100),80(80),55(19),54(15), 53(33),42(34),39(39),38(12),27(12), 15(12)

Table VII–1a (*continued*)

Compound and (Molecular formula)	Ref.	Principal fragments (relative intensity %)[a]
N-Furfurylpyrrole (C_9H_9NO)	[36]	148(5),**147**(42),117(1),81(100),58.5(2), 53(22),39(5),27(9)
2-Acetylpyrrole (C_6H_7NO)	[7]	110(6),**109**(80),95(7),94(100),67(5), 66(50),53(7),43(13),41(5),40(10), 39(35),38(12),37(6)
N-Methyl-2-formylpyrrole (C_6H_7NO)	[36]	110(7),**109**(100),108(85),80(33),64(2), 53(41),42(7),39(33),27(13)
N-Methyl-2-acetylpyrrole (C_7H_9NO)	[36]	124(5),**123**(66),108(100),80(24),65(2), 53(35),43(9),39(26),27(9)
2,5-Dimethylpyrrole (C_6H_9N)	[10a]	96(4),**95**(58),94(100),93(9),80(18), 53(8),51(9),42(15),27(11),26(41),15(9)

a) Molecular ions ($M^{+\cdot}$) are given in bold characters, and isotopic abundances (M + 1 and M + 2 ions) have been calculated.

Methyl substituted- furans, thiophenes and pyrroles yield very abundant M – 1 ions which are the base peaks in the spectra of C-methylated pyrroles. The marked similarity between 2- and 3-isomers spectra indicates a common relatively stable cation having a six-membered pyrilium (X = O), thiapyrilium (X = S) or pyridinium (X = NH) ring structure (*see Scheme 2*).

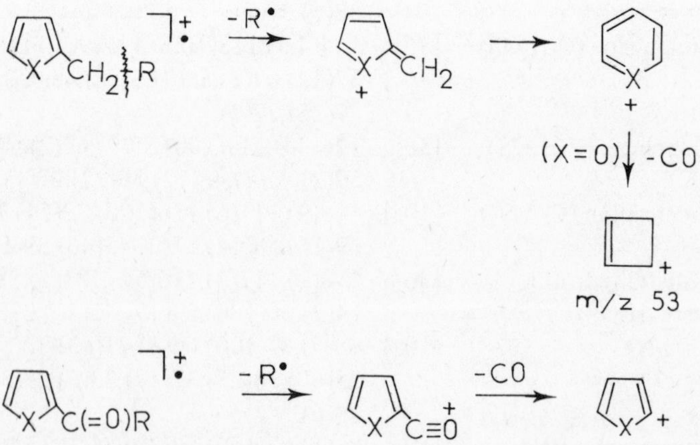

Scheme 2 – Fragmentation of 2- (or 3-) alkyl-, furans, thiophenes and pyrroles.

Table VII–1b Mass spectra of some pyrroles identified in freshly roasted coffee[a)]

Compound and (Molecular formula)	Principal fragments[b)]
N-Ethylpyrrole (C_6H_9N)	80,**95**,67,41,53,39
N-Methyl-2-methylpyrrole (C_6H_9N)	94,**95**,67,39,53,80
N-Methyl-3-methylpyrrole (C_6H_9N)	94,**95**,53,39,67,80
N-Propylpyrrole ($C_7H_{11}N$)	80,**109**,81,67,53,39,94
N-Ethyl-2-methylpyrrole ($C_7H_{11}N$)	**109**,108,94,80
N-Ethyl-3-methylpyrrole ($C_7H_{11}N$)	94,**109**,108,80
N-Methyl-2-ethylpyrrole ($C_7H_{11}N$)	94,**109**,42,39,41
N-Isobutylpyrrole ($C_8H_{13}N$)	80,81,**123**,68,53
N-Propyl-2-methylpyrrole ($C_8H_{13}N$)	94,**123**,95,80
N-Isobutyl-2-methylpyrrole ($C_9H_{15}N$)	94,95,**137**,80,41
N-Isobutyl-3-methylpyrrole ($C_9H_{15}N$)	94,80,95,**137**,41
N-(2-Methylbutyl)-pyrrole ($C_9H_{15}N$)	81,80,**137**,41,53
N-Isoamylpyrrole ($C_9H_{15}N$)	81,80,**137**,53,39
N-(2-Methylbutyl)-2-methylpyrrole ($C_{10}H_{17}N$)	95,94,**151**,53,41
N-Isoamyl-2-methylpyrrole ($C_{10}H_{17}N$)	95,136,**151**,94,53
N-Isoamyl-3-methylpyrrole ($C_{10}H_{17}N$)	94,95,**151**,53,80
N-Isobutyl-2,5-dimethylpyrrole ($C_{10}H_{17}N$)	108,**151**,94,95,109,41,42
N-Isoamyl-2,5-dimethylpyrrole ($C_{11}H_{19}N$)	109,108,**165**,94,41,92
N-Acetonylpyrrole (C_7H_9NO)	109,108,**165**,94,41,92
N-(2-Butanoyl)-pyrrole ($C_8H_{11}NO$)	80,**137**,53,39
N-(5-Methylfurfuryl)-pyrrole ($C_{10}H_{11}NO$)	95,**161**,43,67,57,53
N-Furfuryl-2-methylpyrrole ($C_{10}H_{11}NO$)	81,**161**,53,51,39
N-Furfuryl-3-methylpyrrole ($C_{10}H_{11}NO$)	81,**161**,53,51,39
N-(3-Methylfurfuryl)-pyrrole ($C_{10}H_{11}NO$)	95,**161**,67,43
N-(5-Methylfurfuryl)-2-methylpyrrole ($C_{11}H_{13}NO$)	95,**175**,43,41,81,65
N-Furfuryl-2,5-dimethylpyrrole ($C_{11}H_{13}NO$)	81,**175**,53,51,39
N-Furfuryl-2,3-dimethylpyrrole ($C_{11}H_{13}NO$)	81,**175**,53,51
N-(2-Phenylethyl)-pyrrole ($C_{12}H_{13}N$)	80,**171**,104,53,91
N-Furfuryl-2-ethyl-5-methylpyrrole ($C_{12}H_{15}NO$)	81,174,**189**,53
N-Furfuryl-2,3,5-trimethylpyrrole ($C_{12}H_{15}NO$)	81,**189**,53
N-Furfuryl-2-formylpyrrole ($C_{10}H_9NO_2$)	81,**175**,53,39
N-(5-Methylfurfuryl)-2-formylpyrrole ($C_{11}H_{11}NO_2$)	95,**189**,43
N-Furfuryl-2-acetylpyrrole ($C_{11}N_{11}NO_2$)	81,**189**,53,43,146
N-(5-Methylfurfuryl)-2-acetylpyrrole ($C_{12}H_{13}NO_2$)	95,**203**,53,43,160

a) Reported by Tressl *et al.*, *Chem. Mikrobiol. Technol. Lebensm.*, **7**, 28 (1981). Mass spectra were measured with a CH 5DF Varian MAT. The ionization voltage was kept at 70 eV.

b) Principal fragments are given by decreasing relative intensity.

The tendency towards the formation of this cation increases with alkyl chain length. The spectra of 2-*n*-alkylfurans [18] contain the following peaks: M, M - 1, M - 15 (for R > CH_3), M - 28, M - 29, M - 42 (for R > CH_3), M - 43 and the fragments at m/z 81, 53, 51, 44, 41 and 39.

The fragmentation mode of 2-pentylfuran is shown in *Scheme 3*.

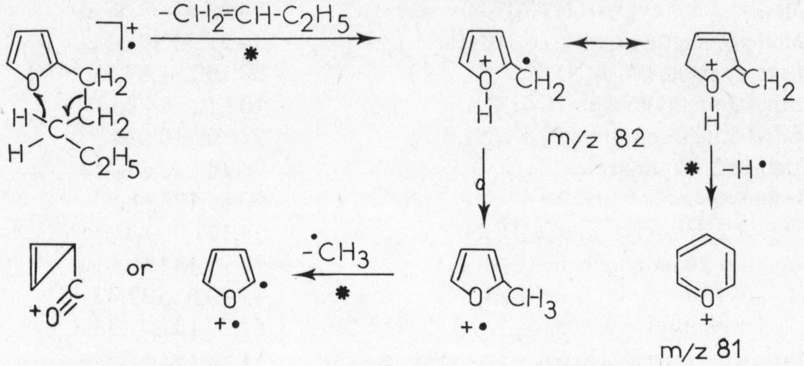

Scheme 3 - Fragmentation of 2-pentylfuran.

The fragmentation of certain long-chain *N*-alkylpyrroles has also been studied in detail by means of labelling and high-resolution techniques [33].

α-Cleavage leading to the carbonyl group is the dominant feature in each of the spectra of 2-acyl-, furan, thiophene and pyrrole (*see Scheme 4*).

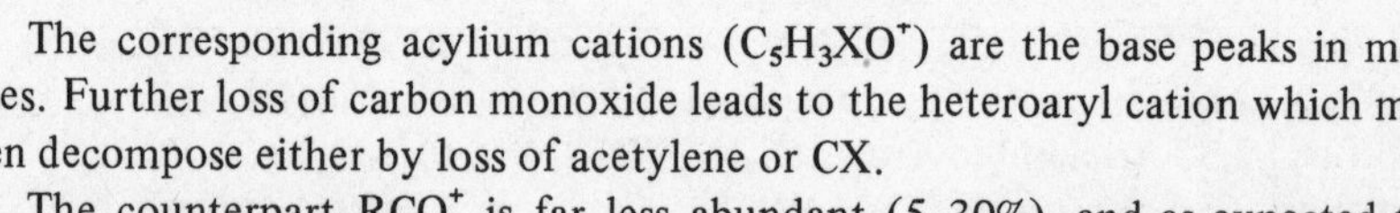

Scheme 4 – Fragmentation of 2-acyl-, furan, thiophene and pyrrole.

The corresponding acylium cations ($C_5H_3XO^+$) are the base peaks in most cases. Further loss of carbon monoxide leads to the heteroaryl cation which may then decompose either by loss of acetylene or CX.

The counterpart RCO^+ is far less abundant (5–30%), and as expected the longer alkyl group decreases the intensity of this ion. The mass spectral behaviour of *N*-acylpyrroles is quite different from that of *C*-acylpyrroles [46]. The mass spectrum of *N*-acetylpyrrole shows a parent peak at m/z 109. The base peak is at m/z 67, indicating a loss of ketene. In comparison with *N*-acylpyrrole, the base peak of 2-acetylpyrrole is the ion at m/z 94 resulting from a loss of a methyl radical from the molecular ion. The only other abundant fragment of 2-acetylpyrrole is at m/z 66 indicating a loss of carbon monoxide.

The mass spectra of *N*-propionylpyrrole and *N*-acetyl-2-methylpyrrole are similar in many respects to the spectrum of *N*-acetylpyrrole.

Carboxylic acid derivatives are characterized by skeletal rearrangement ions due to a loss of CO with accompanying OH migration:

$$\mathrm{HarCOOH}^{+\bullet} \xrightarrow{-\mathrm{CO}} \mathrm{Har{-}OH}^{+\bullet}$$

Nuclear carboxylic esters such as methyl furoate behave like aromatic methyl esters giving a base peak corresponding to a loss of methoxyl radical, then followed by a loss of carbon monoxide.

In the pyrrole series additional peaks corresponding to cleavage due to the N - H group are observed. When a methyl substituent is adjacent to the ester group, the loss of methanol can compete with the loss of the methoxyl radical through '*ortho* effect' (*see Scheme 5*). This effect is also responsible for the loss of a methyl radical from the ester group.

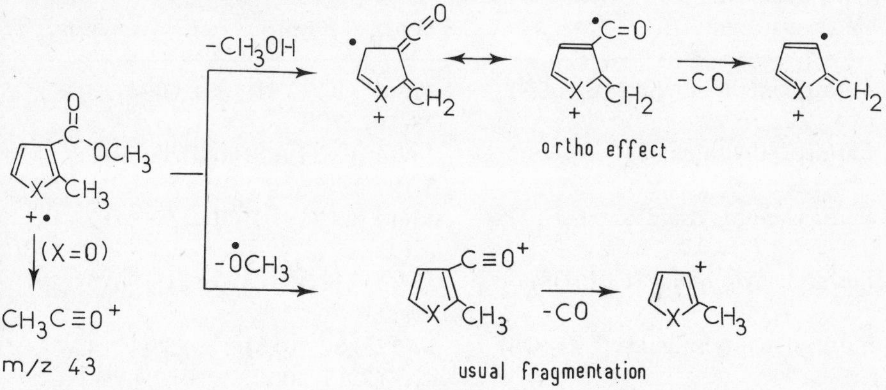

$-CH_3OH$; $-\dot{O}CH_3$; $-CO$; $(X=O)$; $CH_3C{\equiv}O^+$ m/z 43

ortho effect

usual fragmentation

Scheme 5 – "*Ortho* effect" in the fragmentation of nuclear carboxylic methyl esters.

Nuclear ethyl esters behave similarly:

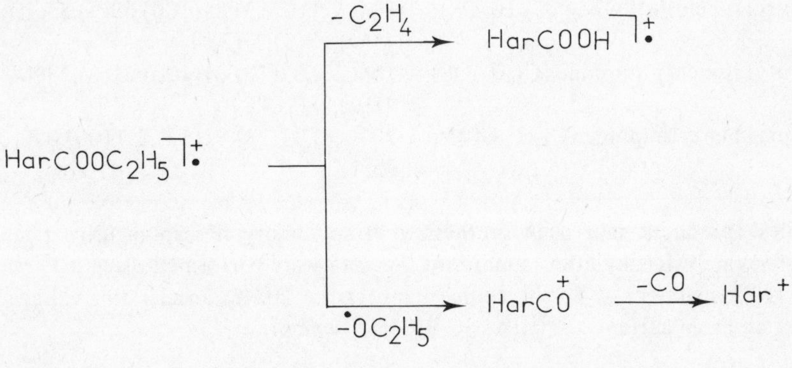

$$\mathrm{HarCOOC_2H_5}^{+\bullet} \xrightarrow{-C_2H_4} \mathrm{HarCOOH}^{+\bullet}$$

$$\mathrm{HarCOOC_2H_5}^{+\bullet} \xrightarrow{-\dot{O}C_2H_5} \mathrm{HarC\overset{+}{O}} \xrightarrow{-\mathrm{CO}} \mathrm{Har}^{+}$$

The base peak of 2-methylthiofuroate found in coffee is at m/z 95 corresponding to the furoyl group. Besides the rather strong parent peak at m/z 142 (26%) the third peak of interest at m/z 39 arises from fragmentation of the furan ring.

The mass spectrum of furfuryl mercaptan shows the characteristic fragmentation pattern of thiols where an SH group is lost from the molecular ion. The base peak at m/z 81 is a very characteristic fragment in furfuryl compounds such as difurfuryl sulphide, difurfuryl disulphide, deoxyfuroin, *N*-furfurylpyrrole and furfuryl thioethers (*see Table VII–2* and *Scheme 6*).

Peak at m/z 53 results from fragment at m/z 81 by loss of carbon monoxide.

2-Methyl-3-tetrahydrofuranone affords four main fragments at m/z 43 (100%), 100 (26%), 72 (37%) and 44 (31%) [43].

Table VII–2 Mass spectra of furfurylthioethers [42] [a)]

Compound and (Molecular formula)	Principal fragments (relative intensity, %)
Furfuryl methylsulphide (C_6H_8OS)	**128**(57),82(6),81(100),53(34),45(11),28(19),27(9)
Furfuryl ethylsulphide ($C_7H_{10}OS$)	**142**(40),82(10),81(100),53(20),28(15),27(9)
Furfuryl-*n*-propylsulphide ($C_8H_{12}OS$)	**156**(29),82(10),81(100),53(14),28(12),27(9)
Furfuryl allylsulphide ($C_6H_{10}OS$)	**154**(14),112(29),82(6),81(100),53(20),45(11)
Furfuryl-*n*-butylsulphide ($C_9H_{14}OS$)	**170**(23),82(10),81(100),53(19),45(6),41(4),39(5),30(3),29(10),28(11)
Furfuryl isopropylsulphide ($C_8H_{12}OS$)	**156**(39),82(6),81(100),53(9),45(6),41(6),28(4),27(7)
Furfuryl isobutylsulphide ($C_9H_{14}OS$)	**170**(24),82(7),81(100),53(11),28(6),27(8)
Furfuryl-*n*-pentylsulphide ($C_{10}H_{16}OS$)	**184**(12),82(12),81(100),69(5),53(10),41(9),28(5),27(5)
Furfuryl isopentylsulphide ($C_{10}H_{16}OS$)	**184**(23),82(10),81(100),61(6),53(14),41(9),28(10),27(6)
Furfuryl benzylsulphide ($C_{12}H_{12}OS$)	**204**(18),91(23),82(13),81(100),65(13),45(13)

a) These products have been synthesized from furfurylmercaptan using phase transfer catalysis alkylation conditions. Spectra were performed using a *Varian* Mat 111 instrument at 80 eV, source temperature 200°C, accelerating voltage : 820 V and trap current : 270 μA, GC-MS inlet system.

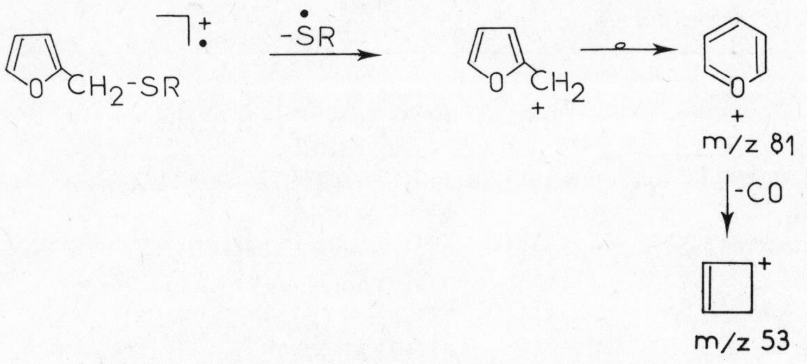

Scheme 6 – Fragmentation of furfuryl thioethers [42].

The mass spectrum of 2,5-dimethyl-4-hydroxy-3-(2*H*)-furanone gives little structural information. In this case other spectral data such as IR and 1H-NMR are required to elucidate the structure of this compound [22].

VII–3 OXAZOLES, THIAZOLES, IMIDAZOLES, 1,3-DIOXOLANES AND ISOTHIAZOLES

The fragmentation processes of oxazoles [4,44], thiazoles [8,9,16,31,37,44], imidazoles [5], 1,3-dioxolanes [24], and isothiazoles [27,32] have already been discussed. The mass spectra of the parent compounds and their *C*-methyl derivatives (*see Table VII–3*) are dominated by the molecular ions which are often the base peaks and by fragments formed mainly by cleavage of the 1,2 and 3,4 bonds with charge retention by the heteroatom-containing portion.

Table VII–3 Mass spectra of some oxazoles, thiazoles, imidazoles, 1,3-dioxolanes and isothiazoles

Compound and (Molecular formula)	Ref.	Principal fragments (relative intensity, %)
Oxazole (C_3H_3NO)	[33]	70(4),**69**(100),68(8),42(11),41(22),40(37), 39(5),38(2)
2,4,5-Trimethyloxazole (C_6H_9NO)	[44]	112(7),**111**(100),110(27),96(7),83(7), 82(29),70(41),69(21),68(46),55(93),54(7), 43(93),42(45),41(25),39(9)
2,4-Dimethyl-5-acetyl-oxazole ($C_7H_9NO_2$)	[44]	140(5),**139**(65),124(18),96(100),68(62), 43(78),42(64),41(45),39(18)
2,4-Dimethyl-2-oxazoline (C_5H_9NO)	[34]	**99**(29),84(54),70(5),69(74),68(47),57(8), 56(44),43(34),42(100),41(15),40(7),39(11), 29(17),28(38),27(31),26(21)

Table VII–3 (*continued*)

Compound and (Molecular formula)	Ref.	Principal fragments (relative intensity, %)
2,4,5-Trimethyl-3-oxazoline ($C_6H_{11}NO$)	[28]	**113**(8),98(23),72(100),71(22),44(87), 43(65),42(59)
Thiazole (C_3H_3NS)	[44]	87(4.5),86(4.6),**85**(100),59(6),58(73), 57(19),56(7),45(9),42(13)
2-Isobutylthiazole ($C_7H_{11}NS$)	[31]	**141**,126(17),113(15),99(100),58(38), 41(17),27(20)
2-Acetylthiazole (C_5H_5NOS)	[31]	**127**(41),112(26),99(40),58(40),57(32), 43(100)
4-Methylthiazole (C_4H_5NS)	[44]	101(4.6),100(5.7),**99**(100),72(57),71(94), 45(32),39(32)
2,4-Dimethylthiazole (C_5H_7NS)	[44]	115(3.9),114(5.8),**113**(85),72(100),71(65), 59(26),45(13),42(17),39(16)
4-Methyl-5-vinylthiazole (C_6H_7NS)	[20] a)	**125**,97,45,58,39,98
2,4,5-Trimethylthiazole (C_6H_9NS)	[44]	129(4.7),128(8),**127**(100),126(8),112(4), 86(80),85(48),71(86),68(12),59(36),53(12), 45(32),39(16)
2,4,5-Trimethyl-3-thiazoline ($C_6H_{11}NS$)	[28]	**129**(45),114(30),88(100),87(22),73(18), 69(60),68(35),59(33),54(47),45(25), 42(83),39
2-Acetyl-2-thiazoline (C_5H_7NOS)	[28]	131(4),130(4),**129**(65),101(9),87(12), 73(4),60(55),59(26),58(5),45(10),43(100), 42(5),41(6),29(11),27(7),15(20)
2-Alkylthio-, 2-thiazolines and thiazoles		*See Table VII–4*
N-Methylimidazole ($C_4H_6N_2$)	[5]	83(5),**82**(100),81(20),55(8),54(14),52(2), 42(13),41(8),40(11),28(8),27(4),26(2)
N-Methyl-2-formylimidazoles ($C_5H_6N_2O$)	[5]	111(6),**110**(100),109(13),83(2),82(39), 81(30),56(11),55(8),54(21),53(3),52(6), 51(1),42(17),41(8),40(10)
1,2-Dimethylimidazole ($C_5H_8N_2$)	[5]	97(6),**96**(100),95(49),55(14),54(29),52(4), 44(2),42(22),41(6),40(9),28(20),27(8), 26(5)
syn-2,4-Dimethyl-1,3-dioxolane ($C_5H_{10}O_2$)	[24]	**102**,87(100),59(43),58(8),45(19),44(14), 43(98),41(57),31(72)
anti-2,4-Dimethyl-1,3-dioxolane ($C_5H_{10}O_2$)	[24]	**102**,87(100),59(62),58(81),45(50),44(45), 43(38),41(58),31(92)
Isothiazole (C_3H_3NS)	[27]	87(4.5),**86**(5),85(100),84(3),59(10),58(40), 57(8),52(2),51(2),47(8),44(3),39(3)

Table VII–3 (*continued*)

Compound and (Molecular formula)	Ref.	Principal fragments (relative intensity, %)
3,4-Dimethylisothiazole (C_5H_7NS)	[32]	115(5),114(8),**113**(100),112(26),98(7), 80(10),73(15),72(43),71(53),45(27)
3,5-Dimethylisothiazole (C_5H_7NS)	[32]	115(5),114(8),**113**(100),112(21),98(12), 73(19),72(17),71(29),45(10),43(11),39(14)
4,5-Dimethylisothiazole (C_5H_7NS)	[32]	115(5),114(8),**113**(100),112(33),98(35), 85(15),71(23),59(14),45(13),39(9)
3,4,5-Trimethylisothiazole (C_6H_9NS)	[32]	129(5),128(9),**127**(100),126(21),112(40), 86(20),85(19),71(32),59(13),45(11),39(7)
5-Hydroxymethylisothia-zole (C_4H_5NSO)	[27]	117(5),116(6),**115**(98),114(24),112(8), 98(8),87(32),86(100),82(40),59(32),58(16), 57(24),55(20),54(10),45(16),44(8)

a) Relative intensities have not been reported. Principal fragments are given by decreasing order of relative intensity.

Thus oxazole (X = O) affords the molecular ion at m/z 69 as base peak, with three main fragment ions at m/z 42, 41 and 40 corresponding to the loss of hydrogen cyanide, carbon monoxide and formyl radical, respectively (*see Scheme 7*).

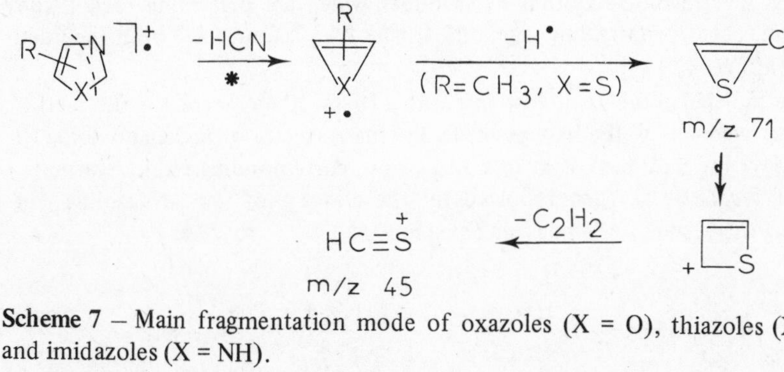

-HCN
*
-H•
(R=CH3, X=S)
CH_2^+
m/z 71
$-C_2H_2$
$HC\equiv\overset{+}{S}$
m/z 45

Scheme 7 – Main fragmentation mode of oxazoles (X = O), thiazoles (X = S) and imidazoles (X = NH).

It should be noted that a loss of hydrogen cyanide is easier in the thiazole series than in the corresponding oxazole and imidazole series. This is due to easier cleavage of the C–S bond relative to C–N bond and C–O bonds.

The substitution pattern of an imidazole greatly affects the elimination of hydrogen cyanide. Mass spectra of 1-methyl-, 2-methyl- and 4- (or 5) methylimidazole show a loss of HCN from the 2,3-, 1,5- and 2,3-positions, respectively. 1,2-Dimethylimidazole shows a negligible loss of HCN from the molecular ion.

The major fragment ions in 2- and 4- alkyl derivatives (three carbon length or longer) are formed by cleavage of the alkyl chain.

As in the case of other nitrogen-containing ring systems, *MacLafferty* rearrangement rather than β-cleavage give rises to an abundant rearrangement ion (*see Scheme 8*).

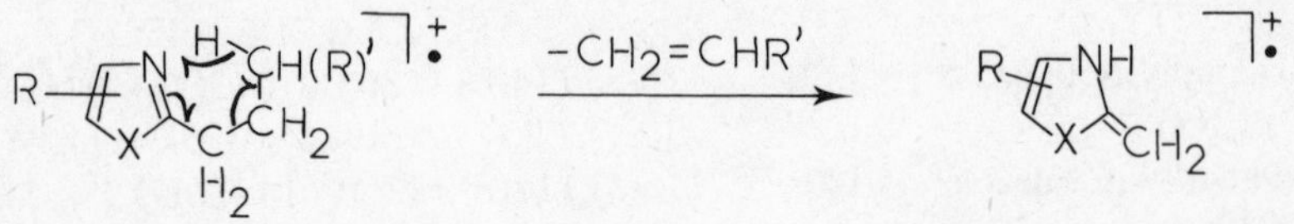

$$\text{R}-\text{[ring N, X]}-\text{CH}_2-\text{CH}_2-\text{CH(R')H}^{+\cdot} \xrightarrow{-\text{CH}_2=\text{CHR}'} \text{R}-\text{[ring NH, X]}=\text{CH}_2^{+\cdot}$$

Scheme 8 – *MacLafferty* rearrangement in 2- (or 4) alkyl-, oxazoles, thiazoles and imidazoles.

The isomeric 5-alkylthiazoles (X = S) give intense ions corresponding at $[M - (R - 14)]^+$. If the heterocyclic compound contains an ester substituent in the 2- position as in ethyl 4-methylthiazole-2-carboxylate, then fragmentation within the ester group is able to compete with the bond cleavages. When methyl and ester groups are in adjacent positions, '*ortho* effect' can operate. The abundance of the resulting ions $(M - C_2H_5)^+$ and $(M - C_2H_5OH)^{+\cdot}$ relative to the normal $(M - C_2H_4)^{+\cdot}$ and $(M - OC_2H_5)^+$ ions is less important than in the related furans, pyrroles and thiophenes.

Acyl and carboxyl substituents are also frequently eliminated to give the corresponding heteroaryl molecular ion.

Mass spectra of 2-alkylthio-2-thiazolines which are present in rocket, rue, dandelion, chicory, tarragon, sage, and truffle like-flavour, have been recorded (*see Table VII–4*) [42].

The molecular ion is of low intensity (10 to 20%) except for the methyl derivative where it is the base peak. In the mass spectra of higher homologues $(R > CH_3)$ the base peak is at m/z 119 or 60, corresponding to the rearrangement of *MacLafferty* then followed by the cleavage of the 2-thiazoline ring according to 1,2 and 3,4 bonds (*see Scheme 9*).

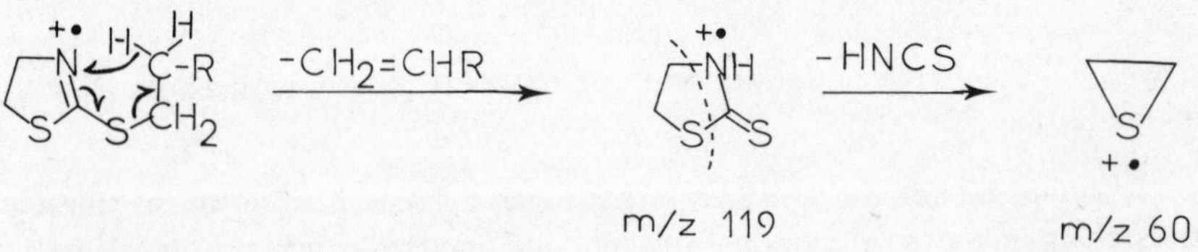

$$\xrightarrow{-\text{CH}_2=\text{CHR}} \quad \text{m/z } 119 \quad \xrightarrow{-\text{HNCS}} \quad \text{m/z } 60$$

Scheme 9 – *MacLafferty* rearrangement in 2-alkylthio-2-thiazolines.

2-Substituted-4-methyl-1,3-dioxolanes [24] on electron impact give a characteristic base peak at m/z 87 corresponding to a resonance stabilized oxonium ion (*see Scheme 10*).

Table VII–4 Mass spectra of some 2-alkylthio-, 2-thiazolines and thiazoles [42] [a)]

Compound and (Molecular formula)	Principal fragments (relative intensity, %)
2-Methylthiothiazoline ($C_4H_7NS_2$)	**133**(100),87(17),74(18),72(28),60(76),59(39), 47(11),45(31),28(15),27(7)
2-Ethylthio-2-thiazoline ($C_5H_9NS_2$)	**147**(22),121(10),119(73),114(15),73(12), 72(22),61(15),60(100),59(42),45(19),29(12), 28(18),27(20)
2-*n*-Propylthio-2-thiazoline ($C_6H_{11}NS_2$)	**161**(5),146(13),120(15),119(92),114(17), 73(15),72(10),61(23),60(100),59(27),45(19), 43(38),41(19),40(13),28(21),27(15)
2-Isopropylthio-2-thiazoline ($C_6H_{11}NS_2$)	**161**(10),128(13),120(13),119(81),72(16), 61(23),60(100),59(29),45(19),43(55),41(16), 28(64),27(32)
2-Allylthio-2-thiazoline ($C_6H_9NS_2$)	**159**(21),158(12),144(100),136(10),131(13), 87(10),73(13),72(22),60(37),39(22),45(22), 41(33),39(22),31(10),28(30),27(9)
2-*n*-Butylthio-2-thiazoline ($C_7H_{13}NS_2$)	**175**(6),146(18),128(39),120(24),119(100), 61(30),60(85),59(24),57(24),45(24),41(18), 29(30),28(24)
2-Isobutylthio-2-thiazoline ($C_7H_{13}NS_2$)	**175**(2),160(17),120(40),119(100),60(67), 57(20),45(17),43(17),41(37),32(10),29(13), 28(37),27(10)
2-*n*-Pentylthio-2-thiazoline ($C_8H_{15}NS_2$)	**189**(9),146(9),142(21),133(24),121(15), 120(21),**119**(100),72(15),69(9),61(15),60(48), 59(15),45(15),43(21),32(15),29(9),28(39), 27(15)
2-Isopentylthio-2-thiazoline ($C_8H_{15}NS_2$)	**189**(22),146(26),133(26),120(26),119(100), 61(22),60(70),59(52),55(13),45(17),43(57), 41(35),28(35),27(35)
2-Isopentylthiothiazole ($C_8H_{13}NS_2$)	**187**(22),144(24),131(42),126(22),119(16), 118(16),117(100),58(16),43(32),42(12), 41(24),32(64),28(75),27(16)
2-*n*-Heptylthio-2-thiazoline ($C_{10}H_{19}NS_2$)	**217**(8),174(8),170(22),146(11),133(22), 121(17),120(47),119(100),61(28),60(50), 59(17),57(17),43(14),41(44),29(14),28(19)

a) These products have been synthesized from the corresponding mercaptans using phase transfer catalysis alkylation conditions. Spectra were performed using a *Varian* Mat 111 instrument at 80 eV, source temperature 200°C, accelerating voltage : 820 V and trap current : 270 μA, GC–MS inlet system.

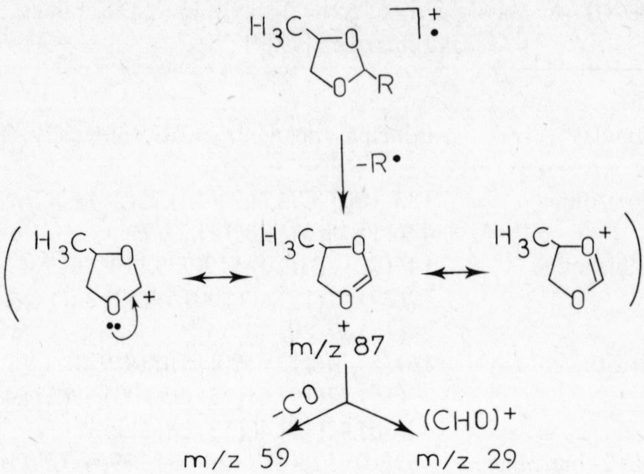

Scheme 10 – Fragmentation of 2-substituted 4-methyl-1,3-dioxolanes.

As yet isothiazoles have not been reported in food flavours, but their presence in *Maillard* reactions is highly probable (*see Chapter V*). The mass spectral behaviour of these compounds has been described [27,32]. The spectrum of the parent compound is quite similar to that of thiazole. The second most intense peak after the molecular ion at m/z 58 is formed by loss of hydrogen cyanide from the molecular ion. This process seems to be typical of nitrogen-containing five- or six-membered heterocyclic compounds which are not highly substituted.

In the mass spectra of the three methyl isomers, a ring extension occurs giving a 1,2-thiazine intermediate (*see Scheme 11*).

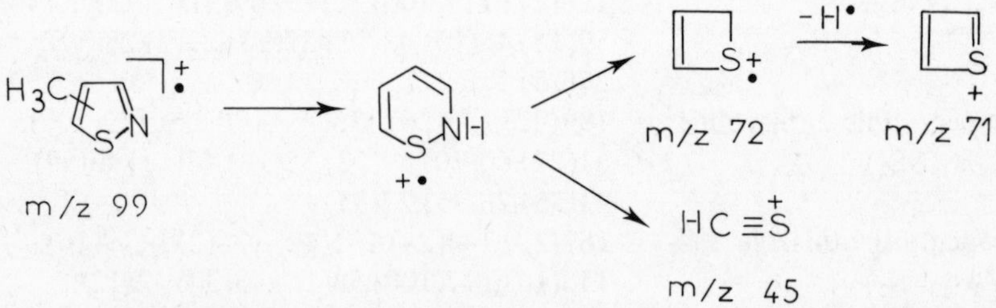

Scheme 11 – Fragmentation of methylisothiazole isomers.

But not all ions are derived from such structures, since the mass spectra for the three compounds are not identical. Indeed 4- and 5-methyl isomers show intense M – 1 peaks, analogous to those of methylpyrroles and methylpyridines, whereas 3-methyl isomer behaves like 4-methylthiazole and gives M – 1 ion of low intensity. Mass spectra of dimethylisothiazole isomers and trimethylisothiazole reported in *Table VII–3* show similar fragmentation patterns. As in the mass spectra of other heterocyclic compounds bearing a carboxylic or ester group adjacent to a methyl group an '*ortho* effect' is observed (*see Scheme 5*) with methylisothiazole carboxylates.

In some respects 5-hydroxymethylisothiazole resembles benzyl alcohol. The M - 1 ion is intense and the latter loses a hydrogen molecule to give M - 3 ion. The loss of the formyl radical from the molecular ion leads to the fragments shown in *Scheme 12*.

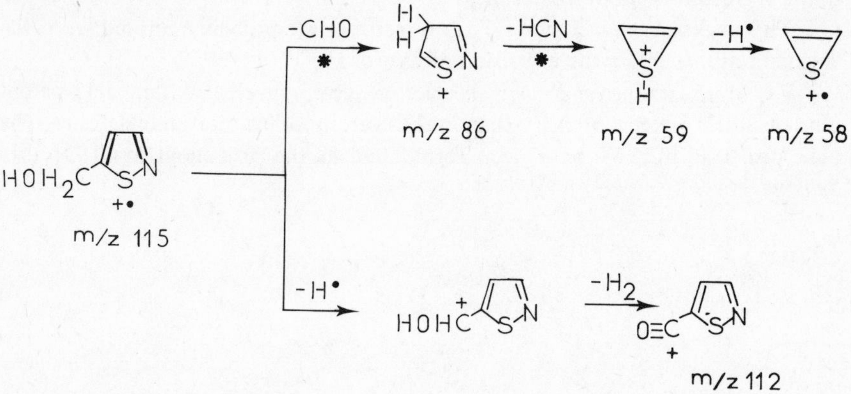

Scheme 12 – Fragmentation of 5-hydroxymethylisothiazole [27].

VII–4 LACTONES, THIOLACTONES AND LACTAMS

The mass spectra of γ- and δ-lactones have been studied in detail [12,19,33]. The main features which characterize these spectra can be summarized as follows:
(i) The abundance of the molecular ion is generally very low, and it is difficult to determine the molecular formula on electron impact alone.
(ii) The predominant fragmentation is the cleavage of the carbon—carbon bond next to the heteroatom leading to the base peaks at m/z 85 for γ-lactones ranging from C_6 to C_{18} and at m/z 99 for the δ-lactones ranging from C_8 to C_{12} (*see Scheme 13*).

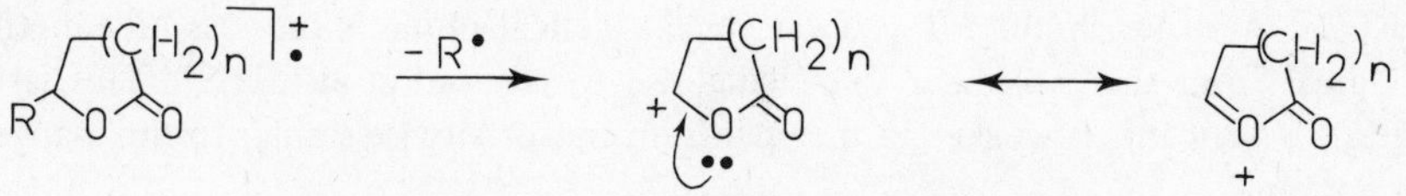

Scheme 13 – Fragmentation of γ- and δ-lactones.

This is due to the stabilization of the positive charge at the 5-position by the free electron of the heteroatom. When R = H or CH_3 this cleavage is less important. In this case the favoured reaction path is loss of 44 mass units from the molecular ions $(M - CO_2)^{+\cdot}$.

(iii) The low peak corresponding to the loss of a molecule of water $(M - 18)^{+\cdot}$ from the higher lactones having a carbon-chain length $n > 3$, may be useful to establish the molecular weight.

(iv) From the smaller lactones an ion at m/z 28 is due to the $C_2H_4^+$ fragment ion. This fragmentation mode is less significant for γ-lactones.

(v) The fragment ion at m/z 42 $(C_2H_2O)^+$ occurs at very low intensity in the spectra of γ-lactones from C_5 to C_{12}, but is however the second strongest peak in the spectrum of γ-butyrolactone.

(vi) The peaks at m/z 70 and 71 for δ-lactones are probably formed from the fragment at m/z 99 by the expulsion of CO (or HCO).$^{\cdot}$

γ-Crotonolactone and β-angelicalactone give, on electron impact, parent ions of much greater intensity than in the corresponding saturated lactones. The base peaks at m/z 55 have been formulated as the propenoyl ion [33] (*see Scheme 14*).

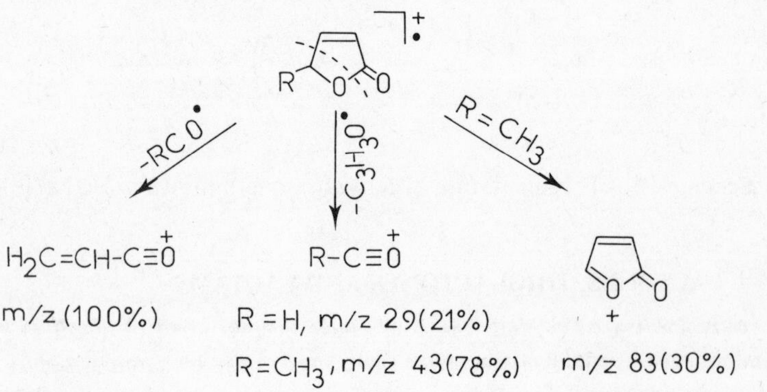

$-RCO^{\cdot}$ | $-C_3H_3O^{\cdot}$ | $R = CH_3$

$H_2C{=}CH{-}C{\equiv}\overset{+}{O}$ m/z (100%)

$R{-}C{\equiv}\overset{+}{O}$ R = H, m/z 29(21%) R = CH_3, m/z 43(78%)

m/z 83(30%)

Scheme 14 – Fragmentation of γ-crotonolactone (R = H) and β-angelicalactone (R = CH_3).

The loss of $C_3H_3O^{\cdot}$ $(M - 55)^+$ as the neutral fragment gives the acyl ions RCO^+. Another important process for β-angelicalactone is the loss of a methyl radical from the molecular ion giving rise to the cation at m/z 83. This latter peak is very much weaker in the spectrum but otherwise similar to the α-angelicalactone.

The mass spectrum of 2-(5*H*)-thiophenone can be rationalized as shown in *Scheme 15*.

In the spectrum of 2-pyrrolidone, the ion at m/z 56 is a triplet composed of the isobaric species $C_3H_4O^+$, $C_3H_6N^+$ and $C_2H_2NO^+$ arising from cleavage of 1,2 and 3,4 bonds (*see Scheme 16*). The M – 1 ion results from the loss of a hydrogen atom from position 5.

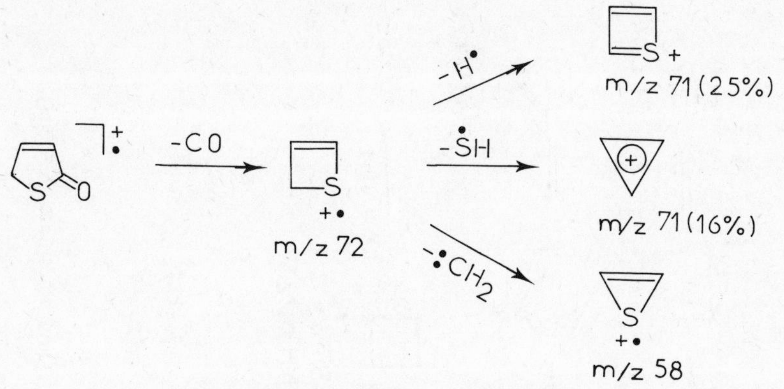

Scheme 15 – Fragmentation of 2-(5*H*)-thiophenone.

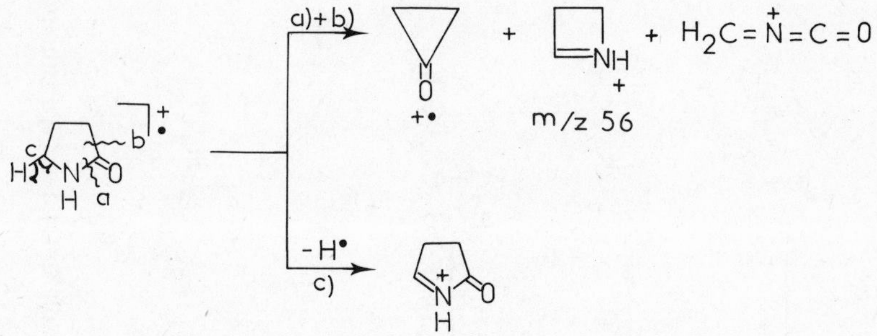

Scheme 16 – Fragmentation of 2-pyrrolidone.

VII–5 PYRANS, THIAPYRANS, PYRIDINES AND PYRAZINES

Pyrans and thiapyrans

The fragmentation of tetrahydropyran and piperidine are very similar. The base peaks are the M - 1 ions arising from a loss of an α-hydrogen atom (*see Scheme 17, way c*).

In the quite dissimilar mass spectrum of tetrahydrothiapyran, the base peak at $(M - CH_3)^+$ involves loss of an α-methylene together with a transferred β-hydrogen atom (*pathway a*). This ion is also the base peak in the spectra of the isomeric methyltetrahydrothiapyrans. Other ions present in these spectra can be explained by 1,2 and 2,3 bond cleavages.

The important fragmentation process of 4-pyrone (X = O) and 4-thiapyrone (X = S) is the 'retro *Diels-Alder*' reaction with loss of neutral acetylene and charge retention on the ketene fragment [2b,33]. Subsequent loss of carbon monoxide gives the ketene radical at m/z 42 (*Scheme 18*).

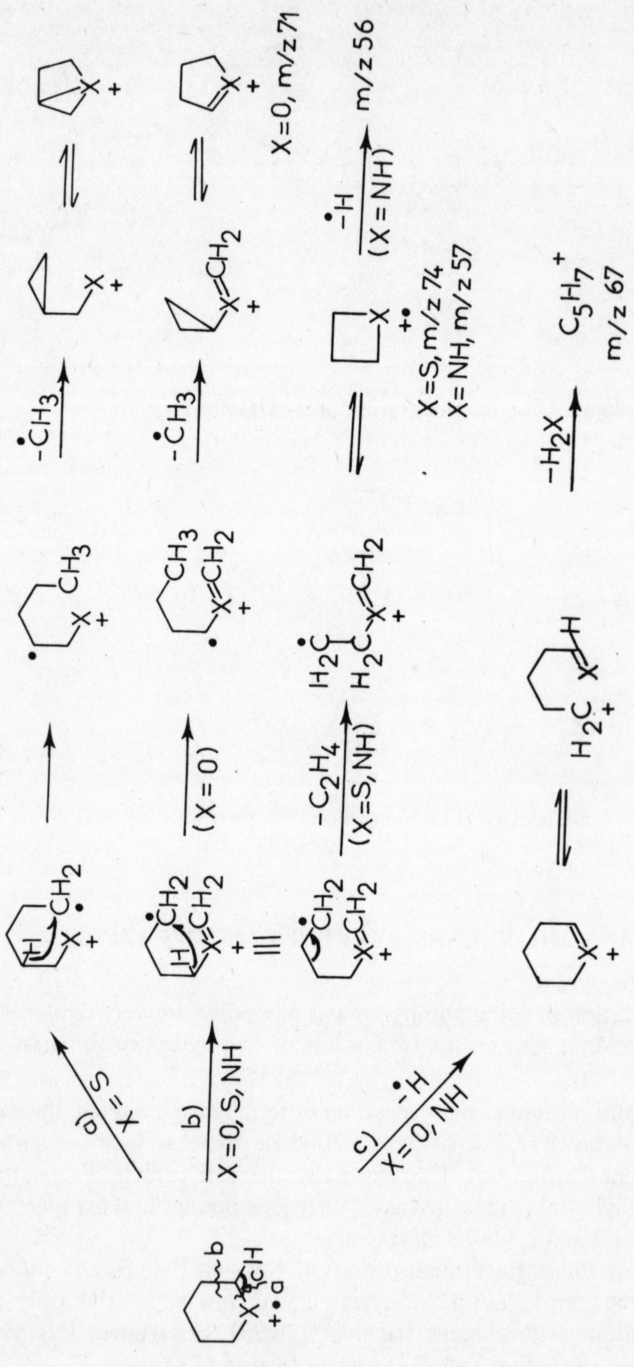

Scheme 17 – Fragmentation of tetrahydropyran (X =O), tetrahydrothiapyran (X = S) and piperidine (X = NH).

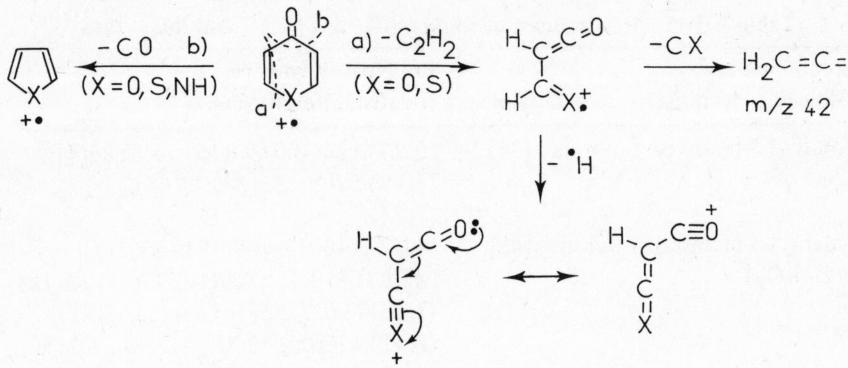

Scheme 18 – Fragmentation of 4-pyrone (X = O), 4-thiapyrone (X = S) and 4-pyridinone (X = NH).

A less important process is extrusion of carbon monoxide from the molecular ion to give species which may be formulated as furan and thiophene radical cations. The latter process also occurs in the mass spectrum of 4-pyridinone, but it is interesting to note that this compound does not lose the elements of acetylene in 'retro *Diels-Alder*' reaction. Thus the fragmentation of 4-pyrone appears to provide an adequate model for the fragmentation of 4-thiapyrone but not for 4-pyridone or its *N*-methyl derivative. Furthermore, it may also serve as the basis for the mass spectrometric differentiation with the 2-pyrones in which the 'retro *Diels-Alder*' fragmentation does not occur.

Mass spectra of recently synthesized 2-methyl-3-alkoxy-4-pyrones have been recorded [42]. They are listed in *Table VII–5*.

The molecular ions are generally of low intensity except for the lower members. These spectra are dominated by an intense acyl ion at m/z 43 which is the base peak in the mass spectra of methyl, ethyl, allyl, and isopentyl derivatives.

The mass spectra of isopropyl, *n*-butyl, isobutyl and *n*-pentyl compounds are quite similar to that of the parent compound since the base peak is at m/z 126. It arises by loss of a molecule of alkene from the molecular ions followed by hydrogen migration (*see Scheme 19*).

As in the case of 4-hydroxy-2-pyrones and 4-hydroxycoumarins, the ion at m/z 71 is probably best envisaged as arising from the keto form by a mechanism of 'retro *Diels-Alder*' reaction with hydrogen transfer [33]. It is by a similar mechanism that the formation of the ion at m/z 69 (from m/z 109) can be explained.

Pyridines

In pyridine and its methyl derivatives (*see Table VII–6*) molecular ions are the base peaks as expected for aromatic rings. Mass spectra of the methyl pyridine

Table VII–5 Mass spectra of some γ-pyrones [42] [a] and thiapyrans

Compound and (Molecular formula)	Ref.	Principal fragments (relative intensity, %)
2-Methyl-3-hydroxy-γ-pyrone ($C_6H_6O_3$)	[36]	127(7),**126**(100),97(15),94(8),85(4), 71(49),57(22),55(37),43(68),31(22), 27(23)
2-Methyl-3-methoxy-γ-pyrone ($C_7H_8O_3$)	[42]	141(7),**140**(92),139(19),125(21), 122(63),110(18),109(12),97(13),82(18), 71(16),69(77),55(18),54(21),53(21), 52(11),43(100),42(18),41(21),39(25), 29(11),27(38),26(27)
2-Methyl-3-ethoxy-γ-pyrone ($C_8H_{10}O_3$)	[42]	155(4),**154**(40),139(57),126(49), 125(11),111(13),110(69),109(12),97(27), 82(20),71(55),69(76),56(13),55(38), 54(19),53(15),44(10),43(100),41(14), 39(8),29(36),27(57),26(30)
2-Methyl-3-propyloxy-γ-pyrone ($C_9H_{12}O_3$)	[42]	**168**(15),126(20),111(100),98(86), 83(20),70(7),69(11),55(29),53(11), 43(53),41(70),39(32),29(13),27(33)
2-Methyl-3-isopropyloxy-γ-pyrone ($C_9H_{12}O_3$)	[42]	**168**(13),126(100),110(11),97(12), 71(30),69(16),56(7),55(20),43(55), 41(20),27(22)
2-Methyl-3-allyloxy-γ-pyrone ($C_9H_{10}O_3$)	[42]	**166**(1.6),110(14),97(24),96(52),95(11), 68(13),67(24),55(5),54(26),53(16), 43(100),42(11),41(27),40(13),39(32), 27(11)
2-Methyl-3-butyloxy-γ-pyrone ($C_{10}H_{14}O_3$)	[42]	**182**(5),139(12),126(100),110(17),71(16), 69(15),55(14),43(36),41(17),29(25), 27(19)
2 Methyl-3-isobutyloxy-γ-pyrone ($C_{10}H_{14}O_3$)	[42]	**182**(1.3),126(100),110(5),71(10),55(10), 43(20),41(15),29(13)
2-Methyl-3-pentyloxy-γ-pyrone ($C_{11}H_{16}O_3$)	[42]	**196**(5),181(6),127(24),126(100),110(15), 71(12),55(13),43(50),41(30),39(10), 29(15),27(17)
2-Methyl-3-isopentyloxy-γ-pyrone ($C_{11}H_{16}O_3$)	[42]	197(2),**196**(22),127(59),126(25),111(55), 98(53),71(14),70(20),55(29),43(100), 41(45),39(17),29(17),27(17)
γ-Thiapyrone (C_5H_4OS)	[2b]	114(4),113(8),**112**(100),86(68),84(36), 58(60),57(24),45(12),39(8)
3,5-Dimethyl-6-ethylthiapyran	[23a]	156(5),155(11),**154**(14),125(100),91(8)

a) The spectra from Ref. [42] were obtained on a *Ribermag R–10* spectrometer at 70 eV as the ionizing potential, GC–MS inlet system.

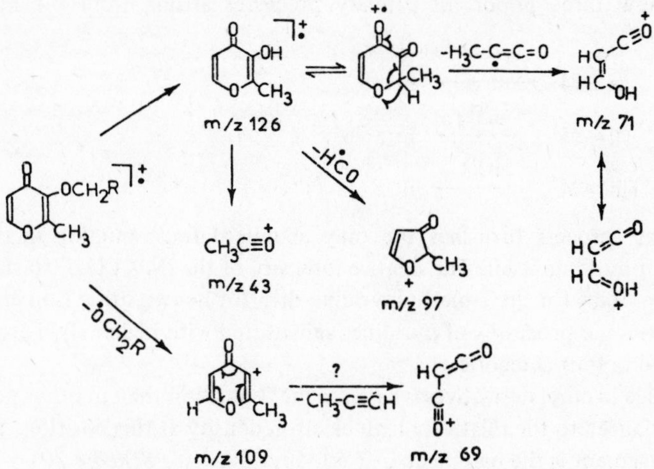

Scheme 19 – Supposed fragmentation patterns for 2-methyl-3-alkoxy-4-pyrones.

Table VII 6 Mass spectra of some pyridines

Compound and (Molecular formula)	Ref.	Principal fragments (relative intensity, %)
Pyridine (C_5H_5N)	[10a]	80(6),**79**(100),78(12),53(9),52(76),51(40), 50(31),39(12),28(7),27(7),26(21)
4-Methylpyridine (C_6H_7N)	[10a]	94(7),**93**(100),92(23),67(11),66(43), 65(22),51(16),50(11),40(15),39(41), 38(15)
2-*n*-Pentylpyridine ($C_{10}H_{16}N$)	[10a]	**149**,120(22),106(23),94(7),93(100),78(7), 65(9),51(7),39(14),29(11),27(16)
2-Acetylpyridine (C_7H_7NO)	[20]	79,78,**121**,43,51,52
3-Acetylpyridine (C_7H_7NO)	[10a]	**121**(44),106(87),79(16),78(100),52(15), 51(55),50(26),43(43),39(14),28(16)
2,6-Dimethylpyridine (C_7H_9N)	[10a]	108(8),**107**(100),106(27),92(19),79(9), 66(22),65(18),42(9),39(37),38(10),27(9)
3,5-Dimethylpyridine (C_7H_9N)	[10a]	108(8),**107**(100),106(49),92(21),79(36), 77(15),52(11),51(15),39(37),38(11), 27(17)
2-Methyl-5-ethylpyridine ($C_8H_{11}N$)	[10a]	122(5),**121**(54),120(16),107(9),106(100), 79(15),77(18),53(9),51(12),39(21),27(15)
Methyl nicotinate ($C_7H_7NO_2$)	[10a]	138(5),**137**(56),106(100),79(21),78(91), 51(62),50(33),44(32),16(38),15(35), 14(51)
N-Methyl-4-pyridone (C_6H_7NO)	[2b]	110(9),**109**(100),82(6),81(27),80(15), 55(9),54(6),53(9),42(18),39(9)

isomers show three important primary processes arising from the molecular ions [6]:

$$\text{(i)}\ M^{+\cdot} \xrightarrow{-H^{\cdot}} m/z\ 92$$

$$\text{(ii)}\ M^{+\cdot} \xrightarrow{-{}^{\cdot}CH_3} m/z\ 76$$

$$\text{(iii)}\ M^{+\cdot} \xrightarrow{-HCN} m/z\ 66$$

The last process furnishes the only abundant fragmentation in the mass spectrum of pyridine itself. The relative intensity of the $(M - CH_3)^+$ fragment ion is more important for the 2-methylpyridine than for its two other isomers.

The cleavage processes of pyridines substituted with higher alkyl groups can be classified in four categories:

– β-Cleavage in ethyl derivatives is easier in the 3 position than in other positions. This is attributed to the relatively high electron density at this position. Thus the resulting fragment is the base peak in 3-ethylpyridine (*see Scheme 20*).

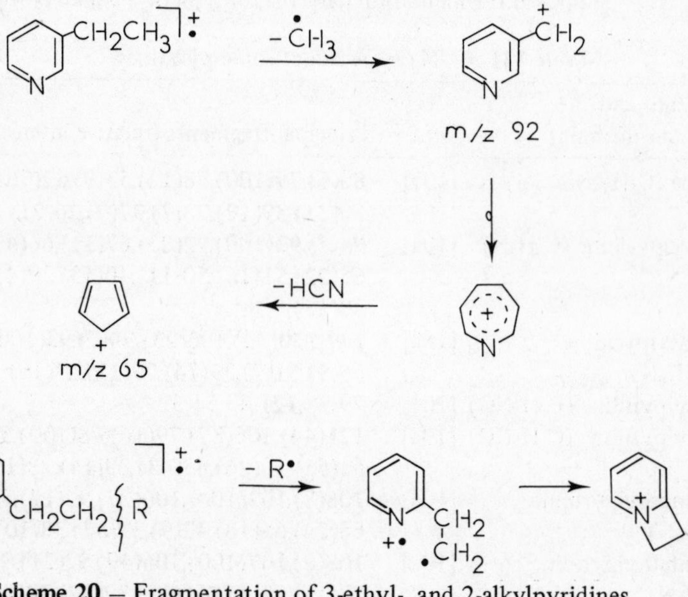

Scheme 20 – Fragmentation of 3-ethyl-, and 2-alkylpyridines.

These fragments undergo further elimination of hydrogen cyanide leading to the peak at m/z 65.

– γ-Cleavage is especially favoured in 2-alkylpyridines. The relative intensity of the resulting fragment ion depends on the nature of the radical lost.

– As previously described, the *MacLafferty* rearrangement takes place when the adjacent position to the heteroatom bears a side-chain with at least three carbon atoms.

– Finally, α-cleavage with hydrogen rearrangement can also occur in pyridines having a C_2-chain in 2-position. Thus 2-ethylpyridine eliminates ethylene.

2-Pyridone and *N*-methyl-4-pyridone in which tautomerism is prevented display the expected expulsion of carbon monoxide. On the other hand, the favoured fragmentation process of 3-hydroxypyridine and 4-pyridone is the loss of hydrogen cyanide giving rise to the furyl ion at m/z 68 (*see Scheme 21*).

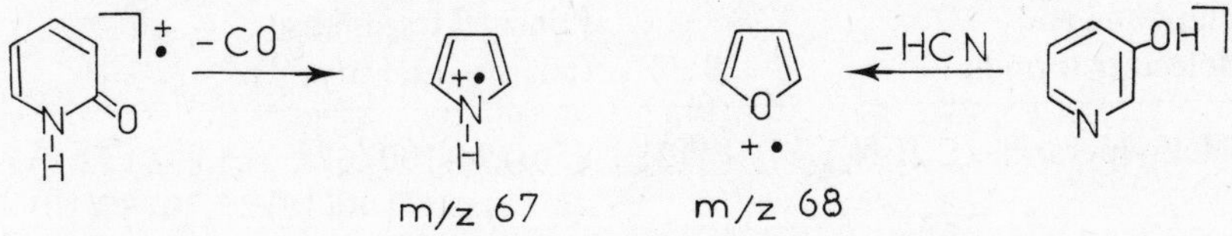

-CO
m/z 67
m/z 68
-HCN

Scheme 21 – Fragmentation of 2-pyridone and 3-hydroxypyridine.

Pyridine-3- and 4-carboxylic acids decompose by consecutive losses of OH and CO from the molecular ion. Mass spectra of methyl nicotinates found in jasmine [39] show the characteristic fragment ions $(M - OCH_3)^+$ and $[M - (OCH_3 + CO)]^+$ of the methyl ester group. In the 4-methyl derivative the '*ortho* effect' corresponding to the loss of methanol is also observed. The molecular ion of methyl pyridine-4-carboxylate loses H_2O in the first and second field-free regions. Exchange of a C(3)-H with a methoxy (CH_3) and operation of an '*ortho* effect' as shown in *Scheme 22* are responsible for the dehydration [29b].

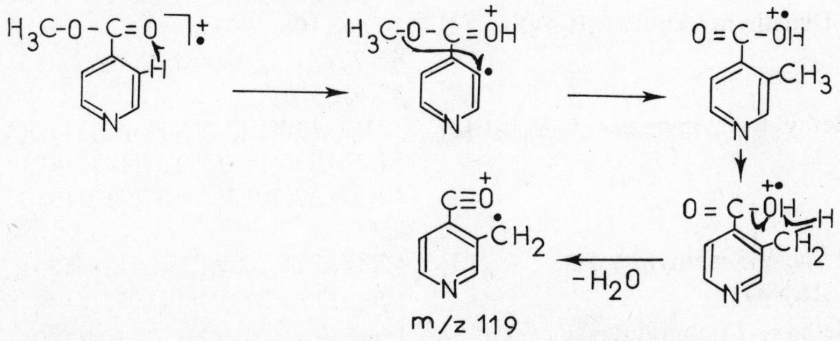

-H₂O
m/z 119

Scheme 22 – Fragmentation of methyl pyridine-4-carboxylate.

However, as a C(2)-H atom was also found to be incorporated with the eliminated H_2O, hydrogen exchange appears to proceed extensively.

Pyrazines

The mass spectrum of the parent compound is dominated by the loss of HCN molecules. The fragmentation of 2-methylpyrazine (*see Table VII–7*) involves losses of HCN and CH_3CN from the molecular ion as shown in *Scheme 23*.

In the spectrum of tetramethylpyrazine, the base peak at m/z 54 is derived by successive and simultaneous elimination of two molecules of acetonitrile from the molecular ion. Subsequent loss of a methyl radical from this ion [6] affords the ($C_3H_3^+$) cation at m/z 39.

Table VII 7 Mass spectra of some pyrazines

Compound and (Molecular formula)	Ref.	Principal fragments (relative intensity, %)
2-Methylpyrazine ($C_6H_5N_2$)	[2]	95(6),**94**(100),67(55),53(16),52(7), 42(13),41(7),40(18),39(20),38(10), 28(7),26(13)
2-Ethylpyrazine ($C_6H_{18}N_2$)	[2]	109(5),**108**(70),107(100),81(12), 80(21),56(9),54(5),53(16),52(16), 51(5),40(4),39(14),38(4),28(6),27(9), 26(12)
2-Acetylpyrazine ($C_6H_6N_2O$)	[47]	123(3),**122**(37),80(45),79(27),63(34), 52(44),43(100)
2,3-Dimethylpyrazine ($C_6H_8N_2$)		109(7),**108**(98),67(100),52(10),51(6), 42(21),41(13),40(19),39(13),28(7), 27(12),26(14)
2,5-Dimethylpyrazine ($C_6H_8N_2$)	[2]	109(8),**108**(100),81(17),52(7),42(86), 40(23),39(35),38(5),28(10)
2,6-Dimethylpyrazine ($C_6H_8N_2$)	[2]	109(8),**108**(100),81(7),67(7),66(6), 42(72),41(5),40(43),39(42),38(13), 37(6),27(6)
2-Methyl-6-vinylpyrazine ($C_7H_8N_2$)	[2]	121(9),**120**(100),94(12),93(5),66(4), 54(20),53(4),52(52),51(14),50(7), 40(12),39(20),38(5),37(3),28(5),27(4), 26(3)
2-Methoxy-3-methylpyrazine ($C_6H_8N_2O$)	[13]	125(8),**124**(100),123(28),109(35), 106(31),95(21),94(14),93(13)
2-Methoxy-3-isobutylpyrazine ($C_9H_{14}N_2O$)	[25]	**166**(8),151(26),125(10),124(100), 123(8),95(10),94(27),93(13),81(12), 80(5),54(9),53(12),52(7),43(10), 42(11),41(18),40(13),39(13)
Trimethylpyrazine ($C_7H_{10}N_2$)	[2]	123(7),**122**(81),81(20),80(5),54(10), 53(7),52(6),51(4),50(2),42(100),41(3), 40(12),39(25),38(3),28(4),27(9),26(2)
2,5-Dimethyl-3-ethylpyrazine ($C_8H_{12}N_2$)	[2]	137(9),**136**(86),135(100),121(7),108(19), 107(12),80(4),67(5),66(3),56(21), 54(5),53(9),52(6),51(3),42(42),41(7), 40(12),39(27),38(4),29(4),28(8),27(9)

Table VII–7 (*continued*)

Compound and (Molecular formula)	Ref.	Principal fragments (relative intensity, %)
2-Ethyl-3,5-dimethylpyrazine ($C_8H_{12}N_2$)	[14]	137(9),**136**(90),135(100),121(7),108(27), 107(7),94(2),80(4),67(6),56(35),42(33) 39(28),27(8)
Tetramethylpyrazine ($C_8H_{12}N_2$)	[6]	137,**136**(60),95(4),54(100),53(22),42(64), 39(28)
2,3-Dimethyl-5-*n*-pentylpyrazine ($C_{11}H_{18}N_2$)	(a)	**178**(24),163(10),149(36),135(58), 122(100),94(8),80(18),42(14)
2,3-Dimethyl-5-isopentyl pyrazine ($C_{11}H_{18}N_2$)	(a)	**178**(11),163(32),149(14),135(59), 122(100),94(8),80(22),42(14)
2,3-Dimethyl-5-(2-methylbutyl)-pyrazine ($C_{11}H_{18}N_2$)	(a)	**178**(16),163(41),149(39),135(23) 123(46),122(100),94(9),80(22),42(18)
2,3-Dimethyl-5-neopentyl-pyrazine ($C_{11}H_{18}N_2$)	(a)	**178**(27),163(45),123(36),122(100), 121(30),94(8),80(18),57(78),42(24)
2,3-Dimethyl-5-(1,2-dimethyl-propyl)-pyrazine ($C_{11}H_{18}N_2$)	(a)	**178**(37),163(64) 150(16),137(46), 136(94),135(100),122(86),108(30), 80(9),42(30)
2,3-Dimethyl-5-(1-methylbutyl)-pyrazine ($C_{11}H_{18}N_2$)	(a)	**178**(15),163(37),149(54),137(40), 136(100),135(95),122(16),108(30), 80(11),42(5)
2,3-Dimethyl-5-(1-ethylpropyl)-pyrazine ($C_{11}H_{18}N_2$)	(a)	**178**(41),163(54),150(93),149(95), 136(45),135(100),122(95),108(8), 80(24),53(8),42(8)
2-Isobutyl-3,5,6-trimethyl-pyrazine ($C_{11}H_{18}N_2$)	(a)	**178**(44),177(32) 163(54),136(100), 135(51),122(24),94(30),80(11),54(32), 53(42),42(38)
2-(2-Methylbutyl)-3,5,6-trimethylpyrazine ($C_{12}H_{20}N_2$)	(a)	**192**(5),177(27),164(16),137(38) 136(100),135(27),94(8),53(16),42(8)
2,3-Dimethyl-5-(1,5-dimethyl-4-hexenyl)-pyrazine ($C_{14}H_{22}N_2$)	(a)	**218**(34),203(8),175(9),149(32) 137(49), 136(100),135(78),108(15),85(9),41(10)
2,3-Dimethyl-5-(3,7-dimethyl-7-heptenyl)-pyrazine ($C_{16}H_{26}N_2$)	(a)	**246**(38),231(9),177(30),163(18),150(35), 136(32),135(86),123(35),122(100), 109(14),69(14),41(8)

(a) Reported by Kitamura & Shibamoto, *J. Agric. Food Chem.*, **29**, 188 (1981). A *Hitachi* Model *RMU 6M* mass spectrometer was used at 70 eV.

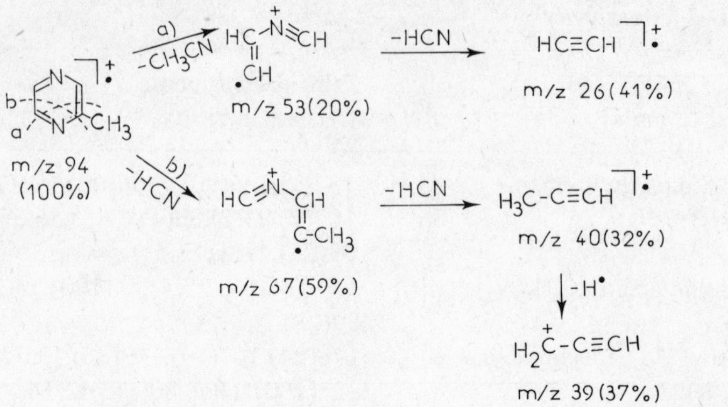

Scheme 23 – Fragmentation of 2-methylpyrazine [33].

Other fragments occur at m/z 41 (M - C_3H_3N) and 27 (M - C_4H_5N). Pyrazines which possess an alkyl side-chain containing three or more carbons undergo the *MacLafferty* rearrangement [14,21,22]. Another important fragmentation of alkylpyrazines is γ-cleavage of side-chains which are either ethyl group or longer ones. The loss of 15 mass units from the molecular ion $(M - CH_3)^+$ is much more abundant from an isobutyl group than from an *n*-butyl group. Methoxyalkylpyrazines are characterized by the loss of the methoxyl group from the molecular ion $(M - OCH_3)^+$.

2,3- and 2,6-Methylmethoxypyrazine isomers can be distinguished because the 2,3 isomer undergoes the expulsion of a molecule of methanol from the molecular ion by means of the '*ortho* effect'. This process does not occur with the 2,6 isomer.

VII–6 POLYSULPHUR-CONTAINING RING SYSTEMS

Mass spectra of some polysulphur-containing ring systems are listed in *Table VII–8.*

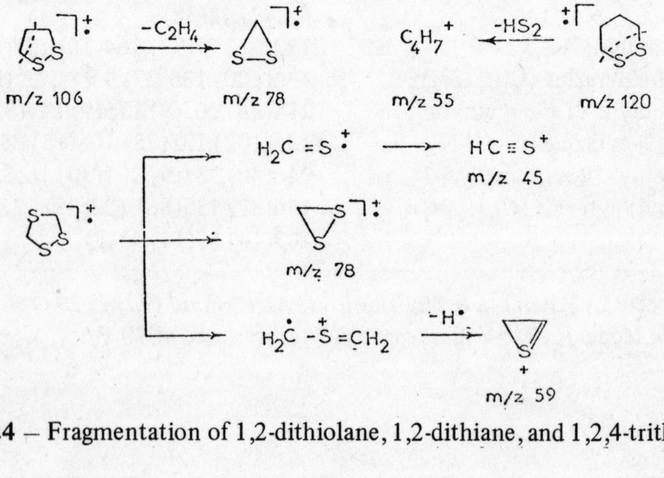

Scheme 24 – Fragmentation of 1,2-dithiolane, 1,2-dithiane, and 1,2,4-trithiolane.

The main feature of the fragmentation of 2-ethyl-4-methyl-1,3-dithiole is the loss of the ethyl radical giving the base peak at m/z 117. Mass spectra of 1,2-dithiolane and 1,2-dithiane are characterized by loss of ethylene and elements of HS_2, respectively (*see Scheme 24*).

Table VII 8 Mass spectra of some polysulphur-containing ring systems

Compound and (Molecular formula)	Ref.	Principal fragments (relative intensity, %)
2-Ethyl-4-methyl-1,3-dithiole ($C_6H_{10}S_2$)	[23a]	**146**(21),117(100),45(27),41(24)
3,5-Dimethyl-1,2,4-trithiolane ($C_4H_8S_3$)	[11]	154(14),153(7),152(100),92(66),88(62), 64(46),60(36),59(63),55(25),45(23)
1,2-Dithiane ($C_4H_8S_2$)	[10c]	**120**(75),87(8),86(8),64(10),56(8) 55(100),45(10),41(9),39(9)
2,6-Dimethyl-1,3-dithiin ($C_6H_{10}S_2$)	[23]	**146**(44),131(8),86(91),85(100),60(43), 59(58),45(83),41(36),39(35)
1,4-Dithiane ($C_4H_8S_2$)	[10b]	122(9),121(6),**120**(100),92(23),64(20), 61(68),60(30),59(17),58(14),46(80), 45(43),27(13)
1,2,4-Trithiolane ($C_2H_4S_3$)	(a)	**124**(82),78(94),64(8),60(47),59(25), 46(77),45(100)
1,2,4-Trithiane ($C_3H_6S_3$)	[23b]	**138**(70),110(29),92(30),73(55),64(56), 46(80),45(100)
2,4,6-Trimethyl-1,3,5-trithiane ($C_6H_{12}S_3$)		**180**(63),120(26),88(15),61(21),60(100), 59(18),58(9),55(23),45(22),27(9)
2,4,6-Trimethyl-1,3,5-dithiazine ($C_6H_{13}NS_2$)	[23a]	**163**(45),103(18),74(18),71(36),70(30), 60(43),56(33),44(100)
3,5-Dimethyltetrathiane ($C_4H_8S_3$)	[11]	186(6),185(3),**184**(34),152(12),124(32), 119(12),92(14),64(24),60(77),59(100), 45(35)

a) Reported by Gil & MacLeod, *J. Agric. Food Chem.*, **29**, 484 (1981).

Mass spectral fragmentations of 1,2,4-trithiolane in egg flavour volatiles were recently proposed by Gil & MacLeod (*see Scheme 24* and *Table VII–8*). The two main fragments are the thioformyl cation at m/z 45 and the radical cation dithiirannium at m/z 78.

The spectral data of *cis* and *trans* 3,5-dimethyl-1,2,4-trithiolane [11] were very similar and had the following characteristics:

(i) A high relative intensity of the molecular ion;
(ii) A fragment at m/z 92 corresponding to the loss of the C_2H_5S element (M - 60);
(iii) The height of the M + 2 peak (e.g., three sulphur atoms).

In the mass spectra of 2,6-dimethyl-1,3-dithiin, 2,4,6-trimethyl-1,3,5-trithiane and 2,4,6-trimethyl-1,3,5-dithiazine the peaks corresponding to the loss of the C_2H_4S fragment (M - 60) are also predominant. The mass spectrum of lenthionine [33] shows a series of ions corresponding to $(CH_2)_2S_x$ (x = 3,4,5) and $(CH_2)S_y$(y = 1,2,3,4,5) as well as the species $(M\text{-}S_2)^+$, $(M\text{-}S_3)^+$, and $(M\text{-}CHS)^+$.

VII–7 *BIS* HETEROARYL COMPOUNDS AND FUSED HETEROCYCLIC RING SYSTEMS

Bis heteroaryl compounds

The effect of extending the conjugation of one five- or six-membered heterocyclic system to a *bis*-heterocyclic system is to increase the stability of the

Table VII–9 Mass spectra of some *bis*-heteroaryl aroma compounds derivatives

Compound and (Molecular formula)	Ref.	Principal fragments (relative intensity, %)
2,2′-Bifuran ($C_8H_6O_2$)	[15b]	135(10),**134**(100),105(27),78(37),77(8), 51(18),39(10)
2-(2′-Thienyl)-thiazole ($C_7H_5NS_2$)	[40]	169(10),168(10),**167**(96),58(100),45(8), 39(5)
5-(2′-Thienyl)-3-methyl-isothiazole ($C_8H_7NS_2$)	[40]	183(11),182(12),**181**(100),140(30), 108(8),96(35),74(9),69(9)
5-(2′-Thienyl)-3,4-dimethyl-isoxazole (C_9H_9ONS)	[40]	181(4),180(10),**179**(100),122(8),111(50), 110(44),109(12),96(9),66(11),45(8),43(8)
2-(2′-Furyl)-thiazole (C_7H_5ONS)	[42]	153(3),152(6),**151**(59),111(32),110(23), 58(100)
2-(2′-Thienyl)-pyridine (C_9H_7NS)	[40]	163(7),162(15),**161**(100),160(42),135(8), 128(10),117(26),116(11),78(12),51(9), 45(5)
2-(2′-Thiazolyl)-pyridine ($C_8H_6N_2S$)	[41]	164(12),163(16),**162**(100),161(20), 136(10),105(10),79(20),78(12),58(100)
2-(2′Thiazolyl)-4-methyl-pyridine ($C_9H_8N_2S$)	[41]	178(7),177(13),**176**(100),161(13),118(46), 100(23),92(20),91(20),73(13),58(100), 44(86)
3-(2′-Thiazolyl)-4-methyl-pyridine ($C_9H_8N_2S$)	[41]	178(8),177(14.5),**176**(80),175(36), 161(27.5),119(19),100(23),93(22),77(20), 58(100),44(60)
(2′-Furyl)-pyrazine ($C_8H_6N_2O$)	[48]	**146**(100),118(18),93(56),64(26),63(25), 53(22),43(22),39(28),38(18)
(2′-Thienyl)-pyrazine ($C_8N_6H_2S$)	[40]	164(6),163(10),**162**(100),135(10),109(55)
2-(2′-Pyrrolyl)-3-methyl-pyrazine ($C_9H_9N_3$)		**159**(100),158(75),92(74),51(57),39(68)

molecular ion which is usually the base peak in the spectra of *bis*-heteroaryl aroma compounds (*see Table VII–9*).

Other fragments are characteristic of the cleavage of heteroaryl substituents.

Fragmentation of 2,2′-bifuran [15b] occurs by loss of a formyl radical to give the benzoyl cation or by loss of $^{\bullet}C_3H_3$ to give the furoyl cation. Both fragmentations occur but the benzoyl cation is the more abundant (*see Scheme 25*).

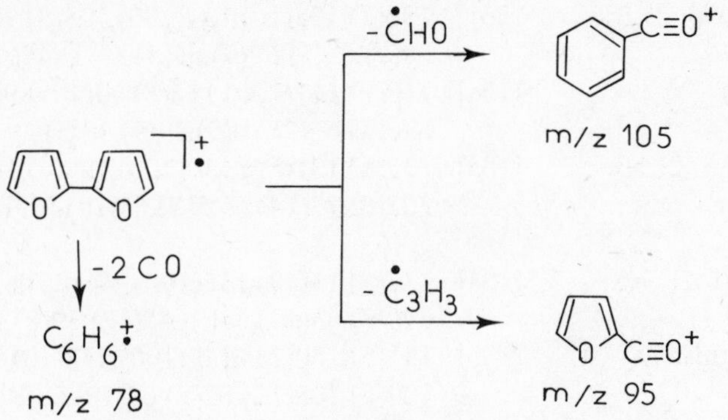

$-\dot{C}HO$

m/z 105

$-2\,CO$

$C_6H_6^{+\bullet}$

m/z 78

$-\dot{C}_3H_3$

m/z 95

Scheme 25 – Fragmentation of 2,2′-bifuran.

In the mass spectra of 2-heteroarylthiazoles [40,41] the abundant fragment at m/z 58 is characteristic of the cleavage of the thiazole ring according to the 1,2 and 3,4 bonds. Another interesting feature of the mass spectrum of 2-(2′-furyl)-thiazole is the formation of the thiofuroyl cation at m/z 111 (*see Scheme 26*).

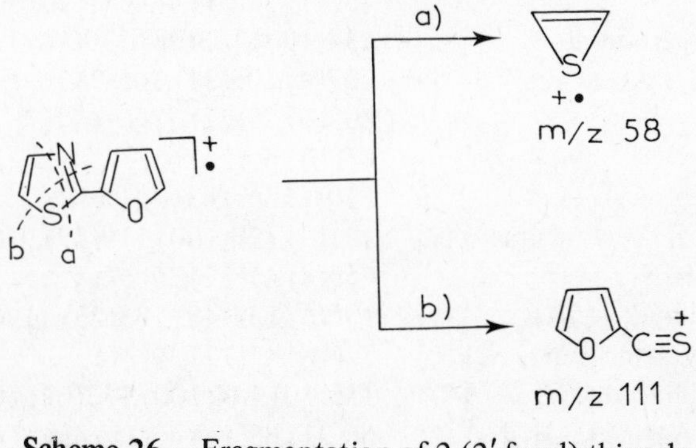

a)

m/z 58

b)

m/z 111

Scheme 26 – Fragmentation of 2-(2′-furyl)-thiazole.

Fused heterocyclic ring systems

The conjugation effect is yet more pronounced in the fused ring systems. Indeed the molecular ions are the base peaks in the spectra of parent compounds. Peaks associated with (M - CX) or $(M - HCX)^{+\bullet}$ ions may also be seen in these spectra.

Table VII–10 Mass spectra of some fused heterocyclic systems

Compound and (Molecular formula)	Ref.	Principal fragments, relative intensity, %)
Benzofuran (C_8H_6O)	[36]	119(9),**118**(100),90(33),89(31),63(16), 59(5),51(3),45(5),39(7)
Benzothiophene (C_8H_6S)	[36]	136(5),135(10),**134**(100),108(3),90(7), 89(9),69(5),67(9),63(5),57(3),45(4),39(3)
Indole (C_8H_7N)	[10a]	118(9),**117**(100),116(8),90(40),89(24), 63(14),62(7),50(5),39(9),38(5)
3-Methylindole (C_9H_9N) (skatole)	[10a]	132(6),**131**(59),130(100),129(6),103(9), 102(6),77(14),65.5(7),65(10),51(10), 50(6)
Benzothiazole ($C_7H_{15}NS$)	[10a]	137(5),136(9),**135**(100),108(35),82(12), 69(27),63(13),54(9),45(12),39(8),38(8)
2-Methylbenzothiazole ($C_8H_{17}NS$)	[10c]	151(5),150(11),**149**(100),148(10),117(4), 109(6),108(12),82(8),69(10),63(8) 55(6), 43(5)
2-Methylthiobenzothiazole ($C_8H_{17}NS_2$)	[10a]	182(13),**181**(100),180(22),148(60), 136(14),135(15),108(32),69(17),45(26), 15(17)
Quinoline (C_9H_7N)	[10a]	130(11),**129**(100),128(17),103(7),102(23), 76(11),75(9),51(18),50(13),39(6)
Isoquinoline (C_9H_7N)	[10a]	130(11),**129**(100),128(17),103(8),102(24), 76(9),75(10),74(7),51(20),50(13)
5,6,7,8-Tetrahydroquin-oxaline ($C_8H_{10}N_2$)	[45]	**134**(100),133(78),130(18),119(27), 107(9),106(31),105(9),103(9),92(13), 80(10),79(22),78(13),77(7),76(10),69(9), 67(16),66(16),65(9),54(11),53(9),52(22), 51(9),50(8),41(7),39(13)
6,7-Dihydro-(5*H*)-cyclopenta-pyrazine ($C_7H_8N_2$)	[47]	121(9),**120**(100),119(87),93(11),66(19), 65(18),41(22),39(27)
5-Methyl-6,7-dihydro-(5*H*)-cyclopentapyrazine ($C_8H_{10}N_2$)	[47]	135(5),**134**(48),133(25),119(100),78(12), 52(16),41(12),39(21)
2-Methyl-6,7-dihydro-(5*H*)-cyclopentapyrazine ($C_8H_{10}N_2$)	[47]	135(10),**134**(100),133(69),107(16), 66(31),65(14),41(15),40(17),39(46)
2-Ethyl-6,7-dihydro-(5*H*)-cyclopentapyrazine ($C_{10}H_{12}N_2$)	[45]	**148**(96),147(100),133(9),121(6),120(17), 119(6),93(5),66(11),65(13),54(5),53(9), 52(5),41(9),39(24)

Simple methyl derivatives give upon electron impact an abundant molecular ion which is usually the base peak. However, some exceptions can be observed (*Table VII–10*). Thus in the mass spectra of 2- and 3-methylbenzofuran, six methyl-benzothiophenes and methylindole isomers (except for the *N*-methyl derivative), the most abundant fragment ion arises from the loss of a hydrogen atom from the molecular ion (*see Scheme 27*).

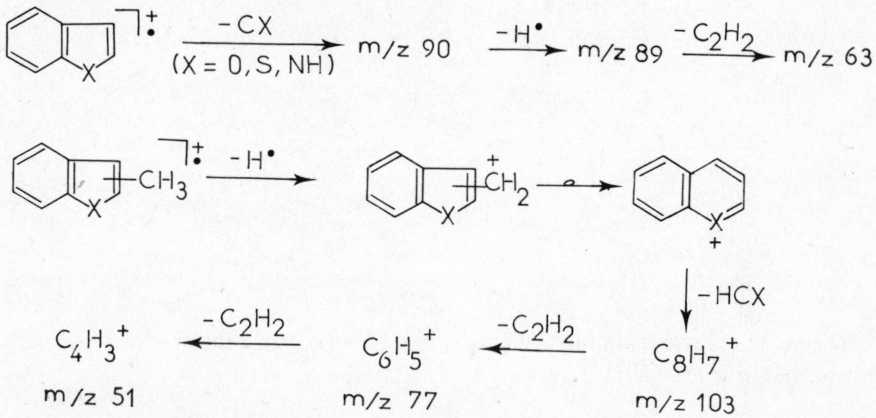

−CX (X = O, S, NH) → m/z 90 −H• → m/z 89 $-C_2H_2$ → m/z 63

$-CH_3$ −H• → CH_2^+ → −HCX → $C_8H_7^+$ m/z 103 → $-C_2H_2$ → $C_6H_5^+$ m/z 77 → $-C_2H_2$ → $C_4H_3^+$ m/z 51

Scheme 27 – Fragmentation of benzofuran, benzothiophene, indole, and their methyl derivatives.

The formation of such ions explains the loss of HCX from the M - 1 ions. A characteristic feature in the mass spectra of 2- and 3-methylindole (X = NH) is the intense double charged ions at m/z 65. For higher homologues, the base peak is due to β-cleavage with loss of the alkyl group.

The mass spectrum of thieno[2,3-c] thiophene is a fairly simple one. The relatively intense M - 1 ion (22%) may result from ring-opening leading to the well stabilized ion at m/z 139.

The only other significant fragment ions result from losses of CS and ${}^{\bullet}$CHS from the molecular ion [33].

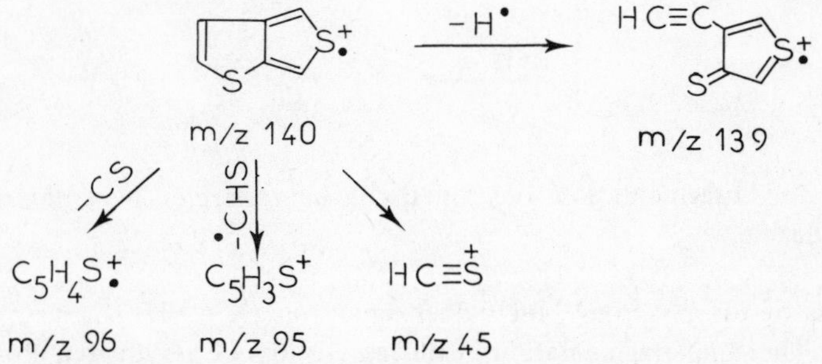

m/z 140 −H• → HC≡C– m/z 139

−CS → $C_5H_4S^{+\bullet}$ m/z 96; −•CHS → $C_5H_3S^+$ m/z 95; → $HC{\equiv}S^+$ m/z 45

Scheme 28 – Fragmentation of thieno [2,3-c] thiophene.

Fragmentations of benzoxazole (X = O), benzothiazole (X = S), and benzimidazole (X = NH) closely parallel those of the corresponding five-membered ring systems. The molecular ions are the base peaks and the most important fragmentation arises by loss of hydrogen cyanide (*see Scheme 29*).

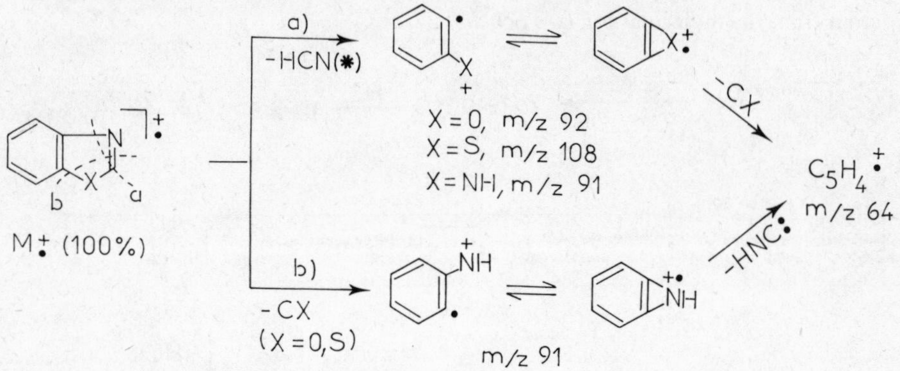

a) −HCN(*)

X = O, m/z 92
X = S, m/z 108
X = NH, m/z 91

−CX

$C_5H_4^{+\bullet}$ m/z 64

−HNC

$M^{+\bullet}$ (100%)

b) −CX (X = O,S)

m/z 91

Scheme 29 – Fragmentation of benzoxazole (X = O), benzothiazole (X = S), and benzimidazole (X = NH).

The loss of another molecule of HCN from the benzimidazole ion leads to the species $C_5H_4^{+\bullet}$ at m/z 64. A second fragmentation process occurs in the mass spectra of benzoxazole and benzothiazole involving losses of CO and CS from the molecular ions, respectively. 2-Methylbenzothiazole (X = S) shows the expected elision of CH_3CN at m/z 108 together with an important loss of $^{\bullet}CH_2CN$ (m/z 109) (*see Scheme 30*).

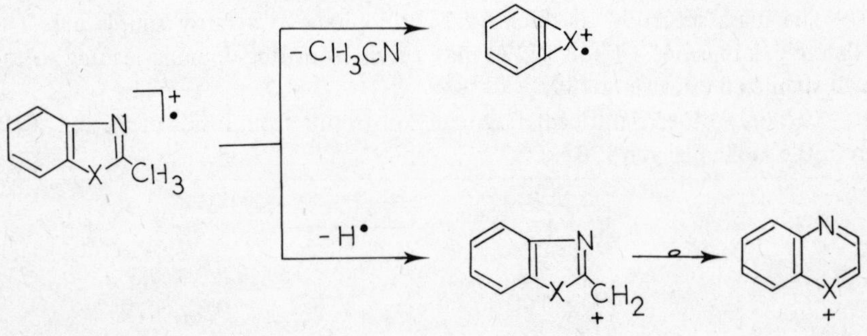

$-CH_3CN$

$-H^{\bullet}$

Scheme 30 – Fragmentation of 2-methyl-, benzoxazole, benzothiazole, and benzimidazole.

Loss of CH_3CN from 2-methylbenzimidazole (X = NH) is a less important process. The chief fragmentation in this case is loss of a hydrogen atom to give an intense M - 1 ion at m/z 131 (60%).

The mass spectra of alkylquinolines and isoquinolines [35a] evidenced an azatropylium ion intermediate leading to the $[M-(H^{\cdot}+HCN)]^{+}$ ion from methyl derivative. The predominant fragmentation modes in compounds having more than three carbon atoms in the chain are β-cleavage and the *MacLafferty* rearrangement. The relative amounts of these cleavages can be correlated with the electron density at the position of substitution.

Alkyl chains at C-4 and C-8 undergo facile γ and δ cleavage, probably because the resulting radicals can be stabilized by cyclization.

Mass spectra of 6,7-dihydro-(5*H*)-cyclopentapyrazines and tetrahydroquinoxalines have been reported by several authors [14,30,45]. Alkyl derivatives of cyclopentapyrazines with no substituents on the alicyclic ring show characteristic peaks at m/z 66 and 52, and the base peak is the molecular ion. However, the 2-ethyl derivative exhibits M – 1 as the most abundant ion, typical of the ethyl substituted alkylpyrazines. The fragmentation pathway for 5-methyl-6,7-dihydro-(5*H*)-cyclopentapyrazine is given in *Scheme 31*.

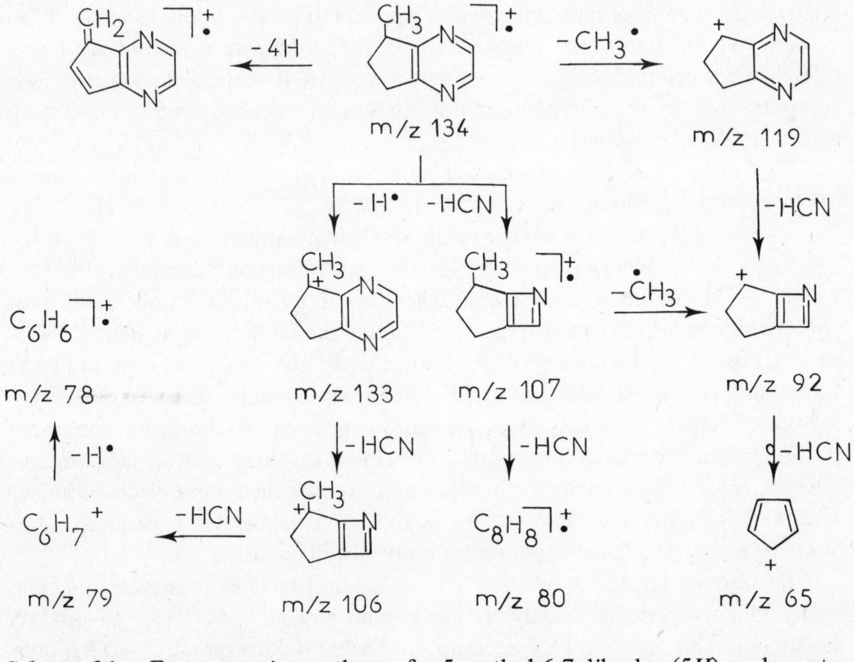

Scheme 31 – Fragmentation pathway for 5-methyl-6,7-dihydro-(5*H*)-cyclopentapyrazine [45].

The most characteristic feature of this spectrum is loss of a methyl radical, giving a base peak at m/z 119. Subsequent loss of hydrogen cyanide leads to an ion of mass 92 which likewise ejects HCN to give the $(C_5H_5)^{+}$ cation at m/z 65.

Of some significance is the M - 1 species which is of approximately equal abundance in comparison with the molecular ion. Two consecutive losses of hydrogen cyanide from m/z 107 and 80 of low abundance are due to the consecutive expulsion of hydrogen cyanide from the molecular ion.

VII–8 BANK MASS SPECTRA DATA: AN INDISPENSABLE TOOL FOR A FAST IDENTIFICATION OF HETEROCYCLIC AROMA COMPOUNDS

The identification of a compound from its mass spectrum remains difficult on account of two factors:

– Owing to the wide range of structures encountered it is not possible to put forward simplifying hypothesis meant to interpret the spectrum.

– General fragmentation rules are not always available for all types of products.

The only reliable method consists in comparing the unknown product spectrum with that of reference compounds spectra and observe when they are the same.

The chances of success being closely dependent on the number of reference compounds, a bank of mass spectra data appears to be absolutely essential. When carried out by hand, the comparison of the unknown spectrum with each reference spectrum rapidly becomes very tiresome. It is then necessary to use a computer and set up a program in order to trace the desired compound as safely and as rapidly as possible.

This method of comparison will be a hundred per cent successful, only if the two following conditions are fulfilled:

Identical compounds ⇒ identical spectra
Identical spectra ⇒ identical compounds

The first condition can be considered as validly fulfilled, only if the compound in question is always fragmented under similar conditions, i.e., with the same apparatus and under similar conditions. In the case when the unknown compound and the standard compound are analysed under identical conditions, reproducibility is excellent, and the various spectra of the same compound hardly differ. Otherwise, very significant differences may appear. Nevertheless, in most cases the available standard spectra are obtained from electron impact around 70 eV, and the same sort of analysis is generally used for aroma components, as a result, the first condition is fairly well admitted.

The second implication is nearly always admitted, the mass spectrum being highly characteristic, and only a few isomers are at times likely to produce indistinguishable spectra. If a spectrum happens to correspond to several compounds, the determination is completed by means of Kováts indexes which characterize retention on a given chromatographic column. It is always necessary to calculate Kováts indexes in order to prove or disprove the results given by mass spectrometry. Besides, these indexes can be fed into the bank computer.

With a view to building up this bank, it is advisable to collect reference spectra, the latter should be as complete as possible without storing useless information, however, such as too low peaks which may be caused by impurities. Furthermore, the conditions under which they are obtained should be very narrowly related to those that will be used for one's own analysis. These reference spectra can be compounds previously identified from aromas, or compounds of parent structures, or compounds predicted in a computer-assisted synthesis (such as the products of the reaction aldehydes + ammonia + hydrogen sulphide) (*see Chapter V*).

Then a spectrum research program must be set up. A numerical criterion is generally selected such as:

$$Y = 0 \Rightarrow \text{identical spectra}$$
$$\text{identical spectra} \Rightarrow Y = 0$$

As the rigorous equality of two different spectra is of rare occurrence it is necessary to replace $Y = 0$ by $Y \neq 0$ but as small as possible. A threshold value is determined such as:

$$Y \leqslant Y_0 \Leftrightarrow \text{identical spectra}$$
$$Y > Y_0 \Leftrightarrow \text{different spectra}$$

If this threshold value is well chosen, i.e. if it takes into account the variations due to the reproductibility of the equipment and all the approximations and uncertainties of the spectra published in the literature, two cases are to be considered:

– Case $Y \leqslant Y_0$ corresponds to spectra that cannot be distinguished by using mass spectrometry (in that case Kováts indexes are calculated in order to find out whether the isomer is the right one), and

– Case $Y > Y_0$ corresponds to a compound still missing from the bank data (or to a spectrum consisting of a mixture of compounds).

The numerical criterion itself must satisfy the following conditions:

– It is imperative that it should cancel out for strictly identical spectra.
– It should make the best use of all the information contained in spectra.
– It must be simple so as to have acceptable calculation times.

As a matter of course, combinations of numerical criteria can be used along with a sensible series of 'or' and of 'and'.

The set represented by the bank of mass spectra, the spectra research program, and Kováts indexes may constitute, if it is properly worked out, a very powerful tool in the analysis of aromas. Besides, our laboratory possesses a bank of that type which already contains about a thousand heterocyclic compounds.† There exist, of course, other sorts of conceptions, such as theoretical methods of interpretation and statistical methods, but at any rate the future looks more promising for the above-described conception. This is true, at least in the particular field of aromas for which theoretical fragmentation rules remain unknown.

† Vernin *et al., Ind. Aliment. agric.*, 9, 741–751 (1981) would like to thank the French Research and Technology Ministry (MIDIST) for financial support of this work.

VII–9 REFERENCES

[1] Boelens, M., L. M. Van Der Linde, P. J. De Valois, H. M. Van Dort & H. C. Takken, *J. Agric. Food Chem.*, **22**, 1071 (1974).

[2a] Bondarovich, H. A., P. Friedel, V. Krampl, J. A. Renner, F. W. Shephard & M. A. Gianturco, *J. Agric. Food Chem.*, **15**, 1093 (1967).

[2b] Bonham, J., E. McLeister & P. Beak, *J. Org. Chem.*, **32**, 639 (1967).

[3] Bowie, J. H., R. G. Cooks, S. C. Lawesson & C. Nolde, *J. Chem. Soc.*, (B), **1967**, 616.

[4] Bowie, J. H., P. F. Donaghue, H. J. Rodda, R. G. Cooks & D. H. Williams, *Org. Mass. Spectrometry*, **1**, 13 (1968).

[5] Bowie, J. H., R. G. Cooks, S. O. Lawesson & G. Schroll, *Austral. J. Chem.*, **20**, 1613 (1967).

[6] Budzikiewicz, H., C. Djerassi & D. H. Williams, in *Mass Spectrometry of Organic Compounds*, Holden-Day Inc. (1967).

[7] Budzikiewicz, H., C. Djerassi, A. H. Jackson, G. W. Kenner, D. J. Newman & J. M. Wilson, *J. Chem. Soc.*, **1964**, 1949.

[8] Buttery, R. G., L. C. Ling & R. E. Lundin, *J. Agric. Food Chem.*, **21**, 488 (1973).

[9] Clarke, G. M., R. Grigg & D. H. Williams, *J. Chem. Soc.*, (B), **1966**, 339.

[10] Cornu, A., R. Massot, *Compilation of Mass Spectral Data*, Section C, Heyden & Sons Limited, London (1966a); *ibid*, Supplement 1 (1967b); *ibid*, Supplement 2 (1971c).

[11] Dubs, P. & M. Joho, *Helv. Chim. Acta*, **61**, 1404 (1978).

[12] Fadden Mc, W. H., E. A. Day & J. M. Diamond, *Anal. Chem.*, **37**, 89 (1965).

[13] Friedel, P., V. Krampl, T. Radford, J. A. Renner, F. W. Shephard & M. J. Gianturco, *J. Agric. Food Chem.*, **19**, 530 (1971).

[14] Goldman, I. M., J. Seibl, I. Flament, F. Gautschi, M. Winter, B. Willhalm & M. Stoll, *Helv. Chim. Acta*, **50**, 694 (1967).

[15a] Grigg, R., M. V. Sargent, D. H. Williams & J. A. Knight, *Tetrahedron*, **21**, 3441 (1965).

[15b] Grigg, R., J. A. Knight & M. V. Sargent, *J. Chem. Soc.*, (C), **1966**, 976.

[16] Haag, A. & P. Werkhoff, *Org. Mass Spectrometry*, **11**, 511 (1976).

[17] Hanus, V. & V. Cermák, *Collect. Czechoslov. Chem. Comm.*, **24**, 1602 (1959).

[18] Heyns, K., R. Stute & H. Scharmann, *Tetrahedron*, **22**, 2223 (1966).

[19] Honkanen, E., T. Moisio & P. Karvonen, *Acta Chem. Scand.*, **19**, 370 (1965).

[20] Kinlin, T. E., R. Muralidhara, A. O. Pittet, A. Sanderson & J. P. Walradt, *J. Agric. Food Chem.*, **20**, 1021 (1972).

[21] Kitamura, K. & T. Shibamoto, *J. Agric. Food Chem.*, **29**, 192 (1981).

[22] Kolor, M. G., in *Mass Spectrometry* (C. Merritt & C. N. McEven, Eds.), M. Dekker Inc., New York, 1979, pp. 67-117.

[23] Ledl, F., *Z. Lebensm.-Unters. Forsch.*, **157**, 28 (1975a); *ibid*, **161**, 125 (1976b).

[24] MacLeod, G., M. S. Ardebili & A. J. MacLeod, *J. Agric. Food Chem.*, **28**, 441 (1980).

[25] Merritt, L., McEven, C. N., in *Practical Spectroscopic Series*, Vol. 3, Part A *Mass Spectrometry*, Chap. 3, 1979.

[26] Metzger, J. V., E. J. Vincent, J. Chouteau & G. Mille, in *Thiazole and its Derivatives* (J. V. Metzger, Ed.), John Wiley & Sons, New York, 1979, p. 83.

[27] Millard, B. J., *J. Chem. Soc.*, (C) **1969**, 1231.

[28] Mussinan, C. J., R. A. Wilson, I. Katz, A. Hruza & M. H. Vock, in *Phenolic, Sulfur and Nitrogen Compounds in Food Flavors* (G. Charalambous & I. Katz, Eds.), ACS Symposium Series No. 26, Washington, D.C., 1976, pp. 133-145.

[29] Nishiwaki, T., *Tetrahedron*, **23**, 2979 (1967a); *idem*, *Heterocycles*, **2**, 473 (1974b).

[30] Pittet, A. O., R. Muralidhara, J. P. Walradt & T. Kinlin, *J. Agric. Food Chem.*, **22**, 273 (1974).

[31] Pittet, A. O. & D. E. Hruza, *J. Agric. Food Chem.*, **22**, 264 (1974).

[32] Poite, J. C., R. Vivaldi, A. Bonzom & J. Roggero, *Compt. Rend. (Paris)*, **268**, 12 (1969); Poite, J. C., *Thesis*, University of Marseille, France, 1972.

[33] Porter, Q. N. & J. Baldas, in *Mass Spectrometry of Heterocyclic Compounds*, Wiley Interscience, New York, 1971.

[34] Powers, J. C., *J. Org. Chem.*, **33**, 2044 (1968).

[35a] Sample, S. D., D. A. Lightner, O. Buchardt & C. Djerassi, *J. Org. Chem.*, **32**, 997 (1967).

[35b] Shigematsu, H., T. Kurata, H. Kato & M. Fujimaki, *Agric. Biol. Chem.*, **35**, 2077 (1971).

[36] Stoll, M., M. Winter, F. Gautschi, I. Flament & B. Willhalm, *Helv. Chim. Acta*, **50**, 628 (1967).

[37] Tabacchi, R., *Helv. Chim. Acta*, **57**, 324 (1974).

[38] Tonsbeek, C. H. T., H. Copier & A. J. Plancken, *J. Agric. Food Chem.*, **19**, 1014 (1971).

[39] Toyoda, T., S. Muraki & T. Yoshida, *Agric. Biol. Chem.*, **42**, 1901 (1978).

[40] Vernin, G., J. Metzger & C. Párkányi, *J. Org. Chem.*, **40**, 3183 (1975).

[41] Vernin, G. & J. Metzger, *J. Chim. Phys.*, **71**, 865 (1974).

[42] Vernin, G., Unpublished results.

[43] Viani, R., F. Mueggler-Chavan, D. Reymond & R. E. Egli, *Helv. Chim. Acta*, **48**, 1809 (1965).

[44] Vitzthum, O. G. & P. Werkhoff, *J. Food Sci.*, **39**, 1210 (1974).

[45] Vitzthum, O. G. & P. Werkhoff, *J. Agric. Food Chem.*, **23**, 510 (1975).

[46] Vitzthum, O. G. & P. Werkhoff, *Z. Lebensm.-Unters.-Forsch.*, **160**, 277-291 (1976).

[47] Walradt, J. P., A. O. Pittet, T. E. Kinlin, R. Muralidhara & A. Sanderson, *J. Agric. Food Chem.*, **19**, 972 (1971).
[48] Yamanishi, T., S. Shimojo, M. Ukita, K. Kawashima & Y. Nakatani, *Agric. Biol. Chem.*, **37**, 2147 (1973).

CHAPTER VIII

The Legislation of Flavours

Gaston VERNIN, University of Aix-Marseilles, France

VIII–1 INTRODUCTION

Most countries have now adopted "permitted lists of flavouring substances or materials, the use of which in foodstuffs is authorized to the exclusion of all others". These substances which must not be added in an amount greater than therein specified, have been subjected to independent toxicological evaluation to assess their safety in use.

We have attempted to list here heterocyclic compounds used in flavours both in the United States and in the European Community. The largest number of such compounds have been found in the FEMA GRAS† reference list [2]. The second source arises from the publication of the Council of Europe in 1974 (CoE list) [5]. In this latter list, flavouring substances are classified into four categories:

– CATEGORY I (No. 1 to 2000) means flavouring substances which may be added to foodstuffs without hazard to public health (≃ 692 products).

– CATEGORY II (No. 2001 to 4000) means flavouring substances which may be added temporarily to foodstuffs without hazard to public health (≃ 293 substances).

– CATEGORY III (No. 4001 to 6000) means flavouring substances not yet authorized owing to the absence of technological and toxicological data (≃ 243 substances).

– CATEGORY IV (No. 6001 to . . .) are flavouring substances forbidden owing to their toxicity (3 substances).

The central theme of the last proposal for a directive of the Commission of the European Community [7,10] is the adoption of the "Positive List System" for flavouring materials (article 5). A number of special terms for different kinds of flavouring are proposed. These terms may not be exactly the same as those used by other organizations (e.g., FAO/WHO Codex Alimentarius) for other purposes.

† Flavoring Extract Manufacturers' Association; Generally Recognized as Safe.

These lists of substances shall include four groups of flavourings:

(a) artificial flavouring substances,
(b) nature-identical flavouring substances,
(c) source materials for the production of natural flavouring preparations and natural flavouring substances;
(d) source materials for the production of artificial flavouring preparations.

The third source of information was derived from the *International Organisation of Flavour Industries* (IOFI list) [3]. Numbers 1, 2 or 3 are defined as Natural-, Nature-identical-, and Artificial flavouring substances, respectively.

In the present publication, flavouring substances authorized for human consumption are defined as follows:

Natural Flavours and Flavouring Substances are preparations or a single substance, respectively, isolated exclusively by appropriate physical extraction (including distillation, solvent extraction) from flavouring raw materials either in their natural state or processed.

Nature-Identical Flavouring Substances are obtained by chemical synthesis or isolated by chemical process from flavouring raw materials. They are chemically identical to substances occurring in natural products either processed or not. They constitute by far the most important category [8].

Artificial Flavouring Substances are those substances which as yet have not been identified in natural products either processed or not ($\simeq$ 357 products according to FEMA GRAS list).

In *Tables VIII–1* to *VIII–6*, numerical references refer to the three above-mentioned lists. When data were not sufficient to categorize a substance, a blank was left at this position. In some cases, foods in which heterocyclic compounds may be added in the USA are also given as well as their usage levels. Finally, another list (UK list) has also been published by the Ministry of Agriculture, Fisheries, and Food of the United Kingdom [1]. It is not included here. In the last table (*Table VIII–7*) we compare the distribution of the various categories of heterocyclic flavouring substances on FEMA and CoE lists. It can be seen that FEMA lists contain a larger number of these compounds ($\simeq$ 231) with $\simeq$ 14% of the listed compounds.

We attempted to make this list as error free as possible. Nevertheless there may be a number of errors that we have missed. Therefore for more accuracy, the original lists should be consulted.

VIII–2 TOXICITY OF HETEROCYCLIC COMPOUNDS

The acute lethal doses have been recently reported in rats or mice for 43 heterocyclic compounds used as flavourings [4] (*see Table VIII–8*).

The test substance was administered by oral gavage as a solution on suspension in corn oil.

Table VIII–1 Furans used as flavouring substances

Monosubstituted furans	FEMA[a)]	Product[b)]	Usage levels (ppm)[c)]	CoE	IOFI
2-Ethylfuran	–	–		+	2
2-Pentylfuran	3317	a, e, f, g, n	3.0	–	2
2-Heptylfuran	3401	a, d, f, i, j	0.02– 0.06	–	2
Isoamyl 4-(2′-furyl)-butyrate	2070	a, d, e, f, i	0.03– 8.0	2080	–
Ethyl 3-(2′-furyl)-propionate	2435	a, d, e, f	–	2091	–
Isobutyl 3-(2′-furyl)-propionate	2198	a, d, e, f, i, j, m	4.0 –30.0	2093	–
Isoamyl 3-(2′-furyl)-propionate	2071	a, d, e, f	0.02– 3.6	2092	–
(2′-Furyl)-2-propanone (Furfurylmethyl ketone)	2496	d, e, f	2.0 –20	–	2
Furfuryl alcohol	2491			2023	2
Furfuryl methyl ether	3159	a, d, e, g	2.0	–	2
Difurfuryl ether	3337	f, g	1.0		2
Furfuryl acetate	2490			2065	
Furfuryl propionate	3346	a,d,e,f,g,i,j,n	0.5 –10.0		2
Furfuryl butyrate	–	–	–	638	2
α-Furfuryl pentanoate	3397	a, e	1.5 – 3.0		2
Furfuryl-3-methyl butanoate	3283	a, d, e, i	5.0 –10.0	–	2
Isopentyl-3-furfuryl acrylate				+	–
α-Furfuryl octanoate	3396	a, i, g, k, l	2.0 –10.0	–	–
Furfuryl mercaptan	2493	a, d, e, f, i, m	0.5 – 2.0	2202	2
Furfuryl methyl sulphide	3160	a, d, e, i	1.0	–	2
Furfuryl isopropyl sulphide	3161	a, d, e, f	0.5	2248	–
Difurfuryl sulphide	3258	a, d, e, f, g	5.0	–	2
2-Furan methanethiol formate (Furfurylthiol formate)	3158	a, d, e	1.0	+	–
Furfuryl thioacetate	3162	a, d, e, i	0.2 – 1.5	2250	2

Table VIII–1 (*continued*)

	FEMA[a)]	Product[b)]	Usage levels (ppm)[c)]	CoE	IOFI
Furfuryl thiopropionate	3347	a, d, e, f	1.0	—	
2-Methyl-3-(5- or 6-furfurylthio)-pyrazine (2-Furfurylthio-3-methyl pyrazine)	3189	a, d, f, h	1.0	2287	—
Furfural (2-furaldehyde)	2489	a,b,c,d,e,f,i,j	0.8 —45.0	2014	2
2-Furyl methyl ketone (2-Acetylfuran)	3163	f, g, h	20.0	+	2
Pentyl 2-furyl ketone	3418	a, d, f, i, j	0.05— 0.15	—	—
1-(2-Furyl)-1,3-butanedione (Furoyl acetone)	—		—	—	—
Methyl 2-furoate	2703	a, d, e, f, h	0.02— 1.3	358	2
Propyl 2-furoate	2946			359	—
Amyl 2-furoate	2072			2109	—
Hexyl 2-furoate	2571	a, d, e, f	—	361	—
Octyl 2-furoate	3518	a, d, e, f	—		2
Allyl 2-furoate	2030	—		360	3
Phenethyl 2-furoate	2865			362	—
Methyl 2-thiofuroate	3311	a, d, e, f, g, h	3.0	—	2
Methyl furfuryl disulphide (Furfuryl methyl disulphide)	3362	a, d	1.0	—	2
2,2′-(Dithiomethylene)-difuran (2-Furfuryl disulphide)	3146	a,d,e,f,g,h,i,j	1.0	—	2
3-(2′-Furyl)-acrolein	2494	—		2252	2
2-Methyl-3(2′-furyl)-acrolein	2704	—		2216	—
2-Furfurylidene butyraldehyde (3-(2′-Furyl)-2-ethylacrolein)	2492	—		—	—
2-Phenyl-3-(2′-furyl)-prop-2-enal (2-Furfurylidene phenyl acetaldehyde)	3586	—		—	—
4-(2′-Furyl)-but-3-en-2-one (Furfurylidene acetone)	2495			2049	2
Methyl 3-(2′-furyl)-acrylate	—			2267	

n-Propyl 3-(2′-furyl)-acrylate	2945			2090	–
Allyl 3-(2′-furyl)-acrylate	–			+	
Polysubstituted furans					
2-Methyl-3-furanthiol	3188	f, g, h	0.3	+	–
Bis-(2-methyl-3-furyl)-disulphide (2-Methyl-3-furyl disulphide)	3259	–		–	–
Bis-(2-methyl-3-furyl)-tetrasulphide	3260			–	–
2,6-Dimethyl-3-[(2′-methyl-3′-furyl)thio]-4-heptanone	3538			–	–
3-[(2′-Methyl-3′-furyl)thio]-4-heptanone	3570			–	–
4-[(2′-Methyl-3′-furyl)thio]-5-nonanone	3571			–	–
2-Phenyl-3-carbethoxyfuran	3468			–	–
Methyl 2-methyl-3-furyl disulphide	3573			–	–
Propyl 2-methyl-3-furyl disulphide	3607				3
2,5-Dimethylfuran	–			+	
3-(5′-Methyl-2′-furyl)-butanal	3307	a, d, e, f, h	0.1 – 2.5		
5-Methylfurfural	2702	a, d, e, f	0.03– 0.1	119	2
2-Acetyl-5-methylfuran	3609				2
5-Methyl-2-furanthiol	–			+	
2-Methyl-5-(methylthio)-furan	3366	f, g	1.5	–	2
3-Acetyl-2,5-dimethylfuran	3391			–	–
2,5-Dimethyl-3-furanthiol	3451			–	–
2,5-Dimethyl-3-thiofuroylfuran	3481			–	–
2,5-Dimethyl-3-thioisovalerylfuran	3482	f, g	0.1	–	–
Bis-(2,5-dimethyl-3-furyl)-disulphide	3476	f, g	0.1	–	–
Bis-(2-methyl-3-furyl)-disulphide	3259			–	–
Bis-(2-methyl-3-furyl)-tetrasulphide	3260			–	–

Table VIII–1 (*continued*)

Reduced furans	FEMA[a)]	Product[b)]	Usage levels (ppm)[c)]	CoE	IOFI
2-Methyl-3-thioacetoxy-4,5-dihydrofuran	3636			–	2
5-Methyl-2-methylthiocarbonyl-2,3-dihydrofuran	–			+	–
2-(3-Phenylpropyl)-tetrahydrofuran	2898			489	–
Tetrahydrofurfuryl acetate	3055	a, d, e, f	1.0 –20.0	2069	
Tetrahydrofurfuryl alcohol	3056	a, d, e, f	0.03–18.0	2029	–
Tetrahydrofurfuryl propionate	3058	a, d, e, f	1.0 –20.0	2096	–
Tetrahydrofurfuryl butyrate	3057	a, d, e, f	0.9 –15.0	2081	–
Tetrahydrofurfuryl cinnamate	3320	a, d, f	20.0	+	–
2-Hexyl-4-acetoxytetrahydrofuran	–			490	–
2,5-Diethyltetrahydrofuran	–			2232	2
6-Hydroxydihydrotheaspirane	3549			–	2
Linalool oxide (2-(α-Hydroxyisopropyl)-5-methyl-5-vinyltetrahydrofuran)	–			2214	2
4-Hydroxyalkanoic acid lactones (γ-Lactones, 1,4-alkanolides or 2-tetrahydrofuranones)					
4-Hydroxybutanoic acid lactone (γ-Butyrolactone)	3291			–	2
γ-Valerolactone	3103			2283	2
γ-Hexalactone	2556			2254	2
γ-Heptalactone	2539			2253	2
γ-Octalactone	2796			2274	2
γ-Nonalactone	2781			178	2
γ-Decalactone	2360			2230	2
γ-Undecalactone	3091			179	2
γ-Dodecalactone	2400			2240	2

γ-Dodecene-6-lactone	—			625	—
4,4-Dibutyl-γ-butyrolactone	2372			2231	—
Erythrobic acid	2410			30	—
Unsaturated γ-lactones (or Dihydrofuran-2-ones)					
4,5-Dimethyl-3-hydroxy-2-(5*H*)-furanone	3634			—	2
5-Ethyl-3-hydroxy-4-methyl-2-(5*H*)-furanone	3153	e, f, i	1.0	+	—
L-Ascorbic acid	2109			27	2
3-Heptyldihydro-5-methyl-2-(3*H*)-furanone	3350	f, g, o	1.5	—	—
3-(2*H*)-Furanones					
2-Methyltetrahydrofuran-3-one	3373			+	2
4-Hydroxy-2-methyltetrahydrofuran-3-one	—			+	—
4-Hydroxy-5-methyl-3-(2*H*)-furanone	3635			+	2
4-Hydroxy-2,5-dimethyl-3-(2*H*)-furanone (Furaneol)	3174			—	2
2-Ethyl-4-hydroxy-5-methyl-3-(2*H*)-furanone	3623				2

a) FEMA: Flavoring Extract Manufacturers' Association; GRAS: Generally Recognized as Safe.

b) Products in which heterocyclic compounds may be used as flavouring additives are: a) nonalcoholic beverages; b) alcoholic beverages; c) syrups; d) ice cream; e) candy; f) baked goods; g) meat sauces, soups; h) condiments, pickles; i) gelatins and puddings; j) chewing gum; k) margarine; l) snacks; m) icing; n) cereals; o) dairy products; p) protein foods.

c) Usage levels have been reported by Hall, R. L. & B. L. Oser, *Food Technol.*, **24**(5), 25 (1970); Oser, B. L. & R. A. Ford, *Food Technol.*, **27**(1), 64 (1973); *ibid.*, **27**(11), 56 (1973); *ibid.*, **28**(9), 76 (1974); *ibid.*, **29**(9), 70 (1975); *ibid.*, **31**(1), 65 (1977); **32**(2), 60 (1978); *ibid.*, **33**(7), 65 (1979).

Table VIII–2 Thiophenes, pyrroles, thiazoles, pyrazoles, dioxolanes, and trithiolanes

Thiophenes	FEMA	Usage levels (ppm)	CoE	IOFI	
2-Formylthiophene (2-Thiophene carboxaldehyde)	–		+	2	
Methyl 2-thienylketone (2-Acetylthiophene)	–		+	2	
2-Thienyl mercaptan	3062		–	2	
2-Thienyl disulphide	3323	0.15			
5-Methyl-2-thiophene carboxaldehyde	3209	0.5	–	2	
4,5-Dihydro-3-(2*H*)-thiophenone	3266		–	2	
2-Methyltetrahydrothiophen-3-one	3512		–	2	
Pyrroles					
Pyrrole	3386		–	2	
1-Methylpyrrole	–		+	2	
1-Furfurylpyrrole	3284	2	2249	2	
1-Methyl-2-acetylpyrrole	3184	10	–	2	
1-Ethyl-2-acetylpyrrole	3147	5		–	
Methyl 2-pyrrolyl ketone (2-Acetylpyrrole)	3202	50	+	2	
Ethyl 2-pyrrolyl ketone (2-Propionylpyrrole)	3614	0.15	–	2	
5-Ethyl-2-formylpyrrole	–		+	2	
2,5-Dimethylpyrrole	–		–		
Pyrrolidine	3523		–	2	
2-Acetyl-1-methylpyrrolidine	–		–	2	(bread)
L-Proline	3319		+	2	
1,2-Pyrrolidine dicarboxylic acid	–		+		
5-Pyrrolidone 2-carboxylic acid	–		+		
N-Ethyl succinimide	–		+	2	(black tea)

Thiazoles						
Thiazole	3615			–	2	
2-Isobutylthiazole	3134	1.0		+	2	
2-(1-Methylpropyl)-thiazole	3372	0.1	–0.5	–		
2-Acetylthiazole	3328	0.6	–1.4	+	2	
2-Propionylthiazole	3611			–	2	
2-Ethoxythiazole	3340	0.2	–2.0	–	3	
2,4-Dimethylthiazole	–		–		2	
2-Isopropyl-4-methylthiazole	3555				2	
4,5-Dimethylthiazole	3274				2	(coffee)
5-(β-Hydroxyethyl)-4-methylthiazole (4-Methyl-5-thiazole ethanol)	3204		–			
4-Methyl-5-vinylthiazole	3313		–		2	
4-Methyl-5-thiazole ethanol acetate	3205					
2,4,5-Trimethylthiazole	3325				2	
2,4-Dimethyl-5-vinylthiazole	3145			2237		
2,4-Dimethyl-5-acetylthiazole	3267			–		
2-Methyl-5-methoxythiazole	3192			+		
Thiamine hydrochloride	3322			+	2	
4,5-Dimethyl-2-ethyl-3-thiazoline	3620			–	3	
2-(*sec*-Butyl)-4,5-dimethyl-3-thiazoline	3619			–	3	
4,5-Dimethyl-2-isobutyl-3-thiazoline	3621			–	3	
2-Acetylthiazolidine	–			+	–	
Pyrazoles						
1-Phenyl-3-(or 5-)-propylpyrazole	–			2277	–	

Table VIII–2 (*continued*)

1,3-Dioxolanes	**FEMA**	**Usage levels (ppm)**	**CoE**	**IOFI**
Cinnamaldehyde ethylene glycol acetal	2287		48	–
2-Phenethyl-1,3-dioxolane	–		+	
4-Methyl-2-phenyl-1,3-dioxolane	–		+	
4-Methyl-2-pentyl-1,3-dioxolane	3630		–	3
2,2,4-Trimethyl-1,3-dioxacyclopentane	3341		–	2
Phenylacetaldehyde 2,3-butylene	2875		669	–
Glycol acetal (2-Benzyl-4,5-dimethyl-1,3-dioxolane)				
4,4-Dimethyl-2-hexyl-1,3-dioxolane	–		+	–
4,5-Dimethyl-2-phenoxymethyl-1,3-dioxolane	–		+	–
Citral propylene glycol acetal	–		+	–
4-(3,4-Methylenedioxyphenyl)-2-butanone	2701		165	–
Phenoxy acetaldehyde butylene glycol acetal (4,5-Dimethyl-2-phenoxymethyl-1,3-dioxolane)	–		+	–
1,2,4-Trithiolane				
3,5-Dimethyl-1,2,4-trithiolane	3541		2236	2
Oxazole				
2,4,5-Trimethyl-3-oxazoline	3525		–	2

Table VIII–3 Oxygen and/or sulphur-containing six or more membered heterocyclic systems

	FEMA	CoE	IOFI
Pyrans and derivatives			
Tetrahydro-4-methyl-2-(2-methylpropen-1-yl)-pyran	3236	2269	2
Tetrahydro-3-hydroxy-2,2,6-trimethyl-6-vinylpyran (Linalool oxide)	–	+	
3-Hydroxy-2-methyl-4-pyrone (Maltol)	2656	148	2
Maltyl isobutyrate	3462	–	–
Ethyl maltol (3-Hydroxy-2-ethyl-4-pyrone)	3487	218	–
3,5-Dihydroxy-2-methyl-4-pyrone	–	+	–
δ-Lactones (2-Tetrahydropyranones or 5-hydroxyalkanoic acid lactones)			
δ-Hexalactone	3167	641	2
δ-Heptalactone	–	+	
δ-Octalactone	3124	2195	2
δ-Nonalactone	3356	–	2
δ-Decalactone	2361	621	2
δ-Undecalactone	3294	–	2
δ-Dodecalactone	2401	624	2
δ-Tridecalactone	–	+	
δ-Tetradecalactone	3590	2196	2
δ-Dodecen-9-lactone	–	+	–
Miscellaneous lactones			
ϵ-Decalactone	3613	–	3
ϵ-Dodecalactone	3610	–	3
6-Hydroxy-3,7-dimethyloctanoic acid lactone	3355	–	–
ω-6-Hexadecen lactone (Ambrettolide or 16-Hydroxy-6-hexadecenoic acid ω-lactone)	2555	180	2

Table VIII–3 (*continued*)

	FEMA	CoE	IOFI
ω-Pentadecalactone	2840	181	2
Ethylene tridecanedioate	–	+	–
Dioxanes			
Heptanal glyceryl acetal (mixed 1,2- and 1,3-acetals)	2542	2016	–
Benzaldehyde glyceryl acetal	2129	36	2
Phenylacetaldehyde glyceryl acetal (5-Hydroxy-2-benzyl-1,3-dioxane)	2877	41	–
2-Butyl-5 (or 6-)-keto-1,4-dioxane	2204	2206	
2-Amyl-5 (or 6-)-keto-1,4-dioxane	2076	–	
2-Hexyl-5 (or 6-)-keto-1,4-dioxane	2574	+	
Oxathiane			
2-Methyl-4-propyl-1,3-oxathiane	3578	–	2
Dithiane			
2,5-Dimethyl-2,5-dihydroxy-1,4-dithiane	3450	–	–
Trioxane			
Paraldehyde	–	+	–
Trithiane			
Trithioacetone (2,2,4,4,6,6-Hexamethyl-*s*-trithiane)	3475	–	2

Table VIII–4 Pyridines

Pyridines	**FEMA**	**CoE**	**IOFI**
Pyridine	2966	604	2
2-Ethylpyridine	–	+	–
2-Pentylpyridine	3383	–	2
2-(2-Methylpropyl)-pyridine (2-Isobutylpyridine)	3370	–	2
2-Acetylpyridine	3251	+	2
2-Pyridine methanethiol	3232	2279	–
2-Pyridyl mercaptan	–	+	–
3-Methylpyridine	–	+	–
3-Ethylpyridine	3394	–	2
3-(2-Methylpropyl)-pyridine (3-Isobutylpyridine)	3371	–	–
3-Acetylpyridine	3424	–	2
2,6-Dimethylpyridine	3540	–	2
Piperidine	2908	675	2
Piperine	2909	492	2

Some effects of structure on acute oral toxicity were observed. For pyrazines the LD_{50} ranged from 456 to 1910 mg/kg body weight. The addition and position of methyl groups affected the toxicity, while the substitution of ethyl for methyl had little effect. Toxicity increased with the proximity of the methyl group. The thiazoles and thiazolines tested had similar toxicities regardless of the side groups. These compounds may compete for the thiamine pyrophosphate enzyme site, and the resulting loss of enzyme activity could be responsible for the observed toxicity. The furan thioesters tested were about equally toxic and slightly less toxic than the parent molecule 2,5-dimethyl-3-furanthiol. These data suggest that hydrolysis to the acid plus thiol is a likely metabolic event. No toxicity was observed for the three lactones studied.

Toxicological data of certain heterocyclic flavouring compounds have been recently reported by Opdyke in *Monographs on Fragrance of Raw Materials* [6]. Substances reviewed include: isobutylfuryl propionate, methylfuroate, γ-hexalactone, γ-nonalactone, γ-methyldecalactone, δ-dodecalactone, and 6-methylquinoline.

Some other data are also available from *Dangerous Properties of Industrial Materials* [9]. Heterocyclic compounds tested include:

– Nine furans (furan, 2-methylfuran, furfural, furfurylic alcohol, furfuryl acetate, tetrahydrofuran, 2-methyltetrahydrofuran, 2-hydroxymethyltetrahydrofuran, 2,5-dimethylfuran).
– Two thiophenes (thiophene, 2,4-dimethylthiophene).
– Eight pyrroles (pyrrole, pyrrolidine, pyrrolidone, pyrroline, *N*-methylpyrrolidine, *N*-methylpyrrolidone, *N*-vinylpyrrolidone, proline).

Table VIII–5 Pyrazines

Pyrazines	**FEMA**	**Usage levels (ppm)**	**CoE**	**IOFI**
Methylpyrazine	3309	10	2270	2
Ethylpyrazine	3281	10	2213	2
2-Mercaptomethylpyrazine (Pyrazine methanethiol)	3299	2–4	–	–
Pyrazine ethanethiol	3230	10	2285	–
Cyclohexylmethylpyrazine (mixture of isomers)	3631	–	–	3
Isopropenylpyrazine	3296	10	–	2
Acetylpyrazine	3126	5	2286	2
Methoxypyrazine	3302	10	–	–
Pyrazinylmethyl sulphide	3231		2288	–
2,3-Dimethylpyrazine	3271	10	–	2
3-Ethyl-2-methylpyrazine	3155	3	548	2
2-Isobutyl-3-methylpyrazine	3133	5	+	2
2-Methyl-3 (5 or 6-furfurylthio)-pyrazine (mixture of isomers)	3189	1	2287	–
2-Acetyl-3 (5 or 6)-methylpyrazine (mixture of isomers)	—	10	–	
2-Acetyl-3-ethylpyrazine	3250		–	–
2-(5 or 6-Methoxy)-3-methylpyrazine (mixture of isomers)	3183	2.4	2266	–
2-(5 or 6-Methoxy)-3-ethylpyrazine (mixture of isomers)	3280	5	–	–
2-Methoxy-3 (5 or 6)-isopropylpyrazine	3358		–	2
2-Methoxy-3-(1-methylpropyl)-pyrazine	3433		–	2
2-Isobutyl-3-methoxypyrazine	3132	0.05	–	2
2-Methyl-3 (5 or 6)-ethoxypyrazine	3569			–
2-Methylthio-3 (5 or 6)-methylpyrazine	3208	2–4		–
2,5-Dimethylpyrazine	3272	10	2210	2
2-Ethyl-5-methylpyrazine	3154	10	2212	2

5-Isopropyl-2-methylpyrazine	3554	–	2268	2
2-Methyl-5-vinylpyrazine	3211	10	–	2
2,6-Dimethylpyrazine	3273	–	2211	2
Trimethylpyrazine	3244	5	2282	2
3-Ethyl-2,6-dimethylpyrazine	3150	5	2246	2
2-Ethyl-3 (5 or 6)-dimethylpyrazine	3149	2–5	2245	2
2,3-Diethyl-5-methylpyrazine	3336	0.1–1	–	2
2-Acetyl-3,5 (and 3,6)-dimethylpyrazine	3327	1–5	–	2
Tetramethylpyrazine	3237	5	2280	2

Table VIII–6 Fused ring systems

	FEMA	Usage levels (ppm)	CoE	IOFI
2-Benzofuran carboxaldehyde	3128	10–20	2247	2
2,3-Dimethylbenzofuran	3535		–	2
4,5,6,7-Tetrahydro-3,6-dimethylbenzofuran (Menthofuran)	3235	10	2265	2
Indole	2593		560	2
Skatole (3-Methylindole)	3019		493	2
Benzothiazole	3256	0.5	–	2
3-Propylidene phthalide	2952		494	–
4-Butylidene phthalide	3333		–	2
3-*n*-Butylphthalide	3334		–	2
5,7-Dihydro-2-methylthieno [3,4-d] pyrimidine	3338		–	–
5-Methyl-6,7-dihydro-(5*H*)-cyclopentapyrazine	3306	0.045–1	–	2
d-Piperitone	2910		2052	2
Piperonal (Heliotropine)	2911		104	2
Piperonyl acetate	2912		2068	–
Piperonyl isobutyrate	2913		305	–

Table VIII–6 (*continued*)

	FEMA	Usage levels (ppm)	CoE	IOFI
Piperine	2909		492	2
Saccharine, sodium salt	2997		–	–
Caffeine	2224		+	2
Theobromine	3591		–	2
Dihydrocoumarin	2381		535	–
6-Methylcoumarin	2699		579	–
3-Methylcoumarin	–		+	–
4-Methylcoumarin	–		+	–
Quinoline	3470		–	2
6-Methylquinoline	2744		+	2
Quinine bisulphate	2975		–	–
Quinine hydrochloride	2976		–	–
Quinine sulphate	2977		–	–
Isoquinoline	2978		487	–
5,6,7,8-Tetrahydroquinoxaline	3321	1–5	–	2
5-Methylquinoxaline	3203	10	2271	2
Spiro-2,4-dithia-1-methyl-8-oxabicyclo [3.3.0] octane-3,3′	3270		–	–
1,5,5,9-Tetramethyl-13-oxatricyclo [8.3.0.0] (4,9)tridecane	3471		–	2
Miscellaneous (Cyclic ethers)				
Eucalyptol	3466		185	–
Ethyl 3-phenyl glucidate	2454		2097	–
Ethyl 3-Methyl-3-*para* tolyl glucidate	–		+	–
Ethylene oxide	2433			
Ethyl 3-methyl-3-phenyl glucidate	–			
Ethyl 2,3-eposy-3-methyl octanoate	–			

Table VIII–7 Distribution of the various categories of heterocyclic flavouring compounds on FEMA and CoE lists

	FEMA	CoE[a)]
Cyclic ethers	3	5 (3)
Furans	70	43 (11)
γ-Lactones	15	13 (1)
3-(2*H*)-Furanones	4	3 (3)
Thiophenes	5	2 (2)
Pyrroles	8	8 (7)
Thiazoles	19	6 (5)
Pyrazole	0	1
1,3-Dioxolanes	5	9 (6)
1,2,4-Trithiolanes	1	1
Oxazole	1	0
Pyrans	4	5 (2)
δ-Lactones	7	8 (3)
Miscellaneous lactones	5	3 (1)
Dioxanes	6	5 (1)
Oxathiane	1	0
Dithiane	1	0
Trioxane	—	1 (1)
Trithiane	1	0
Pyridines	11	8 (3)
Pyrazines	33	17 (1)
Fused ring systems	31	18 (4)
Total	231	156 (54)

a) Values in brackets point out heterocyclic compounds whose toxicological evaluation was not completed in 1974.

One lactone (α-acetylbutyrolactone).

- Three pyrans (dihydropyran, 2,6-dimethyl-1,4-pyran, 6-methyl-4,5-dihydro-2-pyranone).
- Seven pyridines (pyridine, 4-isopropylpyridine, 2-methyl-5-ethylpyridine, α, β and γ-picolines, piperine).
- Two pyrazines (2,5-dimethylpyrazine and piperazine).
- Six fused heterocyclic systems (benzothiazole, quinoline, isoquinoline, quinine, benzimidazole, acridine).
- Two vitamins (thiamin (B_1) and ascorbic acid (C)).

Table VIII–8 Acute oral lethal dose for heterocyclic compounds [4]

Heterocyclic compounds	Species	LD_{50} mg/kg	Confidence interval (95%)
Pyrazines			
2-Methylpyrazine	Rats	1800	1590 – 2010
2,5-Dimethylpyrazine	Rats	1020	780 – 1250
2,6-Dimethylpyrazine	Rats	880	775 – 985
2,3-Dimethylpyrazine	Rats	613	433 – 867
2-Ethyl-5-methylpyrazine	Rats	900	755 – 1045
2-Ethyl-3-methylpyrazine	Rats	600	501 – 699
2-Ethyl-3(5 or 6)-dimethylpyrazine	Rats	456	394 – 527
2,3,5-Trimethylpyrazine	Rats	806	470 – 1381
2,3,5,6-Tetramethylpyrazine	Rats	1910	1675 – 2145
Thiazoles and oxazoles			
Thiazole	Mice	983	553 – 1821
4-Methyl-5-vinylthiazole	Mice	400 – 800[a]	
2,4-Dimethyl-5-acetylthiazole	Mice	975	845 – 1130
Benzothiazole	Mice	900	803 – 1008
2-Propionylthiazole	Mice	2113	1124 – 3750
2-Ethyl-4,5-dimethylthiazoline	Mice	1265	1162 – 1381
2-*sec*-butyl-4,5-dimethylthiazoline	Mice	2827	1943 – 4115
2-Isobutyl-4,5-dimethylthiazoline	Mice	3067	2786 – 3370
2,4,5-Trimethyloxazoline	Mice	4840	4380 – 5300
Furans			
2-Methyl-3-furanthiol	Mice	100	77 – 123
2,5-Dimethyl-3-furanthiol	Mice	360	328 – 392

2-Methyltetrahydrofuran-3-one	Mice	1860	325 – 3395
2-Ethyl-4-hydroxy-5-methyl-3-(2*H*)-furanone	Mice	1932	1783 – 2014
2-Acetyl-5-methylfuran	Mice	438	334 – 576
2-Pentylfuran	Mice	1200	863 – 1672
2,5-Dimethyl-3-thioisovalerylfuran	Mice	625[b]	603 – 647
		720[b]	617 – 823
2,5-Dimethyl-3-thiofuroylfuran	Mice	540	511 – 569
2,5-Dimethyl-3-thiobenzoylfuran	Mice	567	542 – 592
Miscellaneous			
ϵ-Decalactone	Mice	5252	2778 – 7726
ϵ-Dodecalactone	Mice	7898	7141 – 8997
α-Angelicalactone	Mice	2800	2333 – 3365
4,5-Dihydro-3-(2*H*)-thiophenone	Mice	1860	325 – 3395
5-Methylthienopyrimidine	Mice	260	220 – 300
2-Propionylpyrrole	Mice	1620	886 – 2963
N-Furfurylpyrrole	Mice	380	288 – 502
4-Methyl-2-pentyldioxolane	Mice	6802	5382 – 8597
1,3,5-Triisobutyl-2,4,6-trioxane	Mice	935	861 – 1013
Hexamethyl-1,3,5-trithiane	Mice	2400	972 – 5933
2,5-Dimethyl-2,5-dihydro-1,4-dithiane	Mice	560	298 – 1050
3([2-Methyl-3-furyl] thio)-4-heptanone	Mice	425	372 – 479
Methyl-2-methyl-3-furyl disulphide	Mice	142	130 – 154
Propyl-2-methyl-3-furyl disulphide	Mice	284	234 – 345
2-Thienyl disulphide	Mice	473	220 – 726
Bis-(2-methyl-3-furyl) tetrasulphide	Mice	220	134 – 306

a) Insufficient compound available to obtain a more definitive value.

b) Results obtained at two different laboratories.

VIII–3 REFERENCES

[1] *Food Additives and Contaminants Committee*, Report on the Review of Flavouring in Food. HMSO, London, 1976, p. 190.

[2] Ford, R. A., *Flavor Materials 1979*; Reference list of Flavoring Substances in Use in the United States, Allured Publ. Co., Wheaton (1979); Ford, R. A. & G. M. Cramer, *Perf. Flavorist.*, **11**, No. 1, 1 (1977).

[3] Grundschober, F., R. L. Hall, J. Stofberg & C. A. Vodoz, *Survey of World-wide Use Levels of Artificial Flavouring Substances, Flavours*, 1975, pp. 223-230.

[4] Moran, E. J., O. D. Easterday & B. L. Oser, *Drug and Chemical Toxicology*, **13**, 249 (1980).

[5] *Natural Flavouring Substances, their Sources and Added Artificial Flavouring Substances*, Council of Europe; Maisonneuve, Ed. (1974); 3rd edn. (1982). B.P. 39, 57160 Moulins les Metz.

[6] Opdyke, D. L. J., *Food Cosm. Toxicol.*, 1979, **17** (Suppl.), pp. 773, 791, 835, 867, 869, 871, 877.

[7] Proposal for a Council Directive on the Approximation of the laws of the Member States Relating to Flavourings for Use in Foodstuffs and to Source Materials for their Production, COM (80) 286 final, Brussels, 22 May 1980.

[8] Rijkens, F. & H. Boelens, *Proc. Int. Symp. Research*; Zeist (1975).

[9] Sax, N. I., *Dangerous Properties of Industrial Materials*, Van Nostrand, Reinhold (1975).

[10] Vernin, G., *Riv. Ital. EPPOS.*, **62**, 249 (1980).

* A 2nd Ed. of "Flavor Materials" has been recently published by Allured Pub. Co., Weaton, USA (1981).

Index

A

Acetals, odours of cyclic derivatives, 120
1-Acetonylpyrrole, from hydroxyproline and pyruvaldehyde, 175
4-Acetyl-3,5-dimethylpyridine, synthesis of, 230
2-Acetylfuran, as aroma components
of chestnut syrup, 78
of sticks of liquorice, 310
fragmentation of, 310
2-Acetyloxazole, synthesis
from pyruvaldehyde and serine system, 181
2-Acetylpyrazines, synthesis
from 2-amidopyrazines
from α-aminomethyl reductones
and α-aminoketones, 196
2-Acetylpyridine, occurrence in potato chips, 116
in roasted products, 130
synthesis of, 230
2-Acetylpyrroles, from acylation of *N*-substituted pyrroles, 214
fragmentation of, 310
occurrence of, 99
2-Acetyl-1,4,5,6-tetrahydropyridine, as bread flavouring compound, 51, 118
preparation from proline and dihydroxy-acetone, 118
2-Acetylthiazole, from cysteine and pyruvalhyde, 185
as flavour contributor in asparagus, 45
Acetylthiazoles, olfactory properties of, 110
2-Acetyl-2-thiazoline, isolation and identification from beef broth, 295
Acids, as aroma components in asparagus, 43
in milk, 56
in tomato, 42
in wine, 60

Acids – *continued*
auto-oxidation in milk, 55
averages values in wine, 62
carbonyl compounds formation from, 56
enzymatic reduction of oxo derivatives, 169
Aldehydes, aldol condensation of, 159
condensation with 2-oxo acids, 159
formation from α-amino acids and α-dicarbonyl compounds, 157
from α-amino acids and glucose, 158
from oxidation of α-amino acids, 158
from oxidation of fatty acids, 55
from *Strecker* degradation of α-amino acids in cocoa, 34
in meats, 22
in milk, 57
occurrence in potato aroma, 48
Alkaloids, flavour precursors in tea, 31
3-Alkoxy-2-methyl-4-pyrones, fragmentation of, 325
5-Alkoxyoxazoles, synthesis from *N*-α-acylamino acid esters, 216
2-Alkoxypyridines, phenolic odours of, 118
5-Alkoxythiazoles, synthesis from α-acylamino acid esters and phosphorus pentasulphide, 216
Alkylations, by phase transfer catalysis, 241
2-Alkylbenzothiophenes, synthesis from *o*-alkylthiophenols, 239
2-Alkyldihydropyridines, synthesis of, 227
3-Alkyl-2-ethylpyridines, synthesis of, 230
1-Alkyl-2-formylpyrroles, from sugar (or furfural) and α-amino acids, 178
2-Alkylfurans, fragmentation of, 308
occurrence in canned beef, 78
Alkylfuroates, occurrence in coffee aroma, 78

2-Alkylimidazoles, *MacLafferty* rearrangement of, 316
2-Alkyl-3-methoxypyrazines, occurrence in raw vegetable, 122
biosynthesis in vegetables, 197
synthesis of, 233
Alkyloxazoles, occurrence in coffee aroma, 181
2-Alkyloxazoles, *MacLafferty* rearrangement of, 316
Alkylpropenyl disulphides, as precursors of dimethylthiophenes, 171
Alkylpyrazines, extraction of, 273–276
^{1}H-NMR data of, 285
occurrence in glucose-cystine system, 192
Alkylpyridines, extraction of, 273, 274
2-Alkylpyridines, fragmentation of, 326
synthesis from butadiene and nitriles, 229
Alkylpyrroles, retention indices of, 292
1-Alkylpyrroles, formation from sugars (or furfural) and α-amino acids, 178
2-Alkylpyrroles, fragmentation of, 308
Alkylthiazoles, synthesis from dialkylamines and sulphur, 219
2-Alkylthiazoles, effect in tomato soup, 110
MacLafferty rearrangement of, 316
odours of, 110
Alkylthiofuroates, in coffee aroma, 78
2-Alkylthiophenes, fragmentation of, 308
2-Alkylthiothiazoles, mass spectra data of, 317
2-Alkylthio-2-thiazolines, mass spectra data of, 317
Amadori and *Heyns* intermediates, formation of, 153
from proline (or hydroxyproline) and glucose, 175
rearrangement of, 155
retro-aldolization of, 155
Ambrettolide, as food flavouring, 132
Nor-Ambreinolides, in oriental tobaccos, 140
Amines, as flavour precursors in coffee, 38
α-Amino acids, as flavour precursors in asparagus, 45
in bread, 50
in cocoa beans, 33
in coffee, 37
in peanut, 24
in potato, 46
in red meat, 20
in tea extract, 30
in tomato, 41
in wine, 62
Strecker degradation of, 22, 156
thermal decomposition products in red meat 20, 21
1,5-Aminoketones, synthesis of, 228
1-Amino-1-deoxy-2-ketoses, *see Amadori* intermediates, 153
1-Amino-2-deoxy-2-aldoses, *see Heyns* intermediates, 154
α-Angelicalactone, from carbohydrate degradation, 170
β-Angelicalactone, fragmentation of, 320
Acridines, in tobacco, 140
Acrolein, formation from hydroxyacetone, 256
Acylation reactions, 214, 239
Acyloxazoles, in coffee aroma, 181
2-Acylthiophene, fragmentation of, 310
Adsorption, 269
Alcoholic beverages, 59
Alcohols, as aroma components in wine, 60
isolation of, 272
Ascorbic acid, content in tomato, 43
Asparagus, aroma precursors of, 43
carboxylic acids from, 43
composition of, 43
enzyme systems in, 45
sulphur-containing acids of, 44
Asparagusic acid, degradation of, 44

B

Baking, in bread processing, 49
Beef, flavour precursors in, 18
Benzimidazole, fragmentation of, 336
synthesis from *o*-phenylene diamine and acyl chlorides, 240
toxicological data of, 357
in tobacco, 135
Benziosothiazoles, synthesis of, 241
Benzolfuran, from degradation of glucose, 170
fragmentation of, 335
mass spectra data of, 334
Benzofurans, occurrence in food flavours, 132
synthesis of, 237
Benzolquinolines, in tobacco, 140
Benzopyrans, as flavouring agents in perfume compositions, 136
synthesis of 243
Benzopyrimidines, synthesis of, 245
Benzothiazole, as flavouring agents in foods, 357
formation from *Amadori* intermediates, 187
occurrence of, 135
Benzothiazoles, fragmentation of, 337
mass spectra data of, 334
synthesis from intramolecular cyclization of *N*-arylthioamides, 241
from *o*-aminothiophenol and acyl chlorides, 240

Benzothiophene, from *D*-glucose-*L*-cysteine system, 175
fragmentation of, 335
mass spectra data of, 334
occurrence of, 134
sensory properties of, 134
Benzothiophenes, synthesis of, 237, 239
Benzolxazole, fragmentation of, 336
Benzoxazoles, occurrence of, 135
synthesis of, 240
Betaines, as pyridine precursors in foods, 189
Bread, content of, 49
flavour precursors of, 50
ingredients for processing of, 50
processing of, 48
sugars degradation during fermentation, 51
browning reactions, effect of some inorganic ions on, 153
inhibitions by reductones, 154
2,3-Butanediol, in wine, 62
δ-Butenolactone, from degradation of ascorbic acid, 188
Butter, methylketones in, 55
γ-Butyrolactone, formation from oxocids, 169
occurrence in model systems, 167

C

Caffeic acid, as diphenol precursors in coffee, 39
Caffeine, in coffee and tea, 26, 31, 139
as flavouring substance in foods, 358
Carbonyl compounds, isolation of, 272
from β-ketoacids and β-hydroxyacids, 159
in milk and butter aroma, 55
from oxidation of carotenoids, 159
in roasted peanut aroma, 25
carotenoids, biogenesis from leucine, 31
as flavour precursors in tea, 29
in tomato, 42
Casein, α-amino acids content of, 57
as milk protein, 57
Carbazoles, in tobacco, 140
Cinnolines, synthesis of, 245
Cheese, flavour precursors in, 55
polysulpur compounds in, 59
various processing stages of, 53
various types of, 53, 55
Chemical methods, 271
Chicken, flavour precursors of, 18
Chlorogenic acid, degradation during roasting of coffee, 39
Chromans, synthesis of, 244
Chromatographic methods, 270
adsorption, 270, 287
capillary gas, 288, 293
in combination with other physical methods, 288, 293
exclusion, 270
high-performance, 287
ion-exchange, 272
preparative, 270
reverse-phase, 287
Cocoa, α-amino acids content of, 33
flavours precursors of, 32
Coffee, sucrose content of, 37
volatile substances of, 26
Coffee (green), composition of, 35
content of, 35
Coffee (roasted), composition of, 36
sugars content of, 37
Computer, in organic synthesis, 250
program of, 250
Connectivity table, 251
Cooling, in bread processing, 49
Cumarins, as flavouring agents in foods, 358
occurrence of, 136
synthesis of, 243
γ-Crotonolactone, fragmentation of, 320
Cycloalkylpyrazines, ^{1}H-NMR data for, 285
Cycloalliin, as cyclic amino acid in onion, 120
Cyclopentapyran, in essential oils, 136
Cyclopentapyrazines, in cooked and roasted products, 137
extraction from grilled meat, 276
from roasted cocoa, 234
formation from cyclotene, α-diketones and ammonia, 197
fragmentation of, 337
mass spectra data of, 334
occurrence in model systems, 195
in *L*-rhamnose-ammonia system, 195
synthesis of, 244
Cyclopentapyrimidines, as flavouring agents, 139
Cyclotene, from hydroxyacetone, 155

D

Dairy products, 52
δ-Decalactone, sensory properties of, 115
3-Deoxyhexosones, from *Heyns* intermediates, 155
1-Deoxy-1-piperidino-*D*-fructose, pyrolysis of, 165
1-Deoxy-*L*-prolino-*D*-fructose, decomposition of, 166
Detectors, in gas chromatography, 288, 289

Derivatization methods, in gas chromatography, 297
in liquid chromatography, 296
in thin layer chromatography, 296
1,5-Dialdehydes, synthesis of, 225
4,5-Dialkylthiazoles, aroma of, 110
α-Dicarbonyl compounds, from rearranged sugars, 155
Dicyclohexapyrazine, formation of, 197
2,2′-Dufuran, fragmentation of, 333
2,5-Diketo-3,6-dimethylpiperazine, from serine and threonine pyrolisis, 196
Diketopiperazines, as bitter compounds in cocoa, 34
Diketoses, influence on pyrazine formation, 25
Dihydroactinidiolide, occurrence of, 132
2,3-Dihydroindoles, from *D*-glucose-proline system, 180
4,5-Dihydro-2-(3*H*)-furanones, *see* γ-lactones
Dihydropyran, from ring expansion of tetrahydrofurfurylic acid, 188
Dihydropyrans, occurrence of, 112
4,5-Dihydro-α-pyrones, from 1,5-hydroxy acids, 226
4,5-Dihydro-3-(2*H*)-thiophenone, acute lethal dose of, 361
from cysteine and xylose, 173
3,6-Dihydro-3-vinyl-1,2-dithiin, extraction from asparagus, 273
2,3-Dihydroxy-5-vinylfuran, *self*-condensation of, 256
2,3-Dimethylbenzofuran, as flavouring substance, 357
2,5-Dimethyl-2,5-dihydroxy-1,4-dithiane, as flavouring substance, 354
2,4-Dimethyl-1,3-dioxolanes, mass spectra data of, 314
2,6-Dimethyl-1,3-dithiin, mass spectra data of, 331
2,5-Dimethyl-4-hydroxy-3-(2*H*)-furanone, occurrence of, 85
synthesis from 3-hexyne-2,5-diol, 211
4,5-Dimethyl-3-hydroxy-2-(5*H*)-furanone, from 2-oxobutyric acid, 171
2,4-Dimethyloxazole, from α-aminopropanol and acetaldehyde, 181
3,5-Dimethyl-1,3-oxazolidine-2,4-dione, in meat aroma, 106
Dimethylpyrazines, occurrence of, 122
3,6-Dimethyl-2-ethylpyrazine, in chips aroma, 46
3,5-Dimethyl-2-propylpyrazine, charcterization from grilled meat aroma, 299
2,4-Dimethylpyrrole, synthesis from β-ketoester and α-aminoketone, 212
2,5-Dimethylpyrrole, synthesis from acetonylacetone and ammonia, 210
2,6-Dimethyl-γ-pyrone, from dehydration of 1,3,5-pentanetrione, 228
Dimethyl sulphide, as aroma precursors in tea, 31
formation from *S*-methylmethionium salt in asparagus, 45
in tomato, 42
3,5-Dimethyltetrathiane, mass spectra data of, 331
3,5-Dimethyl-1,2,4-trithiolane, as flavouring substance, 352
mass spectra data of, 331
1,3-Dioxane, fused derivative in tobacco
1,4-Dixoane, in lemon, 120
Dioxanes, as flavouring substances, 354
synthesis of, 237
1,3,5-Dioxathianes, from aldehydes and hydrogen sulphide, 200
in model systems, 120
Dioxolanes, bicyclic derivatives in tobacco, 134
1,3-Dioxolanes, as flavouring substances, 352
fragmentation of, 318
occurrence in fruits and wines, 102
synthesis from α-glycols and aldehydes (or ketones), 219
Distillation, 267
in combination with extraction, 271
various techniques of, 267
Disulphides, acute oral lethal doses of, 361
Diterpenic acids, in coffee oil, 38
1,2-Dithiacyclopentene, extraction from asparagus, 273
occurrence in asparagus aroma, 102
1,2-Dithiacyclopentene, as flavour contributor in asparagus, 45
1,2-Dithiane, fragmentation of, 330
mass spectra data of, 331
in model systems, 120
from pyrolysis of cysteine-xylose system, 200
1,3-Dithianes, synthesis of, 237
1,3,5-Dithiazines, in model systems, 120
1,4-Dithiin, synthesis of, 234
Dithiins, in cooked asparagus, 120
odours of, 120
in model systems, 120
synthesis of, 235
Dithiins, from aldehydes and hydrogen sulphide, 199, 200
in synthetic onion aroma, 120
1,2-Dithiolane, fragmentation of, 330
1,2-Dithiolanes, alkyl carboxylate derivatives in cooked asparagus, 102
sensory properties of, 102

1,3-Dithiolanes, odours of, 102
1,2-Dithiolene, from degradation of asparagusic acid, 198
Dithiolium salts, from α-dithioacylketones, 217
δ-Dodecalactone, sensory properties of, 115

E

Edulans, in passion fruit, 136
Enzymes, as flavour precursors in cocoa, 32
  as catalysts in milk, 59
Enzyme systems, specificity of, 158
Esters, *ortho* effect in the fragmentation of heterocyclic derivatives, 311
  occurrence in wine, 60
Ethers, cyclic derivatives as flavouring substances, 358
2-Ethyl-4 (or 5) acetyloxazoline, from glucose-theanine system, 182
Ethylmaltol, as flavouring agent, 116
2-Ethyl-4-methyl-1,3-dithiole, mass spectra of, 321
Ethylpyrazines, from acetaldehyde and dihydropyrazines, 197
3-Ethylpyridine, fragmentation of, 326
2-Ethylthiazole, from cysreine-pyruvaldehyde system, 185
Eucalyptol, in essential oils, 297
Extraction, various methods of, 269

F

FEMA GRAS list, 263, 343
Fermentation, in bread processing, 49
  in dairy and milk products, 58
Flavanols, occurrence in tea, 137
  oxidation in tea, 27
Flavanones, as flavour precursors in tomato, 43
Flavonols, as flavour precursors in cocoa, 33
  occurrence in wine, 61
Flavouring substances, books and reviews dealing with, 73
  classification of, 262, 343, 344
  legislation of, 343
Flour (wheat), content of, 49
2-Formylbenzofuran, as flavouring substance, 357
3-Formylbenzofuran, from acrolein and hydrogen sulphide, 228
2-Formylpyrrole, occurrence of, 100
2-Formylpyrroles, synthesis of, 215
2-Formylthiophenes, formation of, 173
Free radical substitution, 215
Fruits, 40
3-(2*H*)-Furanones, *o*-alkyl derivatives of, 87
  as flavouring substances, 349
  formation from 1-deoxyhexosones, 167
  occurrence of, 86
Furan, fragmentation of, 306
Furans, acute oral lethal doses of, 361
  analysis by GC/MS in brandy, 294
  in red wine, 295
2,5-Disubstituted derivatives in food flavours, 79, 80
  ub model systems, 164
  extraction from food flavours, 272
  flavour characteristics of, 83
  formation from ascorbic acid and *L*-dehydroascorbic acid, 166
  mass spectra of, 306
  mono-unsaturated and saturated derivatives in model systems, 168
  ^{1}H-NMR spectra of some 2-substituted derivatives, 284
  odour threshold values of, 279
  patented as flavouring agents, 85
  predicted for the thermal degradation of glucose, 257
  2-substituted derivatives in food flavours, 74–77
    in model systems, 161–163
  from sugar degradation in wine, 61
  sulphur derivatives in coffee and bread, 85
  synthesis from α-halogenoketones and β-ketoesters, 212
    from parent compounds, 214
  terpenoid- and saturated derivatives in foods, essential oils and tobacco, 82
  used as flavouring substances, 345–348
  variously substituted derivatives in food flavours, 81
Furanthiols, acute oral lethal dose of, 360
Furfural, formation from pentoses, 154, 165, 211
    from rearranged sugars, 165
  fragmentation of, 259
  occurrence in food flavours and essential oils, 78
  odour units of, 279
  as precursors of other heterocyclic compounds, 165
  reaction with actived methylene compounds, 165
    with hydrogen sulphide and ammonia, 165
    with *L*-valine, 165
Furfuryl alcohol, occurrence of, 78
Furfuryl mercaptan, in coffee aroma, 78

N-Furfurylpyrrole, acute oral lethal dose of, 361
Furfurylpyrroles, retention indices of, 292
Furfuryithioethers, fragmentation of, 313
  mass spectra data of, 312
2-(2′-Furyl)-pyrazines, formation of, 196
  mass spectra data of, 392
  occurrence of, 78
2-(2′Furyl)-thiazole, fragmentation of, 333

G

Galbanum oil, pyrazines in, 130
GC/MS coupling, specific structural informations of, 294
D-Glucose, hydrolysis in the presence of α-amino acids, 158
Glutamic acid, formation of 4,5-dihydro-5-ethoxy-2-(3*H*)-furanone from, 62
N-Glycosylamines, from sugar-amine condensations, 152
Guaiapyridine, characterization from the patchouli oil, 299
  extraction of, 273

H

Hazelnut, components of seed, 25
  heterocyclic compounds of, 25
Headspace technique, 289
Bis-Heteroaryl compounds, mass spectra data of, 332
Heterocyclic bases, extraction of, 272
Heterocyclic compounds, acylation of, 214
  analysis of, 262
  bank mass spectra data of, 338
  classification of, 74
  distribution on FEMA and CoE lists, 359
  electronic absorption spectra of, 281
  five membered ring closure, 209
  formation in foods, 152
    from α-amino acids and diacetyl, 156
    from furfural, hydrogen sulphide and ammonia, 258–260
  fused derivatives, 132
    used as flavouring substances, 357
    predicted from 4-hydroxy-3-(2*H*)-furanones, 254
  isolation of, 265
  mass spectrometry of, 305
  predicted from hydroxyacetone, 255
    from the thermal degradation of glucose, 257
  polyfused derivatives predicted by computer, 260
  proton chemical shifts of, 283
  separation by HPLC in model systems, 288
  toxicity of, 344, 355
Heterocyclic compounds – *continued*
  various steps involved in the separation of, 266
  various types of ring closure, 215, 220
  (E)-2-Hexenal, formation in tea aroma, 30
  (Z)-2-Hexenal, from enzymatic isomerization of *cis*-3-hexenal, 158
  Heyns intermediates, formation of, 153
  rearrangement of, 155
  retro-aldolization of, 155
Hexamethyl-1,3,5-trithiane, acute oral lethal dose of, 361
High-performance liquid chromatography, 277, 288
Hydrogen sulphide, as aroma precursors in tea, 31
  formation from cysteine and α-diketones, 160
    from thiamin and glutathione, 160
  reaction with 4-hydroxy-5-methyl-3-(2*H*)-furanone, 167
Homoserine, from methylmethionine sulphonium salt, 31
Hydantoins, 1,3-dialkyl derivatives in beef aroma, 103
Hydroxyacetone, heterocyclic compounds predicted from, 255
γ-(and δ)-Hydroxy acids, as lactones precursors in butter, 55
2-Hydroxy-2-cyclohexenones, dimerization of, 256
1-β-Hydroxyethylpyrrole, from glycol aldehyde and aminoethanol, 179
4-Hydroxy-3-(2*H*)-furanones, 2,5-dialkyl derivatives of, 87
  dimerization of, 253
5-Hydroxymaltol, formation from *Amadori* intermediates, 187
  from glucose-theanine system, 187
3-Hydroxy-2-methylbenzopyranone, from pentoses and ascorbic acid system, 188
4-Hydroxy-5-methyl-3-(2*H*)-furanone, condensation with furfural, 167
  from hydrolysis of ribonucleotides, 166
  from thermal degradation of *D*-xylose, 166
5-Hydroxymethylfurfural, from degradation of sucrose in bread, 51
  determination in tomato paste by HPLC, 297
  increasing concentration in tomato during storage, 78
  formation from 2-ketohexoses, 154, 165
    from sugars in tomato, 43
  synthesis for hexoses, 211

2-Hydroxymethylpyrazines, synthesis from 2-methyl-*N*-oxyde pyrazines, 333
3-Hydroxypyridine, fragmentation of, 327
3-Hydroxy-2-pyrone, from degradation of ascorbic acid, 188
3-Hydroxy-4-(4*H*)-pyrones, *see* γ-pyrones, 115
5-Hydroxymethylisothiazole, fragmentation of, 319
5-(β-Hydroxyethyl)-4-methylthiazole, from thiamin degradation, 185

I

Imadazoles, formation from α-amino carbonyl compounds, 183
  from *L*-rhamnose-hydrogen sulphide system, 182
  2,4-disubstituted derivatives from α-aminoaldhydes-ammonia-carboxylic acid systems, 182
  main fragmentation modes of, 315
  occurrence in food flavours, 103
    in model systems, 183
  synthesis from α-halogenoketones and amidines, 216
    from oxazoles and ammonia, 217
    from reduction of monoacetyl derivavative of α-aminonitriles, 219
Imidazolines, synthesis from β-aminoacyclamines, 218
  from 1,2-diaminoethanes and aldehydes, 219
Indazoles, synthesis of, 241
Indole, as flavouring substance, 135
  fragmentation of, 335
  mass spectra date of, 334
Indoles, occurrence in essential oils, tobacco and food flavours, 135
  synthesis of, 238
Infra-red spectroscopy, 282
Iridans, in essentials oils, 136
N-Isobutylacetamide, in ripened cheese, 59
2-Isobutyl-3-methoxypyrazine, in grapes, 62
  threshold value of, 122
2-Isobutylthiazole, as aroma component of tomato, 40
  from cysteamine and isovaleraldehyde, 185
Isomaltol, as caramelization products of carbohydrates, 78
  heterocyclic compounds predicted from, 253
2-Isopropyl-2-methoxypyrazine, as characteristic flavour compound in raw potatoes, 46
Isoquinoline, as flavouring substance, 358
  mass spectra data of, 334
Isoquinolines, as flavouring agents in nonalcoholic beverages, 137
  synthesis of, 243
Isothiazoles, mass spectra data of, 315
Isoxazoles, synthesis of, 220

K

Ketones, as flavour components in tomato, 42
Kneading, in bread processing, 48
Kovats indices, in gas chromatography, 289

L

δ-Lactams, formation from glucose-theanine system, 180
  from model systems, 192
γ-Lactams, synthesis from γ-amino acids, 210
Lactones,
  acute oral lethal dose of, 361
  extraction of, 271
  as flavouring substances in foods, 348, 349, 353
  formation form δ- and γ-hydroxylation of saturated fatty acids, 169
  fragmentation of, 319
  macrocyclic derivatives in essential oils, 132
  occurrence in milk and butter flavours, 55
  synthesis from γ-hydroxy acids, 210
δ-Lactones, formation of δ-alkylated derivatives from δ-hydroxy fatty acids, 188
  unsaturated derivatives in peach flavour, 115
  occurrence in food flavours, 113, 114
  α, β-unsaturated derivatives in food aromas, 15
  synthesis of bicyclic derivatives, 244
    from cyclopentanone and aldehydes, 227
γ-Lactones, γ-alkyl derivatives from chemical and biochemical pathways, 168
  bicyclic derivatives in spice plants, 134
  α,β-unsaturated derivatives from α,β-unsaturated γ-hydroxy acics, 170
  occurrence in food flavours, 88–91
  synthesis of α, β-unsaturated derivatives from α, β-unsaturated acids of esters, 210
Lactose, as flavour precursor in milk and dairy products, 58
β-Lactoglobulin, as source of hydrogen sulphide and mercaptans in milk, 58

Lamb, flavour precursors in, 18
Lenthionine, in red algae and mushrooms, 132
Linalol oxides (*cis* and *trans*), occurrence in essential oils, 112
in food flavours, 81
in tea aroma, 30
Linoleic acid, oxidation in bread, 52
in potato, 48
Linolenic acid, enzymatic oxydation of the ^{14}C-labelled derivative, 45
in tea, 30
Lipids, as flavour precursors in asparagus, 45
in cocoa, 34
in coffee, 38
in milk, 54
in potato, 47
in tomato, 41
in wheat flour, 52
Maillard reactions, *see also* browning reactions
computer application in, 249
occurrence in cocoa, 34
in coffee, 26
various steps of, 152
Maleimides, in tobacco and meat aromas, 101
Maltol, from *Amadori* intermediates, 187
heterocyclic compounds predicted from, 253
occurrence in food flavours, 116
sensory properties of, 116
Mass spectra data, compilation of, 294
Meat, degradation of precursors in, 19
fatty acids oxidation of, 22
flavour precursors of, 18, 22
flavour sensation of, 23
nucleotide content of, 20
ratios of free α-amino compounds in, 19
Menthofuran, as flavouring substance, 357
in *mentha piperata L.* oils, 132
Menthanethiol, from menthionine and α-diketones, 160
Methional, from *Strecker* degradation of methionine in potato, 46
Methionine, as methional precursor in grapes, 62
Strecker degradation of, 157
Milk, composition of, 53
fat content of, 54
fatty acids content of, 55
flavour precursors in, 53
phenolic compounds as flavour components in, 58
vitamins content of, 53
2-Methylbenzofuran, from laetose and casein, 170
2-Methylbenzoxazole, from *o*-substituted aniline and acetic anhydride, 240
6-Mehtylbenzopyrone, olfactory properties of, 136
2-Methylbenzothiazole, in rice, 136
2-[(Methyldithio)-methyl]-furan, in coffee, 78
5-Methyl-3-(2*H*)-furanone from ribose-5′ phosphate, 46
5-Methylfurfural, in food systems, 78
synthesis from methylpentose, 211
4- or 5)-Methylimidazole, from *D*-glucose-glycine system, 182
N-Methylimidazoles, mass spectra data of, 314
2-Methylindole-3-carboxylate, synthesis of, 239
Methylsothiazole isomers, fragmentation of, 318
S-Mehtylmethionine sulphonium salt, as flavour precursor in tomato, 42
Strecker degradation of, 157
synthesis of, 230
2-Methyl-4-propyl-1, 3-oxathiane, as flavouring substance, 354
2-Methylpyrazine, fragmentation of, 330
Methylpyridine-4-carboxylate, fragmentation of, 327
2-Methylquinazoline, in dried mushrooms, 139
1-Methylpyrroline, from *L*-rhamnose-ammonia system, 180
2-Methyl-thiazoline, from cysteine (or cystine) pyrolysis, 185
2-Methylthiazolidine, from sulphur-containing α-amino acids pyrolysis, 185
2-Methyl-3-tetrahydrofuranone, mass spectra data of, 312
2-Methyl-1,2,4-trithiane, in beef extract, 120
2-Methylthiobenzothiazole, mass spectra data of, 334
in meat flavour, 136
4-Methylthio-1,2-dithiolane, in algae, 295
1-Methylthioethanethiol, from acetaldehyde and methanethiol, 160
2-Methylthipene, from thiamin, 175
5-Methylthio-1,2,3-trithiolane, in algae, 295
Mercaptofurans, from sulphide in 4-hydroxy-3-(2*H*)-furanones, 167
as thiamin degradation products, 166
Model systems, various categories of, 161, 163

N

1,6-Naphthyridines, burnt meat taste of, 136
Nerol oxides, in essential oils, 113
N-Nitrosopiperidine, in cured meat, 118
Non-alcoholic beverages, 25
Non-volatil acids, as flavour precursors in coffee, 39
Norcepanone, from diacetylene alcohols and mercuric salts, 211
Nornicotines, as tobacco flavourings, 118
Nuclear magnetic resonance, 282
Nucleophilic substitution, representation of, 250
Nucleotides, content in potato, 47
  enzymatic decomposition of, 20, 22

O

1-Octen-3-ol, formation in Camembert, 55
1-Octen-3-one, from linoleic oxidation in potato, 48
Odours units, 278
Organoleptic methods, 278
1,3,5-Oxadithianes, from aldehydes and hydrogen sulphide, 200
1,3-Oxathianes, in Passion fruit, 120
1,4-Oxathianes, synthesis of, 234
1,2-Oxazines, synthesis of, 235
1,3-Oxazines, synthesis of, 237
Oxazoles, alkyl- and acyl derivatives in food flavours, 104
  formation from aldehydes and α-aminoketones, 182
    from model sytems, 181
    from sugars and ammonia, 182
  main fragmentation mode of, 315
  mass spectra date of, 313
  synthesis from α-halogenoketones and amides, 216
Oxazolines, mass spectra data of, 313, 314
2-Oxazolines, olfactory properties of, 106
  rearomatization of, 218
  synthesis from β-hydroxyacylamines, 218
3-Oxazolines, as flavouring agents, 105
  olfactory properties of, 105
Oxazolidines, from 1-amino-2-hydroxyethanes and aldehydes, 219
5-(4*H*)-Oxazolinpnes, from α-acylamino carboxylic acids, 217
2-Oxo-1,4-oxathiane, synthesis of, 234

P

Peanuts, α-amino acids content in, 24
  aroma precursors of, 23
  volatile aroma compounds of, 23
Peanut seed, composition of, 23
  lipids and fatty acids in, 24
  sugars in, 24
Petit Grain oil, pyrazines in, 130
2-*n*-Pentylfuran, acute oral lethal dose of, 361
  formation from linoleate auto-oxidation, 166
  fragmentation of, 310
  identification in potato juice, 48
  occurrence in food flavours, 75
  in model systems, 161
  odiur units of, 279
  sensory properties of, 78, 83
Peptides, as flavour precursors in bread, 50
  in cocoa, 33
  in coffee, 37
Phase transfer catalysis, 215
Phenols, classification of, 61
  extraction of, 271
  from phenolic acids degradation, during dough baking, 52
  in wine, 61
Phenols acids, as flavour precursors in cocoa, 33
  in coffee, 39
  in tomato, 42
  in wheat flour, 52
Phenylalanine, reaction with α,β-unsaturated carbonyl compounds, 158
2-Phenyl-2-1lkenals, from aldehydes and phenylacetaldehydes, 159
2-Phenylethanol, formation from *Strecker* degradation of phenylalanine, 45, 46
Physical methods, in combination with chemical methods, 296
  identification of heterocyclic compounds by, 294
Phospholipids, as flavour precursors in cocoa, 34
Phtalides, in celeri, lovage and essential oils, 134
  used as flavouring substances, 357
Piperidine, in fish, 118
  synthesis of, 227
Piperine, amount in oleorsin from pepper, 297
  as component of black pepper, 120
  as flavouring substance, 358
Piperonal, in essential oils, 134
Polyalcohols, as flavour precursors in cocoa, 33
  in potato, 48
  in tea, 27
  in wine, 61, 62
Polyhydroxyphenols, as astringency agents in wine, 61

Polyphenols, predicted from the thermal degradation of glucose, 257
Polysulphides, cyclic derivatives formation in model systems, 131, 199
Polythiepans, from enzymatic degradation of lentinic acid, 200
Pork, flavours precursors of, 18
Potato, composition of, 47
enzyme systems in, 48
flavour precursors of, 46
nucleotides content of, 47
sugar content of, 47
Program, computer assisted organic synthesis, 251
2-Propionylpyrrole, acute oral lethal dose of, 361
Proteins, as flavour precursors in bread, 50
in coffee, 37
in milk, 57, 58
Pyrans, as flavouring substances, 353
predicted from the thermal degradation of glucose, 257
synthesis from 1,5-dicarbonyl compounds, 224
Pyrazines, acute oral lethal dose of, 360
GC/MS analysis in roasted peanuts, 295
as flavouring agents, 131, 356
formation from degradation of dipeptides, 192
from dipeptides-glyoxal systems, 196
from α-dicarbonyl compounds and α-amino acids, 197
from pyrolysis of α-hydroxyamino acids, 192
fused derivatives as flavouring agents, 138
influence of pH on the formation of, 122
induced by fermentation in cocoa, 34
mass spectra data of, 328
occurrence in alcoholic beverages,130
in cheese, 130
in essential oils, 112, 130
in food flavours, 123–129
in glucose-*L*-alanine system, 192
in glucose-theanine system, 192
in model systems, 193, 194, 196
in *L*-rhamnose-ammonia system, 195
in sugars-ammonia systems, 192
odour threshold values of, 130, 280
in perfume compositions, 130, 131
retention indices of, 290
sensory properties of, 130, 278
synthesis from oxidation of α-amino alcohols, 232
from cyclized compounds, 232
Pyrazines – *continued*
synthesis – *continued*
from α-dicarbonyl compounds and aliphatic diamines, 232
from ketone-nitrogen iodide reaction, 232
from oxidation of piperazines
from thermal degradation of hydroxyamino acids, 38
tricyclic derivatives as nut flavourings, 140
Pyrazinium salts, as flavouring agents, 130
Pyrazoles, occurrence in food flavours, 103
synthesis of, 220
1,2-Pyridazines, synthesis of, 235
Pyridines, as flavouring substances, 355
formation from aldehydes and α-amino acids, 189
from aldehydes and ammonia systems
from 1,5-dicarbonyl compounds and ammonia (or α-amino acids), 189
from other heterocyclic compounds, 189, 192
homolytic alkylation, of, 230
mass spectra of, 325
occurrence in jasmine oil, 117
in food flavours, 117, 119
in model systems, 190, 191
odour threshold values of, 280
saturated derivatives in tobacco, 118
sensory properties of, 118
synthesis fron 1,5-dicarbonyl compounds and ammonia, 224
from glutaconic aldehyde and ammonia, 225
from other heterocyclic compounds, 229
from β-ketoesters, aldehydes and ammonia, 228
from 1,5-ketonitriles, 226
from thermal degradation of trigonelline, 39
2-Pyridinemethanethiol, as ingredient in meat flavour, 118
Pyridinium salts, synthesis from 1,5-diketones and amines (or α-amino acids), 225
2-Pyridinone, fragmentation of, 327
4-Pyridinone, fragmentation of, 323
4-(or γ)-Pyridinones, from 1,3,5-triones and ammonia, 229
Pyrilium salts, transformation in pyridines, 224, 230
Pyrimidines, in meat aroma, 139
synthesis of, 236

2-(or α)-Pyrones, formation from unsaturated 1,5-dialdehydes, 188
organleptic properties of, 115
synthesis from 1,5-acetylenic acids, 226
from 1,5-ketoacids, 226
4-(or γ)-Pyrones, fragmentation of, 323
mass spectra data of, 324
Pyrrole, fragmentation of, 306
Pyrroles, from degradation of trigonelline, 39, 179
extraction from roasted peanuts, 275
as flavouring subatances, 350
formation from hydroxyproline, 178
from sugars and α-amino acids, 101
mass spectra data of, 307, 309
occurrence in model systems, 176, 177, 180
odour threshold values of, 280
reduced derivatives of, 101
2-substituted derivatives in food flavours, 99
N-substituted derivatives in food flavours, 98
synthesis from 3-alkyl-2,4-pentanediones and α-aminoketones, 212
from 1,4-diketones and ammonia, 209
from α-halogenoketones, β-ketoesters and ammonia, 212
from pyrolysis of furans in the presence of ammonia, 213
from pyrrylthallium salts and alkylhalide, 215
Pyrrolidine, from from D-glucose-L-alanine system, 180
from tetramethylene dibromide and ammonia, 210
Pyrrolidines, from 1-deoxy-1-*L*-prolino-*D*-fructose, 179
N-substituted derivatives in proline-glucose system, 180
synthesis of 1,2-disubstituted derivatives from *N*-chloroalkylamines, 213
2-Pyrrolidone, fragmentation of, 321
Pyrrolidone carboxylic acid, in tomato, 101
1-Pyrroline, butter-like flavour of, 99
Pyrrolines, from 1-deoxy-1-*L*-prolino-*D*-fructose, 179
3-Pyrrolines, in tobacco, 100
Pyrrolo lactones, from *D*-glucose-α-amino acids systems, 180
as tobacco flavourings, 134

Q

Quinic acid, as phenol precursors in coffee, 38
Quinine, in non-alcoholic beverages, 139
as flavouring substance, 358
Quinoline, mass spectra data of, 334
Quinolines, as flavouring substances, 358
main routes for the preparation of, 242
occurrenc in food flavours, 137
Quinoxalines, as flavouring substances, 358
extraction from roasted cocoa, 274
from filberts
from roasted sesame seeds, 275
synthesis of, 244

R

Ribose-5′-phosphate, from nucleotides, 20, 22
Rose oxides, in essential oils, 112

S

Saccharine, as flavouring substance, 358
Safrole, in spice and cocoa aromas, 134
Salts, in bread, 52
as milk compnents, 59
Sclareol oxide, in oriental tobaccos, 140
Sedanolides, odours of, 134
Skatole, occurrence of, 135
Spectroscopic methods, 281
Steroids, biogenesis from leucine, 31
Strecker degradation of α-amino acids in meats, 22
during roasting of cocoa, 34
Staling, in bread processing, 49
Starch, as flavour precursor in bread, 51
Succinimides, in tobacco, 101
Sugars, dehydration of, 154
as flavour precursors in asparagus, 45
in bread, 51
in cocoa beans, 33
in coffee, 36
in potato, 46
in red meat, 20
in tea, 31
in wine, 61
Sulphides, cyclic derivatives from aldehydes and hydrogen sulphide, 199
fufuryl and furfurylalkyl derivatives of, 215

T

Tannins, in wine, 61
Tea, α-amino acids content in, 30
aroma formation of, 32
composition of, 27
content of, 26

Tea – *continued*
flavour precursors in, 26
volatile heterocyclic compounds of, 31
volatiles from carotene oxidation of, 29
Terpenes, in wine, 61
Terpenoids, biogenesis from leucine, 31
Tetrahydrofuran, from dehydration of 1,4-butanediol, 210
Tetrahydrofurans, flavour characteristics of, 84
occurrence in food flavours, essential oils and tobacco, 82
Tetrahydropyran, fragmentation of, 322
sysnthesis from, 1,5-dibromopentane, 227
Tetrahydropyrans, occurrence in food flavours, 112
Tetrahydropyranspirocyclohexanes, as flavouring ingredients in perfumes and colognes, 113
Tetrahydro-2-pyrones, *see* δ-lactones, 113
Tetrahydroquinoxalines, occurrence in model systems, 195
from *L*-rhamnose-ammonia system, 195
Tetrahydrothispyran, fragmentation of, 322
synthesis of, 227
Tetrahydrothiapyrans, in food flavours, 116
Tetramethylpyrazine, in food flavours, 122
1,2,4,5-Tetrathiane, 3,6 derivative from propionaldehyde, hydrogen sulphide and ammonia, 200
1,2,4,6-Tetrathiepane, from furfural, hydrogen sulphide and ammonia, 200
1,2,5,6–Tetrathiocane, in grape leaves, 132
Theanine, as precursor of *N*-ethylsuccinimide, 180
Theaspirone, characterization from tea, 298
1,3,5-Thisdiazines, in model systems, 120
Thialdine, formation from acetaldehyde, ammonia and hydrogen sulphide, 199
Thiamin, thermal degradation of, 21
4-(or γ)-Thiapyrans, mass spectra data of, 324
4-(or γ)-Thiapyrone, fragmentation of, 323
Thiazoles, acute oral lethal dose of, 360
alcohol and carbonyl derivatives as flavouring agents, 111
cysteine and cystine as source of, 185
from α -dicarbonyl compounds, hydrogen sulphide and ammonia, 185
extraction from food flavours, 276
as flavouring agents, 111, 351
flavour description of, 107–109
formation of, 186
main fragmentation modes of, 315
mass spectra data of, 314

Thiazoles – *continued*
^{1}H-NMR data for mono-, and disubstituted derivatives, 284
occurrence in food flavours, 107, 108
in model systems, 184
odour thrshold values of, 280
retention indices of, 291
sensory properties of, 110
from thermal degradation of cysteine, 38
synthesis from acyl derivatives of α-mercaptoketones, 217
from α-halogenoketones and thioamides, 216
from mercaptoacetaldehyde dimer, ammonia and aldehydes, 219
from oxazoles and hydrogen sulphide, 217
from α-thiocyanoketones, 218
1,3-Thiazines, synthesis of, 237
Thiazolidines, as flavouring agents, 112
formation of, 186
synthesis from 1-amino-2-mercaptoethanes and aldehydes, 219
from α-dicarbonyl compounds, hydrogen sulphide and ammonia, 185
Thiazolines, formation of, 186
mass spectra data of, 314
occurrence in food flavours, 111
2-Thiazolines, from β-mercaptoacylamines, 218
3-Thiazolines, from α-mercaptoketones (or aldehydes) and ammonia, 219
Thiazolylketones, from cysteine/cystine-ribose system, 185
Theobromine, in cocoa beans, 26
as flavouring substance, 358
stimulating effect of, 139
Thienopyrimidines, flavour properties of, 139
Thieno-[2.3-c]-thiopene, fragmentation of
Thienothiophenes, in coffee aroma, 134
occurrence in model systems, 134, 175
2-(2′-Thienyl)-pyrazine, mass spectra data of, 332
3-(2′-Thienyl)-thiazole, mass spectra data of, 332
γ-Thiolactones, from γ-thiolacids, 210
Thiolane, from tetramethylene dibromide and hydrogen sulphide, 210
2-Thiopene carboxyaldehydes, from α-dicarbonyl compounds and hydrogen sulphide, 171
Thiophenes, acylation of, 214
GC/MS analysis in shade oil, 295
from cysteine pyrolysis, 173

Thiophenes – *continued*
di-, and polysubstituted derivatives in food flavours, 94, 95
extraction from food flavours, 273
as flavouring substances, 350
fragmentation of, 306
from furans and hydrogen sulphide, 171
mass spectra data of, 307
occurrence in food flavours, 92–94
in model systems, 172, 174
odour threshold values of methyl-, and dimethyl derivatives, 96
patented as flavouring agents, 96
sensory properties of, 94, 95
synthesis from 1,4-diketones and phosphorus pentasulphide, 209
from α-mercaptoketones, 213
from pyrolysis of alkanes in the presence of sulphur, 211
from pyrolysis of furans in the presence of hydrogen sulphide, 213
from thiamin pyrolysis, 174
from thermal degradation of cysteine, 38
2-(5*H*)-Thiophenone, fragmentation of, 321
Thiophenones, in food flavours, 96, 97
1,3-Thioxanes, synthesis of, 237
Tomato, C_6 aldehyde formation in, 41
content of, 40
flavour precursors of, 40
2,4,5-Trialkylimidazoles, from αaminoketones and iminoesters, 217
2,4,5-Trialkyloxazoles, from α-acylamidoketones and sulphuric acid, 216
2,4,5-Trialkylthiazoles, from acylaminoketones and phosphorus pentasulphide, 216
Triethyl-1,3,5-oxadithiane in model systems, 120
Trigonelline, thermal degradation in coffee, 39
2,4,6-Trimethyl-1,3,5-dithiazine, mass spectra data of, 331
Trimethylethylpyrazine, as flavour contributor in cocoa, 34
2,4,5-Trimethyl-3-oxazoline, acute oral lethal dose, of, 360
as flavouring substance, 352
formation from alanine and diacetyl, 182
occurrence in beef aroma, 104
synthesis from acetaldehyde, acetoin and ammonia, 219
2,4,5-Trimethyl-3-thiazoline, from 3-mercapto-2-butane, acetaldehyde and ammonia, 185
s-Trithiane, from acetaldehyde and hydrogen sulphide, 199
as flavouring substance, 354
from glucose-hydrogen sulphide system, 199
s-Trithianes, in meat aroma, 120
1,2,3-Trithiane-5-carboxylic acid, from degradation of asparagusic acid, 198
1,2,4-Trithiane, as degradation product of cysteine and cystine, 120
1,2,4-Trithianes, synthesis of, 273
1,2,4-Trithiolane, fragmentation of, 330
mass spectra data of, 331
1,2,4-Trithiolanes, from aldehydes and hydrogen sulphide, 223
cis and *trans* 3,5-dimethyl derivatives in food flavours, 102
Tryptamines, as flavour precursors in coffee, 38
Tuberolactone, in tuberose oil, 115

V

δ-Valerolactum, from *N*-formyl-*L*-lysine-*D*-lactose system, 102
vegetables, 40
5-Vinyl-2-oxazolidinone, as degradation product of progoitrin, 182
Vitamin C, *see* ascorbic acid, 43
Vitamins, content in whole cow's milk, 53
as flavour precursors in milk and dairy products, 58
in potato, 48
toxicological data of, 359

W

Wine, composition of, 60